The Theory of Order-Disorder in Alloys

THE THEORY OF ORDER-DISORDER IN ALLOYS

M. A. KRIVOGLAZ
AND
A. A. SMIRNOV

Translated from the original Russian by Scripta Technica, Inc.

Edited by Professor Bruce Chalmers, Gordon McKay Professor of Metallurgy at Harvard University

NEW YORK

AMERICAN ELSEVIER PUBLISHING COMPANY, INC.

U.S.A. EDITION 1965

AMERICAN ELSEVIER PUBLISHING COMPANY, INC.
52 VANDERBILT AVENUE
NEW YORK, N.Y. 10017

Original Russian title:
Teoriya Uporyadochivayushchikhsya Splavov

Library of Congress Catalog Card Number : 64-21728

PRINTED IN GREAT BRITAIN

Contents

Preface

Numerous experimental investigations have shown that the ordering of atoms in alloys has a great influence on their properties. The magnetic, electrical, mechanical and many other properties of the alloys change strongly upon ordering. Consequently, both experimental and theoretical studies of ordering of atoms and its influence on the properties of alloys is of great practical interest for solving the problem of obtaining alloys with desired properties. The study of an order-disorder transformation is important also for the development of a theory of solid state, since such a theory will provide an explanation of the mechanism of a number of phenomena occurring in alloys. A theoretical analysis of ordering is also of interest from the point view of the general theory of phase transitions, because the latter is one of the few phenomena that can be completely analyzed by the method of statistical mechanics.

The ordered state may occur not only in alloys with metallic properties, but also in semiconducting alloys or compounds. It is well known, for example, that the ordering of atoms can alter the properties of compound ferrites.

The study of ordered alloys is of great current interest in the field of solid state physics. Up to the present time hundreds of experimental and theoretical papers devoted to this problem have been published. Among these papers, however, there does not exist a sufficiently complete up-to-date review devoted to ordered alloys. The aim of this book is to fill this gap by presenting a systematic exposition of some of the more completely developed aspects of the theory of ordered alloys.

The study of ordered alloys can be conveniently divided into the study of the order-disorder transformation of alloys and the investigation of the effect of ordering on the properties of an alloy. These two subjects are, of course, closely related to one another, since the experimental study of an order-disorder transformation is possible only after a clarification of the nature of the influence of order in alloys on their properties is obtained, and the latter will, in turn, serve as an indication of the degree of order. We shall therefore consider first the theory of ordering of atoms in alloys and then proceed with analyzing the influence of order on some of their properties. In this manner we will be able to consider not only ordered alloys, but also alloys in a disordered state, with only some degree of ordering of atoms in the crystalline lattice.

The theory of ordering provides a means for determining the mechanism of establishing both the long- and short-range order in alloys of different compositions and at different temperatures. This theory also enables us to explain an item of great practical importance: the influence of composition and heat treatment on the various properties of alloys.

Chapter I gives general information concerning order-disorder transformation and the basic experimental facts. It was not the author's intention to provide an exhaustive review of all experimental studies of ordered alloys, but to cite chiefly those results that are necessary for illustrations of the theoretical discussion in the following chapters. Chapters II and III are devoted to the thermodynamic and the statistical theories of the ordering of alloys. In Chapter IV, we examine the theory of diffusion in alloys both from the macroscopic and the microscopic points of view. Chapter V presents the quantum mechanical theory of the motion of electrons in an ordered crystal, which makes possible a determination of the electron energy spectrum and wave functions. Chapter VI contains the theory of scattering of x-rays and slow neutrons in ordered alloys. In Chapter VII we employ the many-electron theory to determine the dependence of the residual resistance of alloys on composition, and on long- and short-range order (correlation) parameters. Chapter VIII is devoted to an exposition of the theory of magnetic, galvanomagnetic, optical and mechanical properties of ordered alloys.

THE THEORY OF ORDER-DISORDER IN ALLOYS

Chapter I

General Information Concerning the Order of Alloys

1. Order-Disorder Transformation of Atoms in Alloys

Solid solutions of metals are usually of the substitutional type, where the atoms are positioned at sites of some crystal lattice. Experiments show that, in a number of alloys with stoichiometric composition and at sufficiently low temperatures, the atoms are arranged in such a manner that each species of atom occupies only a certain type of site in the crystal lattice. An alloy in such a state is called completely ordered. As the temperature increases, we observe a transition of a portion of the atoms from their sites to foreign sites; such an alloy is called partially ordered. The concentration of atoms of a given type on foreign sites increases with temperature, but the concentration on their own sites decreases.

In some alloys the concentration of atoms on sites of different types becomes identical even before the melting point is reached; such an alloy is called disordered. The temperature at which such a transition occurs is called the order-disorder phase transition temperature or the critical temperature.

Order-disorder phase transitions occur not only in alloys with a stoichiometric composition, but also in alloys with other compositions. The critical temperature is a well defined function of the alloy composition.

Contrary to pure metals with a perfect crystal lattice, alloys do not exhibit translational symmetry. The only exception is the case of complete ordering. However, this symmetry, even in an incompletely ordered alloy, presents a probability that different sites will be occupied (substituted) by different atoms. This symmetry changes during an order-disorder transition. Thus, two cases are possible. In the first case, the probability of site substitution at the transition point changes discontinuously (step-change), and a first-order phase transition occurs. In the second case, the probability changes continuously, and the transition is called a second-order phase transition. The ordering of atoms in the crystal lattice of an alloy may be primarily characterized by the degree to which the different sites (forming a sublattice) are occupied by the different types of atoms. Here, the ordering is considered in relation to the lattice sites, and the degree of order, called the degree of long-range order, is determined by the arrangement of atoms over the entire crystal.

The degree of long-range order may be given quantitatively by different methods. First, we shall present the following method. Let us consider a substitutional binary alloy composed of N_A atoms of A and N_B atoms of B, in which the crystal lattice containing N sites may be divided into two sublattices. These sublattices are formed from $N^{(1)}$ sites of the first type, applying to atoms A, and $N^{(2)}$ sites of the second type, applying to atoms B, respectively. Here, $\nu = \frac{N^{(1)}}{N}$ denotes the relative concentration of sites of the first type, and $c_A = \frac{N_A}{N}$ the relative concentration of atoms of the first type (which must not be equal to ν). Further, let $N_A^{(1)}$, $N_A^{(2)}$, $N_B^{(1)}$, $N_B^{(2)}$ denote the number of atoms A and B on sites of the first and second type, respectively, and

$$p_A^{(1)} = \frac{N_A^{(1)}}{N^{(1)}}, \quad p_A^{(2)} = \frac{N_A^{(2)}}{N^{(2)}}, \quad p_B^{(1)} = \frac{N_B^{(1)}}{N^{(1)}}, \quad p_B^{(2)} = \frac{N_B^{(2)}}{N^{(2)}} \tag{1.1}$$

denote the probabilities of occupation of the first and second sites for atoms A and B. Furthermore,

$$\left.\begin{aligned} N_A^{(1)} + N_A^{(2)} = N_A, \quad N_B^{(1)} + N_B^{(2)} = N_B, \\ N_A + N_B = N^{(1)} + N^{(2)} = N, \\ N_A^{(1)} + N_B^{(1)} = N^{(1)}, \quad N_A^{(2)} + N_B^{(2)} = N^{(2)}. \end{aligned}\right\} \tag{1.2}$$

Consequently,

$$\left.\begin{aligned} p_A^{(1)} + p_B^{(1)} = 1, \quad p_A^{(2)} + p_B^{(2)} = 1, \\ \nu p_A^{(1)} + (1 - \nu)\, p_A^{(2)} = c_A. \end{aligned}\right\} \tag{1.3}$$

The degree of long-range order η is defined by

$$\eta = \frac{p_A^{(1)} - c_A}{1 - \nu}. \tag{1.4}$$

It is easily seen that once the values of η and c_A are known, all the probabilities (1.1) of occupation of lattice sites for atoms A and B will be determined. In fact, from (1.3) and (1.4) we obtain:

$$\left.\begin{aligned} p_A^{(1)} &= c_A + (1 - \nu)\eta, & p_B^{(1)} &= 1 - c_A - (1 - \nu)\eta, \\ p_A^{(2)} &= c_A - \nu\eta, & p_B^{(2)} &= 1 - c_A + \nu\eta, \end{aligned}\right\} \qquad (1.5)$$

where ν is determined by the crystal lattice structure.

The degree of long-range order η introduced by Eq. (1.4) is proportional to the deviation of the probability $p_A^{(1)}$ from its value c_A in a disordered alloy. Therefore, in a disordered alloy (of any composition) $\eta = 0$, while in ordered alloys this magnitude will be the larger, the more perfect the crystal. For a completely ordered alloy (stoichiometric composition) $\eta = 1$. In the case of alloys with non-stoichiometric composition ($c_A \neq \nu$) η will be always less than unity. For $c_A \leqslant \nu$ in a state with the maximum possible order, all atoms A are found at sites of the first type, i.e., the probability $p_A^{(2)}$ equals zero. Thus, from (1.5) we find that the maximum value of η in this case is given by

$$\eta_{\max} = \frac{c}{\nu}. \qquad (1.6)$$

If, however, $c_A \geqslant \nu$, then in a state with the greatest possible ordering all sites of the first type are known to be occupied by atoms A and $p_A^{(1)} = 1$. Hence,

$$\eta_{\max} = \frac{1 - c_A}{1 - \nu}. \qquad (1.7)$$

Thus, the maximum degree of long-range order is a function of the composition of the alloy. This function, given by (1.6) and (1.7), is illustrated in Fig. 1,* for $\nu = \frac{1}{2}$ and $\nu = \frac{1}{4}$. The fact that $\eta_{\max} < 1$ for alloys with a non-stoichiometric composition is due to the impossibility of obtaining perfect periodicity of the crystal lattice (due to alternation of atoms) in such an alloy.

---

*It will be shown later that ordered structures exist within a certain limited range of concentrations.

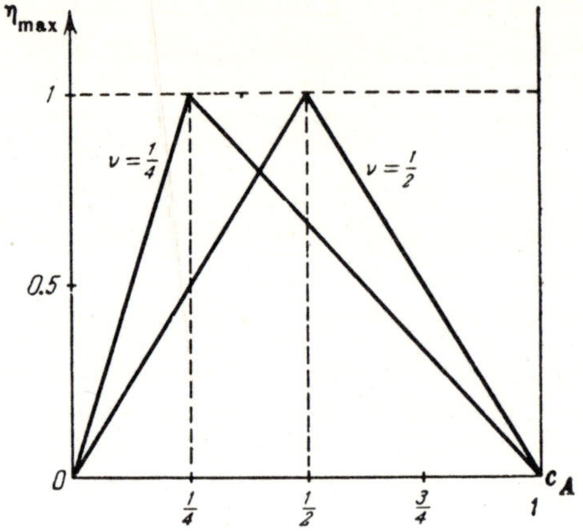

Fig. 1. Dependence of the maximum possible value of the degree of long-range order η_{max} on the concentration c_A of atom A in alloys (for $\nu=\frac{1}{4}$ and $\nu=\frac{1}{2}$).

Another definition of the degree of long-range order is also used in the literature

$$\eta_* = \frac{p_A^{(1)} - c_A}{\nu}, \tag{1.8}$$

where $\gamma = c_A \frac{1-\nu}{\gamma}$ for $c_A \leqslant \nu$ and $\gamma = 1 - c_A$ for $c_A \geqslant \nu$. The degree of long-range order η_* for an alloy of any composition, defined in this manner, varies from zero (for a disordered alloy) to unity (for the maximum possible order in a given alloy). The parameter η_* therefore is more suitable for characterization of the deviation of order from the maximum possible value for an alloy of given composition, whereas the parameter η is more suitable for describing the deviation of order in a given alloy from the order in a completely ordered alloy with stoichiometric composition having a perfectly periodic lattice (in the sense of alternation of the atoms). Since many properties of alloys are related to the latter deviation, it will be more convenient to use the parameter η in the following presentation.

We note that for alloys with stoichiometric composition ($c_A = \nu$) the definitions of η and η_* given by formulas (1.4) and (1.8) are identical. For alloys of non-stoichiometric composition these formulas lead to the following relationship between

η and η_*:

$$\eta = \frac{\gamma}{1-\nu}\eta_*. \tag{1.9}$$

The long-range order in alloys having more than two types of atoms or more than two types of sites can no longer be characterized by a single long-range order parameter η. In these cases one must introduce several analogous long-range order parameters.

Consequently, the degree of long-range order decreases with an increase in temperature and disappears at the critical temperature.

The ordered state may also be characterized by the average number of atoms of a different type surrounding a given type of atom. In this case the degree of order is defined with respect to atoms of the alloy and not with respect to the lattice sites and is called the degree of short-range order. The degree of short-range order may also be defined in different ways. For example, for binary alloys $A-B$ with a 50-percent stoichiometric composition in which sites of the first type are always surrounded by sites of the second type and vice versa, the degree of short-range order may be defined by

$$\sigma = \frac{2N_{AB} - N^*}{N^*}, \tag{1.10}$$

where N_{AB} is the number of pairs of neighboring atoms A and B, and N^* is the total number of pairs of neighboring atoms. It is obvious that in this case the degree of short-range order varies from unity (for complete ordering, when $N_{AB} = N^*$) to zero (for a completely random arrangement, when $N_{AB} = \frac{N^*}{2}$).

In the general case, short-range order may be characterized by another parameter (for instance, by the correlation parameters for the different coordination spheres). For a binary alloy one may choose as such parameters the quantities $\varepsilon_{AB}^{LL'}(\rho_l)$ defined by

$$\varepsilon_{AB}^{LL'}(\rho_l) = p_{AB}^{LL'}(\rho_l) - p_A^{(L)} p_B^{(L')}. \tag{1.11}$$

Here $p_{AB}^{LL'}(\rho_l)$ is a probability that sites L are occupied by the atom A, and that sites L' located a distance ρ_l from atom A (in the l-th coordination sphere) sites are occupied by atoms B,* while $p_A^{(L)}$ and $p_B^{(L')}$ are the probabilities introduced above for the occupation of L and L' sites by atoms A and B, respectively. (If two types of sites exist, $L, L' = 1, 2$.)

If the probabilities of occupation of sites of a given type by atoms A and B were independent of the arrangement of atoms on the other sites, then $p_{AB}^{LL'}(\rho_l)$ would be equal to the product of the a priori probabilities $p_A^{(L)} p_B^{(L')}$ and $\varepsilon_{AB}^{LL'}(\rho_l)$ would be equal to zero. In this case there would be no correlation between the filling of different types of sites by atoms A and B. Therefore, when the parameters $\varepsilon_{AB}^{LL'}(\rho_l)$ are not equal to zero, a correlation will exist in the alloy. The correlation for different spheres of coordination will be a function of the value of ρ_l.

In the special case of an alloy with stoichiometric composition AB, in which sites of the first type are surrounded only by sites of the second type and vice versa, there exists a simple relation between the parameter $\varepsilon_{AB}^{12}(\rho_1)$ for the first coordination sphere and the degrees of short-range and long-range orders σ and η,

$$\sigma = \eta^2 + 4\varepsilon_{AB}^{12}(\rho_1). \tag{1.12}$$

Thus, the degree of short-range order in an ordered alloy is determined by both the degree of long-range order and by correlation. It follows from Eq (1.12) that the correlation parameter $\varepsilon_{AB}^{12}(\rho_1)$ tends to zero both with a temperature increase, when $\eta = 0$ and $\sigma \to 0$, and with a temperature decrease, when almost a completely ordered state exists, i.e., η and σ tend to unity.

For disordered alloys,** when all the sites are equivalent and their probabilities of occupation $p_A^{(L)}$ and $p_B^{(L')}$ are equal to the atomic concentration of components c_A and c_B, respectively, the correlation parameters are independent of the types of sites L and L' and are determined by

$$\varepsilon_{AB}(\rho_l) = p_{AB}(\rho_l) - c_A c_B, \tag{1.13}$$

where $p_{AB}(\rho_l)$ is the probability of occupation of any site by atom A and of another site, located at a distance ρ_l from the former by atom B.

---

*Eq. (1.11) is written for the case where the probability $p_{AB}^{LL'}$ depends only on the modulus of the vector ρ_l. Otherwise $\varepsilon_{AB}^{LL'}$ would also depend on the direction of this vector.

**Alloys in which only short-range order exists and long-range order is absent are also called disordered.

If this probability for neighboring sites in an alloy is greater than the product of the a priori probabilities $c_A c_B$, then $\varepsilon_{AB}(\rho_1) > 0$. This case corresponds to random arrangement of atoms (when atoms of one type are surrounded predominantly by neighboring atoms of the other type). In the opposite case, the atoms are surrounded predomināntly by neighboring atoms of the same type and $\varepsilon_{AB}(\rho_1) < 0$. The short-range order in a disordered alloy is determined entirely by the correlation parameter. For example, when $\eta = 0$ Eq. (1.12) will read $\sigma = 4\varepsilon_{AB}(\rho_1)$

The following parameters are sometimes used to characterize the short-range order in disordered binary alloys:

$$\alpha_l = 1 - \frac{p_{AB}(\rho_l)}{c_A c_B}. \tag{1.14}$$

Equations (1.13) and (1.14) show that the correlation parameter $p_{AB}(\rho_l)$ is related to α_l by

$$\varepsilon_{AB}(\rho_l) = -c_A c_B \alpha_l. \tag{1.15}$$

The degree of short-range order also decreases as the temperature increases; in contrast to the degree of long-range order, however, it does not disappear at the order-disorder phase transition point.

The short- and long-range orders are closely interrelated, because they characterize the ordered arrangement of atoms from different points of view. But these concepts are, of course, not equivalent. Thus, nonequilibrium states of an alloy are possible in which the degree of long-range order is equal to zero, and simultaneously an almost complete short-range order exists (Fig. 2). In describing the state of order in an alloy, one may speak not only of the composition of the first, but also of any coordination sphere surrounding a given atom. In this sense we speak of short-range order at any distance. If the difference in the mean concentration of a given type of atom in different spheres is retained when the radius of the coordination sphere tends to infinity, long-range order will exist in the alloy. If, however, the mean concentrations tend to the same limit, long-range order will not exist. Hence the concept of long-range order can be introduced by considering short-range order in remote coordination spheres whose radii tend to infinity. The physical basis of ordering in alloys capable of being ordered is that, from the energy standpoint, it is more favorable for atoms of any type to be surrounded by atoms of another type. At sufficiently high temperatures (at which the long-range order is still absent) this leads to the appearance of short-range order. (In this case $\sigma > 0$ and $\varepsilon_{AB}(\rho_1) > 0$.) By the same virtue, a temperature decrease will lead to an order-disorder phase transition and to the formation of an ordered structure with long-range order and with a more or less regular alternation of atoms on the lattice sites.

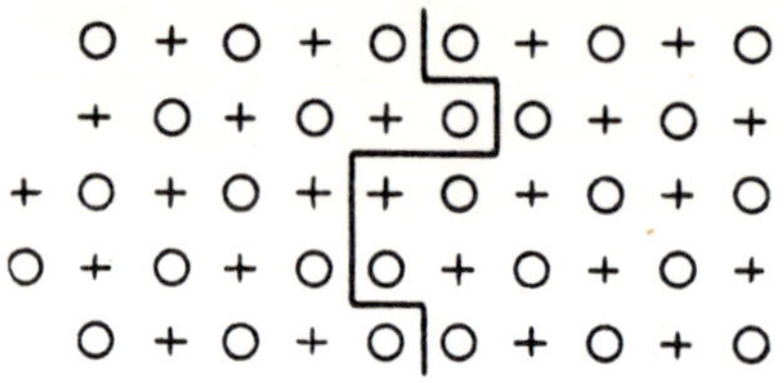

Fig. 2. An example of the arrangement of atoms having almost complete short-range order with the degree of long-range order equal to zero. Circles—atom A, crosses—atoms B.

In some alloys it is energetically favorable that atoms of a given type be surrounded by atoms of the same type. In fact, such a type of short-range order is established at high temperatures. (Thus, $\sigma < 0$ and $\varepsilon_{AB}(\rho_1) < 0$.) It is well known that such alloys may decompose at low temperatures with the precipitation of different phases, i.e., disordered solid solutions of a different composition. Alloys in which long-range order can exist are called ordered, while alloys which can decompose into disordered solutions are called completely disordered.

In a number of cases the experimental investigation of alloys does not reveal either long-range order or complete disorder, and such alloys exist in the form of a solid solution with only short-range order down to the lowest temperatures. This is explained by the fact that the critical or decomposition temperature is quite low, and diffusion of atoms is virtually impossible below this temperature. The alloy therefore can not attain a state of equilibrium.

2. Examples of Crystal Structures in Ordered Alloys

Let us examine certain structures of ordered alloys (sometimes called superlattices). We shall classify alloys according to type of crystal lattice in the ordered state.

1. One of the simplest structures encountered among the ordered alloys is the structure of β-brass. In the disordered state, atoms of one type (A or B) of this alloy are distributed with equal probability over all sites of the body-centered cubic lattice. In the ordered state, atoms of one type, for example, A, occupy sites predominantly at the corners of the cubic unit cells, and atoms B— at their centers (Fig. 3). With stoichiometric composition (alloy AB) in a completely ordered state, all the sites at the corners of the cubic unit cells are occupied by atoms A and the sites at the centers of the cells by atoms B. Obviously the state does not change in this type of alloy if the sites correct for atoms A are centers of the cubic cells, while sites correct for atoms B are at the corners.

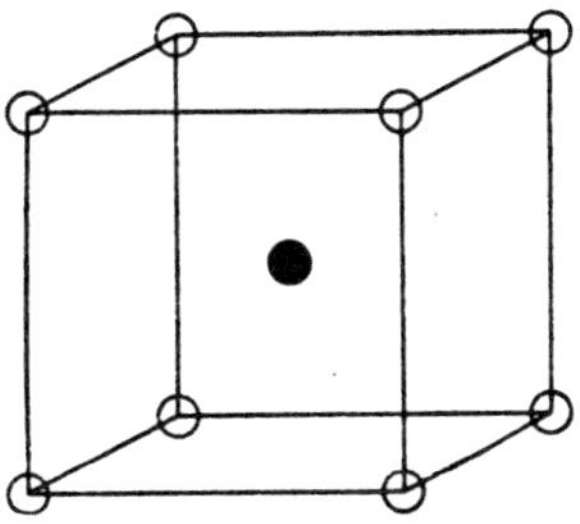

Fig. 3. Arrangement of atoms in the unit cell of completely ordered alloy with a body-centered cubic β-brass lattice. Open circles—atoms A, solid circles—atoms B.

A typical representative of alloys with this structure is the Cu-Zn alloy over the concentration range from 45.8 to 48.9 at.-% Zn (β-brass). The degree of long-range order in such an alloy decreases continuously with increasing temperatures and becomes zero at the critical temperature $T_0 = 742°$ K (for an alloy with a composition of 48.9 at.-% Zn). Hence, a second-order phase transition occurs at this point. The critical temperature depends on the concentration of atoms and decreases with the departure from stoichiometric composition (Fig. 4). An identical type of structure occurs also in the ordered alloys with a composition close to CuBe, CuPd, AgMg, FeAl, AuZn, FeCo etc. Some of these alloys (for example, AuZn) exist in the ordered state up to the melting point.

2. Alloys with a composition close to stoichiometric, corresponding to formula AB_3, may have a structure similar to

$AuCu_3$. In this case the alloy has a face-centered cubic lattice in the disordered state, and atoms A (or B) are encountered with equal probability at all the sites. In the ordered state sites correct for atoms A are the corners of the cubic cells, and for atoms B—the centers of the faces of these cells. Obviously, the number of sites of the second type is three times greater than the number of sites of the first type. In the completely ordered stoichiometric alloy all sites are substituted by atoms A and B, respectively (Fig. 5). In this type of alloy, as the temperature increases the degree of long-range order at first decreases continuously to some nonzero value, and then, by a step change, drops to zero at the critical temperature. Thus, a first-order phase transition occurs at this point. The most completely investigated alloys of this type are the alloys of the AuCu system that are close to the $AuCu_3$ composition. Such alloys with a stoichiometric composition become ordered upon the attainment of an equilibrium state below a temperature of approximately 665°K. Experiments [2] show that the critical temperature decreases (Fig. 6) with departure from the stoichiometric composition in either direction. The ordered phase exists over an approximate range 17 to 37 atomic % of Au.

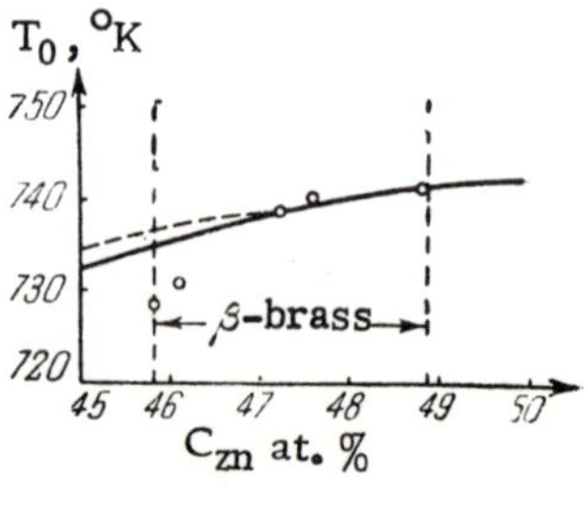

Fig. 4. Dependence of the critical temperature of β-brass on atomic concentration of zinc [1]. Circles—experimental points; dashed and solid curves—theoretical curves obtained by the Gorsky-Bragg-Williams method and from the quasichemical approximation, respectively.

An order-disorder phase transition is also detected [3] for the Au_3Cu alloy, but the critical temperature $T_0 = 516°$ K is lower than that in the $AuCu_3$ alloy. The alloys $PtCu_3$, $MnNi_3$, $FeNi_3$, $GaNi_3$ and some others also exhibit this type of structure in the ordered state.

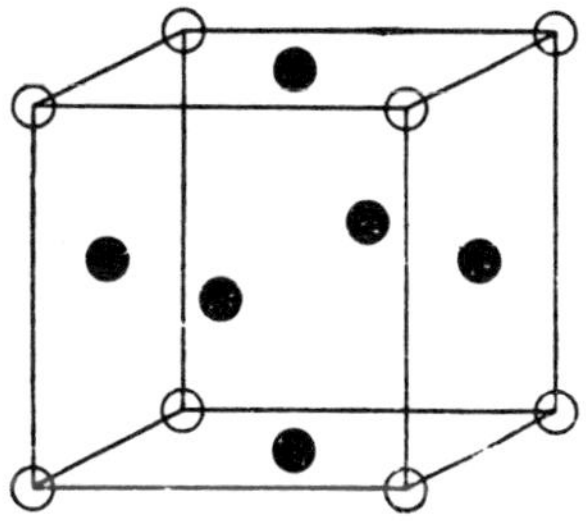

Fig. 5. The arrangement of atoms in the unit cell of a completely ordered alloy with a face-centered cubic $AuCu_3$ lattice.
Open circles—atoms A, solid circles—atoms B.

3. Alloys whose lattice in the disordered state is a face-centered cubic and whose composition is close to 50% possess the AuCu type of crystal structure in the ordered state. In a completely ordered alloy of stoichiometric composition (alloy AB) atoms A and B are located in alternating atomic planes, as illustrated in Fig. 7. Unlike the structures considered above, in which the cubic symmetry is retained in the ordered state, in AuCu type alloys the crystal lattice becomes tetragonal during ordering. In these alloys, as in the alloy $AuCu_3$, the degree of long-range order at the order-disorder transition point undergoes a step-change. In the alloy AuCu, the disordered phase exists at temperatures higher than approximately 685°K. The ratio of the sides of the unit cell $\frac{a}{c}$ (Fig. 7) in a completely ordered alloy is equal to 1.08. Such type of superlattice is observed in CoPt, and also in some other alloys.

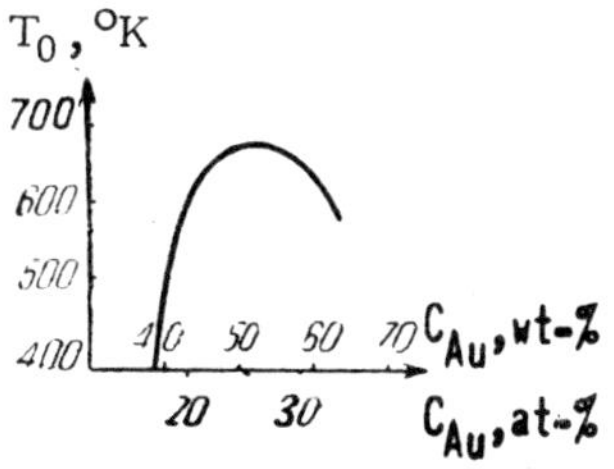

Fig. 6. Critical temperature of the Au-Cu alloy (close to the stoichiometric composition of the alloy $AuCu_3$) as a function of the concentration c_{Au} of gold.

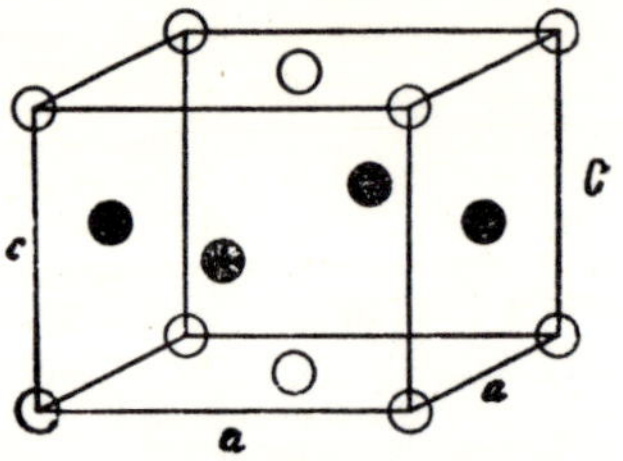

Fig. 7. Arrangement of atoms in a unit cell of a completely ordered alloy with a face-centered cubic AuCu lattice.
Open circles—atoms *A*, solid circles—atoms *B*.

4. Fe_3Al type alloys have a more complex structure. In the disordered state atoms of these alloys are arranged at the sites of a body-centered cubic lattice. Thus, the corners of the cubic cells of the alloy A_3B are permanently occupied by atoms *A*, whereas the centers of the cells are occupied with equal probability by either atoms *A* or *B*. In a completely ordered state the corners of the cell are occupied by atoms *A*, while atoms *A* and *B* occupying the centers of the cell form a NaCl-type ordered structure (Fig. 8). Thus, only one-half the atoms participate in the ordering in such alloys. In this case the probabilities of occupation of the corners of the cubic cells by atoms *A* can be approximately regarded as being independent of temperature over a rather broad range. In Fe-Al alloys this type of ordered phase lies within a certain concentration range near the stoichiometric composition of Fe_3Al below a temperature of approximately 830°K (Fig. 9 [4].)

5. Ordered alloys with a hexagonal close-packed structure have also been encountered. These include: Mg_3Cd, $MgCd_3$, Ni_3Sn and other alloys of similar composition. Thus, the alloy Mg_3Cd in a completely ordered state has the arrangement of atoms depicted in Fig. 10.

6. An interesting superstructure is observed in alloys of the system Cu-Pt. The alloy CuPt has a face-centered cubic structure in the disordered state. In a completely ordered alloy the Cu and Pt atoms are positioned in alternating planes, perpendicular to the body diagonal of the cubic cell [(111) planes, as shown in Fig. 11a]. Thus the cubic lattice is transformed into a rhombohedral (or trigonal) lattice. In ordered alloys having a concentration of copper atoms exceeding 50 atomic %, the excess atoms occupy any site on the planes that are correct for the Pt atoms. If, however, the ordered alloys have a concentration of Pt atoms exceeding 50 atomic %, the

excess Pt atoms are located in planes correct for copper atoms, where they tend to occupy such positions as to be surrounded by copper atoms. This can lead to the formation of the two-dimensional ordered structure, in (111) planes, shown (for the alloy Cu_3Pt_5) in Fig. 11b.

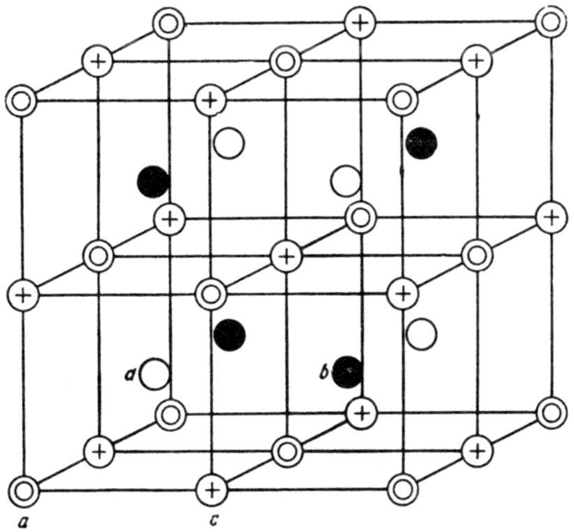

Fig. 8. Arrangement of atoms in the unit cell of a completely ordered alloy with a Fe_3Al type crystal lattice. The sublattices *a*, *c* and *d* in a completely ordered state are occupied by atoms Fe, and the sublattice *b*–by atoms Al.

7. In addition to the structures of ordered alloys mentioned heretofore, we encounter alloys with other types of crystal lattices. In the alloy AgZn, slow cooling produces a transition from a disordered state with a body-centered cubic lattice to an ordered state with a hexagonal lattice. During rapid cooling

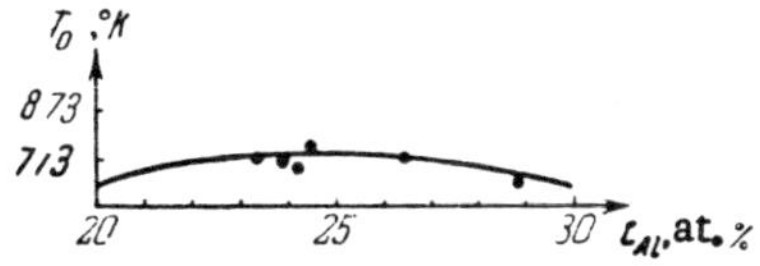

Fig. 9. Dependence of the critical temperature for alloy Fe-Al on the concentration of aluminum (near the stoichiometric composition corresponding to Fe_3Al) according to experimental data of Sykes [4]. The theoretical curve is obtained from the quasi-chemical approximation.

at temperatures of approximately 550°K, an ordered phase with a β-brass body-centered cubic lattice is formed. The addition of several atomic % of Au to the AgZn alloy during slow cooling leads to the formation of a β-brass superlattice and the critical temperature becomes markedly higher.

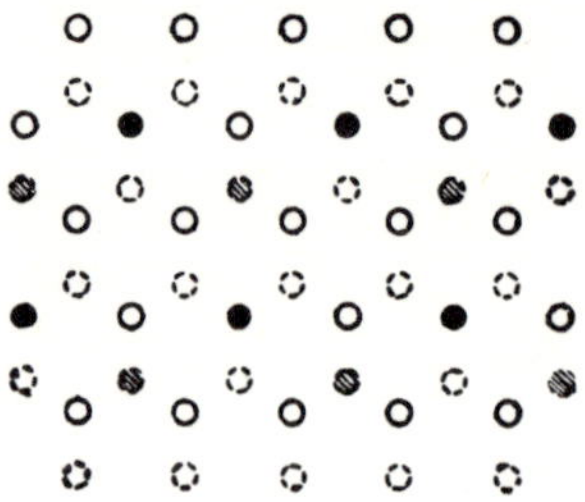

Fig. 10. Arrangement of atoms in a completely ordered alloy having a Mg_3Cd type crystal lattice. Open circles—Mg atoms, solid and cross-hatched circles—Cd atoms, dashed and cross-hatched circles indicate atoms in the odd planes.

An order-disorder transformation may also occur in multicomponent alloys and compounds. Here we refer to the Heusler alloy Cu_2MnAl, which has a Fe_3Al type body-centered cubic lattice in the ordered state (see Fig. 8). Here, the sites at the corners of cubic cells (sublattices *a* and *c*) are correct for Cu atoms, and Al and Mn atoms are properly alternated at the centers of the cells (in sublattices *b* and *d*, respectively).

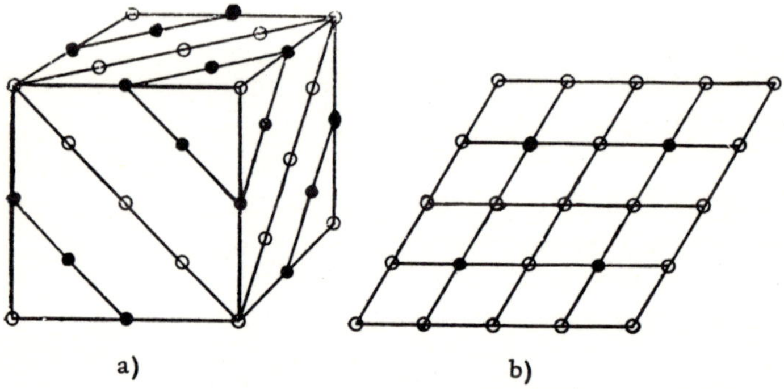

Fig. 11. Arrangement of atoms in ordered Cu-Pt alloys. a—structure corresponding to the composition CuPt (the alternating planes (111), occupied by Cu and Pt atoms are indicated); b—ordered structure in the (111) planes for the alloys Cu_2Pt_5. Dark curves—Pt atoms, light curves—Cu atoms.

An order-disorder transformation also occurs in the compounds Ag_2HgI_4 and Cu_2HgI_4. Here in a completely ordered state the Hg atoms occupy the corners of almost cubic cells (Fig. 12), while the Ag (or Cu) atoms occupy the centers of the vertical faces. The centers of the horizontal faces remain unoccupied. In this manner a layered structure is obtained. The iodine atoms do not take part in the ordering; forming a face-centered lattice, they are located around four of the eight corners of the almost cubic cells. In the disordered state, the atoms Hg, Ag (or Cu) and the vacant lattice sites (vacancies) are arranged in a random manner on sites lying both at the corners and in the middle of the faces of the cells. Thus, we have here an interesting example when vacancies as well as atoms participate in the order-disorder transformation.

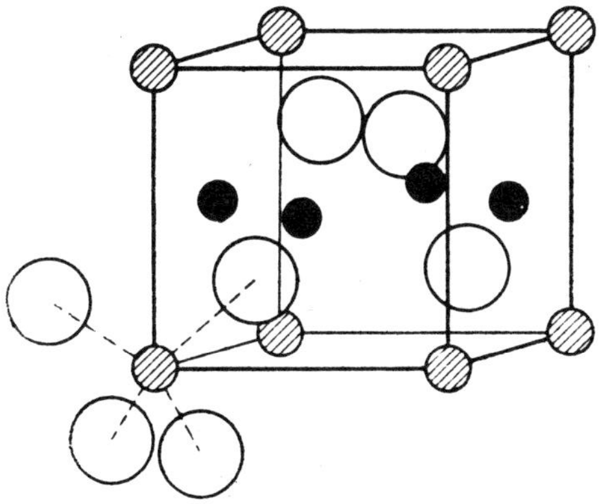

Fig. 12. Arrangement of atoms in the ordered crystal Ag_2HgI_4. Open circles—I atoms, solid circles—Ag atoms, cross-hatched circles—Hg atoms.

Such a transformation is observed in the Ni_2Al_3 alloy which has a hexagonal lattice in the ordered state. This alloy may be considered as a ternary alloy composed of Ni, Al atoms and vacancies, which along with the metallic atoms participate in the ordering process.

We note that the order-disorder transformation of atoms is observed not only in metallic alloys, but also in some non-metallic crystals. Examples include certain oxides having a spinel structure and containing two or more metallic atoms, such as, ferrites. In ferrite crystals the metallic ions may be observed at two types of positions: at the octahedral sites—surrounded by six oxygen ions—and at the tetrahedral sites—surrounded by four oxygen ions. If two species of metallic ions occur at a given type of site, in a number of cases there may occur an ordered arrangement at these sites, which will vanish, however, at a certain critical temperature.*

3. X-Ray Studies of Ordered Alloys

The existence of superlattices in an alloy can be most directly detected by means of x-ray diffraction (and also neutron and electron diffraction methods). There is a large number of experimental papers devoted to the x-ray investigation of ordered alloys. Instead of giving a systematic survey of these studies, we shall only consider some papers as examples of the x-ray method of investigation of order-disorder transformation in alloys.

In order to illustrate the method of investigation of superlattices we shall consider a disordered alloy in which all the parallel atomic planes have the same mean composition and are equivalent for x-ray scattering. Let the x-rays have an angle of incidence ϑ such that the phases of the waves reflected from the neighboring atomic planes differ by π. In this case the neighboring planes give identical scattering amplitudes. However, these amplitudes are of different signs and determine the resultant amplitude of the wave reflected by the planes. Consequently, interference does not produce a reflection corresponding to a given angle ϑ. Hence, in an x-ray photograph only those lines will be visible that correspond to a phase difference which is a multiple of 2π for reflection from neighboring atomic planes, i.e., these lines are determined by the Wulff-Braff equation: $2d \sin \vartheta = n\lambda$ (where λ - wavelength, n — an integer, d — the distance between neighboring atomic planes). The system of obtained lines is identical to the case of a pure metal having the same structure as the disordered alloy. These are called the fundamental lines. The picture changes during passage of the alloy into the ordered state. The parallel atomic planes under consideration may no longer be equivalent, because the number of atoms of any species in different planes may be different. The amplitudes of the radiation scattered from different planes are therefore different in magnitude and do not cancel each other. Consequently, new lines (called superlattice lines), characteristic of the ordered phase (superlattice), appear in the x-ray photo-

*A detailed summary of phenomena occurring in spinels may be found, for instance, in [5].

graph. Difference in amplitude of the radiation scattered from different atomic planes in a binary alloy is proportional to the difference in concentration of any type of atoms in these planes, i.e., to the difference of the probabilities $p_{\alpha}^{(1)} - p_{\alpha}^{(2)}$ $(\alpha = A, B)$, and according to (1.5) is also proportional to the degree of long-range order η. The ability of an atom to scatter x-rays is characterized by the atomic scattering factor, i.e., the ratio of the amplitude of the waves scattered by the atom to the amplitude of the waves scattered by a "classical" free electron under the same conditions. It is obvious that if the atomic scattering factors of different atoms of the alloy are equal, the specified atomic planes are equivalent with respect to the scattering of x-rays, and no superlattice lines will occur. The difference of amplitudes of waves scattered by different atomic planes in a binary alloy is proportional to the difference of the atomic scattering factors of its components. The resultant amplitude of scattered radiation for a superlattice line, therefore, will be proportional to $\eta |f_A - f_B|$, where f_A and f_B are atomic scattering factors of atoms A and B. Hence, the intensity of a superlattice line may be written in the form*

$$I = C |f_A - f_B|^2 \eta^2, \tag{3.1}$$

where C is a coefficient of proportionality. By measuring the intensity of the lines on the x-ray photograph and using (3.1) the degree of long-range order can be determined.

Table 1 gives the values of the indices g_1, g_2, g_3 of atomic planes for some of the simplest crystal lattices, which give fundamental and superlattice lines.

As may be seen from Eq. (3.1), the intensity of the superlattice lines becomes small when atoms A and B have almost the same atomic number Z in the periodic table, because in this case their atomic scattering factors are almost identical. Such a state of affairs occurs in many ordered structures, for example, those formed in the alloys Cu-Zn, Fe-Co, Ni-Mn, etc.

In an x-ray investigation of ordering in such alloys special methods should be used. One may use the fact that the atomic scattering factors depend on the wavelength of x-rays and decrease near the characteristic radiation lines of components of the alloy. Such a decrease is different for different atoms of the alloy, hence the difference $|f_A - f_B|$ may be increased appreciably at these wavelengths. The method was first used in [6] for the determination of the superlattice in the Heusler

Table 1

Values of the indices g_1, g_2, g_3 of atomic planes corresponding to the fundamental and superlattice lines, for the simplest crystal structures.

Structure type	Fundamental lines	Superlattice lines
NaCl cubic lattice	All numbers g_1, g_2, g_3 are even	All numbers g_1, g_2, g_3 are odd
β-brass bcc lattice	The sum of $g_1+g_2+g_3$ is even	The sum $g_1+g_2+g_3$ is odd
$AuCu_3$ fcc lattice	Numbers g_1, g_2, g_3 are either all even or all odd	The numbers g_1, g_2, g_3 are not all even nor all odd
AuCu face-centered lattice	The numbers g_1, g_2, g_3 are either all even or all odd	The numbers g_1 and g_2 are even, and g_3 is odd, or g_1 and g_2 are odd and g_3 is even (the planes filled in the ordered state with atoms of one type are perpendicular to the z axis).

alloy, Cu_2MnAl, in which the atomic numbers of magnesium ($Z=25$) and copper ($Z=29$) are nearly equal. The ordered phases in the alloys CuZn [7, 8], Fe_3Ni and $FeNi_3$ [9, 10] and FeCo [11] have been investigated in the same manner. Another method which may be employed to exhibit the superlattices in alloys $A-B$ is based on the fact that the impurity atoms of the third element are sharply and nonuniformly redistributed over the sites of the first and second types if the interaction energy of these atoms with atoms A and B is greatly different. Consequently, even a small proportion of atoms of a third element (1 atomic %), having an atomic scattering factor substantially different from f_A and f_B, may substantially enhance the intensity of the superlattice line, and consequently a superlattice is detected [12]. A drawback of the method is the fact that the addition of an impurity alters somewhat all the properties of the alloy (critical temperature, degree of long-range order corresponding to a given temperature, etc.). For a quantitative determination of the properties of a pure binary alloy one must therefore conduct measurements at several concentrations of impurities and then extrapolate the results to zero concentration.

We will now consider the results of several x-ray investigations of ordering in alloys. The results of investigations of the structures of alloys in the disordered and the ordered states have already been given in Sec. 2; we will therefore discuss here only the dependence of the parameters characterizing the long-range and short-range order on the temperature and composition of the alloy.

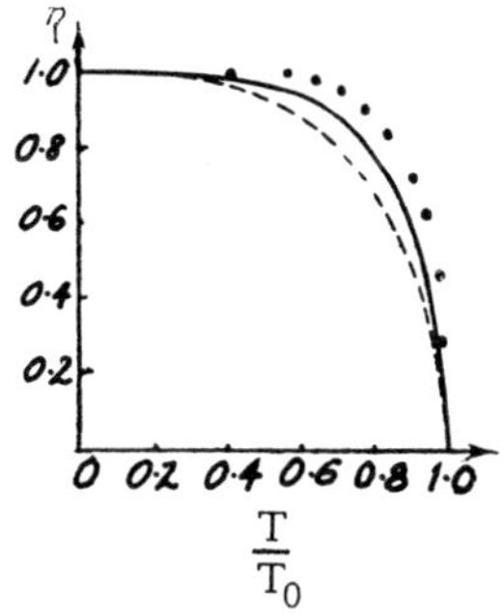

Fig. 13. Dependence of the degree of long-range order of β-brass on the ratio of the absolute temperature T to the critical temperature $T_0 =$ 742°K (experimental points according to Chipman and Warren [8]). Theoretical curves were plotted from the approximation of Gorsky, Bragg and Williams (dashed curve) and from the quasi-chemical approximation (solid curve).

The investigation of the temperature dependence of the degree of long-range order in β-brass has been conducted [8] by the method explained above, based on the use of characteristic x-rays of the metals composing the alloy. CuK_α radiation [8] with the ionization method of recording was used. The dependence of the degree of long-range order η on the ratio of temperature T to the critical temperature T_0, depicted in Fig. 13, was obtained. It is apparent that in β-brass almost complete ordering is already established at $\frac{T}{T_0} = 0.6$, i.e., at $T \approx 440$°K.

A β-brass type ordered structure is also formed in the alloy AgZn [13] upon quenching from temperatures a little above 280°C. Above this temperature the alloy is in a disordered state and has a body-centered cubic lattice. During slow cooling a more stable complex hexagonal structure (ζ-phase) arises, which makes it impossible to investigate the order-disorder transformation in the body-centered cubic lattices of the alloy AgZn. In order to make possible such investigation, alloys were prepared [14] in which relatively small amounts of silver atoms were substituted by gold atoms (up to 5 atomic %). In the alloy AuZn no hexagonal-ζ-phase is formed, and at all temperatures up to the melting point there exists an ordered body-centered cubic β-brass lattice [15]. Thus, the critical temperature in this alloy is very high and lies above the melting point of 725° C. Consequently, in AgZn alloys in which small quantities of Ag atoms were substituted by Au atoms the critical temperature was elevated, and for an atomic concentration of Au above 3% the presence of ζ-phase has not been detected [14]. Thus, in this type of alloy we succeed in investigating the ordering at sites of the body-centered cubic lattice. By varying the concentration of gold in alloys, the dependence of the critical temperature on the atomic concentration of gold could be determined (Fig. 14).

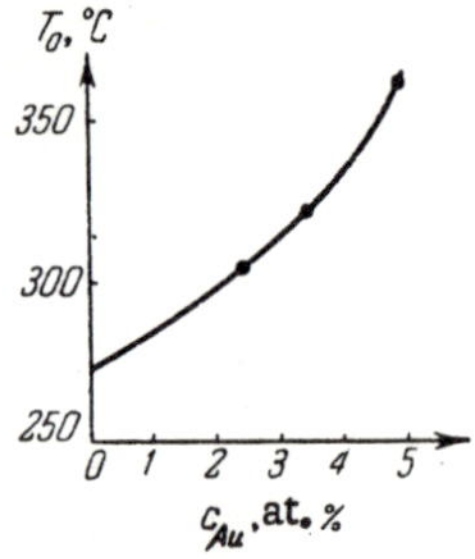

Fig. 14. Dependence of the critical temperature T_0 of the alloy Ag-Zn-Au on the atomic concentration of gold c_{Au} at constant (0.5) atomic concentration of zinc [14].

It is noteworthy that even a small amount of gold impurity (5 atomic %) changes substantially the critical temperature of the alloy (by ∼ 80° C). An investigation of the temperature dependence of the degree of long-range order in these alloys has shown that this dependence is of the same type as for β-brass, i.e., their order-disorder transition is a second-order phase transition. Ordering of alloys in the AuCu system has been studied in detail in a number of papers (cf. [16–22]).

We cite the results of x-ray studies of the temperature dependence of the degree of long-range order of alloys with stoichiometric composition $AuCu_3$ carried out by Komar and Buinov [21]. Figure 15 shows the curve obtained by these authors. It is evident that the step-change in the degree of long-range order during passage through the critical temperature T_0 is, for this alloy, approximately 0.83. The dependence of the degree of long-range order on concentration in the alloys Au-Cu was by Ageev and Shoikhet [16], [17] and Komar and Buinov [21]. The dependence of η on c_{Au} obtained by these authors revealed that the maximum degree of long-range order occurs when the composition is close to the stoichiometric concentration for $AuCu_3$.

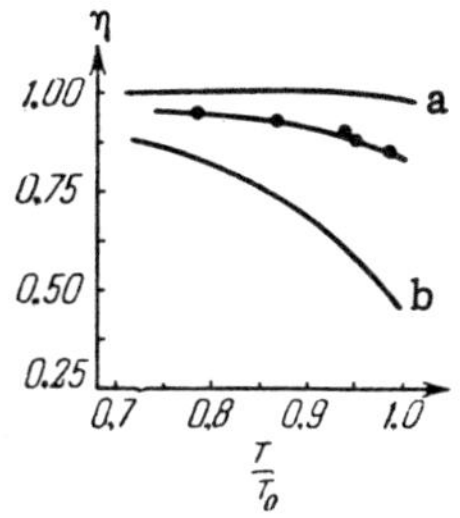

Fig. 15. Dependence of the degree of long-range order in the alloy $AuCu_3$ on the ratio of the absolute temperature T to the critical temperature $T_0 = 665°K$, according to Komar and Buinov [21] (experimental points). Theoretical curves are constructed in the quasi-chemical approximation (curve *a*) and in the approximation of Gorsky, Bragg and Williams (curve b).

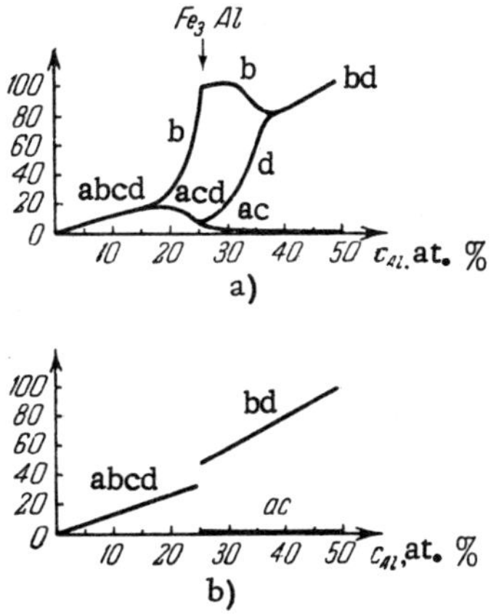

Fig. 16. Concentration of Al atoms in the different sublattices of the alloy Fe-Al as a function of the concentration c_{Al} in the alloy [23]. a—annealed alloy, b—alloy quenched at a temperature of 600°C.

In alloys of the system Fe-Al, ordering was studied by Bradley and Jay using x-ray methods [23]. These authors investigated the change in concentration of aluminum atoms in sublattices, both in alloys quenched at a temperature of 600°C and in slowly annealed alloys, as a function of alloy composition. Figure 16a gives the dependence of the concentration of aluminum atoms in the sublattices *a*, *b*, *c* and *d* (see Fig. 8) on the atomic concentration of aluminum c_{Al} in annealed Fe-Al alloys. At values of c_{Al} less than 18 atomic % the concentrations of aluminum atoms in all the sublattices are the same, i.e., the alloy occurs in a disordered state. The sharp rise in concentration of aluminum atoms in the sublattice *b*, beginning with 18 atomic % Al, is related to the Fe_3Al ordered structure which is formed in these compositions. Beginning with 25 atomic % Al, the sublattice *b* is almost completely occupied by aluminum atoms, and with a further growth of c_{Al} the excess Al atoms are substituted into the sites of the sublattice *d*, which is expressed as a rapid rise in curve *d* near the specified composition. For $c_{Al} \approx 37$ atomic % Al, the concentration of aluminum atoms in the sublattices *b* and *d* become identical. This means that another brass superlattice, corresponding to the composition FeAl, is formed in the alloy. Thus, the concentrations of aluminum atoms in the sublattices *a* and *c* are also identical and differ little from zero.

For the Fe–Al alloys quenched at a temperature of 600° C, Fig. 16b gives the dependence of the concentration of Al atoms in different sublattices on c_{Al}. It is apparent that for $c_{Al} < 25$ atomic % the alloy is in a disordered state. For $c_{Al} < 25$ atomic %, the probability of occupation of the sites of the sublattices *b* and *d* by aluminum atoms differs from the probability of occupation of the sites of sublattices *a* and *c*. This is associated with the appearance of a β-brass type ordered structure in this range of concentration at a temperature of 600° C.

We shall now consider the results of x-ray studies of short-range order in alloys. The experimental determination of the short-range order parameter in different coordination spheres is based on the x-ray analysis of the distribution of background intensities caused by disorder in the alternation of atoms. The difficulty of determining the short-range order in alloys from an analysis of the distribution of background intensity is due to the fact that in addition to the background caused by disorder in alternation there is also some background due to other causes: Compton scattering, scattering by thermal vibration of the crystal lattice and others. The theory and method of determination of short-range order from the intensity of the background associated with an incompletely ordered arrangement of atoms in the crystal lattice is expounded in Chapter VI. Here, however, we shall give only experimental results.

A careful investigation of short-range order in the first ten coordination spheres in single crystals of the alloy AuCu at temperatures above the critical temperature was conducted by Cowley [22]. The scattering of monochromatic CuK_α-radiation by the crystal lattice of this alloy has been investigated by means of the ionization technique. The background caused by Compton and thermal scattering was eliminated. A relation was obtained for the intensity of the scattered x-rays as the function of the difference of the wave vectors of the scattered and incident waves, and the intensity distribution in the reciprocal lattice space was thus determined.

Table 2

Experimental values of the coordination parameter $\varepsilon_{AuCu}(\rho_l)$ for the alloy $AuCu_3$ at temperatures above the critical temperature, determined for the first ten coordination spheres [22] .

Number of the coordination sphere l	Coordination number z_l	Coordinates of one of the atoms in the coordination sphere	$\varepsilon_{AuCu}(\rho_l)$		
			At T = 405°C	At T = 460°C	At T = 550°C
1	12	110	0.0285	0.0278	0.0246
2	6	200	−0.0349	−0.0322	−0.0197
3	24	211	−0.0017	−0.0036	−0.0049
4	12	220	−0.0178	−0.0128	−0.0084
5	24	310	0.0099	0.0092	0.0060
6	8	222	−0.0047	−0.0013	−0.0017
7	48	321	0.0030	0.0015	0.0006
8	6	400	−0.0090	−0.0079	−0.0036
9a	12	330	0.0049	0.0041	0.0021
9b	24	411	−0.0021	−0.0038	−0.0013
10	24	420	−0.0049	−0.0047	−0.0013

Figure 17 presents curves of equal intensity (in arbitrary units) in the (100) planes of the reciprocal lattice of the alloy $AuCu_3$ at a temperature of 405°C. In this figure the corners of the square correspond to the positions of the structural maximum, while the center of the square and the middle of the sides correspond to the positions of the superlattice maxima that appear during cooling below the critical temperature. It is apparent from Fig. 17 that the background is concentrated near the positions of the superlattice maxima.

Analysis of the distribution of background intensity [22] produced the parameters α_l describing the short-range order for ten coordination spheres of the disordered alloy $AuCu_3$ at temperatures of 405, 460 and 550°C. Using the values of these parameters, one may calculate from Eq. (1.15) the values of the correlation parameters $\varepsilon_{AuCu}(\rho_l)$ introduced in

Sec. 1 [cf. Eq. (1.13)]. Table 2 gives the experimental values of $\varepsilon_{AuCu}(\rho_l)$ obtained in this manner (for $l = 1, \ldots, 10$). The second column of this table gives the values of the coordination number z_l for the different coordination spheres. The third column lists the coordinates of one of the atoms of the given coordination sphere, where one half the length of the edge unit cell is chosen as a unit. The data of Table 2 can be used to determine (for each coordination sphere drawn around the central gold atom) the excess or deficiency of the average number of copper atoms compared with their number in the case of a completely random distribution of atoms. Figure 18 illustrates the dependence of the deficiency ΔN^l_{Cu} of the number of copper atoms on the radius of the coordination sphere ρ_l both in the case of a complete (long-range) order and for the disordered state of the alloy at temperatures of 405 and 550°C.

Fig. 17. Curves of equal intensity (in arbitrary units) of the scattered x-rays in (100) plane of the reciprocal lattice of the alloy $AuCu_3$ at a temperature of 405° [22].

The cited experimental data show that even in a disordered state the distribution of copper atoms around the gold atoms differs appreciably from a random distribution, which indicates the existence of appreciable short-range order in the alloy. For example, among the 12 atoms of the first coordination sphere at a temperature of 405°C there is an average of 1.4 excess copper atoms, and even the 24 atoms in the tenth coordination sphere lack an average of 0.5 atom. Thus, at

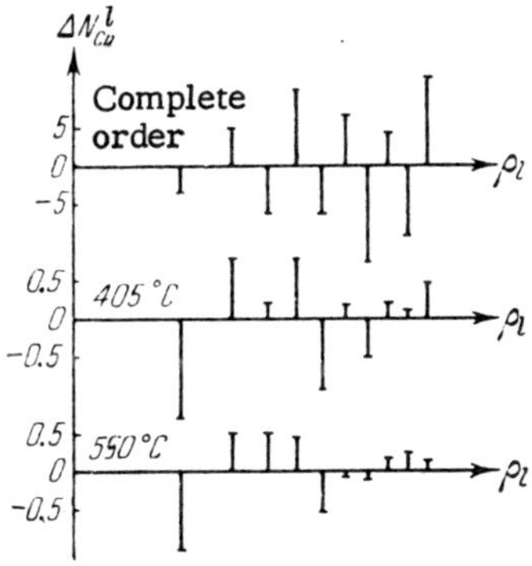

Fig. 18. Deviation of the number of copper atoms ΔN^l_{Cu} from the value corresponding to completely random distribution, in different coordination spheres. Alloy: $AuCu_3$ [22].

temperatures slightly higher than the critical temperature, the correlation in the alloy is significant even in rather remote coordination spheres. It is apparent from Fig. 18 that as the temperature is increased to 550° C, the average correlation effect in the nearest coordination spheres decreases somewhat, but still remains appreciable, whereas at more remote spheres the average correlation decreases strongly.

Table 3

Value of the correlation parameter in the alloy AuCu at a temperature of 425° C for the first six coordination spheres [26]

Number of the coordination sphere l	Coordination number z_l	Correlation parameter $\varepsilon_{AuCu}(\rho_l)$
1	12	0.033
2	6	−0.016
3	24	−0.005
4	12	−0.0165
5	24	0.008
6	8	−0.009

We again note that for a completely ordered state of the alloy the quantity ΔN^l_{Cu} changes sign during transition from one coordination sphere to the neighboring one. In a disordered state a regular alternation of signs of the quantity ΔN^{l}_{Cu} is not observed. For example, at 550° C in the second, third and fourth coordination spheres a deficiency in the average number of copper atoms is observed and in the fifth, sixth and seventh spheres an excess; in the eighth, ninth and tenth spheres a small deficiency is again detected. This may be explained by the fact that in an ordered alloy a given type of atom has a tendency to be surrounded by the other type. Therefore, even in a disordered state an excess over the average number of copper atoms is observed at sites neighboring the central gold atom. Since the nearest neighboring sites of the first coordination sphere are sites of the second, third and fourth spheres, all the spheres will have a tendency to be enriched by gold atoms. A similar situation also occurs for the next coordination sphere.

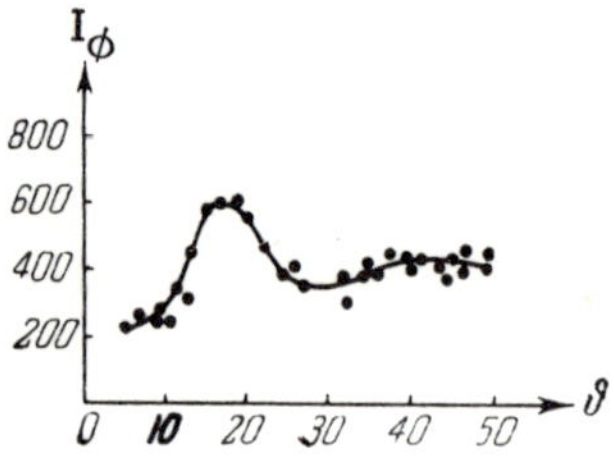

Fig. 19. Dependence of the intensity of diffuse scattering of x-rays by the alloy Ni-Au (30 atomic % Au) quenched from 900°C on the value of ϑ (2ϑ angle of scattering). Taken from the data of [28]. Intensity is given in electron units.

In this manner the correlation parameters in the first three coordination spheres in the ordered alloy $AuCu_3$ were determined at a temperature of 389° C [24].

An x-ray study of short-range order in the $AuCu_3$ alloys was also conducted in polycrystalline samples [20, 25].

In AuCu alloys the short-range order was investigated in [26] using the diffuse scattering of x-rays by a single crystal of the alloy. The correlation parameters $\varepsilon_{AuCu}(\rho_l)$ for the first six coordination spheres in the disordered alloy at a temperature of 425° C are given in Table 3. X-ray analysis of the disordered alloys [26] has established the existence of short-range order corresponding to the same distribution of atoms in small regions of the crystal (containing up to 100 atoms) as in the case of long-range order, i.e., in these regions the atoms AuCu filled predominantly the alternating atomic (100) planes. In these regions a distortion of the crystal lattice, caused by the appearance of a tetragonal structure (the type illustrated in Fig. 7), has been discovered. Thus, in a disordered alloy we encounter equal probability regions where the direction of the shortened edges of the cell c are parallel to the various edges of the cell of the original cubic lattice.

The short-range order in Au-Ag alloys has been determined by diffuse scattering of a monochromatic radiation by single crystals of AgAu and Ag_3Au [27]. In this case the same technique as in [22] was used. The samples were annealed for one week at 450° C, for two weeks at 350° C and four weeks at 300° C. Thus, the measured values of the short-range order correspond to a temperature of 300° C (or somewhat lower).

The correlation parameters $\varepsilon_{AgAu}(\rho_l)$ for the first two coordination spheres in the case of the alloy AgAu were found to have the following values:

$$\varepsilon_{AgAu}(\rho_1) = 0.02, \quad \varepsilon_{AgAu}(\rho_2) = -0.0025, \tag{3.2}$$

and in the case of the alloy Ag_3Au

$$\varepsilon_{AgAu}(\rho_1) = 0.01, \quad \varepsilon_{AgAu}(\rho_2) = -0.002. \tag{3.3}$$

The signs of the correlation parameters $\varepsilon_{AgAu}(\rho_1)$ indicate that in Ag-Au alloys any type of atoms has a tendency to be surrounded by the other type of atoms. Consequently, these alloys belong to the class of ordered alloys. The magnitudes of the correlation parameters Ag-Au in alloys were found to be appreciably less than in $AuCu_3$, in spite of the somewhat lower annealing temperature. Thus, in the first coordination sphere in the alloys AgAu and Ag_3Au, there are found an average of only approximately 0.5 excess atoms in the alloys as compared with a completely random distribution of the Ag atoms. It may be expected therefore, that the critical temperature in these alloys is appreciably less than in $AuCu_3$ (possibly below room temperature). Consequently, at temperatures lower than the critical temperature, diffusion processes leading to an order-disorder phase transition proceed extremely slowly, and the alloy remains in a nonequilibrium disordered state. This explains the fact that ordered phases in the Ag-Au alloys have not been observed.

Short-range order in the system Ni-Au has been determined from the background intensity distribution on a Debye diagram, [28] and [29]. An alloy with 30 atomic % of gold was quenched from a temperature of 900° C (at which the alloy is a homogeneous solid solution), after which the intensity of the diffuse scattering was measured at -190 °C. The dependence obtained for the intensity of duffuse scattering on the scattering angle (expressed in electron units*) is illustrated in Fig. 19. According to these data, the magnitude of the correlation parameter $\varepsilon_{NiAu}(\rho_1)$ for the first coordination sphere is equal to 0.006. The positive sign of $\varepsilon_{NiAu}(\rho_1)$ means that in the investigated alloy Ni-Au, any type of atom is predominantly surrounded by atoms of the other type. Such a type of short-range order was observed [29] at 900° C for alloys of different compositions (Table 4).

Table 4

Variation of the mean number of Ni atoms in the first coordination sphere around the Au atom during the establishment of short-range order in Au-Ni alloys

Content of Ni in the alloy, atomic %	Number of Ni atoms in the first coordination sphere around Au atoms	
	In a random distribution	Measured (with short-order)
30	3.6	3.7
50	6.0	6.2
70	8.4	8.7
75	9.0	9.4
90	10.8	11.2

It is noteworthy that no ordered phase is formed in these alloys as the temperature is reduced, but that the alloy decomposes into two disordered solid solutions with different

*The electron unit of intensity of x-ray scattering is taken as the intensity of scattering by a "classical" free electron under identical conditions.

compositions. Such unusual behavior of an alloy may be explained by the fact that atomic volumes of the Au and Ni atoms differ appreciably (the separation between the nearest neighbor atoms in pure Au is 2.88 Å, while in pure Ni it is 2.49 Å). Consequently, large lattice distortions arise in the solid solution that are associated with an appreciable energy of elastic deformation. The latter decreases if the alloy consists of regions that are closer in composition to the pure metals Ni and Au. Such a state, therefore, is energetically more favorable, and at sufficiently low temperatures total disordering occurs.

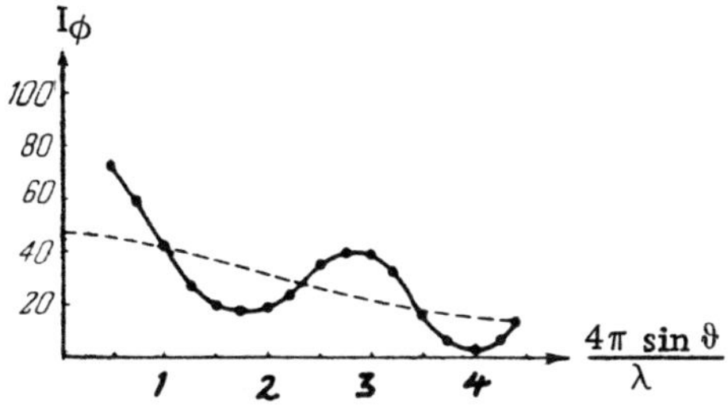

Fig. 20. Dependence of the intensity of diffuse scattering of x-rays by the alloy AlZn at a temperature of 400°C on the quantity $\frac{4\pi \sin \vartheta}{\lambda}$ [28]. Intensity is given in electron units.

The short-range order was also investigated in the disordered alloy CuPt [30]. We shall now consider alloys in which any type of atom is predominantly surrounded by the other type of atoms, i.e., in which correlation parameters $\varepsilon_{AB}(\rho_1)$ are negative. This type of short-range order was investigated in the alloy Al–Zn with 30 atomic % Zn [28]. Diffuse scattering of monochromatic radiation by a polycrystalline sample of this alloy has been investigated at 400°C. Fig. 20 illustrates the dependence of the intensity of scattering, expressed in electron units, on the quantity $\frac{4\pi \sin \vartheta}{\lambda}$, where λ is the wavelength, and 2ϑ is the scattering angle. The dashed curve indicates the monotonic distribution of intensity of scattered radiation of x-rays, which would occur in the case of a completely random distribution of atoms in the first alloy (without short-range order). The correlation parameter for the first coordination sphere was equal to $\varepsilon_{ZnAl}(\rho_1) = -0.04$ in this case. This means that each zinc atom in this alloy has, among its twelve nearest neighbors, an average of 1.7 excess zinc atoms (in comparison with a completely random distribution). Such types of regions enriched by zinc are formed in Al-Zn alloys of other compositions [31].

Thus, in an alloy containing 20 atomic % Zn, at 400°C $\varepsilon_{ZnAl}(\rho_1) = -0.022$, i.e., each Zn atom has an average of 1.3 excess neighboring atoms of Zn. Thus, in contrast with the previously considered alloys Au-Cu, Au-Ag and Ni-Au, the atoms in the system Al-Zn are predominantly surrounded by the same type of atom.

The small regions enriched by one type of atom exist in totally disordered alloys at high temperatures. When the temperature is lowered below the initial temperature, larger regions with an enhanced concentration of atoms of the dissolved component may appear. In fact, at sufficiently low temperatures the free energy of this system is lowered if regions with an increased concentration of one type of atom are formed. Consequently, such regions may be augmented as a result of uphill diffusion. In the case in which the mechanism of concentration pressures proposed by Konobeyevskiy [32] plays a leading role, the growth of a region with enhanced concentration atoms of the dissolved component has been theoretically examined by Lyubov and Maksimov [33].

With rapid quenching of totally disordered alloys a new phase does not precipitate, and the system remains in a nonequilibrium state. The region of enhanced concentration of impurity atoms in such an alloy, as in the equilibrium solution, may be studied by means of diffuse scattering of x-rays.

Walker and Guinier [34] performed an x-ray investigation of the aging process of Al-Ag with 5 atomic % Ag. The samples were annealed for 24 hours at a temperature of 520°C (the disordering temperature of the alloy of such a composition is 450°C) and subsequently quenched in water. The investigation was conducted with monochromatic CuK_α-radiation on polycrystal and single crystal samples. It was discovered that in the reciprocal space there are formed regions having almost spherical shape, which correspond to intense diffuse scattering. Each of these regions has as its center some site of the reciprocal lattice of the mother phase (and also the (000) point of this reciprocal lattice). Walker and Guinier [34] regarded this fact as proof that the nuclei atoms, which appear in the initial stages of aging, occur at sites of the crystal lattice of the mother phase (solid solution). The dependence of the intensity of radiation scattered by a polycrystal on the angle of scattering 2ϑ, as may be seen in Fig. 21, has a pronounced maximum in the region of small angles (at $2\vartheta = 2^\circ 20'$). Treating the results in the same manner as in [22], one obtains the probability $p_{AgAg}(\rho)$ that the silver atom is at a distance ρ from the central site if it is reliably known that this site is a silver atom. The dependence of this probability on the distance between the atoms is depicted in Fig. 22; the values on the curve have meaning only at specific points, corresponding to distances that are equal to the radii of the different coordination spheres. As may be seen from the figure, as ρ increases the probability $p_{AgAg}(\rho)$ decreases to a minimum at $\rho = 20$ Å, which is appreciably smaller than the value corresponding to a completely random distribution of atoms (dotted line in Fig. 22). This value is reached in practice at large distances ($\rho \sim 36$ Å). Such a probability distribution means that a quenched alloy contains regions substantially enriched by silver atoms (containing about 100 Ag atoms) and having a radius of about 10 Å. These regions are surrounded by layers that are impoverished of silver atoms. High-temperature investigations show that above the disordering temperature the regions enriched by silver atoms are appreciably smaller. With these data Walker and Guinier [34] concluded that during the quenching, uphill diffusion of silver atoms occurs into regions enriched by silver, which further increases their enrichment. This process proceeds so rapidly that the average silver concentration is not attained in the regions of enriched silver concentration. Thus, layers impoverished of silver atoms are formed.

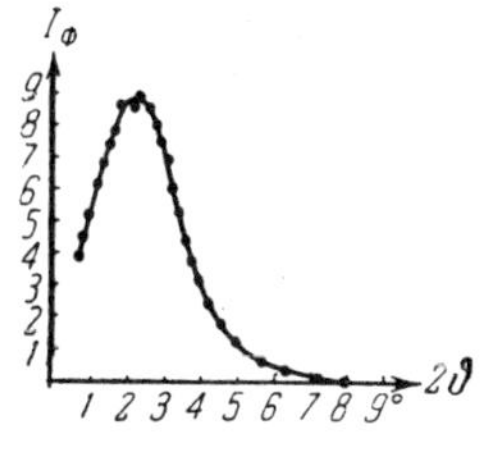

Fig. 21. Relative intensity of diffuse scattering of x-rays (CuK_α radiation) by a polycrystalline sample of the alloy Al-Ag (5 atomic % Ag), quenched from 520° C, as a function of the angle of scattering 2ϑ [34].

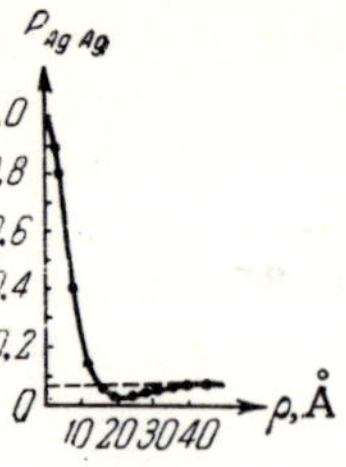

Fig. 22. Probability p_{AgAg} in the alloy Al–Ag (5 atomic % Ag), quenched from 520°C, as a function of the distance ρ between the atoms [34].

In the region of a homogeneous solid solution (at a temperature of 525°C) the short-range order in the alloy Al–Ag of the same composition has been investigated in [35]. The dependence of the probability p_{AgAg} on the distance between the atoms is given in Fig. 23. At a distance equal to the radius of the first coordination sphere (ρ = 2.8 Å), p_{AgAg} = 0.4. This means that in the first coordination sphere the neighborhood of the Ag atom contains an average of 4.8 atoms of Ag instead of 0.6 Ag atoms as in the case of a completely random distribution.

The first stages of the total disordering of a number of aging alloys were studied by x-rays by Bagaryatskiy [36]–[38]. He also reached the conclusion that at the beginning of the aging process, regions with short-range order are formed in some alloys corresponding to an increased concentration

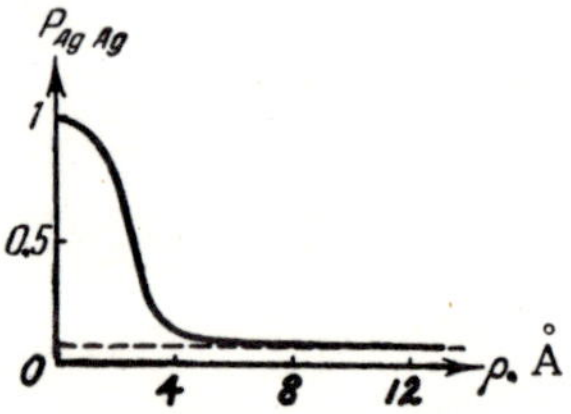

Fig. 23. Dependence of the probability p_{AgAg} in the alloy Al–Ag (approximately 5 atomic % Ag) on the distance ρ between atoms at a temperature of 525°C [35]. The dashed line indicates the value of p_{AgAg} corresponding to a completely random distribution of atoms.

of atoms of one or more species. Thus, for example [36, 38], the alloy Al-Cu-Mg with 1.3 atomic % Cu and 1.3 atomic % Mg was investigated. The author reached the conclusion that at early stages of natural aging in the alloy there arise regions enriched with CuMg, in which a slightly distorted crystal lattice of the mother solid solution is preserved.

At the present time it has not been definitely established whether the arrangement of atoms of the enriched region in the investigated early stages of aging resembles the crystal lattice of the mother phase or a new precipitating phase. It is possible that in different alloys the initial stage of disordering occurs differently. Elistratov [39-43], using an original technique based on the simultaneous investigation of scattering of both characteristic and white radiation by macrocrystalline samples, succeeded in some cases in observing diffusion spots near points of the reciprocal lattice corresponding to sites not only of the mother but also of the precipitated phases. In a number of cases he succeeded in explaining the diffusion maxima near the points of the reciprocal lattice of the mother phase by the presence of regions in which no mother phase exists (owing to the presence of a nucleating new phase) and, in other cases, by the occurrence of microscopic cracks. Consequently, the interpretation of Guinier's experiments [34] given above may require further experimental confirmation.

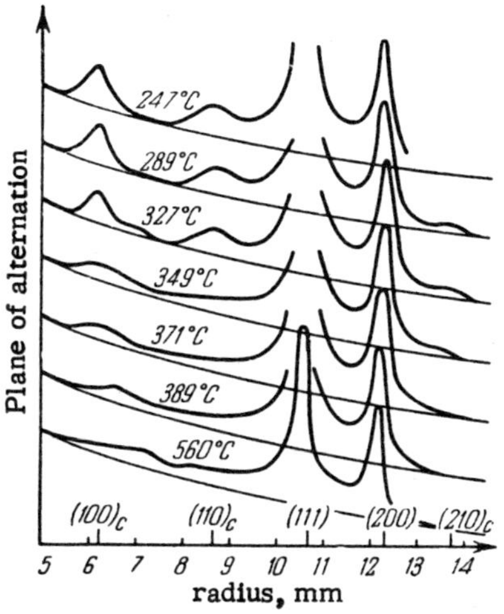

Fig. 24. Microphotometric curves obtained from electron diffraction photographs of the alloy $AuCu_3$ at different temperatures [44]. The subscript *c* denotes superlattice lines.

Results similar to those given in this section have also been obtained by electron diffraction studies of ordered alloys. In the alloy $AuCu_3$ the ordering was studied by Germer, Haworth and Lander [44], who used the electron diffraction method. The samples of alloy were thin films 300 and 400 Å thick. The intensity and width of the rings of the diffraction pattern were investigated as functions of the temperature of the alloy. Microphotometric curves of a number of such electron diffraction photographs are given in Fig. 24 for different temperatures. As in the x-ray method of investigation described above, disappearance of the superlattice lines is also observed here with a temperature increase. Moreover, an increase in width of the diffraction rings with temperature is also observed. Above the critical temperature a weakening of the maxima, caused by the short-range order, is noted.

4. Neutron Diffraction Studies of Ordering in Alloys

Ordering in alloys may also be investigated by scattering of neutrons. The velocity of the neutron used for this purpose is chosen so that the de Broglie wavelength $\lambda = \frac{h}{m_n v}$ (where $h =$ Planck's constant, $m_n =$ the neutron mass, $v =$ neutron velocity) was of the order of the lattice constant. The wavelength of thermal neutrons is of the order of the interatomic separation. Thermal neutrons are those which have been slowed down to thermal energies corresponding to room temperature (of the order $kT \sim \frac{1}{40}$ ev), or lower.

Neutron scattering is characterized by the scattering amplitude, i.e., by the amplitude $f^{(n)}$ of a spherical wave diverging from the scattering center, $f^{(n)} = \frac{e^{ik'r}}{r}$ when a plane wave e^{ikz} incident on the center in the direction of the z axis has an amplitude equal to unity (k and k' are the absolute values of wave vectors of the incident and the scattered wave). The scattering amplitude of neutrons in neutron diffraction studies plays the role of the atomic scattering factor in x-ray analysis.

Neutron scattering exhibits certain features differing from x-ray scattering and attributable to the fact that the scattering center for neutrons is the nucleus rather than the electronic shell. Since the dimensions of the atomic nucleus are several orders of magnitude smaller than the wavelength of thermal neutrons, their scattering is isotropic, i.e., the scattering amplitude $f^{(n)}$ and, consequently, the intensity of neutrons scattered by a single nucleus are independent of the scattering angle. On the other hand, the wavelength for x-rays is of the same order of magnitude as the dimensions of the electronic shells and, consequently, there is interference between waves scattered by different parts of the shell. This is known to lead to angular dependence of the atomic scattering factors of x-rays, which decrease as the angle increases. Thus, all other conditions being equal, in a neutron diffraction pattern the lines corresponding to a large scattering angle will be relatively more intense (in comparison with lines at smaller angles) than in an x-ray pattern.

Whereas the atomic scattering factors for x-rays are positive, for neutron scattering cases are known in which the scattering amplitudes are negative (for example, in the case of Mn and Li). In alloys containing elements with negative scattering amplitudes (averaged over isotopes), one may observe an unusual (in comparison with x-ray scattering) distribution of intensity between the fundamental and superlattice lines. This results in the fact that the intensity of the superlattice line on a neutron diffraction pattern, which is proportional to $|f_A^{(n)} - f_B^{(n)}|^2$, may in the case of different signs of $f_A^{(n)}$ and $f_B^{(n)}$ and at large values of the degree of long-range order turn out to be greater than the intensity of the fundamental line. It is well known that the latter is proportional to $|c_A f_A^{(n)} + c_B f_B^{(n)}|^2$ (see Chap. VI).

The feature of neutron scattering that is most significant for neutron diffraction analysis of alloys is that the neutron scattering amplitudes do not change monotonically with the atomic number, as was the case with x-rays. Moreover, the amplitudes are found to be different even for different isotopes of the same chemical element. Values of the scattering amplitudes of slow neutrons for some isotopes of different elements are given in Table 5 [45]. If the mass number of a given isotope is not indicated alongside the chemical symbol of the element in this table, then the given scattering amplitude is the mean value over the isotopes of the element. For comparison, Table 6 gives values of the atomic scattering factor of x-rays for several elements at a scattering angle 2ϑ and

Table 5

Amplitudes of coherent scattering of slow neutrons for some elements and their isotopes.

Element or isotope	Scattering amplitudes of neutrons $f^{(n)} \cdot 10^{13}$, cm	Element or isotope	Scattering amplitudes of neutrons $f^{(n)} \cdot 10^{13}$, cm
H	−3.78	Ni^{62}	−8.7
H^2	6.6	Cu	7.5
Be	7.74	Zn	5.9
C	6.61	Ge	8.4
C^{13}	6.0	Zr	6.3
N	9.4	Nb	6.9
O	5.81	Mo	6.7
Mg	5.35	Rh	6.0
Al	3.5	Pd	6.3
Si	4.0	Ag	6.1
Ti	−3.3	Ag^{107}	8.3
V	−0.51	Ag^{109}	4.3
Cr	3.52	Sn	6.1
Mn	−3.7	Ta	7.0
Fe	9.52	W	4.67
Fe^{54}	4.18	Pt	9.5
Fe^{56}	10.1	Au	7.62
Fe^{57}	2.3	Hg	13.2
Co	2.82	Tl	9
Ni	10.2	Pb	9.6
Ni^{58}	14.4	Bi	8.63
Ni^{60}	3.0	Th	10.0

Table 6

Atomic factors of scattering of x-rays f for some metals, at $\frac{\sin\vartheta}{\lambda} = 0.2 \text{Å}^{-1}$

Element	f
Mn	18.2
Fe	18.9
Co	19.8
Ni	20.7
Cu	21.6
Zn	22.4
Ag	36.9
Au	65.0

wavelength λ, corresponding to $\frac{\sin \vartheta}{\lambda} = 0.2 \text{Å}^{-1}$. For many alloys composed of elements with similar atomic numbers, the atomic scattering factors of x-rays are similar and the superlattice lines on the x-ray pattern are very weak. As is apparent from Table 6, such a case is encountered, for example, in the alloys Fe-Co, Ni-Mn and Fe-Ni. At the same time, the isotope-averaged scattering amplitudes of neutrons for these alloys differ appreciably (see Table 5). The superlattice lines on a neutron diffraction pattern will therefore be rather intense. Cases are also possible where the isotope-averaged scattering amplitudes of neutrons for the components of an ordered alloy are similar. This occurs, for example, in the Cu-Au system, where the scattering amplitudes of neutrons equal, respectively, 0.76×10^{-12} and 0.77×10^{-12}, so that the neutron diffraction method should be used in studying the ordering in these alloys. We note that when the isotope-averaged values of the scattering amplitudes of neutrons are almost equal, the superlattice lines on the neutron diffraction pattern may be appriciably strengthened, changing in this manner the isotopic composition of the alloy components.

The neutron diffraction method of investigation is preferred to the x-ray method in structures containing atoms with small atomic numbers (for instance, hydrogen), since these atoms have small scattering factors for x-rays and may have neutron scattering amplitudes that are comparable with other elements.

In analyzing the background on a neutron diffraction pattern one should consider that many more causes giving rise to diffuse scattering exist here than in the case of x-rays. Since the amplitudes of neutron scattering by nuclei of different isotopes of any alloy component are different, diffuse scattering arises due to the completely random distribution of isotopes on the sites of this type. Even the scattering of neutrons from a pure metal gives rise to a background identical to that obtained with a "disordered alloy of isotopes," in which no correlation exists.

Another cause of additional diffuse scattering of neutrons (in comparison with x-ray scattering) is the dependence of the forces of interaction between the neutrons and the nucleus on the direction of nuclear spin. The existence of different spin directions and the possibility of neutron scattering with a change in spin direction cause the appearance of the additional background on the neutron diffraction pattern. Consequently, the distribution of isotopes and the possible values of projections of nuclear spins over the lattice sites may, with a high degree of accuracy, be regarded as random (possibly with the exception of cases at extremely low temperatures). The intensity of the diffuse scattering from a single crystal is independent of the scattering angle. Knowing the constants that characterize interaction between the neutron and the nuclei, the background intensity can be computed (see Sec. 33).

Owing to their magnetic moment, neutrons may interact not only with nuclei, but also with the electronic shells, if the latter have a nonzero magnetic moment. For this reason, magnetic scattering of neutrons, as distinguished from x-rays, is not caused by the inner filled electronic shells but is due to the unfilled electronic shells. Diffuse scattering of neutrons may be appreciable, particularly near the Curie point of a ferromagnetic substance, where it has the same order of magnitude as diffuse scattering by nuclei [46]. This should be borne in mind in the case of neutron diffraction studies of magnetic alloys. Magnetic scattering by electron shells should depend on the scattering angle, since the dimensions of the scattering center are now comparable to the wavelength.

Scattering by thermal vibrations of the crystal lattice is observed both with x-rays and with neutrons. Neutron scattering has certain pecularities. Absorption or emission of phonons occurs as a result of an interaction between the neutron and thermal vibrations of the lattice. During these processes the velocity of the neutron can be altered appreciably, because its energy change, equal to the energy of the absorbed or emitted phonon, is of the same order of magnitude as the mean energy of a thermal neutron ($\sim kT$). Thus, thermal scattering of neutrons is essentially inelastic. The variation in the neutron velocity and the corresponding wavelength could be used to eliminate thermal scattering. In the case of thermal scattering of x-rays, the ratio of the energy of the x-ray quantum to the energy of a phonon is of the order of 10^6, so that it is virtually impossible to detect a change in the wavelength caused by thermal scattering.

The application of neutron diffraction methods to investigation of ordered alloys was first limited by the low power of the neutron sources. In [47] neutrons, obtained from a radioactive Rn-Be source, were slowed down in paraffin, following which their mean wavelength was approximately 1.7Å. Because of the low power of the neutron source, only a very crude experimental technique could be employed. Fe-Ni alloys of

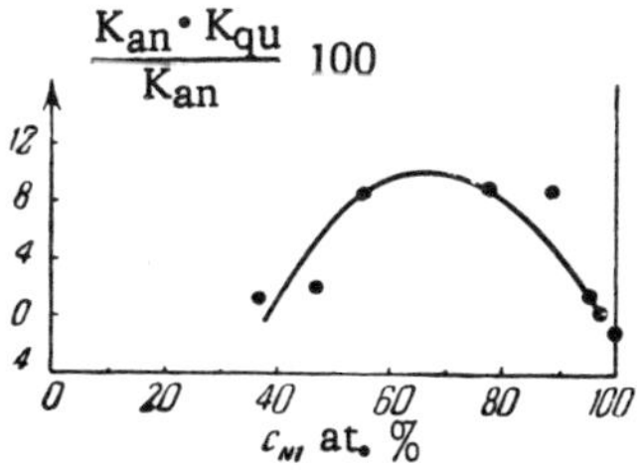

Fig. 25. Dependence of $\frac{(K_{an}-K_{qu})\cdot 100}{K_{an}}$ on the concentration of nickel in Fe-Ni alloys [47].

different compositions were investigated. The samples were annealed at 1,000°C, then each was divided into two samples. After annealing in vacuo at 900°C (to obtain identical grain structure) one series of samples was rapidly quenched, while the other was maintained at 490°C for 700 hours and then cooled down to 360°C (over a period of 340 hours). The passage of neutrons through the quenched and annealed samples has been investigated, and the corresponding transmission coefficients K_{qu} and K_{an} were determined from the attenuation of the intensity of the beam transmitted through the sample. It is apparent from Fig. 25 that over the concentration range from approximately 45 to 90 atomic % of Ni, the transmission coefficient for the annealed samples is higher than for the quenched samples. The difference in the transmission coefficients was related to the existence of pronounced ordering in the annealed samples. The highest difference in transmission coefficients $K_{an}-K_{qu}$ was observed for the stoichiometric alloy Ni_3Fe. Investigations by this method were found to be not very reliable, because a strong dependence of the transmission coefficient on the grain size was detected [48].

However, when more intense neutron fluxes become available from atomic reactors, more modern methods of neutron diffraction study of ordering became possible—similar to the well-known method of x-ray analysis. These methods made possible the determination of the angular dependence of the intensity of the scattered neutrons and, in particular, to find the intensity of the lattice and superlattice lines of the neutron diffraction pattern. A method of this type was used to investigate the long-range order in FeCo, Ni_3Mn and Cu_3Au alloys [48]. For the first two alloys, as is apparent from Table 6, the atomic scattering factors of x-rays by atoms of their components are similar, and the application of the

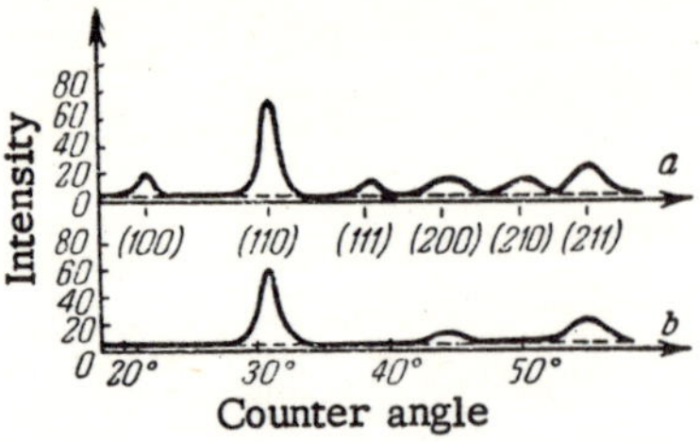

Fig. 26. Angular distribution of intensity; number of neutrons scattered by the ordered (*a*) and disordered (*b*) samples of the FeCo alloy [49].

x-ray structure analysis entailed some difficulties. The neutron diffraction method for these alloys, as indicated above, were found to be more effective. In fact, for the completely ordered alloy FeCo, having a β-brass type body-centered cubic structure in the ordered state, the ratio of the intensity of the superlattice line (110) should be equal to 1/1,390 with x-rays, whereas this ratio in the case of neutron scattering should be equal to 1/6.

In [49] samples of FeCo alloys were prepared in both disordered and ordered conditions. The disordered sample was quenched in water after annealing at a temperature of 850°C. To produce an ordered structure the sample was slowly cooled from a temperature of 750°C for 100 hours. In both cases the scattering of a monochromatic beam of neutrons (with a wavelength of 1.06Å) from a polycrystalline powder was studied. Figure 26 gives the angular dependence of the intensity of scattered neutrons obtained by this method. A comparison of the curves for the ordered and disordered states of the alloy shows that in the ordered state superlattice lines appear at positions corresponding to an odd sum of the indices [(100), (111) and (210) lines], i.e., at the places where they should also appear for β-brass type ordered structures (see Table 1). The intensity of the diffuse scattering of neutrons in the case of a disordered sample was found to be approximately 20% greater than for an ordered sample.

For $AuCu_3$ alloys, Fig. 27 shows that there was no appreciable difference between the neutron diffraction patterns obtained from ordered and disordered samples.

Ordering in the Ni_3Mn alloy was also studied by the neutron diffraction method, [49] [50]. The samples of the ordered alloy were prepared by means of a long-period stepwise cooling from a temperature of 600°C [50]. The angular dependence of the intensity of the neutrons scattered by samples of the ordered and disordered alloy are illustrated in Fig. 28. The superlattice lines were obtained at positions corresponding to the indices (100), (110), (210), (211), (300), and (221)* i.e., at those places where they should appear for the ordered structure with a face-centered cubic lattice. The lower curve of Fig. 28, corresponding to the disordered alloy, displays a diffuse maximum caused by the presence short-range order.

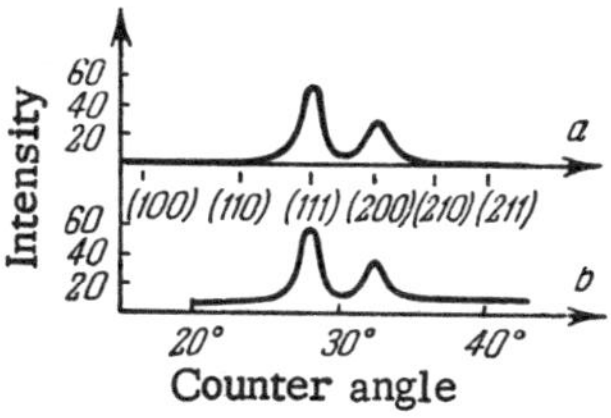

Fig. 27. Angular distribution of intensity of the neutrons scattered by (*a*) ordered and (*b*) disordered $AuCu_3$ samples [49].

The ordering in alloys of the NiFe system was investigated from the variation in the intensity of superlattice lines by

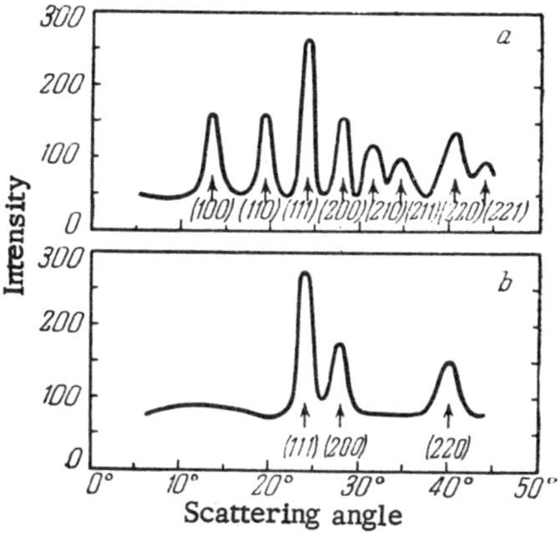

Fig. 28. Angular distribution of the intensity of the neutrons scattered by an ordered (*a*) and disordered (*b*) samples of a Ni Mn alloy [50].

*The positions of the lines (221) and (300) coincide.

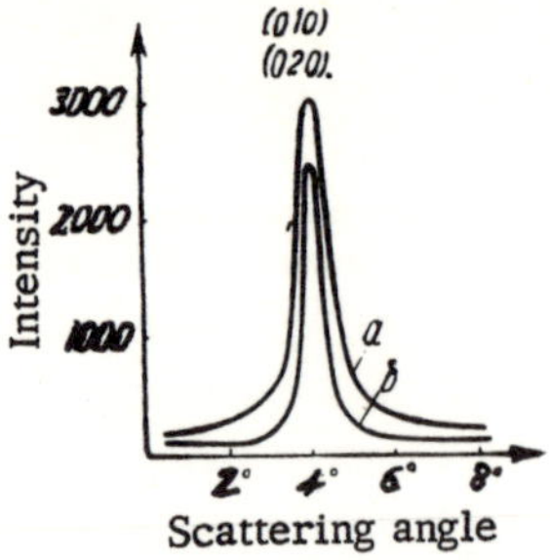

Fig. 29. Angular dependence of the intensity of the neutrons (number of neutrons/min), scattered by an ordered (*a*) and disordered (*b*) samples of Ni_3Fe [51].

[50] and [51]. In these alloys, as apparent from Tables 5 and 6, the amplitudes of the nuclear scattering of neutrons, as well as the atomic scattering factors of x-rays, are almost equal. However, since Ni and Fe atoms have uncompensated magnetic moments of the electronic shells, appreciable magnetic scattering of the neutrons should be observed. The amplitude of the magnetic scattering of neutrons for Fe atoms is several times greater than for Ni atoms; consequently, magnetic scattering makes possible the observation of superlattice lines in the Ni-Fe alloys. In [51], the scattering of nonmonochromatic neutrons by polycrystalline samples of Ni-Fe alloys of different compositions was studied over the range from 50 to 82 atomic % Ni, and also in Ni-Fe alloys with small additions of Mo and Cr. The ordered alloys were prepared by means of a 600-hr step-wise annealing from 600 to 300° C, while the disordered alloys were prepared by quenching from 600° C. Figure 29 shows curves of the intensity of scattered neutrons as a function of the scattering angle for the (010) superlattice reflection. The neutron diffraction pattern for scattering nonmonochromatic neutrons shows that the maximum of this reflection coincides with the maximum of the lattice reflection (020) from a face-centered cubic lattice and caused by neutrons with one half the wavelength. The lower curve corresponds to a disordered alloy, while the upper one corresponds to an ordered alloy with almost a stoichiometric composition Ni_3Fe. It is apparent from Fig. 29 that the appearance of a superlattice reflection in the case of an ordered alloy leads to an increase in the total intensity of the maximum which we are considering. No ordering was observed in the Ni-Fe alloy with 50 atomic % Ni. The

measurement of the intensity of superlattice reflections for alloys annealed at different temperatures, enabled us to determine the temperature dependence of the degree of long-range order and to find the critical temperature of the Ni_3Fe alloy. The latter is approximately 530° C.

5. Changes in Specific Heat and Energy of Alloys During Ordering

During experimental determination of the thermal properties of ordered alloys it was discovered that the specific heat of such alloys changes very rapidly with temperature near the critical temperature. Such a change usually accompanies phase transformation in solids. To determine the temperature dependence of specific heat in ordered alloys measurements of the small energy changes, corresponding to small temperature changes must be made. Such measurements are carried out in a special calorimeter, described in [52, 53, 54], which permits the determination of the heat capacity with a sufficient degree of accuracy.

The curve of the temperature dependence at constant pressure of specific heat (calculated for a single atom) C_p/R of β-brass, containing 49% Zn, is shown in Fig. 30 (curve *a*). Curve *b* in this figure depicts the temperature dependence of the mean specific heat, computed from the experimental values for pure Cu and Zn according to the miscibility rule. It is evident from Fig. 30 that both curves practically coincide below 400°K. As the temperature is raised and the alloy begins to be disordered, a more and more rapid rise in the specific heat is observed, until a maximum is reached at the order-disorder transition temperature. This increase in specific heat is due to the heat added during disordering of the alloy by the disruption of the energetically more favorable ordered structure. At the critical temperature point, the specific heat drops rapidly. Nevertheless, the specific heat above the critical temperature exceeds the maximum calculated theoretical value. This discrepancy is caused by the short-range order remaining in the alloy after the transition to the disordered state. In this case the additional quantity of heat required during heating is necessary for a reduction of the number of pairs of neighboring atoms of opposite species and for the formation of pairs of similar atoms. During subsequent heating the fraction of the specific heat associated with short-range order decreases.

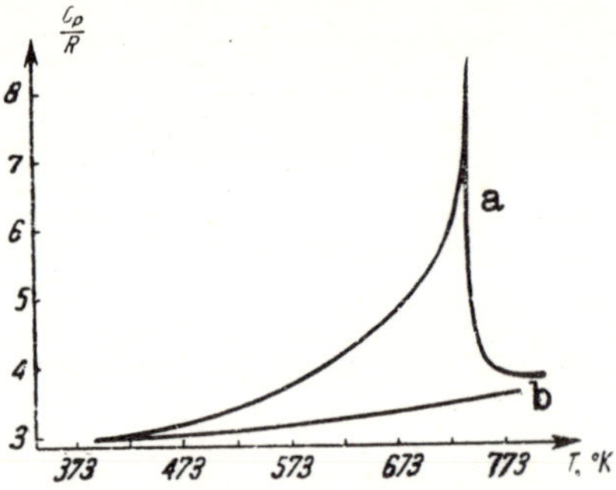

Fig. 30. Temperature dependence of heat capacity (per atom) of β-brass [1].

In the $AuCu_3$ alloy, as distinguished from the case of β-brass, the order-disorder transition is a first-order phase transition. Sykes and Jones [53] determined the specific heat of $AuCu_3$ as a function of temperature. This alloy was first cooled down to 450°C at a rate of 30° per hour to reduce it to the ordered state, and then the heat capacity was determined. Curve (*a*) of Fig. 31 displays the temperature dependence of the specific heat of this alloy, curve (*b*) gives the temperature dependence of the mean heat capacity found from the miscibility rule and the heat capacities of Au and Cu. The heat capacity of alloys of the system Mg-Cd was reported by [55] and [56]. The investigation of the temperature dependence of the heat capacity in the alloys MgCd and $MgCd_3$ presented in [55] has shown that a first-order phase transition is observed in MgCd at a temperature of 251°C, while in $MgCd_3$ a second-order phase transition occurs near 80°C. As shown in [56], a second-order phase transition occurs in the Mg_3Cd alloy at a temperature of 153°C.

The energy $E(T)$, liberated when the alloy is being ordered

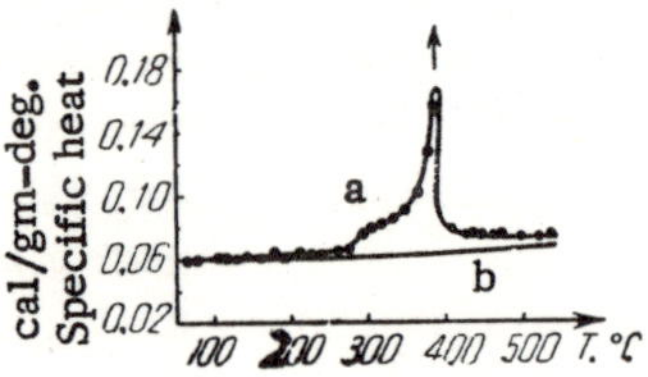

Fig. 31. Temperature dependence of the specific heat of the alloy $AuCu_3$ [53].
The maximum value of specific heat, obtained in these experiments near the transition temperature, equal to 2.0 cal/gm-deg, is not given.

at a given temperature T, may be found either by direct measurements, or from the area bounded by the curve of heat capacity plotted as a function of temperature. The area should be taken over the temperature range from the critical temperature T_0 up to T. The energy $E(T)$, found by this method for the $AuCu_3$ alloy, is illustrated in Fig. 32. When $T = T_0 - 0$, the energy $E(T_0)$ in an alloy with the minimum possible long-range order (in the ordered state) is not zero, i.e., heat of transition is present. This agrees with the fact that the order-disorder transition is a first-order transition in $AuCu_3$ alloys. The heat of transition, as shown by Fig. 32, is equal to 1.3 cal/gm.

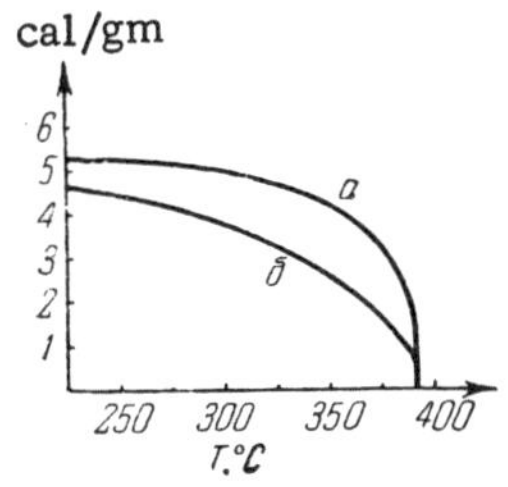

Fig. 32. Energy E of the $AuCu_3$ alloy as a function of temperature T. a-experimental curve, b-theoretical curve obtained by the Gorsky-Bragg-Williams approximation.

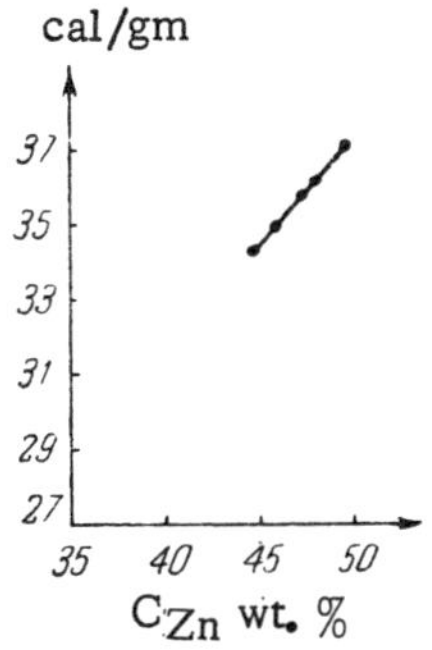

Fig. 33. Energy $E(T)$ over the temperature range 240°-500°C as a function of concentration of zinc in β-brass [1].

The dependence of the energy $E(T)$ on the composition of the Cu-Zn alloy over the temperature range from 240° to 500°C (Fig. 33) was investigated in [1]. In the region of the β-phase, the energy $E(T)$ is a linear function of the atomic concentration of the alloy components.

6. Effect of Ordering of Atoms on the Electrical Resistance of Alloys

The basic laws of electrical resistance of metals and alloys can be qualitatively understood by taking into account the wave properties of conduction electrons. The electron wave travels without scattering in a perfect crystal lattice, which produces a potential that is a periodic function of the coordinates in three dimensions. Such an ideal crystal lattice exhibits no electrical resistance. If, however, the crystal lattice of a metal or an alloy contains any distortion, leading to a disruption of the periodicity of the potential, the electron waves will scatter and thus cause the appearance of electrical resistance. There exist three principal forms of distortion of the crystal lattice giving rise to electrical resistance: (1) thermal motion of atoms, (2) disruption of the periodicity caused by either the disordered alternation of atoms of different types or the existence of vacant lattice sites (holes), and also by the presence of interstitial atoms, and finally, (3) static distortions of the lattice, caused by displacements of the centers of vibrations of the atoms from their correct position.

In pure metals having no static distortions of the lattice or vacancies, only the first of the above reasons should apply. The electrical resistance of the metal will depend on the temperature and should disappear at the absolute zero. The second cause of electron scattering in alloys or metals containing vacancies or interstitial atoms leads to the appearance of an additional resistance which remains even at the absolute zero. The presence of static distortions of the lattice leads to the same result. The electrical resistance remaining at the absolute zero is called residual electrical resistance. The residual electrical resistance may be determined by measuring the electrical resistance at low temperatures and then extrapolating the results to absolute zero. The effect of lattice inhomogeneities, caused by disorder in the alternation of atoms and by static distortions, on the electrical resistance at high temperatures can be determined by removing the electrical resistance associated with the thermal motion of atoms. This can be accomplished by quenching down to low temperatures.

In binary disordered alloys $A-B$ of the substitutional type with unlimited mutual solubility of the components, the residual electrical resistance should depend greatly on the composition of the alloy. When atoms B are added to the pure metal A, and when atoms A are added to the pure metal B, the regularity of the crystal lattices is disturbed and the residual electrical resistance should increase from zero in both cases (for metals which have no static distortion) and reach a maximum in the central part of the phase diagram. In fact, electrical resistivity of the Ag-Au alloys of different compositions [59] shows a maximum at $c_{Ag} \approx 0.5$ (Fig. 34). Similar curves are observed for other disordered alloys of this type.

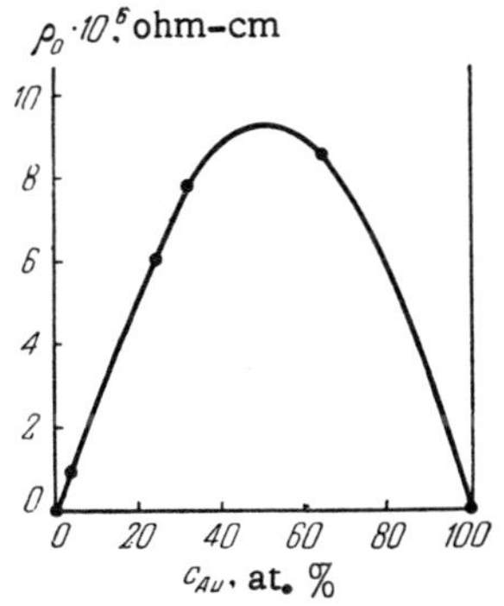

Fig. 34. Residual electrical resistivity of the alloy AuAg as a function of the concentration of gold.

The emergence of long-range order in alloys should decrease the extent of deviation from the periodicity of the potential, produced by the crystal lattice, and hence should also decrease the electrical resistance. This effect should be more pronounced, the closer the alloy lattice approaches a state of complete ordering. In stoichiometric alloys free from lattice defects and with a degree of long-range order equal to unity, perfect periodicity of the lattice is produced, and the residual electrical resistance caused by the irregular alternation of atoms should decrease to zero. In such alloys annealed at higher temperatures, a lower degree of long-range order is established than in an alloy annealed at a lower temperature. This shows that the higher the annealing temperature the higher the residual electrical resistance after quenching. When the composition is not stoichiometric, then, even with maximum possible degree of long-range order, the crystal lattice of the alloy will not show periodicity and the residual resistance may not be equal to zero. Therefore, pronounced minima occur on the curves showing concentration as a function of the residual electrical resistance of highly ordered alloys. Such minima occur at points corresponding to the stoichiometric composition close to where the formation of superlattices is observed. This type of minimum has been detected in alloys of the Au-Cu system by Kurnakov, Zhemchuzhnyi and Zasedatelev [58] at compositions equivalent to AuCu and $AuCu_3$. As a result of such evidence, the authors [58] found that the appearance of phases in such alloys is possible. This discovery was the fundamental development in the physics of ordered alloys. The problem of ordering of atoms arose

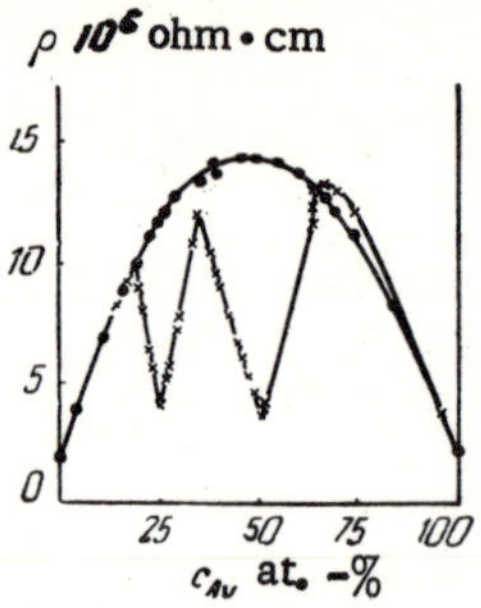

Fig. 35. Electrical resistivity of the alloys Au–Cu as a function of the concentration of gold [60]. ×–denote an ordered alloy, •–denote a disordered alloy.

after this fundamental work was done, and after Tammaun [59] speculated about the existence of ordered structures.

Figure 35 gives data concerning the electrical resistance of these alloys obtained later by Johansson and Linde [60]. Measurements of the electrical resistance have been carried out at room temperature.* The samples, quenched at 650°C (i.e., from a temperature above the critical temperature), were found to be in a disordered state. Therefore, the dependence of the electrical resistivity of these samples on composition has the same form as for Au–Ag alloys. The other series of samples were subjected to continuous annealing at a temperature below the critical temperature. During annealing of the alloys, whose compositions were close to $AuCu_3$ and AuCu, the ordered state was reached, and minima appeared close to these compositions on the electrical resistivity curves.

The ordered phase, as indicated above, does not exist over the entire range of concentration c_A, but only over a certain restricted region of the phase diagram. At concentrations c_0 ($c_0 = c_0'$, c_0''), corresponding to passage from the disordered to the ordered phase, the characteristic features should be observed on the curve $\rho_0(c_A)$ for alloys annealed at the same temperature. Since the degree of long-range order changes

---

*Here, the electrical resistance of even pure metals is not equal to zero, because the measurements were carried out at room temperature, when not only the residual resistance, but also the resistance due to thermal vibrations, are significant.

continuously with temperature in the case of a second-order phase transition, it should also change continuously with composition of alloys annealed at the same temperature. Consequently, electrical resistivity should also change continuously, except at a concentration $c_A = c_0$ the curve of electrical resistivity vs composition should have a discontinuity. If the order-disorder transition is a first-order phase transition, the degree of long-range order exhibits a step-change. Moreover, the electrical resistivity, during passage into the ordered state at $c_A = c_0$, also decreases by a step-change [61]. This type of discontinuity of the curves $\rho(c_A)$ for Au-Cu alloys has been observed experimentally; it is noted on the curves given in Fig. 35. This effect is exhibited more clearly on the curves $\rho(c_A)$ obtained by Pǒspisil [62] for ordered Au-Cu alloys.

Additional features of the dependence of the residual electrical resistivity on concentration emerge when the energy spectrum of the electrons changes appreciably with a change in composition of the alloy. Such a variation of the energy spectrum is encountered, for instance, in alloys of the transition metals with nontransition metals. In this case (in terms of the one-electron approximation) an unfilled d-band of the energy spectrum of electrons with an appreciably higher density of electron levels than in the s-band exists in the alloy over a definite concentration range. The scattering probability of s-electrons by the inhomogeneities of the crystal lattice increases substantially because there arises the possibility of transitions not only into the s-level, but also into the unfilled d-level. Consequently, the number of quantum transitions of conduction electrons per unit time, and thus the electrical resistivity, become higher. As a result, the residual electrical resistivity of alloys, increases appreciably over the range of concentration corresponding to the unfilled d-levels, and the curve $\rho_0(c_A)$ for disordered alloys becomes asymmetrical. Such curves are found, for example, in plots of concentration versus residual electrical resistivity of disordered alloys Ag-Pd, Ag-Pt, Au-Pd, Cu-Pd and Cu-Pt [63-66] (Figs. 36, 37 and 38). During ordering of such alloys, the residual electrical resistivity, just as in the case of alloys of nontransition metals, decreases sharply (Figs. 37, 38).

An increase in temperature causes the appearance of additional distortions of the crystal lattice of the alloy, associated with thermal vibrations. Therefore, even if the distribution of atoms at the crystal lattice sites of the alloy does not change with temperature, the electrical resistivity

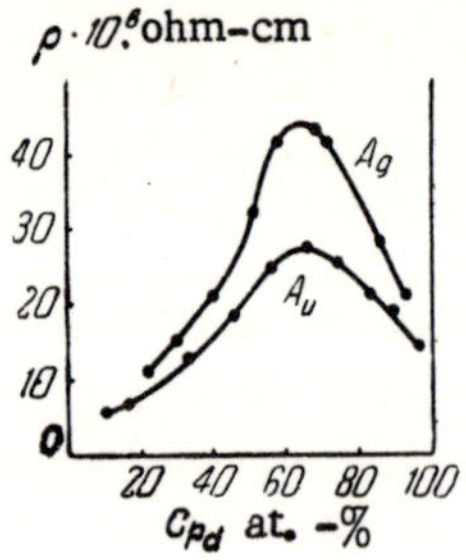

Fig. 36. Electrical resistivity of Ag-Pd and Au-Pd alloys as a function of the concentration of palladium, at 0°C.

will increase with temperature. At high temperatures ($T \gg \theta$, where θ is the Debye characteristic temperature) the electrical resistivity of alloys, like that of pure metals, increases linearly with temperature. Near the melting point we observe small deviations from the linear behavior of $\rho(T)$ which are related chiefly to the nonharmonicity in the vibrations of atoms [67] and the manifestation of vacancies

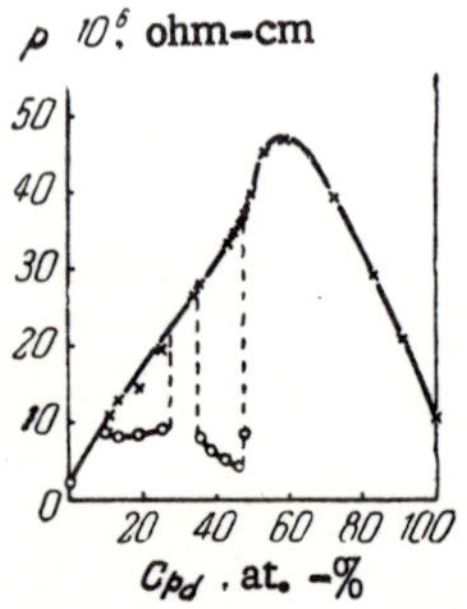

Fig. 37. Electrical resistivity of the alloys Cu-Pd as a function of the concentration of palladium.
x denote alloys rapidly cooled from a temperature above the critical temperature; o denote annealed alloys.

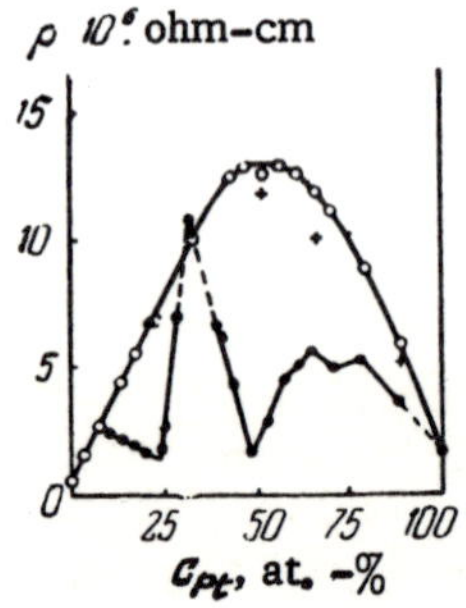

Fig. 38. Electrical resistivity of the alloys Cu-Pt as a function of the concentration of platinum.
Open circles denote rapidly cooled and cold-worked (disordered) alloys, solid circles denote alloys annealed at 300°C; crosses denote alloys rapidly cooled from 900°C.

[68]. At low temperatures ($T \ll \theta$) the temperature variation of electrical resistivity obeys a different law. For pure metals, $\rho(T)$ varies roughly as $\rho \sim T^n$, where n takes values from 2 to 5 for different metals. In the simplest theory of the electrical resistivity of metals employing the one-electron approximation and the isotropic elastic continuum approximation for lattice vibrations, Bloch obtained the law $\rho \sim T^5$. But, taking into account collisions between electrons, departure from the Debye law of the frequency of the lattice vibrations vs wave vector, departure from thermal equilibrium, etc., may lead to a more complex dependence of ρ on T. This relation is approximated by an exponential law with a different exponent for a given temperature range. As $T \to 0$, the resistivity of "perfect" metals tends to zero, while for imperfect metals, or alloys, it tends to the residual electrical resistivity, which we have already examined.

If experiments on the temperature dependence of the electrical resistivity are conducted with the alloy remaining in an equilibrium state at each temperature, an additional variation of the resistivity, associated with the redistribution of atoms at the lattice site, will appear. As the temperature increases, the degree of long-range order decreases, additional scattering appears, and the resistivity increases more rapidly than the resistivity of a sample in which the state of ordering remains unchanged. At the order-disorder transition temperature a discontinuity should be observed in the $\rho(T)$ curve, if the transition is a second-order phase transition. In a first-order phase transition, both the temperature coefficient of resistance and the resistance itself undergo a step-change. These arguments are supported by the results of investigations of the temperature dependence of the electrical resistivity of β-brass [69] and $AuCu_3$ [70] (Figs. 39 and 40). It is apparent from Fig. 39 that the curve of the temperature dependence of the electrical resistivity of β-brass does in fact have a bend at the critical temperature T_0. For $T > T_0$, near the critical temperature, a rapid decrease of ρ is observed as T decreases, which is related to the increase of the degree of long-range order η. At higher temperatures where η differs little from unity, the curve $\rho(T)$, is almost a straight line. In the case of a $AuCu_3$ alloy at the critical temperature T_0, as apparent from Fig. 40 (curve a), one finds a pronounced discontinuous change both of the electrical resistivity and of its temperature coefficient.

In some alloys these changes on passage through the critical temperature are found to be considerably less

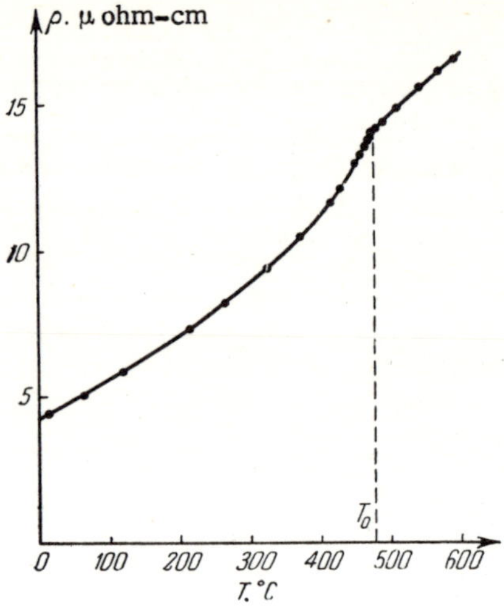

Fig. 39. Resistivity of β-brass containing 47.7 atomic % Zn as a function of temperature.

pronounced. Thus, in the alloy FeCo [71] the break in the curve $\rho(T)$ at the point $T = T_0$ is not noticeable (see curve *b* in Fig. 40).

The effect of ordering on the electrical resistivity of the Ni_3Mn alloy and alloys having a composition close to

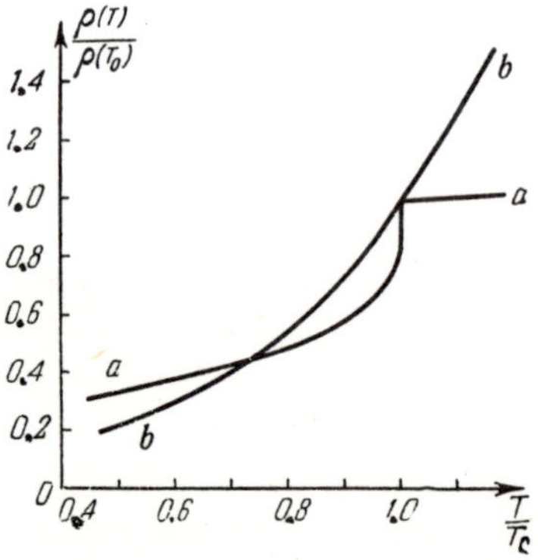

Fig. 40. Ratio of the electrical resistivity $\rho(T)$ of the alloys $AuCu_3$ and FeCo to their electrical resistivity at the critical temperature $\rho(T_0)$ (for the disordered state) as a function of $\frac{T}{T_0}$.

a-alloy $AuCu_3$ (from the data of [70]; b-alloy FeCo (from the data of [71].

Ni_3Mn doped with small amounts of molybdenum was studied by Livshits et al. [72]. In the alloys we first established the initial state by quenching from 1,000°C. Afterwards, prolonged annealing was conducted at temperatures of 250°, 350°, 450° and 550°C, to reduce the alloys to an ordered state. Figure 41 depicts curves of the specific electrical resistivity of the alloy Ni_3Mn, and also alloys doped with 0.64 to 4.1 weight percent Mo, as a function of the annealing temperature. At the annealing temperature of 350°C, the alloy Ni_3Mn (dashed curve) exhibits a minimum, related to the establishment of an ordered state. An increase of molybdenum concentration causes the minimum to become less and less pronounced and to disappear in alloys with a molybdenum concentration of approximately 2.65% by weight. However, the electrical resistivity of an alloy quenched from 1,000°C remains higher than that of alloys annealed at 350°C. Thus, the addition of molybdunum into the alloy Ni_3Mn hinders the establishment of long-range order.

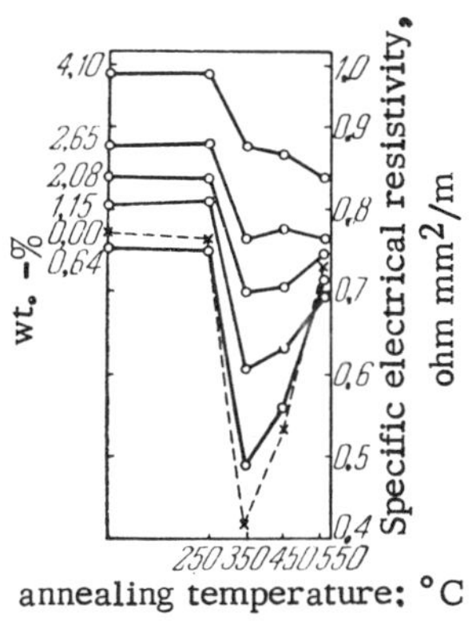

Fig. 41. Electrical resistivity of the alloys Ni_3Mn with a different concentration of molybdenum as a function of the temperature of prolonged annealing [72].

The establishment of short-range order in disordered alloys also changes the residual electrical resistivity, although not as strongly as the establishment of long-range order. This change was detected in β-brass during an investigation of the influence of annealing on electrical resistivity

[73]. The annealing of samples, which leads to the establishment of the equilibrium value of short-range order, was carried out for a period of six weeks at a temperature of 190°C under vacuum. The electrical resistivity of an alloy containing 20% Zn, measured at 80°K, decreased 1.1% after such annealing. After additional annealing at 370°C for a period of 1.5 hours, the resistance of this alloy reached its minimum value (at 80°K), which was reached at 190°C before annealing. This indicates a reduction in the degree of short-range order at temperatures higher than 370°C.

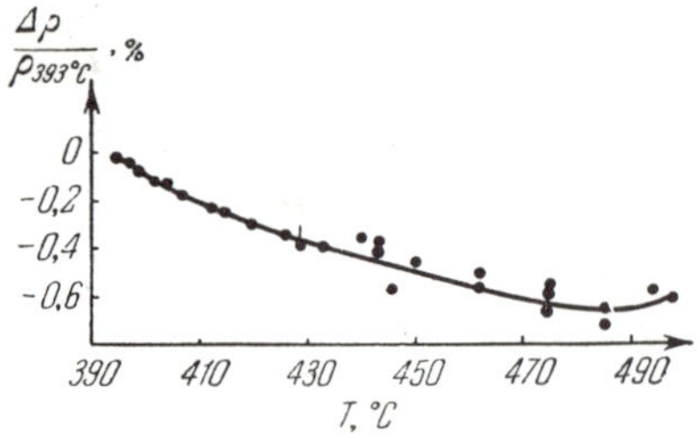

Fig. 42. Relative variation in electrical resistivity of the alloy $AuCu_3$ at 77°C as a function of the quenching temperature [74].

An investigation [74] has been made of the influence of short-range order on the residual electrical resistivity of the alloy $AuCu_3$. To determine the various degrees of short-range order, the alloy was annealed at different temperatures above the critical temperature T_0, from 393° to 500°C. After each annealing the alloy was quenched and the electrical resistivity was measured at a temperature of 77°K, when practically only the residual resistivity remained. In this manner a relation was found for the residual electrical resistivity as a function of annealing temperature at which a corresponding degree of short-range order was established (see Fig. 42). It may be seen from the figure that with a decrease in the annealing temperature from 500° to 393°C, i.e., with an increase in the degree of short-range order, the residual electrical resistivity of the alloy first decreases (down to a temperature of 485°C), and then increases.

7. Kinetics of Ordering

Each temperature corresponds to a certain equilibrium value of the parameters describing the long- and short-range orders. However, these values are not established instantaneously, but over a certain time interval needed for redistribution of atoms on the sites of the crystal lattice. Hence, any physical quantity Q, that depends on the ordering of the alloy, will take the equilibrium value Q_e associated with a given temperature not immediately, but strictly speaking, after an infinitely long continuous annealing at this temperature.

In a number of cases the rate at which Q attains the equilibrium value Q_e may be characterized by a certain relaxation time τ. If the time rate of change of Q, $\frac{dQ}{dt}$ is proportional to the difference between this value and the equilibrium value, i.e.

$$\frac{dQ}{dt} = \frac{1}{\tau}(Q_e - Q), \tag{7.1}$$

then the quantity τ is the relaxation time. In fact, from (7.1) it follows that

$$Q = Q_e - (Q_e - Q_0)\, e^{-\frac{t}{\tau}} \tag{7.2}$$

(where Q_0 is the initial value of Q), i.e. the deviation of Q decreases by a factor e during the time $t = \tau$. Generally, the time dependence of $Q(t)$ is more complicated and there exists no single relaxation time for the entire process. However, we may sometimes approximately describe the rate of change of Q by a certain time, during which Q changes by a definite amount for each stage of the process (or by an average time for the various stages). We note that the relaxation times for the various physical quantities may differ appreciably, because the variation of these quantities may be related to different types of processes, occurring at different stages of ordering. By investigating the time dependence of physical quantities during isothermal annealing, one may draw conclusions concerning the rates at which the ordering occurs, and determine the time after which the ordering is essentially complete.

The most detailed study of the kinetics of ordering was made on the alloy $AuCu_3$, in which the order-disorder transition is a first-order phase transition. If an $AuCu_3$ alloy is rapidly cooled to a temperature $T < T_0$, after prolonged annealing at a temperature above the critical temperature T_0 (i.e., in the disordered state), the establishment of an equilibrium state with long-range order will occur in two stages. In the first stage, at individual points of the crystal (where the short-range order was greatest) centers of the new ordered phase are formed and grow until they become contiguous. At this time a long-range order that is close to the equilibrium condition is established. Since all the sites of the crystal lattice in the disordered state are equivalent, then in the different regions that become ordered, sometimes called domains, the distribution of copper and gold atoms on the sites of the original face-centered cubic lattice may be different. If, for example, the gold atoms in one of the lattices occupy the corners of cubic cells, while the copper atoms occupy the centers of their faces, then in the other lattice one of the three sites, lying at the centers of the faces of a unit cell, may be found to be correct for gold atoms in each cell. Such domains are sometimes called anti-phase domains. The anti-phase domains in a two-dimensional lattice of an alloy A-B are illustrated in Fig. 2. The existence of anti-phase regions leads to a decrease in the average degree of long-range order over the entire crystal in comparison with its value in each domain.

The second, a more slowly progressing stage of the establishment of equilibrium of the ordered state in the $AuCu_3$ alloy, is the growth of some domains at the expense of others, and the resulting passage of the alloy into a state with homogeneous ordering. Inasmuch as there exist two such stages, for the establishment of order, the order-disorder transformation may be approximately described by two relaxation times.*

With rapid heating of a previously ordered alloy to a temperature above the critical temperature, and with isothermal exposure to this temperature, the disordering process occurs

---

*A more detailed treatment of order-disorder transformations in the alloys in which a first-order phase transformation occurs will uncover a greater number of relaxation times. Thus, the first stage of the transformation may be separated into two stages proceeding at different rates (formation of the new ordered phase by nucleation and its subsequent growth up to contiguity). Two relaxation times can therefore be attributed to such a transformation.

only in a single stage. In this case even an alloy having a domain structure before heating becomes homogeneous (disordered) after the disappearance of order in each domain, and no unification of the different regions is required for the attainment of equilibrium.

The passage of the alloy into an ordered state is related to the diffusion processes of the redistribution of atoms on the sites of the crystal lattice. These processes are described by definite activation energies and become extremely slow at low temperatures. With an increase in temperature, the relaxation time initially decreases rapidly. However, when this temperature approaches the critical temperature, another factor, in addition to the rate of diffusion, becomes important: the fact that the order-disorder process is retarded by a decrease in the difference of the free energies of the ordered-disordered phases. At temperatures near the critical temperature this difference is very small, and therefore in spite of the appreciable rate of diffusion, ordering progresses slowly, i.e., the relaxation time again reaches a high value. A reduction of temperature in this region accelerates the order-disorder transformation. Thus, the relaxation time will decrease both with an increase of temperature from low temperatures, and with a decrease of temperature from the critical temperature. Consequently, a minimum exists on curves of the temperature dependence of relaxation time for different quantities.

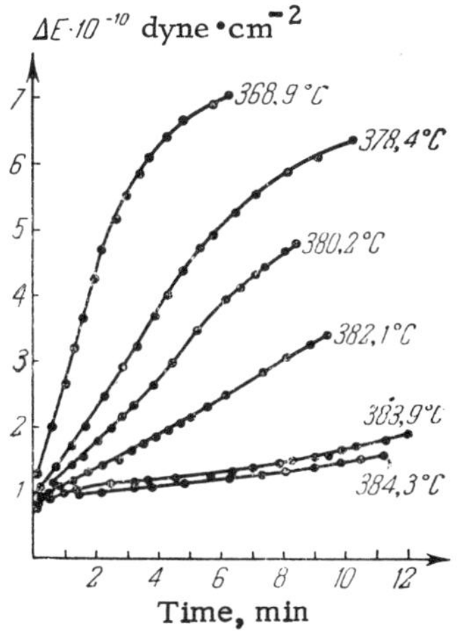

Fig. 43. Variation of Young's modulus with time during isothermal annealing of the $AuCu_3$ alloy [76].

We note that if the relaxation process occurs over an appreciable time interval, the temperature dependence of the different properties of the alloy, measured during heating or cooling of the sample will be different, i.e., hysteresis type curves should be obtained. With a decrease of the rate of heating and cooling, the hysteresis phenomena should appear to a lesser degree.

An experimental investigation of the kinetics of ordering of the $AuCu_3$ alloy was conducted by means of a study of the time dependence of the different properties of the alloy. Siegel [75] and Lord [76] have investigated the time variation of Young's modulus during isothermal exposure of the sample. In [76] the samples were quenched from a temperature of 414.2°C, corresponding to a disordered state, after which the time dependence of Young's modulus E was investigated at various constant temperatures from 279.3° to 384.3 °C. The results of the measurements over a temperature range, near the critical temperature, are given in Fig. 43, where the ordinate axis indicates the difference ΔE between the measured value of the modulus and the value obtained by extrapolation of the temperature dependence of the modulus of a disordered alloy to the temperature of the measurement. The initial sections of the isothermal curves, related to the first stage of the order-disorder transformation, could have been approximated by a type (7.2) exponential time function, which will yield the relaxation time for this stage of transformation. The relaxation time of the first stage of transformation obtained in this manner over a wide temperature range is illustrated in Fig. 44. In accordance with the foregoing discussion this curve exhibits a minimum. The minimum value of the relaxation time ($\sim$ 0.6 min.) is located at a temperature approximately 20° lower than the critical temperature.

The time dependence of Young's modulus for the second stage of the order-disorder transformation was found to differ appreciably from an exponential function. The rate of establishment of the equilibrium state at this stage was appreciably lower than that for the first stage of the transformation. The effective relaxation time for the second stage of the transformation, as shown in [76], has a minimum at approximately the same temperature as τ_1.

An x-ray investigation of the kinetics of ordering of $AuCu_3$ was conducted by [22]. The disordered sample was cooled to a temperature of 360° C (lower than the critical temperature). Thus, the integral intensity of the superlattice lines reached

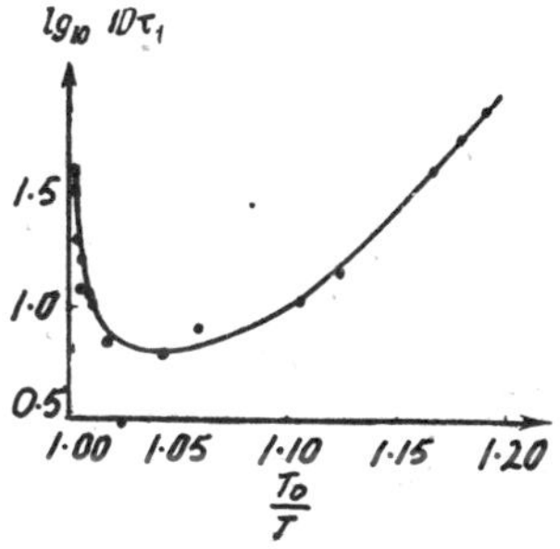

Fig. 44. Temperature dependence of the relaxation time τ_1 (τ_1-in minutes) of the first stage of ordering of the $AuCu_3$ alloy [76].

one half their maximum value, corresponding to the equilibrium state at the temperature, over a period of 15 min., and then increased more and more slowly. The line width was appreciably greater than in the case of alloys annealed continuously at this temperature. This is evidence of the fact that domains were formed in the alloy after such annealing.

The existence of domain structure in ordered alloys could also be explained by the fact that the time required to bring the alloy from a disordered state to an equilibrium state (~ 70 hours) was appreciably greater than the time required to bring the alloy to a disordered state from an ordered state (~ 15 min.). This is due to the fact that in the latter case, the second, slowly occurring stage of the transformation is not present.

The regions of the ordered phase which are formed in the first stage of the order-disorder transformation are extremely small and do not exert any appreciable influence on the electrical resistivity of the alloy, and also do not cause the emergence of superlattice lines. The course of the order-disorder transformation at this stage may, nevertheless, be investigated by a study of the heat capacity of the alloy, which is sensitive to a redistribution of atoms within small volumes whose linear dimensions are of the order of several atomic spacings. In [53] the temperature dependence of the heat capacity of $AuCu_3$ alloys, quenched from temperatures corresponding to a disordered state to a low temperature was determined. Measurement of the electrical resistivity of the alloy did not reveal the existence of an ordered phase, and no superlattice lines were evident on the x-ray pattern. A variation of resistance due to ordering was detected only at a temperature of approximately 300°C. Figure 45 depicts the temperature dependence of the heat capacity of a sample heated from room temperature. At a temperature of 60°C we already observe a decrease of the heat capacity, which reached a minimum at 130°C. This decrease could be explained by the emergence of ordered domains, accompanied by the liberation of energy. When the domains formed during the growth come into contact with each other, their growth and the accompanying liberation of energy ceases, and, consequently, the heat capacity again reaches its normal value. At a temperature somewhat greater than 300°C, the second stage of ordering begins to proceed at a marked rate, during which the domains unite. Energy is liberated when the domains combine (as a result of a disappearance of disordered layers at the boundaries of the domains). As a result the heat capacity again decreases. Upon further heating, as the critical temperature is approached, the equilibrium value of long-range order within the domain decreases. In this connection an additional amount of heat is consumed by the disordering, and the heat capacity begins to increase (in conformity with Fig. 31). Therefore, a sharp minimum is observed in the curve of Fig. 45 at 314°C.

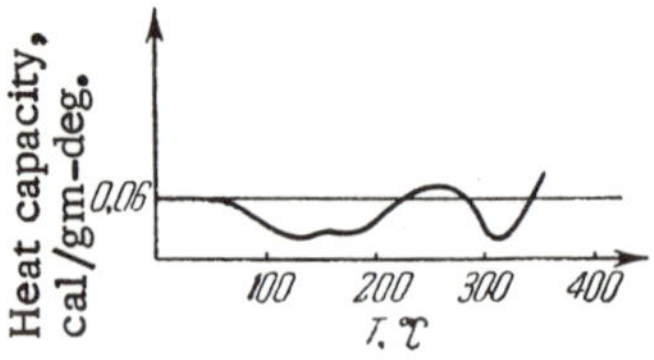

Fig. 45. Temperature dependence of the heat capacity of the $AuCu_3$ alloy, initially quenched in water, from temperatures corresponding to the disordered state.

These experiments show that in alloys whose ordering is not detectable either by the x-ray method, or by the method of electrical resistivity, an appreciable degree of ordering is possible in separate domains. The formation of these domains can lead to liberation of energy greater than 40% of the configurational energy. A measurement of the width of the superlattice lines manifested during the ordering makes possible an estimation of the linear dimensions of the domains, which were found to be of the order of ten interatomic distances.

Dugdale [77] has investigated the kinetics of the variation in the degree of ordering inside individual domains (first stage of ordering) in the $AuCu_3$ alloy. For this purpose the alloy was first annealed for a period of 5-8 days at 370°C, resulting in the formation of rather large domains. Thus, the alloy was in a partially ordered state within each domain. Next, the samples were rapidly cooled to room temperature, and the time dependence of the electrical resistivity was studied at different temperatures, lower than 370°C. Curves of the time dependence of the electrical resistivity of alloys, subjected to such a heat treatment, are given in Fig. 46 for temperatures of 250° and 290°C. It is apparent from the figure that equilibrium is established at 290°C during the period of about one hour, while at 250°C ρ continues to change even after a 50-hour annealing. The pronounced temperature dependence of the rate of the relaxation process is due to the fact that the variation of the degree of long-range order, causing the change in the electrical resistivity, occurs by a diffusion mechanism. Consequently, the rate of change of the degree of order is the larger, the higher is the concentration of vacancies and the higher is their mobility, which,as is well known, increases rapidly with temperature. The change of resistance of a sample quenched from 380°C with heating from room temperature has also been investigated. As is apparent from Fig. 47 (curve a), at temperatures above 200°C, one observes an appreciable decrease in the electrical resistivity, related to the presence of a large number of equilibrium vacancies. The presence of the vacancies leads to the establishment of a long-range order, whose degree is in equilibrium at the temperature under consideration but greater than its value in the quenched sample. With a further increase of temperature, the equilibrium value of η decreases, and consequently, the resistance, passing through a minimum, again begins to increase. In curve (a) of Fig. 47 we also note that the variation of resistivity near the temperature of 170°C is more rapid than that at somewhat higher temperatures. This variation is due to the fact that a certain concentration of vacancies are frozen into the alloy during the quenching of the sample, and they acquire a marked mobility at a temperature of about 160°C. Annealing for a period of 16 hours leads to the disappearance of the frozen vacancies and, consequently, the degree of long-range order remains unchanged up to a temperature of the order of 200°C (see Fig. 47, curve b).

In investigating the kinetics of ordering in the $AuCu_3$ alloy at low temperatures, the activation energy for the

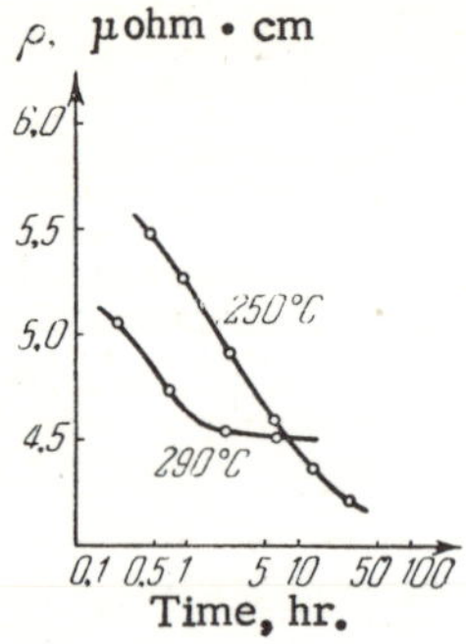

Fig. 46. Electrical resistivity of the $AuCu_3$ alloy, quenched from a temperature of 370°C, vs time of isothermal annealing at 250° and 290°C [77].

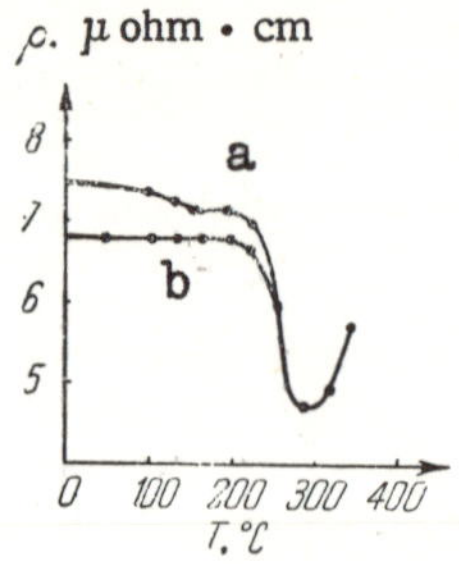

Fig. 47. Electrical resistivity of the partially ordered alloy $AuCu_3$ vs heating temperature of samples quenched from 380°C [77].

motion of vacancies in this alloy was found to be equal to approximately 0.9ev [78]. The energy of vacancy formation in the ordered $AuCu_3$ alloy according to [77] is approximately equal to 1 ev.

Substantially larger relaxation times for the electrical resistivity of the $AuCu_3$ alloy were obtained by [19]. Apparently, the second stage of the order-disorder transformation was investigated.

The initial stages of the order-disorder transformation of the $AuCu_3$ alloy have been studied in [79]. The alloy was quenched from the disordered state (from a temperature of 418°C) to various temperatures corresponding to the ordered state. Then, the variation of the electrical resistivity ρ of the alloy with time t was investigated at constant temperatures. Figure 48 presents a curve ρ as a function of t at a temperature of 384.9°C. We note that electrical resistivity initially rises and only after some time (about 4 minutes) passes through a maximum and begins to decrease. This increase of electrical resistivity can be attributed to the additional scattering of the electrons by the boundaries of the small regions emerging in the ordered phase, whose number initially increases rapidly with time. After an increase of dimensions of these regions, the scattering by their boundaries begins to play a smaller role and the electrical resistivity decreases with the ordering.

The kinetics of ordering of the AuCu alloy was also investigated by the electrical resistivity method [80]. The alloy was first disordered at temperatures corresponding to the disordered state, for a period of 20 hours. The time dependence

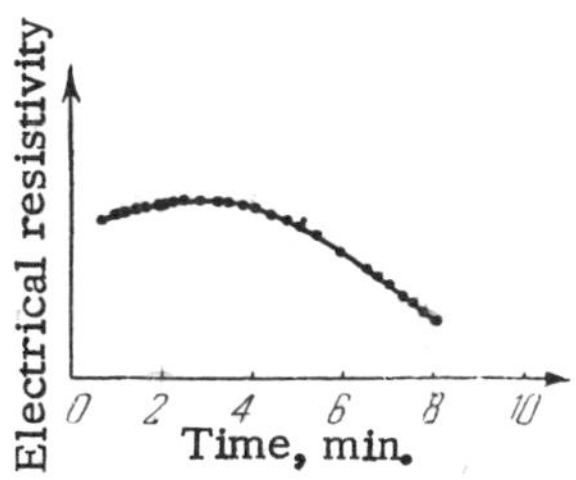

Fig. 48 Time dependence of the electrical resistivity of the $AuCu_3$ alloy quenched from 418°C at a temperature of 384.9°C [79].

of the electrical resistivity was investigated, during isothermal annealing, for different temperatures over the range of existence of the ordered phase. These curves, plotted for previously annealed alloys in the disordered state at 415° C, are given in Fig. 49. In this work a significant dependence of the relaxation time on the temperature of preliminary annealing was also discovered in the disordered phase. It was found that the higher the annealing temperature, the more rapid the order-disorder transformation occurs at temperatures below the critical temperature.

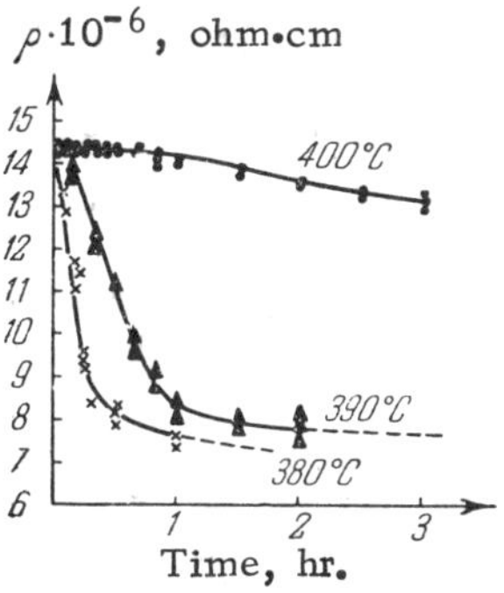

Fig. 49. Electrical resistivity of a disordered alloy AuCu vs time, at different temperatures of isothermal annealing [80].

The process of establishment of the equilibrium state in the AuCu alloy was also investigated by the method of damping of elastic vibrations [80, 81].

The relaxation time may change very much during a transition from one ordered alloy to another. For instance, the relaxation time in β-brass is so small that this alloy does not reach a nonequilibrium state even by quenching to room temperature.

In the FeCo alloy, which has the same structure as β-brass, the relaxation time is also small. This relaxation time was investigated by Kadykova and Selisskiy [82], who measured the electrical resistivity of samples quenched from temperatures corresponding to the disordered state, at various rates of cooling. With an increase in the rate of cooling the electrical resistivity (measured after quenching) initially increases, and then, when the rate of cooling becomes so large that ordering of the alloy does not occur during quenching, saturation sets in. The rate of cooling, which is accompanied by the onset of saturation, is about 6,000 deg/sec., which also describes the time necessary for establishment of the ordered state.

In the Fe_3Si alloy due to the very short relaxation time, just as in β-brass, no ordered state was obtained by quenching [83].

Recently, an interesting new method has been developed for the study of the kinetics of processes occuring in ordered alloys. This new method is based on the study of the effects of fast particles on the alloy. This subject will be treated in the next section.

The theory of the kinetics of ordering of alloys, which will not be examined in this book, was developed in a number of papers (see, for example [84-88]).

8. Influence of Irradiation by Fast Particles on the Ordering of Alloys

New possibilities for the investigation of kinetics of order-disorder transformations were unfolded by the study of the action of beams of fast particles on ordered alloys. The influence of irradiation is found to be different for different types of fast particles. Irradiation may be produced by charged heavy particles (protons, deutrons, α-particles, fission fragments, etc.), neutrons, electrons and γ-quanta.

Penetration by a fast charged heavy particle into a metal or alloy leads to the following phenomena [89-93]. The particle, first, expends its energy in the excitation of the conduction electrons and the inner electrons of the atomic shells and secondly, transfers energy to atoms by collision. The relative role of both processes depends substantially on the magnitude of the parameter

$$\varepsilon = \frac{m}{M}E, \tag{8.1}$$

where m = electron mass, M = mass of the heavy particle, and E = kinetic energy of the particle. For sufficiently large values of the parameter ε, the mechanism of energy transfer to the electrons predominates; the energy given to the electrons is many times greater than the energy transferred to the atoms. At sufficiently low values of ε, lower than ε' (for different metals and different types of particles ε' ranges from 0.1 to 10 ev), the second mechanism of energy transfer predominates, i.e., the energy of the particles is transferred chiefly to the atoms of the substance.

Having obtained energy from the particles travelling through the metal, the electrons transfer energy to it via interaction with the crystal lattice, thus heating the crystal.

Atoms of the alloy, that have obtained energy from the particles, participate in different types of processes depending on the amount of energy transferred to them. If this energy is insufficient to dislodge an atom from the equilibrium position, i.e., for the formation of vacancies and of an interstitial atom (requiring energy of the order of 25 ev), then only vibrations of the crystal lattice atoms are excited. For large energies, the atoms form pairs of crystal lattice defect, vacancies and interstitials. Where a very large energy is transferred to the atoms they move rapidly through the crystal, exciting only the same type of defects as the original particle.

In the region of the crystal, adjoining the trajectory of the particles, a large number of crystal defects (interstitial atoms and vacancies) are formed in this manner. Moreover, the liberation of energy in this region causes a strong local increase of temperature, which may even lead to melting of the alloy in the given region. Subsequent rapid cooling of these parts of the crystal due to heat exchange with the surroundings, may lead to effects similar to the quenching of metal in these regions from the high temperatures attained during the passage of the particle. When the ordered alloy is exposed to irradiation, a disordered phase may be formed in regions of local heating and retained by quenching.

During irradiation of metals by fast neutrons, the latter transfer their energy to the metallic ions by collision with nuclei. Ions that have experienced a collision with a neutron receive a rather large amount of energy and may in turn promote local heating and the formation of crystal lattice defects. Since neutrons interact very weakly with electrons, they do not expend energy by excitation of electrons. Moreover, since neutrons transfer a comparatively small fraction of their energy during collisions with atoms (because the mass of the nucleus is usually substantially larger than that of the neutron), the parameter ε for the ejected atom is not large and its energy is transferred mainly to other atoms of the alloy, rather than to electrons. Neutrons, therefore, expend an appreciably greater fraction of their energy in collisions with atoms, than do charged particles, and eject a comparatively larger number of atoms along their paths.

During irradiation of metals by electrons, the latter transfer the greater part of their energy to electrons, and not to heavy particles. As a result a relatively smaller number of lattice defects are formed.

During irradiation of a metal by γ-rays, fast electrons are formed as a result of the Compton effect and of the excitation of inner electrons (internal photoelectric effect). The fast electrons which emerge in this manner affect the crystal in the same manner as the primary electrons during irradiation by electrons.

The difference in the action of different irradiations on a metal and an alloy lies in the fact that the penetrating radiation (γ-rays and neutrons) produce crystal lattice defects in a comparatively large volume of the crystal, where they may be distributed more or less uniformly; whereas charged particles produce nonuniformly distributed defects in a comparatively thin layer of the irradiated sample surface.

The emergence of a large number of excess (nonequilibrium) lattice defects may lead to a noticeable increase in the rate of the different types of processes associated with the migration of atoms in a solid. In the course of time the interstitial atoms annihilate the vacancies, which leads to additional liberation of heat. The decrease in the number of defects with time leads to slowing-down of the specified processes.

The influence of radiation on ordered alloys is due to two effects. First, the local heating may lead to a change in the state of ordering, and secondly, the lattice defects which appear at such temperatures, possessing sufficient mobility, will accelerate the process of approaching the state of equilibrium (at the given temperature).

The most detailed study of the influence of different forms of radiation was made on the alloy $AuCu_3$. In [94] and [95], the effect of irradiation by neutrons in a nuclear reactor was studied in samples of ordered and disordered $AuCu_3$ alloys. The temperature was maintained at 80° C. The ordering state was ascertained from the electrical resistivity of the samples, measured during irradiation. As is apparent from Fig. 50, irradiation has a different effect on ordered and disordered alloys. The electrical resistivity of an initially ordered sample when the number of neutrons n (calculated per cm² of the transverse cross section of the neutron beam incident on the sample) increased from 0 to $6.5 \cdot 10^{19}$, increased by 23% (Fig. 50a). When n was increased to $2 \cdot 10^{19}$ [95], the electrical resistivity increased by 60%. This means that during irradiation of the ordered sample a disordering occurs, which may be explained by the local heating of the individual spots in the crystal and their subsequent quenching. The electrical resistivity of an initially disordered alloy (Fig. 50b) during irradiation first decreases, and at $n \sim 5 \cdot 10^{20}$ reaches a minimum value, 7.5% smaller than the initial value. This decrease of electrical resistivity may be explained by the acceleration of the diffusion processes when vacancies appear at the lattice sites during irradiation. By decreasing the relaxation time we enable a partial ordering of a previously disordered nonequilibrium alloy even at a temperature of 80° C. The subsequent resistivity increase for long irradiation times discovered by [95], may be related to the formation of a small number of mercury atoms, which appear as a result of absorption of neutrons by the nuclei of gold atoms. Under the conditions of this experiment (for $n \sim 2 \cdot 10^{20}$) there appears

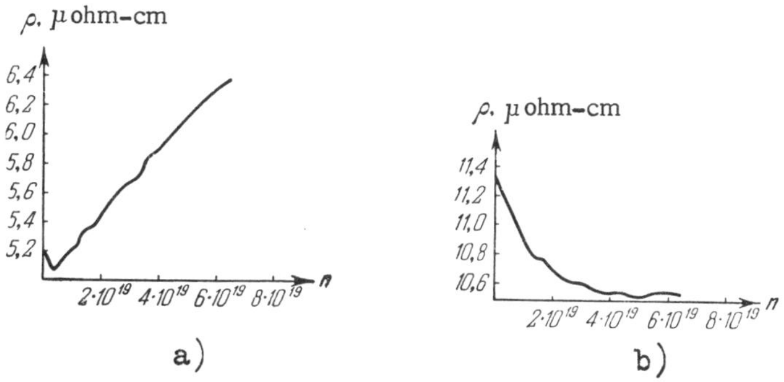

Fig. 50. Variation of the electrical resistivity of the alloy $AuCu_3$ during irradiation by neutrons at a temperature of 80° C [94]. a-initially ordered sample, b-initially disordered sample.

about 0.4 atomic % of mercury, which should increase the electrical resistivity by acting as an impurity (it is also possible that the presence of mercury may lead to a change in the state of ordering).

The effect of neutron irradiation on the ordering of $AuCu_3$ was investigated at different temperatures, where a characteristic variation of this effect was discovered both with an increase and with a decrease in temperature.

An investigation of $AuCu_3$ at a temperature of 140°C [95] has shown that neutron irradiation has an appreciably weaker effect on the electrical resistivity of an initially ordered sample; for $n = 1.4 \cdot 10^{20}$, the electrical resistivity exceeds its initial value only by 4%. For a disordered sample the electrical resistivity at a temperature of 140°C decreases much more rapidly than at a temperature of 80°C (by 17% at $n = 2.5 \cdot 10^{19}$).

Measurements at 200°C were conducted [96] on a disordered sample of $AuCu_3$ with a neutron flux of 10^{12} neutrons per cm^2 per second. The dependence of electrical resistivity on the irradiation time is shown in Fig. 51. Here one observes an appreciable ordering of the sample caused by the irradiation by neutrons, due to a decrease of the relaxation time. Under the conditions of this experiment the relaxation time for a non-irradiated sample was quite substantial. In fact, in Fig. 51, the intitial part of the curve corresponded to the absence of irradiation. The curve is horizontal here and the electrical resistivity did not reveal an order-disorder transformation.

In [97] ordered and disordered samples of the $AuCu_3$ alloy were irradiated by neutrons in a nuclear reactor at 40°C.

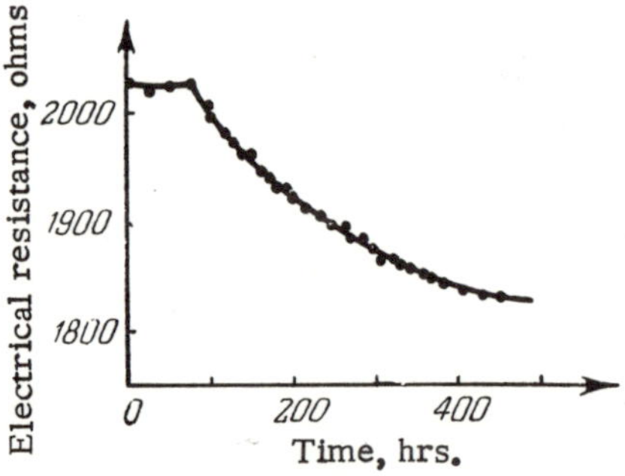

Fig. 51. Time dependence of the electrical resistance of initially disordered sample of $AuCu_3$ at 200°C, irradiated by a neutron flux of 10^{12} neutrons per cm^3 per second [96].

The ordered samples were prepared by slow cooling from 400°C for a period of 50 hours. The disordered samples were obtained by quenching in water from 550°C. It is apparent from Table 7 that at a temperature of 40°C no change of electrical resistivity of the disordered sample was detected, whereas the electrical resistivity of an ordered sample increased substantially, attaining a value corresponding almost to the disordered state. X-ray studies of an initially ordered sample (which could be conducted only after irradiation had ceased for several months) showed that the sample was in a disordered state. Calculations made in [89] have shown that only several per cent of the atoms may be displaced from the lattice sites by collisions. This is insufficient to cause the observed effect of almost completely disordering. The latter may be explained only by considering the regions of local overheating, where the alloy changes to a disordered state.

Table 7

Variation of the electrical resistivity of ordered and disordered samples of the $AuCu_3$ alloy caused by neutron irradiation

Total neutron flux, number of neutrons per cm^2	Electrical resistivity, microohm-cm	
	Initially ordered sample	Initially disordered sample
0	4.60	11.20
$0.4 \cdot 10^{19}$	5.71	11.25
$0.6 \cdot 10^{19}$	6.25	11.25
$1.0 \cdot 10^{19}$	7.54	11.17
$1.5 \cdot 10^{19}$	8.36	11.21
$3.3 \cdot 10^{19}$	10.10	11.30

A study of the ordering of $AuCu_3$ irradiated by neutrons at different temperatures was reported in [98]. Samples with a degree of long-range order corresponding to 376°C were studied. Annealing of the samples (before irradiation) for a period of 100 hours at 150°C did not cause a change of their electrical resistivity. Neutron irradiation of samples in a nuclear reactor at 150°C caused a decrease of electrical resistivity to a certain minimum value, related to further ordering of the samples (because the equilibrium value of the degree of long-range order at 150°C is higher than at 376°C), which now became possible because of the presence of lattice defects. Thus, one of the samples was irradiated in a flux of 10^{12} neutrons per cm^2 per sec, the other in a flux having one-fourth the intensity. Consequently, the rate of attainment of the minimum value of electrical resistivity for the first sample was found to be four times greater than that for the second. A third sample was irradiated by a relatively low neutron flux (approximately $2 \cdot 10^{10}$ neutrons per cm^2 per sec) at -160°C for a period of one week. The electrical resistivity of the alloy increased by 0.6%, i.e., the alloy became slightly disordered. Inasmuch as the alloy was no longer ordered the vacancies formed during irradiation at this temperature were frozen in the crystal. Their mobility was so low, that they did not cause diffusive motion of atoms in the crystal, leading to the establishment of an equilibrium state. Even at a higher temperature of -80°C, which was maintained for a period of 3.5 hours, the third sample (previously irradiated) did not exhibit a change of electrical resistivity. Only at temperatures of approximately 0°C does the vacancy mobility become appreciable. The third sample of the alloy exhibited a decrease of the electrical resistivity after being exposed to temperatures higher than 0°C.

From the rate of change of resistivity one may evaluate the relaxation time of the processes in which the alloy approaches a more ordered state, which is related to the disappearance of lattice defects. For temperatures of 0°, 100° and 150°C these relaxation times were found to be equal to 50 hours, 25 min., and 10 min., respectively.

The change of ordering of $AuCu_3$ was also detected as a result of irradiation by α-particles [90, 91]. The irradiation of the ordered alloy both at -180°C, and at 220°C led to disordering. Since the vacancy mobility at 220°C was rather high, it could restore the breakdown of ordering caused by irradiation. Hence, the irradiation led to a substantially lower disordering, than at -180°C, when the vacancies had practically zero mobility.

The influence of irradiation by protons on the state of ordering of the alloy $AuCu_3$ was studied in [99], in which an appreciable variation of electrical resistivity and consequently of the degree long-range order of ordered samples was also discovered.

Reference [100] gives the results, obtained from irradiation of ordered $AuCu_3$ by 1 Mev electrons at low temperatures.

After irradiation a variation of the electrical resistivity was detected which was substantially smaller than might be expected by accounting for the appearance of a definite quantity of vacancies during irradiation. This could be attributed to the fact that during collisions the electrons transfer appreciably less energy to the atoms of the crystal, than the heavy particles. Therefore, the atoms ejected by electrons from the crystal lattice sites during irradiation travel on the average a much shorter distance than in the case of irradiation by heavy particles, remaining near the vacancies formed by their removal. Hence, even at -195°C a greater part of the interstitial atoms and vacancies which appear during irradiation are able to recombine rapidly. Only the remaining small number of vacancies, from which the interstitial atoms have moved an appreciable distance may, after annealing at room temperature, lead to greater ordering. This was also detected from a certain decrease of the electrical resistivity.

The influence of irradiation by γ-rays on the kinetics of ordering of $AuCu_3$ was investigated in [77]. The variation of

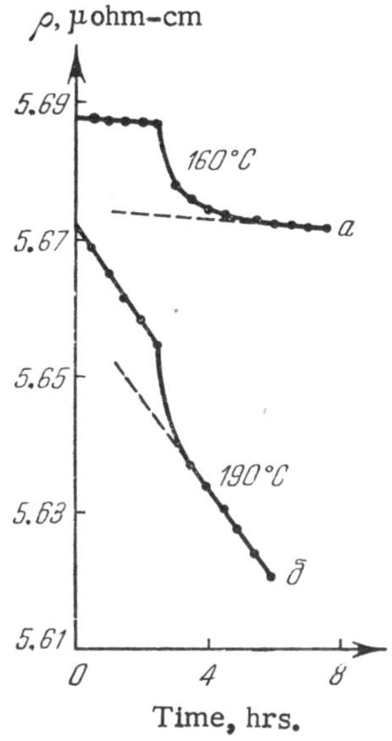

Fig. 52. Influence of γ-irradiation on the electrical resistivity of a quenched, partially ordered $AuCu_3$ alloy during isothermal annealing [77].

the electrical resistivity of a sample of $AuCu_3$ quenched from 370°C as a function of the time of annealing was investigated at 160°C and 190°C.

Figure 52 (curve a) shows that in the absence of the irradiation (the first 2.5 hours) the electrical resistivity at 160°C (and hence the degree of long-range order) changed very slowly during annealing. After γ-ray irradiation (performed after 2.5 hours), the concentration of vacancies at the crystal lattice sites increased sharply and became substantially greater than for the equilibrium concentration. Consequently, the rate of change of ρ increased many times. At 190°C the number of equilibrium vacancies is substantially greater and even in the absence of irradiation it changes markedly during annealing* (Fig. 52 curve b). The time for disappearance of the excess vacancies after irradiation, corresponding to the range of rapid variation of ρ, is in this case smaller than at 160°C.

The disordering effect, occurring in the ordered AuCu (also $AuCu_3$) by irradiation by α-particles, accelerated in a cyclotron, was investigated [101] at -150°C. As a result, the electrical resistivity in the alloy increased substantially, and its value was retained for some time upon heating to room temperature.

The disordering of AuCu upon irradiation by α-particles with energy of 33 Mev was detected by [102] also from the change of the lattice parameter. It may be seen from Fig. 53 that during irradiation of an ordered sample of this alloy, the tetragonality of the crystal lattice diminishes, the lattice constant a decreases somewhat, whereas the constant c increases rapidly. When the disordered state is reached, the values of constants a and c become identical, tetragonality disappears and a face-centered cubic crystal lattice appears. It was also discovered by [102], that during irradiation by α-particles leading to disordering, the hardness of the alloy increased (as distinguished from the case of thermal disordering).

The influence of irradiation by fast neutrons on ordering of the Ni-Mn alloys was studied in [103]. It was discovered that the electrical resistivity of the alloy and the saturation magnetization changed greatly during irradiation, with the

---

*Preliminary annealing at 370°-380°C, which leads to the formation of comparatively large regions of the ordered phase, substantially reduces the time for attainment of equilibrium, compared with alloys quenched from temperatures above the critical temperature. This may be explained by the fact that in the curve of Fig. 51, corresponding to 200°C, the resistance before irradiation remains unchanged.

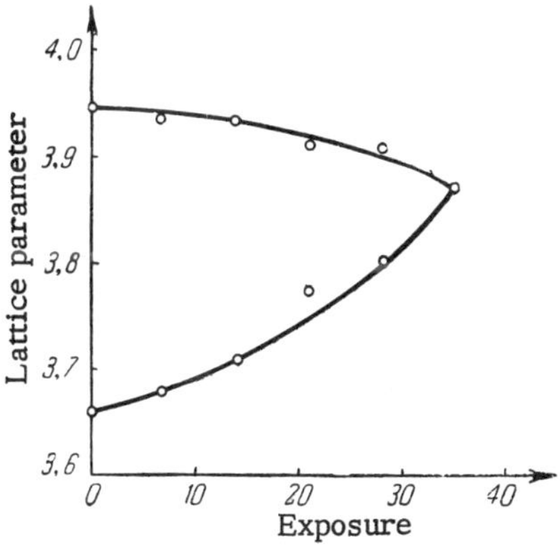

Fig. 53. Influence of α-irradiation on the lattice parameters of a previously ordered AuCu alloy [102].

maximum change found for alloys of stoichiometric composition Ni_3Mn. Figures 54 and 55 give the specific electrical resistivity and 4 π times the saturation magnetization of alloys, previously subjected to a different type of treatment, as a function of composition (from 16.5 to 31.9 atomic % Mn). The curves for irradiated, initially ordered alloys, approach the curves for disordered alloys when the neutron dose is increased. This indicates the disordering effect of neutron irradiation on Ni-Mn alloys.

Analysis of experimental data shows that the mean number of atoms displaced from their positions by one neutron with an energy of approximately 0.5 Mev in the Ni-Mn alloys is approximately 5,000.

As indicated in the preceding section, the relaxation time of β-brass at room temperature is so short that it is practically impossible to preserve the nonequilibrium state of the alloy. This means that the mobility of vacancies is very high, and irradiation by fast particles may have a significant effect on the ordering only at very low temperatures. Irradiation of β-brass samples by α-particles accelerated up to an energy of 33 Mev in a cyclotron, was carried out by [104] at a temperature not exceeding -100°C. The electrical resistivity of the irradiated samples was determined at -146°C for various amounts of radiation. The results of the measurements are given in Fig. 56, where a curve for the dependence of the

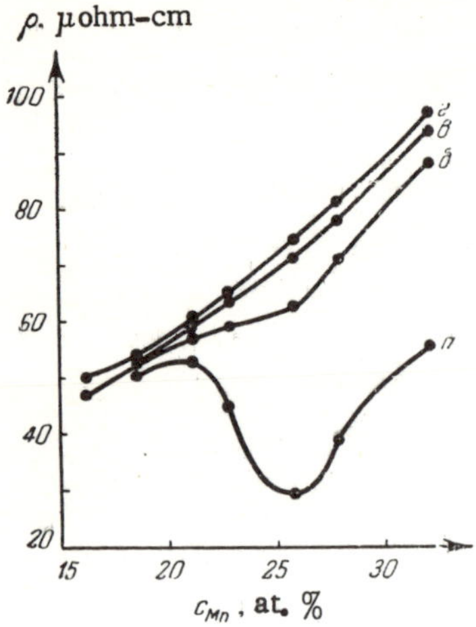

Fig. 54. Influence of irradiation by neutrons on the electrical resistivity (at 20°C) of the NiMn alloys of various compositions [103].
a-nonirradiated ordered alloys; b-previously ordered alloys, subjected to irradiation by a flux of 0.55 • 10^{20} neutrons per cm^2; c-previously ordered alloys, subjected to irradiation by a flux of 0.92 • 10^{20} neutrons per cm^2; d-nonirradiated disordered alloys (quenched from 1,000°C).

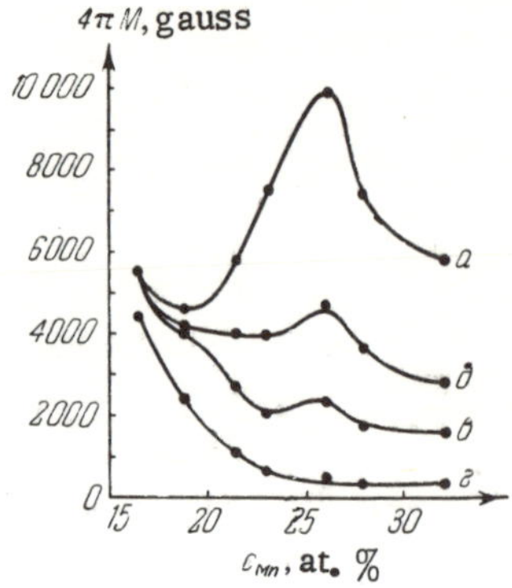

Fig. 55. Influence of the radiation by neutrons on the saturation magnetization M of Ni-Mn alloys of different compositions [103]
a-nonirradiated ordered alloys; b-previously ordered alloys, subjected to irradiation by a flux of 0.55 • 10^{20} neutrons per cm^2; c-previously ordered alloys, subjected to irradiation by a flux of 0.92 • 10^{20} neutrons per cm^2; d-nonirradiated disordered alloys (quenched from 1,000°C).

relative variation of resistivity on the number of α-particles striking the alloy is presented. It is apparent that the radiation causes the increase in electrical resistivity to be almost two-fold. Such a large increase in resistivity is not observed in pure metals (e.g., for copper it should not exceed 25%) and is due both to the appearance of lattice defects and to a decrease in the degree of long-range order under the effect of radiation. After their radiation ceased, the samples were subjected to annealing for a period of 5 minutes at successively increasing temperatures. In the intervals between the annealing, the samples were again cooled down to -146°C and the electrical resistivity was measured at this temperature. It is apparent from Fig. 57, that the decrease in regularity of the lattices, which occurred during irradiation, disappeared quickly during such annealing at low temperatures (approximately -100°C), and vanished almost completely after annealing at room temperature.

Irradiation by fast particles may also change the electrical resistivity of alloys, in which no long-range order exists at a given temperature in the equilibrium state. Such a variation of electrical resistivity can, in a number of cases, be related to the variation of short-range order caused by irradiation. This effect was observed during irradiation of β-brass in a reactor for a period of one week at 50°C [73]. The number of neutrons, striking one cm of the sample during irradiation, was $7 \cdot 10^{17}$. During such an irradiation the resistivity of an alloy, containing 10% Zn, measured at 80° K, decreased by 1.4%; an alloy with 20% Zn-by 1.9%, and an alloy with 30% Zn-by 2.1%. During a month-long irradiation of samples, at a temperature of 80°K, when the lattice defects are virtually immobile and no variation in the short-range order occurs in the alloy, the electrical resistivity increased, because of the lattice defects which were created in the sample. It may be concluded therefore, that the decrease of electrical resistivity, which was observed after irradiation at 50°C, is caused by the increase of short-range order in the alloy during irradiation.

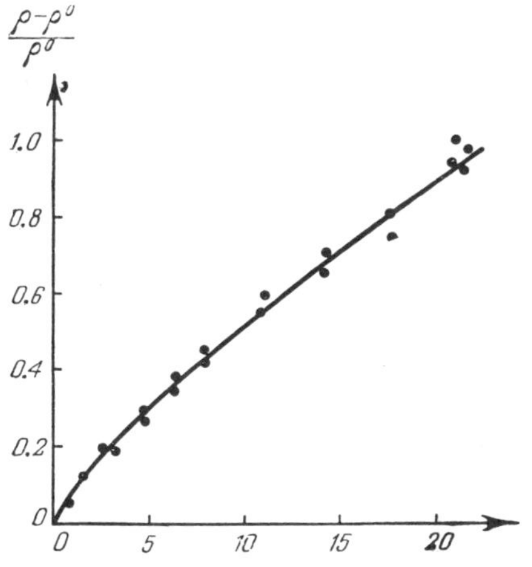

Fig. 56. Relative variation in electrical resistivity $\frac{\rho - \rho^0}{\rho^0}$ of β-brass as a result of irradiation by α-particles at temperatures below -100° C [104].

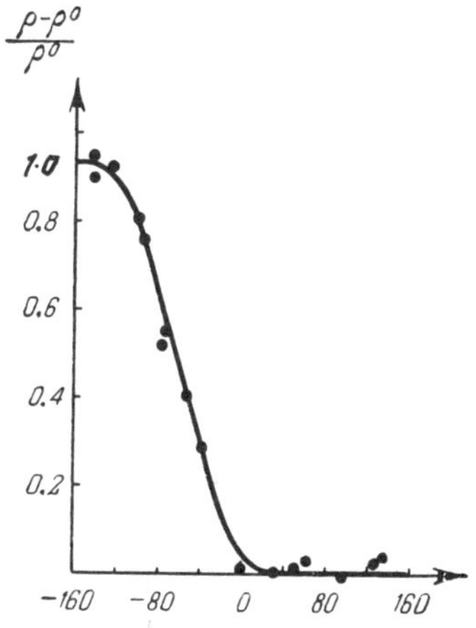

Fig. 57. Relative variation of electrical resistivity of β-brass irradiated by α-particles as a result of consecutive 5-minute annealings at different temperatures [104].

The effect of radiation on several completely disordered alloys was also investigated [105], [90] and [102]. For example, the electrical resistivity of a Cu-Be alloy increased markedly during the radiation by neutrons.

9. Studies of the Correlation Between Ordering and Mechanical, Magnetic, Galvano-Magnetic and Other Properties of Ordered Alloys

Mechanical properties

A change in the order of the distribution of atoms should exert an influence on the elastic properties of alloys. During passage through the critical temperature T_0 the appearance of special features on curves of the temperature dependence of quantities describing the elastic properties of alloys can be expected.

Indeed, a special variation of the elastic constants s_{11}, s_{12} and s_{44} was experimentally detected [106] during an investigation of their temperature dependence in a single crystal of ordered $AuCu_3$. Figure 58 shows curves giving the variation of s_{11}, s_{12} and s_{44} with the temperature of an alloy, in which the degree of long-range order changed with temperature. The initial parts of the curve, lying in the range from 0 to 250° C, where the degree of long-range order is virtually unchanged, show an almost linear increase of the elastic constants. With a subsequent temperature increase, when the decrease of the degree of long-range order becomes marked, a more rapid increase of the elastic constants is observed. At the critical

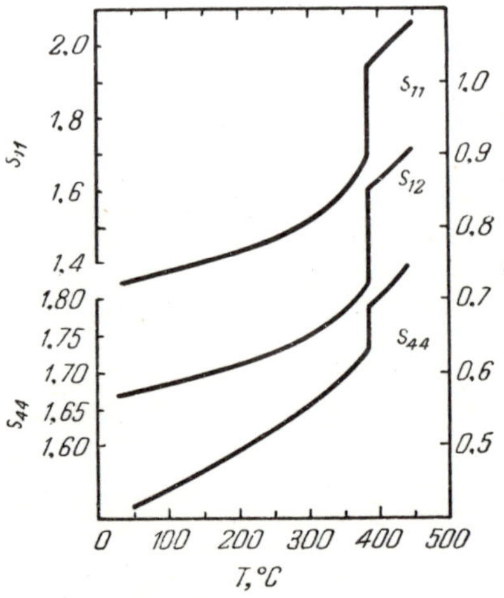

Fig. 58. Temperature dependence of elastic constants (in units of 10^{-12} cm^2/dyne) of a $AuCu_3$ single crystal [106].

temperature the constants undergo a step-change and we again detect an almost linear increase with temperature in the disordered state. The temperature variation of Young's modulus [76] in an ordered sample of $AuCu_3$, cut from a single crystal parallel to the [100] direction, is shown by the upper curve in Fig. 59. At the critical temperature, a step-change in Young's modulus is observed. The lower curve gives an extrapolated temperature dependence of Young's modulus in a disordered state (assuming that the degree of long-range order does not change with temperature and always remains equal to zero). A similar increase of Young's modulus during ordering was detected [107] in Cu_3Pd. In the alloys AuCu and CuPd, however, a decrease in Young's modulus was observed during ordering [107].

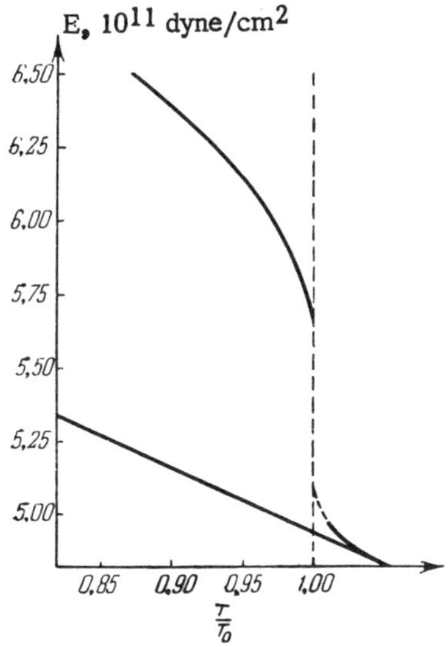

Fig. 59. Temperature dependence of Young's modulus in the $AuCu_3$ alloy [76].

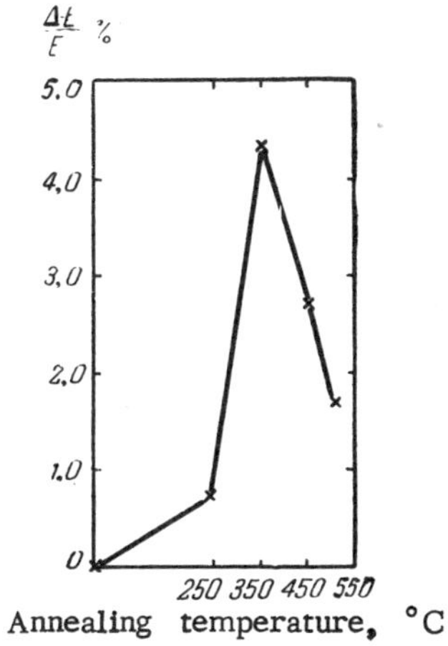

Fig. 60. Relative variation in Young's modulus of the Ni_3Mn alloy as a function of the temperature of prolonged annealing [72].

The relative variation in Young's modulus of annealed, initially disordered, Ni_3Mn with the annealing temperature [72] is illustrated in Fig. 60. The increase of Young's modulus caused by ordering was at a maximum at the annealing temperature of 350°C. This temperature coincides with the annealing temperature at which a minimum in the electrical resistivity was observed (see Fig. 41) in the same alloy, subjected to the same preliminary heat treatment.

The temperature variation of the reciprocal of Young's modulus for β-brass samples cut parallel to the [100], [110] and [111] directions, during ordering of the alloy, is illustrated in Fig. 61 [108]. The temperature dependence of the elastic constants s_{11}, s_{12} and s_{44} in β-brass is illustrated in Fig. 62 [109]. It is apparent from Figs. 61 and 62 that the behavior of the curves changes abruptly during transition into the ordered state; the reciprocal of Young's modulus and also the values of s_{11}, s_{12} and s_{44} decrease rapidly.

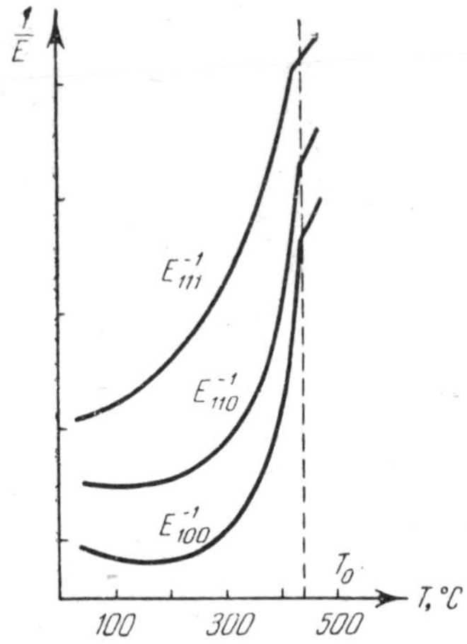

Fig. 61. Relative variation of the reciprocal of Young's modulus for β-brass, in the principal directions, as a function of temperature [108]. Each curve begins at a value of 1 at 25°C. One division on the ordinate axis corresponds to a 10% variation of the given quantity.

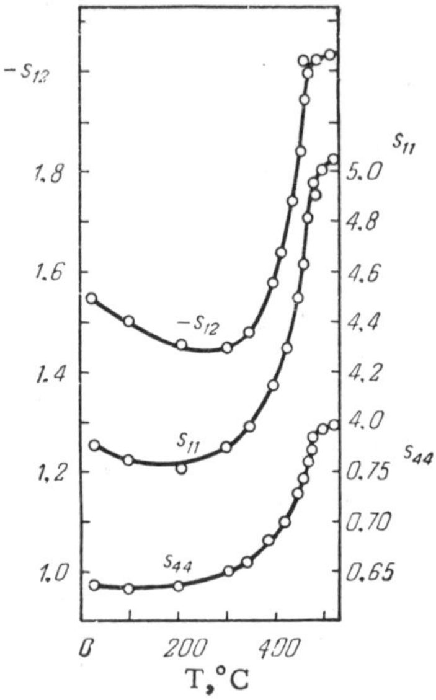

Fig. 62. Temperature dependence of the elastic constants s_{11}, $-s_{12}$ and s_{44} (in units 10^{-12} cm/dyne) for β-brass [109].

Ordering may lead to an appreciable change in the hardness of a number of alloys. Thus, for instance, in CuPt, the ordering leads to a more than two-fold increase in the Brinnell hardness [110]. An alloy quenched from a temperature above T_0 was then annealed at 500°C and in this manner reached the ordered state. A curve obtained for the Brinnell hardness as a function of the annealing time is given in Fig. 63. The sharp increase in hardness at the beginning of the annealing is evidently related to lattice strains, that appear during the formation and growth of domains in the first stage of ordering. These strains appear because a rhombohedral crystal lattice is formed during ordering in this alloy as described in Sec. 2. In the second stage of ordering, when domains combine and

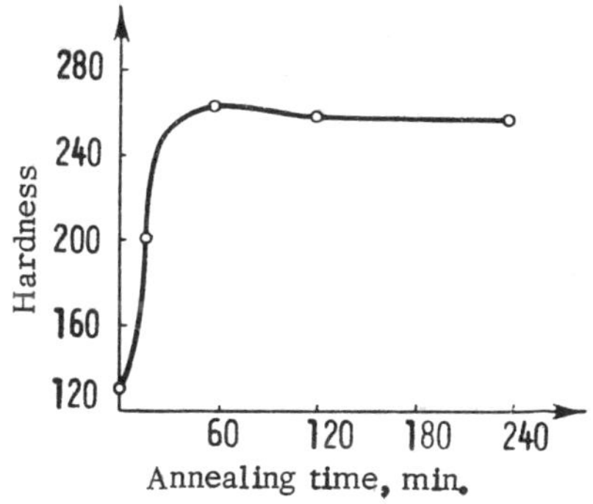

Fig. 63. Dependence of the Brinnell hardness of a previously quenched CuPt alloy on the annealing time at 500° [110].

the alloy reaches a state of homogeneous ordering, these strains disappear, and hence the hardness should diminish. This behavior was observed experimentally (Fig. 63). During ordering of CuPt, the tensile strength also increases.

In AuCu, [111] and [112], an increase of hardness during ordering was also observed, due to the appearance of strains during the formation of the tetragonal lattice in the ordered state.

In $AuCu_3$, which has a face-centered cubic structure both in the ordered and in the disordered states, there is no reason for the emergence of such large ordering strains. In fact, the critical shear stress in the ordered state is equal to 2.3 kg/mm^2 and in the disordered state to 4.3 kg/mm^2 [113]. The influence of ordering on the hardness of the Fe-Co alloys of different composition was investigated by Kadykova and Selisskii [114]. To bring the alloys to an ordered state they were quenched, and then cold-worked. Afterward the alloys were annealed at different temperatures ranging from 150° to 800°C. Over the temperature range 250° to 550°C (the critical temperature of Fe-Co is approximately 730°C) the hardness of the alloys increased substantially. The electrical resistivity of FeCo decreases during annealing within this temperature range, which indicates an increase in the degree of long-range order. During annealing at higher temperatures, when the alloys are disordered, the hardness decreases again.

Ordering also exerts an influence on the rate of creep in alloys. This effect was discovered by Herman and Brown [115] in β-brass. Measurements of the rate of creep were conducted at constant temperatures over the range from

330° to 500°C. The logarithm of the rate of creep $\dot{\gamma}$ was found to be a linear function of the applied stress σ:

$$\ln\dot{\gamma} = \alpha + \beta\sigma,$$

where α and β are functions of temperature. It follows from this formula that the greater $1/\beta$, the better the creep resistance of the material. Figure 64 gives a curve of the dependence of $1/\beta$ on the temperature of a β-brass sample. At the critical temperature the curve has a pronounced bend, and in the ordered state $1/\beta$ becomes considerably greater than in the disordered state. The presence of the bend shows that ordering leads to an increase of creep resistance in β-brass.

It has been discovered that the ordering of an alloy influences not only its elastic and plastic properties, but also that plastic deformation significantly changes the state of ordering. If the ordered alloy is subjected to plastic deformation, its ordering is found to decrease to a great extent. There are two mechanisms of disordering action of a plastic deformation. The first and apparently the chief mechanism is the appearance of a large number of displaced regions (for instance, additional antiphase domains) which appear as a result of the displacement of individual sections of the crystal during slip on slip planes by an appropriate nonintegral number of lattice periods. The second disordering mechanism of plastic deformation consists in the creation of numerous lattice defects, whose motion in the completely ordered crystal causes a displacement of atoms from their proper sites.

The disordering caused by plastic deformation can be readily detected by an increase in the electrical resistivity of

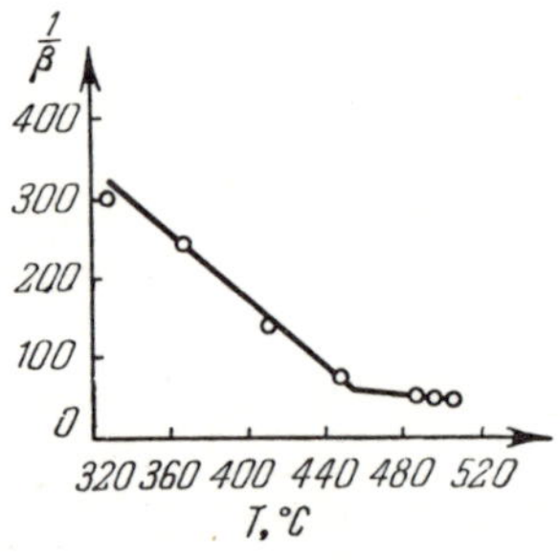

Fig. 64. Temperature dependence of $1/\beta$ (in arbitrary units) describing creep in β-brass [115].

the alloy. This increase should be considerably greater than that in pure metals and disordered alloys. In fact, in pure metals the increase of electrical resistivity during plastic deformation is related only to the geometrical distortion of the crystal lattice and, at room temperature, does not usually exceed 2-3% (excluding some metals with a high recrystalization temperature, e.g., tungsten and molybdenum, where such increase is substantially higher). In disordered alloys, plastic deformation may also cause a change in the short-range order, which does not, however, cause as large a change in electrical resistivity as the change of long-range order in ordered alloys.

The change of the electrical resistivity during plastic deformation of ordered and disordered $AuCu_3$ has been investigated by Dahl [116]. Figure 65 gives curves of the relative variation of electrical resistivity of a disordered alloy (curve a) and an ordered alloy (curve b) as a function of the relative decrease in the cross section of a cold-drawn $AuCu_3$ alloy wire. The electrical resistivity of an undeformed (disordered) sample, rapidly cooled from high temperatures, is taken as 100%. Thus, the electrical resistivity of ordered $AuCu_3$ at sufficiently large deformations (the degree of reduction in area was greater than 60%) becomes the same as for a disordered sample, whose resistance was virtually unchanged during plastic deformation. The superlattice lines on the x-ray diffraction pattern of this alloy also disappeared during deformation, which resulted in a decrease of the cross section of the samples by more than 60%.

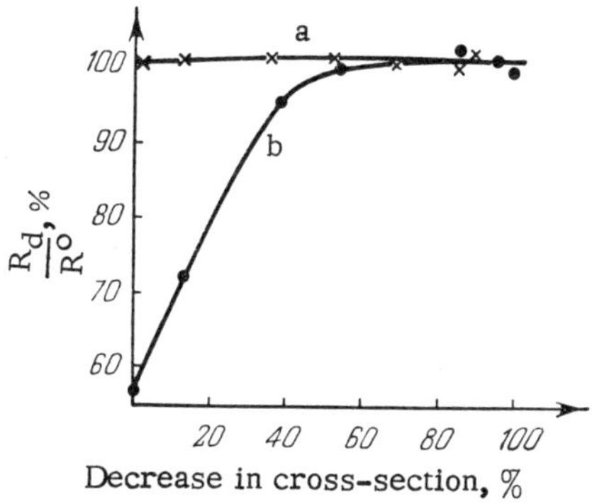

Fig. 65. Ratio of electrical resistivity R_d of a deformed sample of the $AuCu_3$ to the electrical resistivity of a disordered undeformed sample R^0 as a function of the degree of deformation a-disordered alloy, b-ordered alloy.

Similar results were obtained [117] in an investigation of the variation of electrical resistivity with deformation of Ag_3Mg.

A strong effect of plastic deformation, leading to disordering, was also observed in the alloys Ni_3Mn [116], Ni_3Fe [116], AuCu [118] and Cu-Pt (80 atomic % Pt) [119].

The effect of plastic deformation on the electrical resistivity of Ni_3Fe, containing small amounts of molybdenum impurities, was investigated by Lifshits and Ravdel' [120]. The alloys were slowly cooled (for one week) from 550° to 200°C, then plastically deformed. It is apparent from Fig. 66 that the relative variation of the electrical resistivity during cold deformation depends substantially on the concentration of molybdenum in the alloy. In the Ni_3Fe alloy, free of molybdenum, 90% extension leads to appreciable increase (35%) of the electrical resistivity, due to disruption of the ordered structure. An increase in the small amount of molybdenum leads to a considerable lessening of the electrical resistivity increase during compression. It is interesting that for molybdenum concentrations higher than 1%, the sign of the effect changes, i.e., the plastic deformation diminishes the resistivity of the sample. The authors of [120] explain the

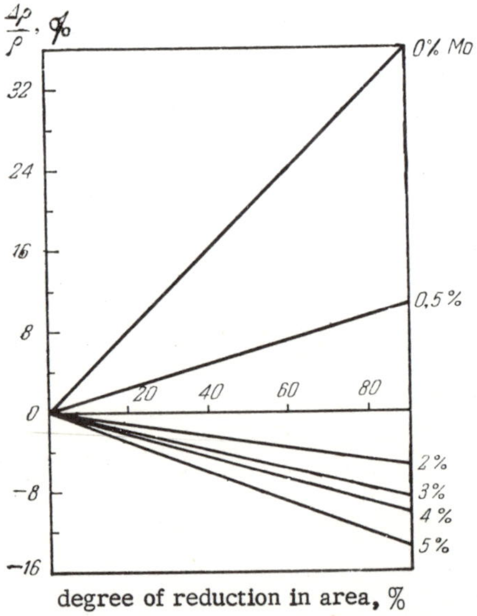

Fig. 66. Relative variation of the electrical resistivity of the Ni_3Fe alloy with different concentrations of molybdenum as a function of the degree of deformation during extension.

strong effect of molybdenum impurities on the change in the electrical resistivity by the fact that the addition of molybdenum leads to a disruption of the long-range order in the alloy, and simultaneously to a segregation of the different types of atoms, leading to the formation of clusters. It is assumed that these clusters lead to additional scattering of electrons, and hence to an increase in the electrical resistivity. Plastic deformation, by disrupting these clusters, decreases the electrical resistivity of alloys containing large amounts of molybdenum impurity.

The increase in number of vacancies caused by the plastic deformation in an alloy quenched from a higher temperature may lead to an increase in the rate of ordering. This was observed in [77], which reports on the variation of the electrical resistivity of $AuCu_3$ during annealing at 100°C after quenching from 380°C. A weak plastic deformation, at intervals of 500 min., caused a small amount of disordering of the type considered above and an increase in electrical resistivity. The vacancies that emerged thereafter appreciably accelerated the approach to equilibrium, i.e., to a more ordered state. The process was retarded at the end of each period because of the loss of vacancies (Fig. 67).

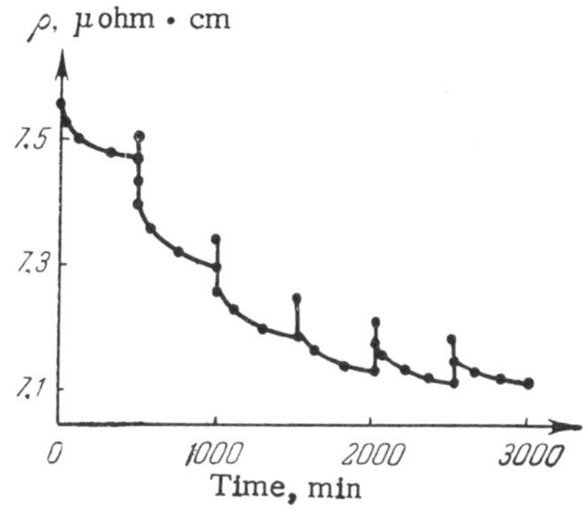

Fig. 67. Electrical resistivity of an ordered $AuCu_3$ alloy as a function of time during plastic deformation occurring at 500 min. intervals [72].

Magnetic properties

The ordering of atoms in a crystal may significantly influence the magnetic properties of alloys. In a number of ferromagnetic alloys the saturation magnetization, Curie temperature, coercive force, magnetic permeability and some other properties are found to be strongly dependent on the degree of order. Particularly noticeable is the influence of ordering on the magnetic properties of Ni_3Mn which is paramagnetic in the disordered state, while in the ordered state it is a pronounced ferromagnetic with saturation magnetization, approximately one and a half times as large as that of pure nickel. The dependence of the saturation magnetization M on the annealing temperature of a sample of Ni_3Mn, obtained by Komar and Vol'kenshteyn [121–123], is given in Fig. 68. Annealing of this alloy at temperatures above the critical temperatures (approximately 510°C), leads to a strong increase of the saturation magnetization. Similar results for the dependence of the saturation magnetization of Ni_3Mn were obtained by Kaya and Nakayama [124]. The dependence of saturation magnetization on composition, for ordered and disordered Ni_3Mn, was also obtained by Kaya and Nakayama [125].

The influence of molybdenum impurities on the saturation magnetization of alloys, close to the stoichiometric composition of Ni_3Mn, was investigated by Lifshits et al [72]. It is apparent from Fig. 69 that the addition of small amounts of molybdenum impurity to Ni_3Mn leads (after continuous annealing) to a strong decrease of the saturation magnetization. Since molybdenum impurity leads to an appreciable decrease of long-range order in Ni_3Mn, the observed effect is related to the influence of ordering on saturation magnetization.

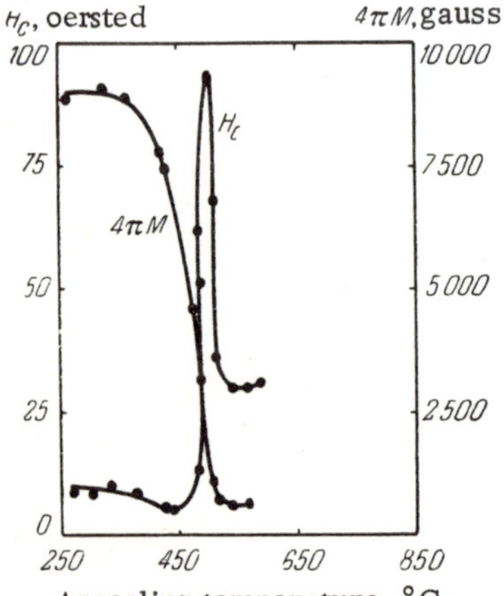

Fig. 68. Dependence of saturation magnetization M and coercive force H_C of Ni_3Mn on the annealing temperature

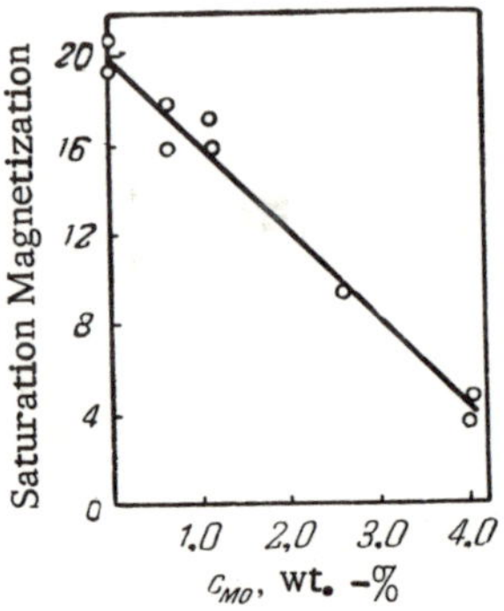

Fig. 69. Saturation magnetization (in arbitrary units) of continuously annealed Ni_3Mn with different percentages of molybdenum impurity as a function of concentration of the impurity [72].

The influence of ordering on saturation magnetization has also been detected in Fe-Si [126], Fe-Ni [127], Ni Pt [128], in Heusler alloys [129] and in some other systems.

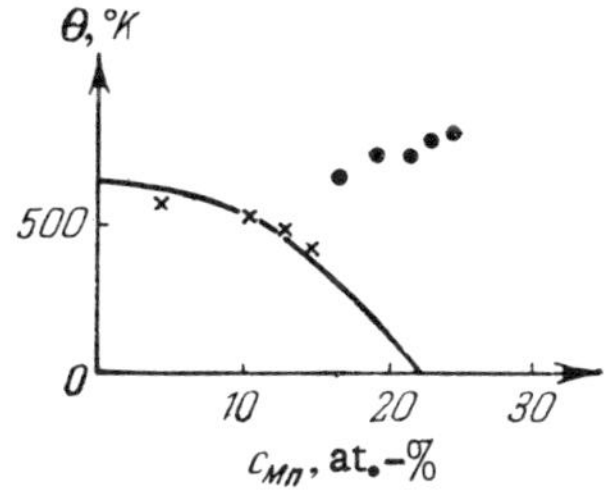

Fig. 70. Curie temperature of disordered (crosses) and ordered (circles) Ni-Mn as a function of concentration of Mn [123-125].

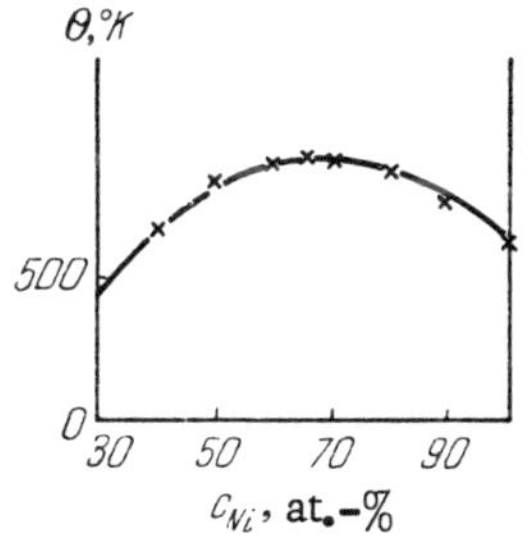

Fig. 71. Curie temperature of disordered Fe-Ni as a function of concentration of Ni [134].

Dekhtyar [130] investigated the temperature dependence of saturation magnetization and coercive force of a molybdenum-doped Ni_3Fe alloy. The influence of plastic deformation on the temperature dependence of these properties was also investigated. The similar temperature dependences obtained for both the saturation magnetization and the coercive force has been attributed to the influence of the order-disorder transformation on the magnetic properties. Figure 68 gives the dependence of the coercive force H_c of Ni_3Mn on the annealing temperature [121-123]. At annealing temperatures close to the critical temperature one observes a sharp maximum of the

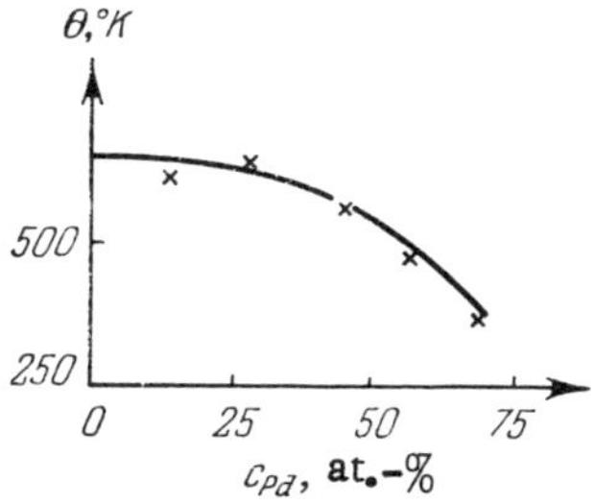

Fig. 72. Curie temperature of disordered Ni-Pd as a function of concentration of Pd [134].

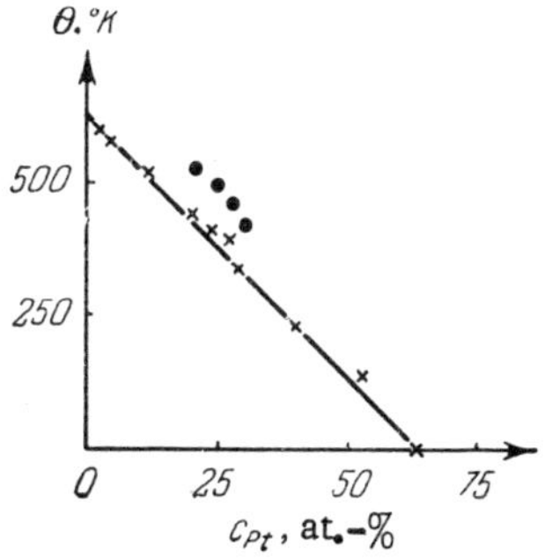

Fig. 73. Curie temperature of disordered (crosses) and ordered (circles) Ni-Pt as a function of the concentration of Pt [128, 135].

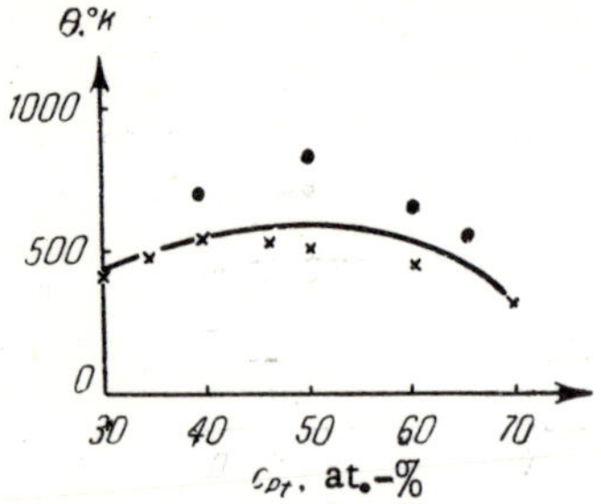

Fig. 74. Curie temperature of disordered (crosses) and ordered (circles) Fe-Pt as a function of the concentration of Pt [134].

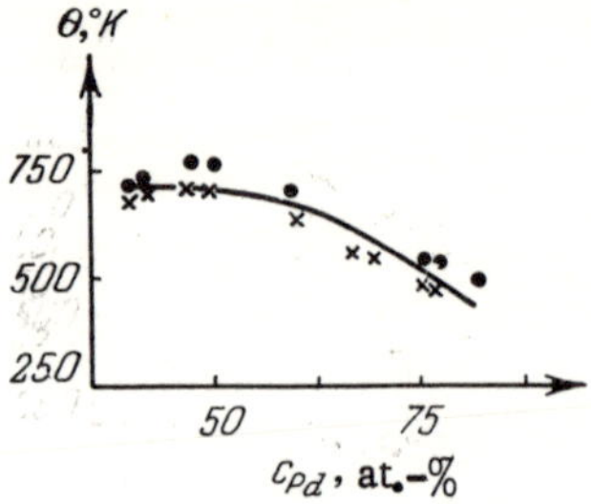

Fig. 75. Curie temperature of disordered (crosses) and ordered (circles) Fe-Pd as a function of the concentration of Pd [136].

coercive force which is evidently caused by the inhomogeneities of the crystal lattice appearing in this temperature range. This type of variation of the coercive force during ordering was detected by Ivanovskiy [131] in Fe_3Al. An even more pronounced rise in the coercive force H_c (reaching a value of 3,000 gauss) during a transition into the ordered state was observed in Co-Pt alloys [132]. In this alloy, during ordering from a structure with a face-centered cubic lattice, a tetragonal structure with a ratio of axes $\frac{c}{a} = 0.953$ is formed [133]. Consequently, large stresses arise in the lattice accompanied by the appearance of a high coercive force.

The effect of composition and degree of ordering on the Curie temperature of a ferromagnetic transformation is illustrated in Fig. 70-75, which give the dependence of the composition in binary alloys: Ni-Mn, Fe-Ni, Ni-Pd, Ni-Pt, Fe-Pt and Fe-Pd. The solid line in these figures is a theoretical curve for the disordered alloys (see Sec. 39). The dependence of θ on the composition was also investigated in ternary alloys. Figure 76 gives the dependence of the Curie temperature on the concentration of manganese in Heusler alloys Mn-Cu-Al annealed at 110°C for a period of 4,000 hours and quenched at temperatures close to the melting point [137].

Ordering may substantially alter other properties of ferromagnetics; magnetic permeability (permalloy [138, 139]), magnetic anisotropic constant (Ni_3Fe [138]), etc.* It is

*Magnetic properties of ferromagnetic ordered alloys are treated in detail by [140], [141] and others.

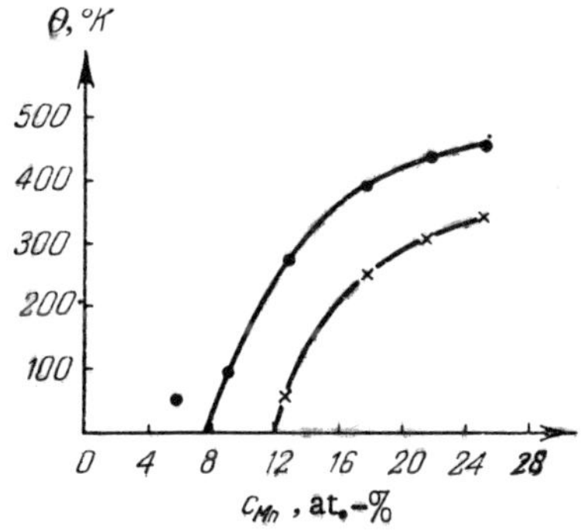

Fig. 76. Curie temperature of Heusler alloys, containing 25 atomic % Al, as a function of the concentration of Mn atoms for an ordered (upper curve) and disordered (lower curve) state [137].

interesting that a change of the state of magnetization may, in turn, influence the ordering of ferromagnetic alloys. Thus, for example, as shown in the experiments of Glazer and Shur [142], in Fe-Ni alloys (near the composition Ni_3Fe) ordering changes as a result of cooling in the magnetic field.

The diamagnetic susceptibility of alloys is also found to be different in the ordered and disordered states. As may be seen from Fig. 77, ordering of $AuCu_3$ leads to an increase in the absolute value of the diamagnetic susceptibility (approx. 18%), while in AuCu it decreases [143]. Figure 78 gives the concentration dependence of the diamagnetic susceptibility of Cu-Pd cooled from high temperatures (disordered) and annealed (ordered) [144]. In this case ordering leads to an increase in the absolute value of diamagnetic susceptibility.

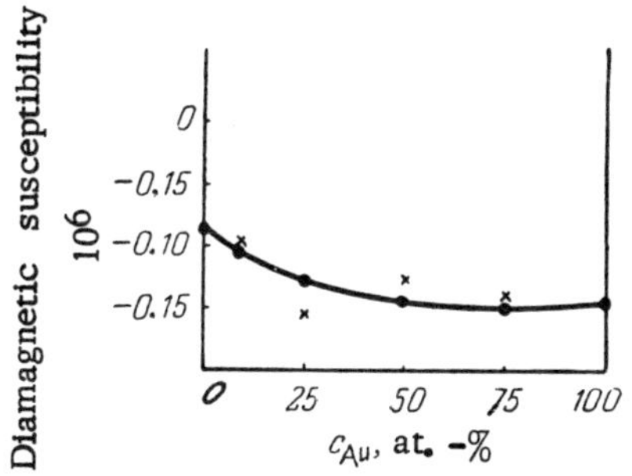

Fig. 77. Diamagnetic susceptibility (per gm) of disordered (circles) and ordered (crosses) $Au-Cu_3$ as a function of the concentration of Au [143].

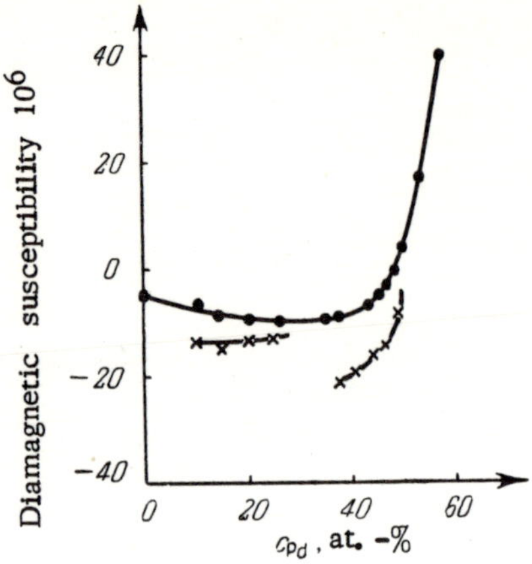

Fig. 78. Diamagnetic susceptibility (per gram-atom) of disordered (circles) and ordered (crosses) Cu-Pd as a function of the concentration of Pd [144].

Galvanomagnetic properties

The influence of ordering on the galvanomagnetic effect in $AuCu_3$, AuCu, $PdCu_3$ and other alloys has been investigated by [145-151]. Interesting results were discovered during an investigation of the Hall effect in $AuCu_3$. In the disordered state the alloy has a negative (normal) sign of the Hall coefficient.

With an increase in the degree of long-range order, the Hall coefficient becomes zero and with a further increase η becomes positive (anomalous Hall effect). The dependence of the Hall coefficient on the degree of long-range order is given in Fig. 79.* The value of the Hall coefficient in the disordered state of $AuCu_3$ equals approximately $-650 \cdot 10^{-6}$ cgs units. For AuCu, in a disordered state, as the degree of long-range order increases, the Hall coefficient remains almost unchanged. For disordered AuCu the Hall coefficient is a substantially larger quantity (in absolute value) and is equal approximately to $-7 \cdot 10^{-4}$ in cgs units. In $PdCu_3$, just as in $AuCu_3$, the Hall coefficient depends strongly on the degree of long-range order, but no change of its sign is observed.

The influence of ordering on the Hall coefficient in β-brass was investigated by Frank [69]. Figure 80 gives the curve

---

*Figs. 79 and 81 were obtained from Prof. A. P. Komar, who plotted them on the basis of data from his doctoral dissertation [145].

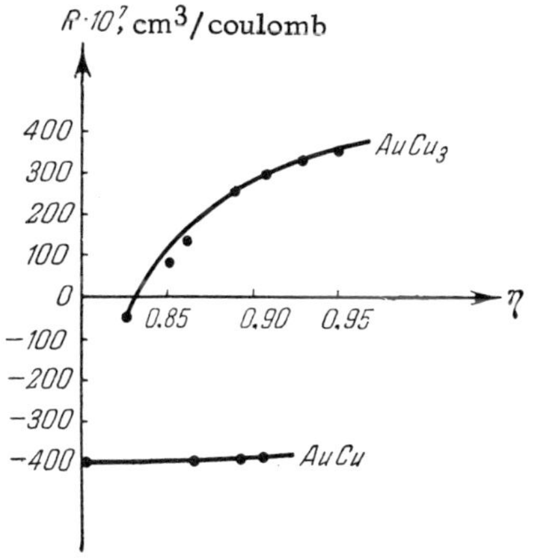

Fig. 79. Dependence of Hall coefficient on the degree of long-range order of $AuCu_3$ and AuCu alloys [145].

obtained by Frank for the dependence of the Hall coefficient R on temperature. This curve reveals a comparatively weak variation of R during ordering. At the critical temperature, a bend in the curve $R(T)$ is observed. In previously mentioned papers [145-151] it was also established that a change in composition and in the degree of long-range order produced an appreciable change in the relative electrical resistivity in a

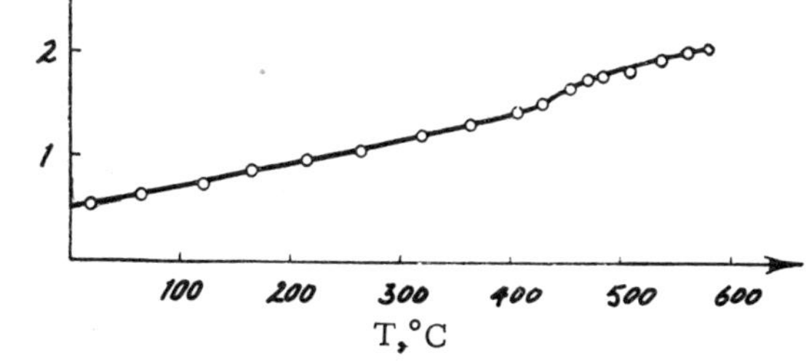

Fig. 80. Dependence of the Hall coefficient of β-brass (47.7 atomic % Zn) on temperature [69].

transverse magnetic field $\frac{\Delta\rho}{\rho^0}$. The relative resistivity may, therefore, serve as a sensitive indicator of ordering. For Au-Cu_3 alloys, whose composition is close to stoichiometric composition $AuCu_3$, the concentration dependence of $\frac{\Delta\rho}{\rho^0}$ is given in Fig. 81. The maxima on the curves correspond to the stoichiometric composition, when the greatest ordering occurs.

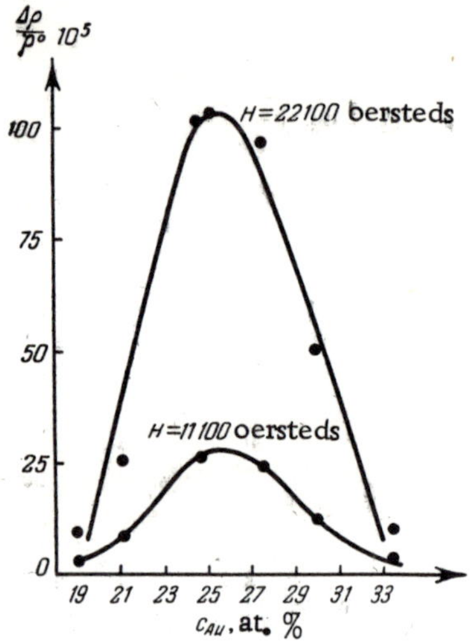

Fig. 81. Concentration dependence of the relative variatiion in the electrical resistivity $\frac{\Delta\rho}{\rho^0}$ in a transverse magnetic field (strength H) for alloys with composition close to $AuCu_3$ [145].

Lattice constants

The lattice constant also changes during ordering of the alloys. In the Au-Cu system this variation was investigated by Ageyev and Shoikhet [16] for alloys of different compositions. The obtained results are given in Fig. 82. It is apparent that during ordering of alloys with composition close to AuCu, the constant a of a tetragonal lattice increases, whereas the constant c decreases. In alloys with composition close to $AuCu_3$, however, the lattice parameter decreased during ordering. In analyzing the change of the ratio of axes $\frac{a}{c}$ (the degree of tetragonality) with a change in the degree of order in

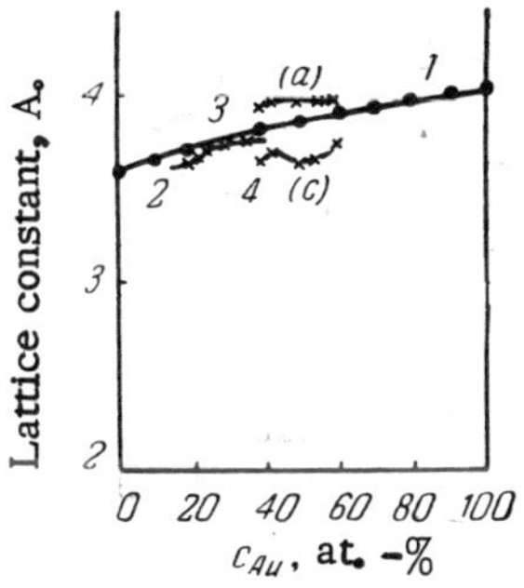

Fig. 82. Lattice constants of disordered (1) and ordered (2, 3, 4) Au-Cu alloys as a function of concentration of Au [16]. Curve 2-ordered alloys with composition close to $AuCu_3$, Curve 3 and curve 4-ordered alloys with composition close to AuCu for the lattice constants *a* and *c*, respectively.

AuCu, Buynov [152] also discovered an increase in this ratio during ordering of the alloy. The temperature dependence of the lattice constant of $AuCu_3$ during ordering was investigated by Betteridge [153]. As is apparent from Fig. 83, a step-change in the lattice constant occurs at the phase transition point.

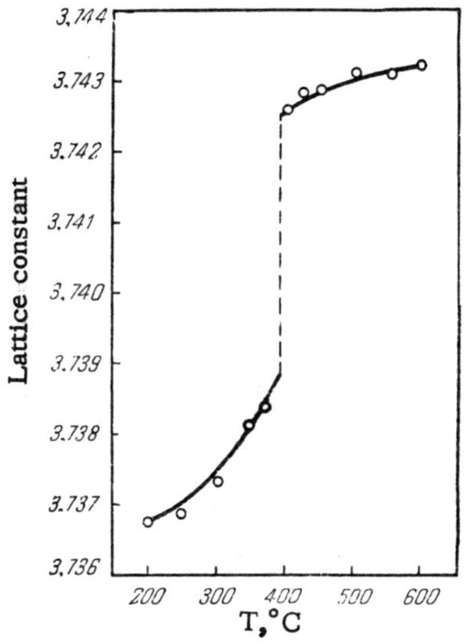

Fig. 83. Temperature dependence of the lattice constant of $AuCu_3$ [153].

The degree of tetragonality (ratio of axes $\frac{c}{a}$) of the ordered alloy Cu_3Pd at different annealing temperatures was measured by Jones and Owens [154]. The influence of annealing temperature on the magnitude of $\frac{c}{a}$ is illustrated in Fig. 84. The dependence of the lattice constants of Fe_3Al on the quenching temperature was investigated by Selisskiy [155]. The curve of the temperature dependence of the lattice constant is illustrated in Fig. 85. It is apparent that a rise of temperature during disordering of Fe_3Al produces an increase in the lattice constant. At the order-disorder transition temperature (approx. 550°C) the curve of the temperature dependence of the lattice constants shows a bend, rather than a step-change observed with Cu_3Au and CuAu.

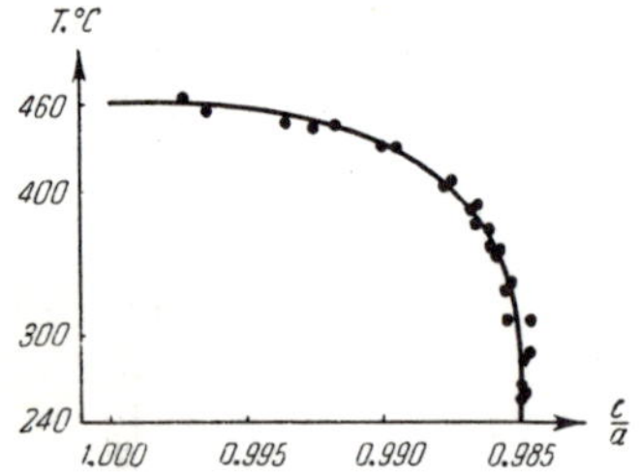

Fig. 84. Ratio of the axes $\frac{c}{a}$ of a unit cell of Cu_3Pd at different annealing temperatures [154].

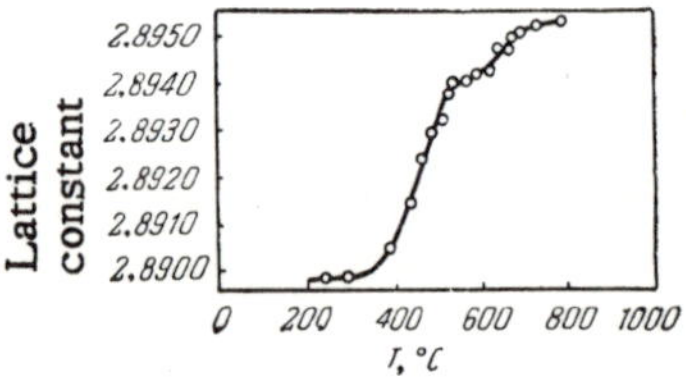

Fig. 85. Dependence of the lattice constants of Fe_3Al on the annealing temperature [155].

Diffusion

It was established by [253, 249, 254 and 257] that ordering exerts an influence on the diffusion of atoms in alloys. Thus, analysis of the experimental data obtained by Gertsriken and Dekhtyar [156], in a study of diffusion in β-brass by the evaporation method, shows that the effective activation energy should change discontinuously at the critical temperature. The absence of a sufficient number of data points above the critical point, however, made it impossible to draw a definite conclusion. The influence of ordering on the diffusion of copper, zinc and antimony in β-brass was investigated by Kuper, Lazarus, Manning and Tomizuka [157] by means of tracer atoms. Diffusion was studied in single crystal samples of β-brass over a wide temperature range both in the disordered and in the ordered states. The diffusion coefficient is given as a function of the reciprocal temperature in Fig. 86. The lower curve corresponds to the diffusion of copper, and the upper one to dffusion of zinc and antimony. The bend in the curves in Fig. 86 in the vicinity of the critical temperature is well noted. Below the critical temperature, when the degree of long-range order changes rapidly with a change of temperature T, the curves differ markedly from straight lines, which should be obtained with diffusion in pure metals, where the diffusion coefficient D is an exponential function of the reciprocal temperature.

$$D = D_0 e^{-Q'/RT}$$

(the parameters D_0 and Q' are called the pre-exponential factor and activation energy, respectively).

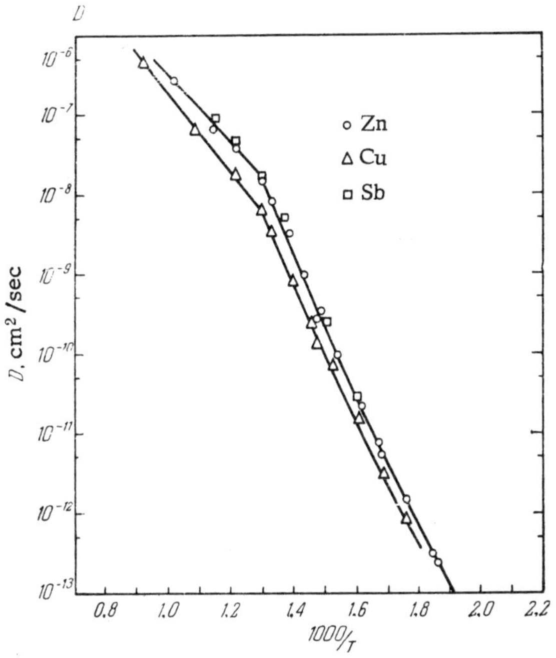

Fig. 86. Dependence of the diffusion coefficients of Cu, Zn and Sb in β-brass as a function of 1,000/T/ (where T — absolute temperature) [157].

Some deviation from linearity is also observed above the critical temperature in a disordered state. In the transition from a disordered to an ordered state of the alloy, the slope

Table 8
Values of the effective factor in front of the exponential coefficient and of the effective activation energy for the diffusion of copper and zinc in β-brass [157].

Element	Range of values of the degree of long-range order	Temperature range, °C	D in cm^2/sec	Q' in kcal/mole
Cu	0—0	817—497	0.011	22.04
Cu	0.55—0.78	442—381	180	37.09
Cu	0.78—0.92	381—292	80	36.02
Zn	0—0	718—499	0.0035	18.78
Zn	0.48—0.79	450—376	78000	44.23
Zn	0.79—0.96	376—264	163	36.30

of the curves increases, which corresponds to an increase of the activation energy and of the pre-exponential factor. The values of D_0 and Q' for Cu and Zn, measured at different degrees of long-range order, are given in Table 8.

Other properties of alloys, such as chemical activity [59], thermoelectromotive force [158] etc., also change during ordering.

Chapter II

Thermodynamic Theory of Order-Disorder Transition

10. Classification of Phase Transitions

In developing a theory of order-disorder transitions, two methods of treatment are possible. A concrete model of an alloy can be considered, and the energy corresponding to different arrangements of atoms on the crystal lattice sites calculated. Then, the statistical partition function and the free energy of the alloy can be determined. From the minimum free energy one may then determine the equilibrium properties of alloys, in particular, the values of the degree of long-range order and the correlation parameters at given temperatures. This treatment of order-disorder transitions employs the statistical theory, which will be expounded in the next chapter. Because of serious mathematical difficulties, arising in the construction of a statistical theory of order-disorder transitions, one must choose a very crude model and approximate methods of calculation. Such procedure produces some results that agree with experimental data, but in other cases no reliable conclusions can be reached.

Another approach to the development of a theory of order-disorder transitions may be based on general thermodynamical relations, on the calculation of the properties of a crystal, and by assuming that certain thermodynamic quantities can be expanded into a series. Thus, a theory of order-disorder transition can be developed without the use of a specific model of an alloy. Such a thermodynamic theory will not contain any drawbacks caused by the choice of a simple model of an alloy, but the results obtained in this theory will not pertain to the entire range of the order parameters. A thermodynamic theory of order-disorder transitions will be examined in this chapter.

Two types of order-disorder phase transitions were described in Chap. I. In a number of alloys (for instance,

$AuCu_3$) the degree of long-range order exhibits a finite step-change at the critical temperature T_0. In such a transition absorption (or liberation) of heat is observed (see Fig. 32), i.e., the entropy of the alloy undergoes a step-change. The energy and volume of the body also change in a similar manner. In this respect the phase transition is similar to the melting of a solid or to allotropic transformations in pure metals.

In another type of transition (e.g., in β-brass) when the temperature is lowered below the critical temperature, the degree of long-range order changes, increasing continuously from zero. In this case no heat of transition is observed, and the change in energy and volume is equal to zero. However, for $T = T_0$ a step-change in the specific heat is observed (Fig. 30).

These two types of order–disorder transition are special cases of the two general types of phase transitions, the first and second-order phase transitions. The classification of phase transitions was given by Ehrenfest [159]. According to this classification, first-order phase transitions are those exhibiting a step-change of the first derivatives of the thermodynamic potential $\Phi(P, T)$ with respect to the temperature T and pressure P, i.e., entropy $S = -\frac{\partial \Phi}{\partial T}$ and volume $V = \frac{\partial \Phi}{\partial P}$, and hence, the energy E of the alloy (the thermodynamical potential $\Phi = E - TS + PV$) remains continuous during the phase transition. Thus, during first-order phase transitions a heat of transition is either liberated or absorbed. Second-order phase transitions are those in which the thermodynamic potential and its first derivatives (i.e., entropy and volume) remain continuous, while the second derivatives of Φ with respect to T and P change discontinuously. Consequently, the discontinuous change will have different thermodynamic quantities, expressed in terms of the second derivatives, i.e., specific heat

$$C_P = T\frac{\partial S}{\partial T} = -T\frac{\partial^2 \Phi}{\partial T^2}, \tag{10.1}$$

compressibility

$$\varkappa = -\frac{1}{V}\frac{\partial V}{\partial P} = -\frac{1}{V}\frac{\partial^2 \Phi}{\partial P^2}, \tag{10.2}$$

coefficient of thermal expansion

$$\alpha = \frac{1}{V}\frac{\partial V}{\partial T} = \frac{1}{V}\frac{\partial^2 \Phi}{\partial T\,\partial P} \tag{10.3}$$

etc. It is understood, of course, that these quantities may also change discontinuously during first-order transitions. Let us note that with second-order phase transitions, neither absorption or liberation of heat occurs, because the entropy remains continuous.

Some general relations may be established between the discontinuities of different thermodynamic quantities, which are expressed in terms of derivatives of the thermodynamic potential. Let us consider a binary ordered alloy $A-B$, in which the order-disorder transition is a second-order phase transition. We shall describe its state by specifying the thermodynamic variables P, T and the relative atomic concentration c_A of component A. Each value of P and c_A is given at the temperature of the phase transition T_0, i.e., an equilibrium surface $T=T_0(P, c_A)$ exists in the space P, c_A and T. Let $\Phi_1(P, T, c_A)$ denote the thermodynamic potential of an alloy in the ordered and $\Phi_2(P, T, c_A)$ the thermodynamic potential in the disordered state. The step-changes of the first derivatives of the thermodynamic potential for a second-order phase transition are

$$\left.\begin{aligned} \frac{\partial\Phi_1}{\partial P}-\frac{\partial\Phi_2}{\partial P}&\equiv\Delta\frac{\partial\Phi}{\partial P}=0,\\ \frac{\partial\Phi_1}{\partial T}-\frac{\partial\Phi_2}{\partial T}&\equiv\Delta\frac{\partial\Phi}{\partial T}=0,\\ \frac{\partial\Phi_1}{\partial c_A}-\frac{\partial\Phi_2}{\partial c_A}&\equiv\Delta\frac{\partial\Phi}{\partial c_A}=0. \end{aligned}\right\} \qquad (10.4)$$

The above equations are applicable at each point of the equilibrium surface $T=T_0(P, c_A)$. Therefore, differentiating along the indicated surface we obtain

$$\left.\begin{aligned} \Delta\frac{\partial^2\Phi}{\partial P^2}dP+\Delta\frac{\partial^2\Phi}{\partial P\,\partial T}dT+\Delta\frac{\partial^2\Phi}{\partial P\,\partial c_A}dc_A&=0,\\ \Delta\frac{\partial^2\Phi}{\partial T\,\partial P}dP+\Delta\frac{\partial^2\Phi}{\partial T^2}dT+\Delta\frac{\partial^2\Phi}{\partial T\,\partial c_A}dc_A&=0,\\ \Delta\frac{\partial^2\Phi}{\partial c_A\,\partial P}dP+\Delta\frac{\partial^2\Phi}{\partial c_A\,\partial T}dT+\Delta\frac{\partial^2\Phi}{\partial c_A^2}dc_A&=0, \end{aligned}\right\} \qquad (10.5)$$

where Δ denotes the step-changes of the corresponding quantities during the transition, and a relation among dP, dT and dc_A is given by

$$dT=\frac{\partial T_0}{\partial P}dP+\frac{\partial T_0}{\partial c_A}dc_A. \qquad (10.6)$$

Using (10.1), (10.2) and (10.3), we obtain from (10.5)

$$-V\,\Delta\varkappa\,dP + V\,\Delta\alpha\,dT + \Delta\frac{\partial V}{\partial c_A}\,dc_A = 0, \tag{10.7}$$

$$V\,\Delta\alpha\,dP - \frac{1}{T_0}\Delta C_P\,dT - \Delta\frac{\partial S}{\partial c_A}\,dc_A = 0, \tag{10.8}$$

$$\Delta\frac{\partial V}{\partial c_A}\,dP - \Delta\frac{\partial S}{\partial c_A}\,dT + \Delta\frac{\partial^2\Phi}{\partial c_A^2}\,dc_A = 0. \tag{10.9}$$

By considering the phase transition in an alloy of specified composition ($c_A = \text{const}, dc_A = 0$) at various pressures and critical temperatures, and taking into account (10.6), we find from (10.7) and (10.8) the Ehrenfest relations

$$\left.\begin{array}{c} \dfrac{\partial T_0}{\partial P} = \dfrac{\Delta\varkappa}{\Delta\alpha}, \quad \dfrac{\partial T_0}{\partial P} = T_0 V\dfrac{\Delta\alpha}{\Delta C_P}, \\ T_0 V(\Delta\alpha)^2 = \Delta\varkappa\cdot\Delta C_P. \end{array}\right\} \tag{10.10}$$

Equations (10.10) correlate the step-changes of specific heat at constant pressure ΔC_P, of the compressibility $\Delta\varkappa$, of the coefficient of volumetric thermal expansion $\Delta\alpha$ with the derivative $\frac{\partial T_0}{\partial P}$, describing the variation of the temperature T_0 with pressure. These formulas enable us to calculate the remaining two quantities from two of these four quantities.

Next, by considering phase transitions in a series of alloys of different compositions, at constant pressure ($P = \text{const}$, $dP = 0$) or temperature ($T = \text{const}$, $dT = 0$), and using, Eqs. (10.7), (10.8) and (10.9), one may determine a number of additional relations [160] containing the derivatives of V, S, Φ and T_0 with respect to concentrations c_A. We shall cite some of these relations

$$\left.\begin{array}{c} \dfrac{\partial T_0}{\partial c_A} = -\dfrac{\Delta\dfrac{\partial V}{\partial c_A}}{V\,\Delta\alpha}, \quad \dfrac{\partial T_0}{\partial c_A} = -\dfrac{T_0\Delta\dfrac{\partial S}{\partial c_A}}{\Delta C_P}, \\ \Delta C_P\cdot\Delta\dfrac{\partial V}{\partial c_A} = T_0 V\,\Delta\alpha\cdot\Delta\dfrac{\partial S}{\partial c_A}, \\ \dfrac{\partial T_0}{\partial P} = \dfrac{\Delta\dfrac{\partial V}{\partial c_A}}{\Delta\dfrac{\partial S}{\partial c_A}}, \quad \dfrac{\partial T_0}{\partial c_A} = \dfrac{\Delta\dfrac{\partial^2\Phi}{\partial c_A^2}}{\Delta\dfrac{\partial S}{\partial c_A}}, \\ \Delta\varkappa\cdot\Delta\dfrac{\partial S}{\partial c_A} = \Delta\alpha\cdot\Delta\dfrac{\partial V}{\partial c_A}. \end{array}\right\} \tag{10.11}$$

Some of the relations (10.10) and (10.11) are not independent. However, writing them in different form may be convenient for different cases.

We note that the quantity $\frac{\partial \Phi}{\partial c_A}$ may be expressed in terms of the chemical potentials $\mu_A = \left(\frac{\partial \Phi}{\partial N_A}\right)_{PTN_B}$ and $\mu_B = \left(\frac{\partial \Phi}{\partial N_B}\right)_{PTN_A}$. Recalling that the total number of atoms of the alloy $N = N_A + N_B$ is regarded as constant, i.e., the derivative $\frac{\partial \Phi}{\partial c_A}$ should be calculated at constant P, T and N, we get

$$\left(\frac{\partial \Phi}{\partial c_A}\right)_{PTN} = N(\mu_A - \mu_B). \tag{10.12}$$

It follows from (10.12) that the quantity $\frac{\partial S}{\partial c_A}$ in Eq. (10.11) is equal to

$$\frac{\partial S}{\partial c_A} = N \frac{\partial}{\partial T}(\mu_B - \mu_A). \tag{10.13}$$

For first-order phase transitions the following conditions of phase equilibrium must be fulfilled:

$$\left.\begin{aligned} \mu_{A1}(P, T, c_{A1}) &= \mu_{A2}(P, T, c_{A2}), \\ \mu_{B1}(P, T, c_{A1}) &= \mu_{B2}(P, T, c_{A2}). \end{aligned}\right\} \tag{10.14}$$

These consist of the equality of the chemical potentials of components A and B in the first and second phase. For given P and T these equalities can not be simultaneously fulfilled for equal phase compositions ($c_{A1} = c_{A2}$). Hence, during ordering, when it is a first-order phase transition, the composition of the ordered and disordered phases must be different, i.e., a two-phase region should exist in the phase diagram. This type of phase diagram is illustrated in Fig. 87. For the concentration $c_A = c_{A\,\max}$, corresponding to the maximum critical temperature, the order-disorder transition should take place without passing through the two-phase region.

For second-order phase transitions, the functions μ_{A1} and μ_{B1} change continuously into μ_{A2} and μ_{B2}, i.e., equation (10.14) does not impose any restriction on composition, and the transition is possible without the appearance of a two-phase region.

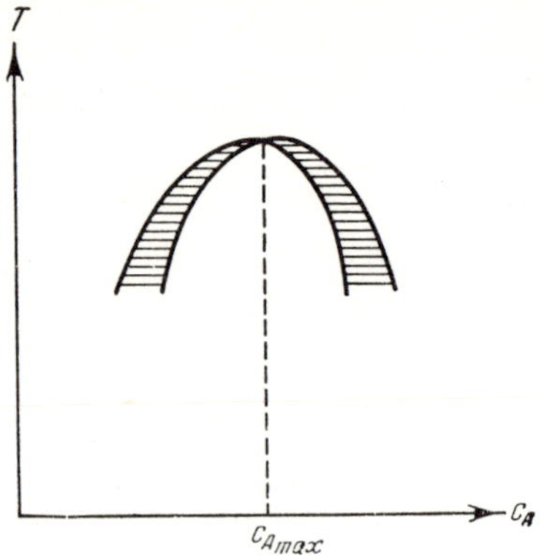

Fig. 87. Typical phase diagram near the maximum critical temperature for first-order phase transition (the dotted line denotes $c_{A\,\max}$).

11. Thermodynamic Theory of Second-Order Phase Transitions

The general thermodynamic theory of second-order phase transitions was developed by Landau [161-163] and Lifshits [164-166] (see also [167]). Essential to this theory is the establishment of a relation between the second-order phase transitions and the changes of symmetry of the body occurring during such transition. Concrete examples of the change of symmetry of the crystal lattices of ordered alloys were given in the previous chapter. By symmetry of the crystal lattice we understand the symmetry of the density probability function $\rho(\mathbf{r})$ for finding any type of atom at the point $\mathbf{r}$. The symmetry of the function $\rho(\mathbf{r})$ is characterized by assigning all possible transformations (symmetry elements), under which this function is invariant. Obviously the symmetry of a crystal during the second-order phase transition undergoes a step-change, because some symmetry elements disappear or appear. Thus, for example, during a transition of β-brass into a disordered state, there appears a new symmetry element—translation along the body diagonal of the cubic cell by a distance equal to one half of this diagonal, i.e., the symmetry increases. An increase of the symmetry during a transition into a disordered state occurs in all the cases considered in the previous chapter.

Although the symmetry of the lattice undergoes a step-change during a second-order phase transition, the function $\rho(\mathbf{r})$ and, hence, the probabilities of substitution of the lattice sites by atoms of a different type change continuously. Changes of these probabilities during a transition from a disordered into an ordered state are conveniently described by the quantity η, determined by Eq. (1.4). In the case of a second-order phase transition, the degree of long-range order increases continuously from zero during passage into an ordered state. One may therefore analyze the state of a crystal in a region close to the transition curve using a suitable small parameter η.

Two approaches to the development of a thermodynamic theory of second-order phase transitions are possible. In the first, the possibility of such a phase transition is postulated and the symmetry of the ordered and disordered phase is regarded as known. An investigation of the temperature and concentration dependence of the degree of long-range order near the phase transition point is carried out, and the variation of different thermodynamic quantities during the transition is considered. Another possible approach does not assume the existence of a second-order phase transition. By using the general symmetry properties of the density probability function, it is possible to predict whether second-order phase transitions will occur in an alloy with a given structure.

Let us analyze the first statement of the problem by considering a binary ordered substitutional alloy $A—B$ (in which the long-range order is determined by the first power of the parameter η) at a temperature $T = T_0(c_A, P)$ with a second-order phase transition. We expand the thermodynamic potential of the alloy $\Phi(T, P, c_A, \eta)$ as a series in powers of η, by considering Φ as a function of temperature, pressure, concentration of component A, and also of the degree of long-range order η. We assume that the latter may take not only its equilibrium value (uniquely determined by the assignment of T, P and c_A), but also some nonequilibrium values. By limiting ourselves to small values of the degree of long-range order, we retain in the expansion only the first few terms:

$$\Phi(T, P, c_A, \eta) = \Phi_0 + A_1\eta + \frac{A_2}{2}\eta^2 + \frac{A_3}{3}\eta^3 + \frac{A_4}{4}\eta^4, \quad (11.1)$$

where the coefficients Φ_0, A_1, A_2, A_3, A_4 are functions of T, P and c_A. Using the symmetry properties, one may show that the coefficient A_1 will always turn zero. By considering the symmetry, one may also show that the coefficient A_3 is also equal to zero for a number of structures. Thus, for example in the case of alloys with a β-brass crystal structure, the replacement of η by $-\eta$ corresponds simply to renaming of the first and second types of sites, which, evidently, cannot alter the thermodynamic potential. Consequently, in the expansion of Φ in powers of η all terms with odd powers of η should be absent, i.e., $A_1 \equiv A_3 \equiv 0$.

By considering next only those alloys in which not only $A_1 \equiv 0$, but also $A_3 \equiv 0$, we write the expansion of Φ in powers of η in the form

$$\Phi = \Phi_0 + \frac{A_2}{2}\eta^2 + \frac{A_4}{4}\eta^4. \tag{11.2}$$

The equilibrium value of the degree of long-range order in an alloy for given T, P and c_A is determined from a minimum value of the thermodynamic potential with respect to a change in η:

$$\frac{\partial\Phi}{\partial\eta} = \eta\,(A_2 + A_4\eta^2) = 0, \tag{11.3}$$

$$\frac{\partial^2\Phi}{\partial\eta^2} = A_2 + 3A_4\eta^2 > 0. \tag{11.4}$$

Equation (11.3) has the solution $\eta = 0$, corresponding to a disordered state of the alloy, and the solution

$$\eta^2 = -\frac{A_2}{A_4}, \tag{11.5}$$

corresponding to the ordered state. To determine which of these two solutions corresponds to the minimum thermodynamic potential, one must substitute the values obtained for η into (11.4). For disordered alloys (at $\eta = 0$) the sign of A_2, which is equal to $\left.\frac{\partial^2\Phi}{\partial\eta^2}\right|_{\eta=0}$, is on the basis of (11.4), positive. Substituting (11.5), corresponding to an ordered state, into Eq. (11.4), we obtain $A_2 < 0$. Thus, during passage through a second-order phase transition point A_2 must change its sign, i.e., at the transition point

$$A_2(T, P, c_A) = 0, \tag{11.6}$$

which is the equation of the surface of the order-disorder transition points in T, P, c_A space.

Since $A_2 < 0$ for an ordered alloy, we find from Eq. (11.5) that in the vicinity of the transition point

$$A_4 > 0. \tag{11.7}$$

For those crystals in which the coefficient A_3 is not identically equal to zero, this coefficient should, because of symmetry, equal zero at a second-order phase transition point.* This gives the auxiliary conditions $A_3(T, P, c_A) = 0$, which must be fulfilled at the transition points. This condition together with (11.6) determines the line of second-order phase transition points in (T, P, c_A) space, i.e., if A_3 does not vanish identically for an alloy of a definite composition, no line of second-order phase transition points is possible, as in the case $A_3 \equiv 0$, but only an isolated point is possible. An investigation of this case was conducted in [161, 162]. Isolated points of this type have not yet been observed experimentally; therefore, in the following, only the case of a curve of second-order phase transition points in the T, P plane (for alloys of known composition) will be considered.

Since the coefficient $A_2 = 0$ at the same point on the second-phase transition line, the expansion of A_2 in powers of $T - T_0$ for given P and c_A near the transition temperature T_0 may be written as

$$A_2 = a(T - T_0). \tag{11.8}$$

Thus, the constant a is positive, since $A_2 > 0$ for the disordered phase, which always lies above the transition temperature in the case of alloys. Substituting (11.8) in (11.5) and remembering that only the zeroth term of the expansion A_4^0 may be retained, in the expansion of the coefficient A_4 in powers of $T - T_0$ near the transition point, we obtain

$$\eta^2 = \frac{a}{A_4^0}(T_0 - T). \tag{11.9}$$

---

*Conditions (11.3) and (11.4) in this case are

$$\frac{\partial \Phi}{\partial \eta} = \eta\,(A_2 + A_3\eta + A_4\eta^2) = 0,$$

$$\frac{\partial^2 \Phi}{\partial \eta^2} = A_2 + 2A_3\eta + 3A_4\eta^2 > 0.$$

Hence, it again follows that $A_2 < 0$ for the disordered phase, and $A_2 > 0$ for the ordered phase. Therefore, a necessary condition for $\frac{\partial^2 \Phi}{\partial \eta^2}$ to be positive at the transition point where $A_2 > 0$, is that $A_3 = 0$ and $A_4 > 0$.

Thus, in ordered alloys, in which the order–disorder transition is a second-order phase transition, the degree of long-range order near the transition temperature for known P and c_A, is proportional to $\sqrt{T_0-T}$.

Substituting the expression for the equilibrium value of the degree of long-range order (11.9) into (11.1) and using (11.8), we obtain an equation for the thermodynamic potential, at the equilibrium ordered state of the alloy

$$\Phi(T, P, c_A)=\Phi_0-\frac{a^2}{4A_4^0}(T_0-T)^2 \qquad (\text{for } T\leqslant T_0). \qquad (11.10)$$

For a disordered alloy the equilibrium thermodynamic potential $\Phi(T, P, c_A)$ is equal to $\Phi_0(T, P, c_A)$.

It becomes apparent that the thermodynamic potential and its first derivative change continuously at the transition point while the second derivative $\frac{\partial^2\Phi}{\partial T^2}$ changes discontinuously. The entropy of the ordered alloy $S=-\frac{\partial\Phi}{\partial T}$ will be

$$S=S_0-\frac{a^2}{2A_4^0}(T_0-T), \qquad (11.11)$$

where $S_0=-\frac{\partial\Phi_0}{\partial T}$. The step-change of the specific heat of the alloy at constant pressure [according to (10.1) and (11.10)] is

$$\Delta C_P=C_P|_{T_0-0}-C_P|_{T_0+0}=\frac{a^2T_0}{2A_4^0}, \qquad (11.12)$$

where $C_P|_{T_0-0}$ and $C_P|_{T_0+0}$ are the heat capacities of ordered and disordered phases for $T=T_0$. It follows from (11.7) that ΔC_P is always positive, i.e., the heat capacity of the ordered alloy is greater than that of the disordered alloy. Equation (11.12) defines a finite step-change of the heat capacity. From the condition $\Delta C_P>0$ and Eqs. (10.10) and (10.11), we see that the step-change of the compressibility during transition from an ordered to a disordered state $\Delta\varkappa>0$ and $\Delta\frac{\partial^2\Phi}{\partial c_A^2}<0$. Hence, if the coefficient of thermal expansion α decreases by a step-change during a transition into the disordered state, then the critical temperature, as follows from (10.10), increases with an increase of pressure, and conversely. It follows from (10.11) that $\Delta \frac{\partial S}{\partial c_A}$ and $\frac{\partial T_0}{\partial c_A}$ have different signs.

It is also easy to obtain the concentration dependence of the degree of long-range order at constant temperature and pressure near the order-disorder transition point [160]. To do this we differentiate η^2, considered to be a function of the variables T and c_A along the curve of the second-order phase transition in the T, c_A plane and at constant P. Since η^2 equals zero at each point of this curve, the corresponding differential is also equal to zero, i.e.,

$$\frac{\partial \eta^2}{\partial T} dT + \frac{\partial \eta^2}{\partial c_A} dc_A = 0. \tag{11.13}$$

Hence, it follows that the derivative of η^2 with respect to concentration c_A is related to the derivative of η^2 with respect to temperature by

$$\frac{\partial \eta^2}{\partial c_A} = - \frac{\partial \eta^2}{\partial T} \frac{\partial T_0}{\partial c_A}. \tag{11.14}$$

By determining $\frac{\partial \eta^2}{\partial T}$ from Eq. (11.9) we obtain

$$\frac{\partial \eta^2}{\partial c_A} = \frac{a}{A_4^0} \frac{\partial T_0}{\partial c_A}. \tag{11.15}$$

It is apparent from Eq. (11.15) that the derivative $\frac{\partial \eta^2}{\partial c_A}$ equals zero at the transition point, where the composition is $c_{A\,\max}$, and where the curve $T_0(c_A)$ (for a given P) has a maximum. For compositions which are not too close to $c_{A\,\max}$, we may obtain the dependence of η on c_A, by considering that for small variations of c_A the right-hand side of Eq. (11.15) may be regarded as constant. By considering a concentration c_A, corresponding to the order-disorder curve (for a given T and P), $\eta = 0$, and integrating (11.15) or (11.14), we obtain

$$\eta^2 = \frac{a}{A_4^0} \frac{\partial T_0}{\partial c_A} (c_A - c_{A0}) = - \frac{\partial \eta^2}{\partial T} \frac{\partial T_0}{\partial c_A} (c_A - c_{A0}). \tag{11.16}$$

Thus, for small $|c_A - c_{A0}|$ the degree of long-range order is proportional to $\sqrt{|c_A - c_{A0}|}$.

In a similar way, we differentiate $\eta^2(T, P)$ for $c_A = \text{const.}$ along the transition curve, to obtain the dependence of the degree of long-range order on pressure near the transition point (at constant T and c_A)

$$\eta^2 = \frac{a}{A_4^0}\frac{\partial T_0}{\partial P}(P - P_0) = -\frac{\partial \eta^2}{\partial T}\frac{\partial T_0}{\partial P}(P - P_0), \tag{11.17}$$

where P_0 is the pressure at the transition point of the curve. Thus, $\eta \sim \sqrt{|P - P_0|}$.

Let us mention still another feature of second-order phase transitions, which is absent in first-order phase transitions. It has been shown at temperatures below the critical temperature that the coefficient A_2 in the expansion (11.2) is negative. Thus, when $\eta = 0$ for a disordered alloy, then in the region of the ordered phase $\frac{\partial^2 \Phi}{\partial \eta^2} = A_2 + 3A_4\eta^2 = A_2 < 0$. This means that the disordered phase for $T < T_0$ corresponds to the maximum of the thermodynamic potential, i.e., this phase cannot possibly exist at temperatures below the critical temperature. This shows the impossibility of supercooling of the disordered alloy in the region where the ordered phase exists. It must be borne in mind that for low atomic mobility, the formation of a new phase may occur rather slowly. As distinguished from the case of second-order transitions, it is well known that both supercooling and superheating are possible in first-order phase transitions, because metastable states having an additional minimum of the thermodynamic potential Φ, considered as a function of η, are present in this case. Therefore, even at a very high mobility of atoms such phenomena as supercooling may be attained.

Let us compare some of the conclusions with the experimental data for β-brass, cited in Secs. 3 and 5. The proportionality of the degree of long-range order η to the square root of $T_0 - T$, which follows from Eq. (11.9) for small values of η, is confirmed by x-ray analysis of the temperature dependence of long-range order in β-brass [8]. From a comparison with experiment one may find following value of the constant a/A_4^0 in Eq. 11.9 for β-brass:

$$\frac{a}{A_4^0} \approx 0.011 \text{ per } {}^\circ\text{C}. \tag{11.18}$$

Thus, the dependence of η on T for β-brass near the transition temperature is expressed by

$$\eta \approx 0{,}105\sqrt{T_0 - T}. \tag{11.19}$$

In Fig. 88b, the solid curve shows this relation plotted on the basis of the thermodynamic theory. The points correspond to experimental values of [8]. As is apparent from the figure, the experimental points agree well with the theoretical curve up to rather large values of η (about 0.5). The agreement between theory and experiment becomes apparent if we plot the dependence of η^2 on $T_0 - T$, because in this case one should obtain a straight line for small values of η. This is confirmed by Fig. 88a.

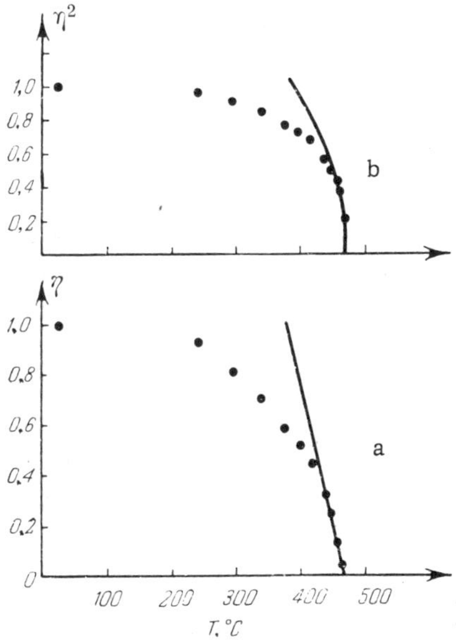

Fig. 88. Comparison of η and η^2 as a function of temperature obtained from the thermodynamic theory, with experimental data [8] for β-brass.

The theoretically predicted step-change in the specific heat at $T = T_0$ was also observed experimentally (see Fig. 30). As is apparent from Fig. 30, the step-change of the specific heat C_P of β-brass per one gram-atom is equal to

$$\approx 3.3 \cdot 10^8 \frac{\text{erg}}{\text{deg}^2 \cdot \text{g-at}}. \tag{11.20}$$

Using Eqs. (11.12, (11.18) and (11.20), we may find the values of the constants a and A_4^0 for one gram-atom of β-brass:

$$a \approx 8.1 \cdot 10^7 \frac{\text{erg}}{\text{deg}}, \; A_4^0 \approx 7.4 \cdot 10^9 \,\text{erg}. \tag{11.21}$$

It also follows from theory that the compressibility, Young's modulus etc. must also undergo step-changes at the second-order phase transition point. Figures 61 and 62 illustrate the temperature dependency of Young's modulus and the elasticity constants of β-brass. There is no evidence, however, of such step-changes. This may be due either to the smallness of the relative variation of Young's modulus during the transition or to the fact that the period of the elastic vibrations used during the measurements is considerably shorter than the relaxation time of the order-disorder transition (see also Chap. VII).

We note that the line of the second-order phase transition points cannot terminate at any point of the (T, P) plane, as is the case with the transition line for liquid-gas transitions which terminates at the critical point of the first-order phase transformation. This happens because the line of second-order phase transition points divides the phase regions of different symmetry, where no continuous transition in symmetry is possible. Landau has shown [161] that the line of second-order phase transition points for ordered solid solutions can, at the critical point, be changed into a decomposition curve. This decomposition curve is a line of first-order phase transition. The phase diagram in the T, c_A plane near such a point is illustrated in Fig. 89. In this Figure AK is the line of second-order phase transition points, which changes at the critical point K into a decomposition curve KB. Region II of the disordered phase is located above the curve AB. Region I shows the area of the ordered phase; the shaded region between curves KD and KB is a two-phase region and corresponds to the coexistence of ordered and disordered phases.

We shall now find the conditions which are satisfied by the temperature and composition of a binary alloy (for a given pressure) at the critical point K. Let us consider a two-phase region. The equilibrium conditions of two phases, in addition to the equality of temperature and pressure, are Eq. (10.14) for the chemical potentials of the components A and B

$$\left.\begin{aligned} \mu_{A1}(T, P, c_{A1}) &= \mu_{A2}(T, P, c_{A2}), \\ \mu_{B1}(T, P, c_{A1}) &= \mu_{B2}(T, P, c_{A2}), \end{aligned}\right\} \tag{11.22}$$

where c_{A1} and c_{A2} are the concentrations of component A in the ordered and disordered phases, respectively. Since the

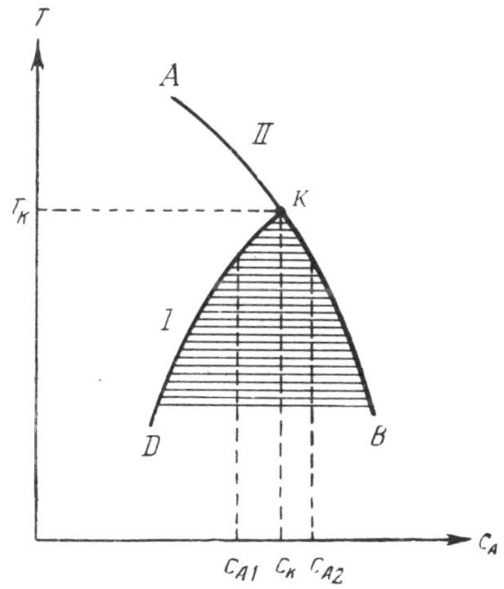

Fig. 89. Phase diagram in the vicinity of the critical point K. AK-line of second-order phase transitions, KB and KD-lines of first-order phase transitions, I-ordered phase, II-disordered phase.

thermodynamic potential of each phase is an additive function for the amount of mass, it can be written in the form $\Phi_j = N_j f_j(T, P, c_{Aj})$, where j denotes the number of the phase ($j = 1, 2$), N_j - the total number of atoms of a given phase, while $f_j(T, P, c_{Aj})$ is independent of N_j. By differentiating this expression with respect to N_A and N_B and allowing for the fact that $N_j = N_{Aj} + N_{Bj}$ and $c_{Aj} = \frac{N_{Aj}}{N_{Aj} + N_{Bj}}$, we obtain

$$\left.\begin{aligned} \mu_{Aj} &= \left(\frac{\partial \Phi_j}{\partial N_{Aj}}\right)_{N_{Bj}} = f_j + c_{Bj}\frac{\partial f_j}{\partial c_{Aj}}, \\ \mu_{Bj} &= \left(\frac{\partial \Phi_j}{\partial N_{Bj}}\right)_{N_{Aj}} = f_j - c_{Aj}\frac{\partial f_j}{\partial c_{Aj}}. \end{aligned}\right\} \qquad (11.23)$$

We shall introduce the thermodynamic potential Φ'_j, per one gram-atom. The coefficients of expansion of Φ'_1 in powers of η, similar to the coefficients of expansion in Eq. (11.2), will be denoted by Φ'_0, A'_2 and A'_4. Substituting Eq. (11.23) into the equilibrium condition (11.22), multiplying the equations obtained by Avogadro's number, expressing the thermodynamic potential per gram-atom of the ordered phase in the form of an expansion in powers of η (similarly to expansion

(11.2)*), and allowing for the fact that for a disordered phase $\Phi'_2 = \Phi'_0$, we obtain equations for the curves OD and OB in the phase diagram.

$$\Phi'_0(c_{A1}) + \frac{1}{2} A'_2(c_{A1})\eta^2 + \frac{1}{4} A'_4(c_{A1})\eta^4 + c_{B1}\left(\frac{\partial \Phi'_0}{\partial c_A}\bigg|_{c_A=c_{A1}} + \right.$$
$$\left. + \frac{1}{2}\frac{\partial A'_2}{\partial c_A}\bigg|_{c_A=c_{A1}} \eta^2 + \frac{1}{4}\frac{\partial A'_4}{\partial c_A}\bigg|_{c_A=c_{A1}} \eta^4\right) = \Phi'_0(c_{A2}) + c_{B2}\frac{\partial \Phi'_0}{\partial c_A}\bigg|_{c_A=c_{A2}}, \tag{11.24}$$

$$\Phi'_0(c_{A1}) + \frac{1}{2} A'_2(c_{A1})\eta^2 + \frac{1}{4} A'_4(c_{A1})\eta^4 - c_{A1}\left(\frac{\partial \Phi'_0}{\partial c_A}\bigg|_{c_A=c_{A1}} + \right.$$
$$\left. + \frac{1}{2}\frac{\partial A'_2}{\partial c_A}\bigg|_{c_A=c_{A1}} \eta^2 + \frac{1}{4}\frac{\partial A'_4}{\partial c_A}\bigg|_{c_A=c_{A1}} \eta^4\right) = \Phi'_0(c_{A2}) - c_{A2}\frac{\partial \Phi'_0}{\partial c_A}\bigg|_{c_A=c_{A2}}. \tag{11.25}$$

For brevity we also omit the variables T and P, that are identical in both phases, and in calculation of the derivatives

$$\frac{\partial \Phi_1}{\partial c_{A1}} = \frac{\partial \Phi_1}{\partial c_A}\bigg|_{c_A=c_{A1}} = \left(\frac{\partial \Phi_1}{\partial c_{A1}}\right)_\eta + \left(\frac{\partial \Phi_1}{\partial \eta}\right)_{c_{A1}} \frac{\partial \eta}{\partial c_{A1}}$$

we take into account the equilibrium condition $\frac{\partial \Phi_1}{\partial \eta} = 0$, so that the dependence of the degree of long-range order on c_A can be neglected. Multiplying (11.24) by c_{A1}, and (11.25) by c_{B1}, adding the results (remembering that $c_B = 1 - c_A$) and then subtracting (11.25) from (11.24), we obtain

$$\Phi'_0(c_{A1}) + \frac{1}{2} A'_2(c_{A1})\eta^2 + \frac{1}{4} A'_4(c_{A1})\eta^4 =$$
$$= \Phi'_0(c_{A2}) + (c_{A1} - c_{A2})\frac{\partial \Phi'_0}{\partial c_A}\bigg|_{c_A=c_{A2}}, \tag{11.26}$$

*It is obvious that this expansion, derived from the symmetry of the system, must be valid near the critical point.

$$\left.\frac{\partial \Phi_0'}{\partial c_A}\right|_{c_A=c_{A1}}+\frac{1}{2}\left.\frac{\partial A_2'}{\partial c_A}\right|_{c_A=c_{A1}}\eta^2+\frac{1}{4}\left.\frac{\partial A_4'}{\partial c_A}\right|_{c_A=c_{A1}}\eta^4=\left.\frac{\partial \Phi_0'}{\partial c_A}\right|_{c_A=c_{A2}} \qquad (11.27)$$

Since the concentrations c_{A1} and c_{A2} near the critical point are nearly equal in the expansion of $\frac{\partial \Phi_0'}{\partial c_A}$ at the point c_{A2} in terms of $c_{A1}-c_{A2}$, we can ignore all but the linear terms

$$\left.\frac{\partial \Phi_0'}{\partial c_A}\right|_{c_A=c_{A1}}=\left.\frac{\partial \Phi_0'}{\partial c_A}\right|_{c_A=c_{A2}}+\left.\frac{\partial^2 \Phi_0'}{\partial c_A^2}\right|_{c_A=c_{A2}}(c_{A1}-c_{A2}).$$

Substituting this expansion into (11.27) and discarding the small term proportional to η^4, the difference $c_{A1}-c_{A2}$ can be found

$$c_{A1}-c_{A2}=-\frac{\left.\frac{\partial A_2'}{\partial c_A}\right|_{c_A=c_{A1}}\eta^2}{2\left.\frac{\partial^2 \Phi_0'}{\partial c_A^2}\right|_{c_A=c_{A2}}}. \qquad (11.28)$$

Next, representing $\Phi_0'(c_{A1})$ in (11.26) in the form of the expansion

$$\Phi_0'(c_{A1})=\Phi_0'(c_{A2})+\left.\frac{\partial \Phi_0'}{\partial c_A}\right|_{c_A=c_{A2}}(c_{A1}-c_{A2})+ +\frac{1}{2}\left.\frac{\partial^2 \Phi_0'}{\partial c_A^2}\right|_{c_A=c_{A2}}(c_{A1}-c_{A2})^2,$$

and substituting Eq. (11.28) for $c_{A1}-c_{A2}$ and Eq. (11.5) for η^2 (where A_2 and A_4 should be replaced by A_2' and A_4'), we obtain a relation between the quantities $\frac{\partial A_2'}{\partial c_A}$, $\frac{\partial^2 \Phi_0'}{\partial c_A^2}$ and A_4', which should be fulfilled at the critical point*

$$A_4'\frac{\partial^2 \Phi_0'}{\partial c_A^2}=\frac{1}{2}\left(\frac{\partial A_2'}{\partial c_A}\right)^2. \qquad (11.29)$$

*Equation (11.29) is obtained from the phase equilibrium condition, corresponding to curves *DK* and *KB* in the vicinity of *K*, as a result of neglecting higher terms in the expansion. As the point *K* is approached, the values of $c_{A2}-c_{A1}$ and η^2 decrease, so that the error tends to zero, and (Eq. (11.9) at the actual critical point is exact.

Equation (11.29), where the quantities A_4', $\frac{\partial A_2'}{\partial c_A}$ and $\frac{\partial^2 \Phi_0'}{\partial c_A^2}$ at a given pressure must be regarded as a function of temperature and concentration c_A, jointly with equation $A_2'(T, c_A)=0$, determine the values $T=T_C$ and $c_A=c_{AC}$, at the critical point. Considering the line of second-order phase transition points, where $A_2=0$, $\frac{\partial A_2}{\partial c_A}=-\frac{\partial A_2}{\partial T}\frac{\partial T_0}{\partial c_A}=-a\frac{\partial T_0}{\partial c_A}$, we find from Eq. (11.29) (where we can set $A_4=A_4^0$) and from Eq. (11.12), that $\frac{\partial^2 \Phi_0}{\partial c_A^2}$ at the critical point can be expressed as

$$\frac{\partial^2 \Phi_0}{\partial c_A^2}=\frac{\Delta C_P}{T_0}\left(\frac{\partial T_0}{\partial c_A}\right)^2, \tag{11.30}$$

where ΔC_P is the maximum value of the step-change in the specific heat during a second-order phase transition, when the transition point on the curve AK approaches point K.

As is well known from thermodynamics (see, for instance, [167], Sec. 94), the inequality $\frac{\partial^2 \Phi}{\partial c_A^2}>0$ must always be fulfilled in solutions. By considering this inequality, we find from (11.29) that $A_4^0>0$ at the critical point. Hence, at the critical point A_4^0 does not become zero, and the step-change of the specific heat, determined from (11.12), remains finite. Substituting (11.29) and (11.5) into (11.28), we obtain

$$A_2(c_{A1})+(c_{A2}-c_{A1})\left.\frac{\partial A_2}{\partial c_A}\right|_{c_A=c_{A1}}=0. \tag{11.31}$$

The left-hand side of Eq. (11.31) represents the value of $A_2(c_A)$ at the point $c_A=c_{A2}$. Thus, the equation of the curve $A_2(T, P, c_A)=0$ is satisfied not only for second-order phase transition points, but (neglecting terms of the second and higher orders in $c_{A1}-c_{A2}$) also for points of the curve KB, which separates the two-phase region from the disordered phase. Hence, the curve of the second-order phase transitions AK changes at the point K into curve KB, as illustrated in Fig. 89.

We will now consider the second statement in the theory of second-order phase transitions, i.e., we shall not postulate beforehand that such a transition is possible and we shall determine in what cases it may be achieved. We shall restrict ourselves to a statement of the fundamental concepts and results of the theory.*

As mentioned above, the crystal symmetry may be described by a set of those coordinate transformations in which the density probability functions $\rho(\boldsymbol{r})$ are invariant. This set of transformations forms a symmetry group, the symmetry groups being different for the disordered and ordered states of the alloy. The function $\rho(\boldsymbol{r})$ near the second-order phase transition point may be represented as the sum of functions $\rho_0(\boldsymbol{r})$, whose symmetry corresponds to the transition point, and a small term $\delta\rho(\boldsymbol{r})$

$$\rho(\boldsymbol{r}) = \rho_0(\boldsymbol{r}) + \delta\rho(\boldsymbol{r}). \tag{11.32}$$

The function $\delta\rho(\boldsymbol{r})$, as group theory shows, may be represented as

$$\delta\rho(\boldsymbol{r}) = \sum_n \sum_i C_i^{(n)} \varphi_i^{(n)}(\boldsymbol{r}). \tag{11.33}$$

Here, the functions $\varphi_i^{(n)}(\boldsymbol{r})$ for a given n are the basis of an nth irreducible representation of the symmetry group** which the function $\rho_0(\boldsymbol{r})$ possesses; i is the number of the basis function of the given irreducible representation and $C_i^{(n)}$ are the expansion coefficients. Since the functions $\varphi_i^{(n)}(\boldsymbol{r})$ (for a given n) transform under these symmetry transformations only into one another, after each such transformation expansion will realized from the same set of functions, but with different coefficients. One may therefore assume that the expansion coefficients $C_i^{(n)}$ and not the functions $\varphi_i^{(n)}(\boldsymbol{r})$ will change during such transformations in the expansion of (11.33) with respect to $\varphi_i^{(n)}(\boldsymbol{r})$.

The thermodynamic potential of the crystal for given T, P and c depends on the form of the function $\rho(\boldsymbol{r})$ and, consequently, on the coefficients $C_i^{(n)}$. These coefficients evidently must become zero at the second-order phase transition point, but near it they should have only small values. Therefore, in expansion of the thermodynamic potential in a series with respect to $C_i^{(n)}$ near the transition point, one can retain only the first terms. The thermodynamic potential of the alloy is independent of the coordinates and therefore does not change

*A detailed statement of these problems may be found in the original papers [161, 164, 165], and also in [167].

**In all transformations of this group, the functions $\varphi_i^{(n)}(\boldsymbol{r})$ transform only into one another (the transformed functions are linear combinations of the remaining ones).

during symmetry transformations. Hence, in its expansion with respect to $C_i^{(n)}$ (that change during such transformations), these quantities may occur in each term of the expansion only in combinations that are invariant under the transformations of the symmetry group under consideration. In order to exhibit the form of the expansion of Φ in $C_i^{(n)}$, it is necessary to establish whether invariant combinations of different orders can be formed from the quantities $C_i^{(n)}$. Using group theory, one can show that it is impossible to construct an invariant of the first order from the quantities $C_i^{(n)}$, whereas an invariant of the second order for each possible representation always exists and may be written in form $\sum_i (C_i^{(n)})^2$. By analogy with the derivation of Eq. (11.6), one may prove that all the coefficients $C_i^{(n)}$ on one side of a point, lying on the second-order phase transition line, become zero, i.e., $\delta\rho(\boldsymbol{r}) = 0$, and the symmetry of this phase coincides with the crystal symmetry at the transition point. Moreover, on the other side of this point some $C_i^{(n)}$ for any of the values n are not equal to zero, i.e., $\delta\rho(\boldsymbol{r}) \neq 0$, and hence, the crystal has a lower symmetry than at the transition point. In the first case the crystal is in a disordered state (with higher symmetry); in the second case it is in an ordered state.

If we neglect thermal motion and static distortion of the crystal lattice, the assignment of the density probability function of the location of any type of atom $\rho(\boldsymbol{r})$ may be accomplished by assigning the probability $p_\alpha^{(L)}$ ($\alpha = A, B;\ L = 1, 2$) of substitution of type L sites by atoms A and B. The additional probability density $\delta\rho(\boldsymbol{r})$, appearing when the alloy transforms into the ordered state, must be proportional to the deviation of the probability $p_\alpha^{(L)}$ of substitution of the lattice sites by atoms A and B from its value at the disordered state, i.e., according to Eq. (1.5) $\delta\rho(\boldsymbol{r})$ must be proportional to the degree of long-range order η. Therefore, the quantities $C_i^{(n)}$ are proportional to η, i.e., $C_i = \gamma_i \eta$ (where the superscript n is discarded, because in the ordered state only the coefficients $C_i^{(n)}$ remain for a single value. Thus, the expansion given above for the thermodynamic potential Φ in powers of C_i is, in essence, an expansion of the type (11.1) in powers of η, the expansion coëfficients being invariants of different orders, composed of the quantities γ_i. As already mentioned, first-order invariants do not exist, which also gives grounds for equating the coefficients A_1 to zero in the expansion (11.1).

According to these arguments concerning the condition of existence of a second-order phase transition line (but not of an isolated point) on the T, P plane, this condition can be formulated more rigorously from the point of view developed here, i.e., the requirement stipulating the impossibility of formation of third-order invariants of the quantities γ_i. For crystals, in which this requirement is fulfilled, $A_3 \equiv 0$. Thus, we obtain for Φ the expansion (11.2) that was used previously. This requirement is not fulfilled for all crystals, and this fact diminishes the number of structures in which second-order phase transition lines are possible.

The second condition, which is imposed on the symmetry change of a crystal during a second-order phase transition, results from the requirement of stability of a homogeneous state, which, according to [167], leads to the absence of linear invariants, composed of quantities of the type $C_i \frac{\partial C_k}{\partial x} - C_k \frac{\partial C_i}{\partial x}$, where both the coefficients C_i and their derivatives with respect to the coordinates will appear.

By investigating the possibility of attaining the two conditions for alloys of different structures, we can determine in which of these alloys the second-order phase transitions are possible. Lifshits [164, 165] has made such an analysis for substitutional alloys, which, in the disordered state, have body-centered cubic, face-centered cubic and hexagonal close-packed structures. Lifshits succeeded in establishing that second-order phase transitions are possible only when such alloys transform to the ordered state with the crystal lattices cited in Tables 9, 10 and 11 [165]. Column I of these Tables lists symbols for the space groups for the ordered alloy. The letters, given in column II, indicate the type of the Bravais lattice* for the ordered crystal. Thus, in accordance with the adopted notation the letters P, I, F, H, R denote simple, body-centered cubic, face-centered cubic, hexagonal and rhombohedral lattices, respectively. Column III lists the ratios of the volume of the unit cell of the ordered crystal to the volume of the unit cell of the disordered crystal (at the transition point). Column IV lists the coordinates of the ends of the fundamental vectors (edges) of the Bravais cell of the ordered alloy. For lattice types I and F column IV also gives the coordinates of the atoms, located at the centers of the cells and in the middle of three

*The Bravais lattice is the set of all equivalent sites which can be superimposed on one another by a translation of integral number of periods of the crystal lattice. Thus, the sublattice of equivalent sites of the ordered alloy may include several Bravais lattices. For example, in an ordered AuCu one may separate four simple cubic Bravais lattices, in such a manner that the sublattice sites proper for the Cu atoms consist of three such Bravais lattices.

of their faces, respectively. In the latter two cases we consider the cubic cells (not the unit cells) of the Bravais lattice of the ordered alloy. The lengths of the edges of the cell of the original disordered alloy were chosen as the unit of measurement. For the disordered alloys with body-centered cubic and face-centered cubic lattices we chose as a unit the cubic cells, along whose edges the coordinate axes x, y, z are directed. For a hexagonal lattice, the axes of the coordinates x and y are selected for the sites of the rhombic unit cells, which form an angle of 120°, and the z axis is the height of the cell. The sign $\supset$ denotes cyclic permutations, which should be carried out over the indices, placed within the brackets. Column V gives formulas corresponding to the stoichiometric composition of an ordered alloy. The coordinates of the atoms in the cell of the ordered crystal are given in column VI. The length of the edges of the cell of the ordered crystal are chosen as the unit of measurement [168]. The letters in parentheses [(a), (b) etc.] form groups of the position of atoms, indicated in these tables. It is also shown in [168] that second-order phase transitions are possible during ordering of alloys with face-centered cubic lattices which change in the ordered state to a structure with the space groups O^6 and O^7 in the stoichiometric composition ABC_3D_3. It should be emphasized that these tables indicate only the possibility of a second-order phase transition in alloys of the specified type. For certain chosen values of the constants of interatomic interaction and concentration of the components of the alloy, which determine the coefficients of expansion of Φ in powers of η, first-order phase transitions for alloys with the structures indicated in the tables are possible.

Among the ordered structures listed in Table 9, which may appear during ordering of alloys with a body-centered cubic lattice, superlattices have been discovered experimentally in β-brass type alloys (space group O_h^1), and also Fe_3Al and Heusler alloys Cu_2AlMn (space group O_h^5). In β-brass, as indicated above, the existence of a second-order phase transition has been well established experimentally. In Fe_3Al phase transitions do not occur according to the scheme cited in Table 9, because a more symmetrical phase in this case does not have a body-centered cubic lattice, in which all the sites are equivalent, but the structure illustrated in Fig. 8. As shown in [165], second-order phase transitions are also possible in alloys of such a structure. Of the structures listed in Table 10, which arise in the ordering of alloys with a face-centered cubic lattice, superlattices were found to occur in CuPt, which has a rhombohedral lattice (space group D_{3d}^5; see Fig. 11a).

Table 9

Superlattices which may arise in second-order phase transitions of disordered alloys with a body-centered cubic lattice

I	II	III	IV	V	VI
O_h^1	P	2	(100, ↄ)	AB	1A(000) (a), 1B $\frac{1}{2}\frac{1}{2}\frac{1}{2}$ (b)
O_h^5	F	4	(200, ↄ) (110, ↄ)	ABC_2	4A(000) (a), 4B $\frac{1}{2}\frac{1}{2}\frac{1}{2}$ (b), 8C $\frac{1}{4}\frac{1}{4}\frac{1}{4}$ $\frac{3}{4}\frac{3}{4}\frac{3}{4}$ (c),
				AB_3	4A (a), 12B (b) (c)
O_h^7	F	4	(200, ↄ) (110 ↄ)	AB	8A(000) $\frac{1}{4}\frac{1}{4}\frac{1}{4}$ (a), 8B $\frac{1}{2}\frac{1}{2}\frac{1}{2}$ $\frac{3}{4}\frac{3}{4}\frac{3}{4}$ (b)

Table 10

Superlattices which may arise in second-order phase transitions of disordered alloys with a face-centered cubic lattice

I	II	III	IV	V	VI
D^5_{3d}	R	2	$\left(\frac{1}{2}\frac{1}{2}1,\ \circlearrowright\right)$	AB	1A (000) (a), 1B $\left(\frac{1}{2}\frac{1}{2}\frac{1}{2}\right)$ (b)
O^5_h	F	8	$(200,\ \circlearrowright)\ (110,\ \circlearrowright)$	ABC_6 AB_7	4A (000) (a), 4B $\left(\frac{1}{2}\frac{1}{2}\frac{1}{2}\right)$ (b), 24C $\left(0\frac{1}{4}\frac{1}{4},\ \circlearrowright\right)\left(0\frac{1}{4}\frac{3}{4},\ \circlearrowright\right)$ (d), 4A (a), 28B (b) (d)
O^7_h	F	8	$(200,\ \circlearrowright)\ (110,\ \circlearrowright)$	AB	16A (000) $\left(0\frac{1}{4}\frac{1}{4},\ \circlearrowright\right)$ (c), 16B $\left(\frac{1}{2}\frac{1}{2}\frac{1}{2}\right)\left(\frac{1}{2}\frac{3}{4}\frac{3}{4},\ \circlearrowright\right)$ (d)
D^{17}_{4h}	I	4	(100, 010, 002) $\left(\frac{1}{2}\frac{1}{2}1\right)$	ABC_2 AB_3	2A (000) (a), 2B $\left(00\frac{1}{2}\right)$ (b), 4C $\left(0\frac{1}{2}\frac{1}{4}\right)\left(\frac{1}{2}0\frac{1}{4}\right)$ (d), 2A (a), 6B (b) (d)

D_{4h}^{19}	I	4	(100, 010, 002) $\left(\frac{1}{2}\frac{1}{2}1\right)$	AB	4A (000) $\left(0\frac{1}{2}\frac{1}{4}\right)$ (a), 4B $\left(00\frac{1}{2}\right)$ $\left(0\frac{1}{2}\frac{3}{4}\right)$ (b)
O_h^1	P	32	$(200, \circlearrowright)$	$ABC_3D_3E_{12}F_{12}$	1A (000) (a), 1B $\left(\frac{1}{2}\frac{1}{2}\frac{1}{2}\right)$ (b), 3C $\left(0\frac{1}{2}\frac{1}{2}, \circlearrowright\right)$ (c), 3D $\left(\frac{1}{2}00, \circlearrowright\right)$ (d), 12E $\left(0\frac{1}{4}\frac{1}{4}, \circlearrowright\right)$ $\left(0\frac{3}{4}\frac{3}{4}, \circlearrowright\right)$ $\left(0\frac{1}{4}\frac{3}{4}, \circlearrowright\right)$ $\left(0\frac{3}{4}\frac{1}{4}, \circlearrowright\right)$ (i), 12F $\left(\frac{1}{2}\frac{1}{4}\frac{1}{4}, \circlearrowright\right)$ $\left(\frac{1}{2}\frac{3}{4}\frac{3}{4}, \circlearrowright\right)$ $\left(\frac{1}{2}\frac{1}{4}\frac{3}{4}, \circlearrowright\right)$ $\left(\frac{1}{2}\frac{3}{4}\frac{1}{4}, \circlearrowright\right)$ (j)
O_h^3	P	32	$(200, \circlearrowright)$	$A_3B_3C_4D_6$	6A $\left(0\frac{3}{4}\frac{1}{4}, \circlearrowright\right)$ $\left(\frac{1}{2}\frac{3}{4}\frac{1}{4}, \circlearrowright\right)$ (c), 6B $\left(0\frac{1}{4}\frac{3}{4}, \circlearrowright\right)$ $\left(\frac{1}{2}\frac{1}{4}\frac{3}{4}, \circlearrowright\right)$ (d), 8C (000) $\left(0\frac{1}{2}\frac{1}{2}, \circlearrowright\right)$ $\left(\frac{1}{2}\frac{1}{2}\frac{1}{2}\right)$ $\left(00\frac{1}{2}, \circlearrowright\right)$ (e), 12D $\left(\frac{3}{4}\frac{3}{4}0, \circlearrowright\right)$ $\left(\frac{3}{4}\frac{3}{4}\frac{1}{2}, \circlearrowright\right)$ $\left(\frac{1}{4}\frac{1}{4}\frac{1}{2}, \circlearrowright\right)$ $\left(\frac{1}{4}\frac{1}{4}0, \circlearrowright\right)$ (f)

Table 11

Superlattices which may arise in second-order phase transitions of disordered alloys with hexagonal close-packed lattice

I	II	III	IV	V	VI
D_{3h}^{1}	H	1	(100, ɔ)	AB	1A $\left(\frac{1}{3}\frac{2}{3}\frac{1}{4}\right)$ (a), 1B $\left(\frac{2}{3}\frac{1}{3}\frac{3}{4}\right)$ (d)
D_{2h}^{5}	P	2	(110, 110, 001)	AB	2A $\left(\frac{1}{2}\frac{1}{6}\frac{1}{4}\right)$ $\left(\frac{1}{2}\frac{5}{6}\frac{3}{4}\right)$ (f), 2B $\left(0\frac{2}{3}\frac{1}{4}\right)$ $\left(0\frac{1}{3}\frac{3}{4}\right)$ (e)
D_{3h}^{1}	H	4	(200, 020, 001)	ABC_3D_3	1A $\left(\frac{1}{6}\frac{1}{3}\frac{1}{4}\right)$ (a), 1B $\left(\frac{5}{6}\frac{2}{3}\frac{3}{4}\right)$ (f), 3C $\left(\frac{1}{3}\frac{1}{6}\frac{3}{4}\right)$ $\left(\frac{5}{6}\frac{1}{6}\frac{3}{4}\right)$ $\left(\frac{1}{3}\frac{2}{3}\frac{3}{4}\right)$ (k), 3D $\left(\frac{2}{3}\frac{1}{3}\frac{1}{4}\right)$ $\left(\frac{1}{6}\frac{5}{6}\frac{1}{4}\right)$ $\left(\frac{2}{3}\frac{5}{6}\frac{1}{4}\right)$ (j)

Thus, in alloys having ordered structures which differ from those considered above but with body-centered cubic, face-centered cubic and hexagonal close packed lattices in the disordered state, second-order phase transitions are impossible. In particular, such transitions were found to be impossible in alloys with very extended structures of the type $AuCu_3$ and AuCu. As was already noted in the preceding chapter, a first-order phase transition occurs during ordering of alloys of this type.

The theory of second-order phase transitions expounded in this section is based on the assumption that the thermodynamic potential for small η may be represented in the form of an expansion (11.1) in powers of η and that the coefficients of this expansion may be represented as a power series with respect to $T - T_0$, $c_A - c_{A0}$ and $P - P_0$. These assumptions, as emphasized by Landau, are not obvious because the thermodynamic potential has a singular point at the second-order phase transition temperature. These assumptions may be checked, either by performing a statistical calculation on the basis of the atomic model (for details see the following Chapter), or by comparing the theory with experimental data.

12. Thermodynamics of Almost Completely Ordered Alloys

In the previous sections we considered the case in which the temperature of the alloy was close to the critical temperature. The transition to the ordered state was a second-order phase transition, consequently the degree of long-range order could be treated as a small quantity. The thermodynamic treatment may also be employed in the opposite limiting case of an almost completely ordered alloy, whose temperature is below the critical temperature [167] and [169]. Such a treatment obviously would be equally applicable to alloys in which a second-order phase transition occurs, and to alloys in which the transition to the ordered state is a first-order phase transition. The treatment of an almost completely ordered solution may be carried out within the framework of a thermodynamic theory because the concentrations of "foreign" atoms on each type of site are small and may be represented by small parameters. Thus, each sublattice formed by any type of site may be considered as a weak solution of "foreign" atoms, and the thermodynamic potential is not difficult to determine, using the same reasoning as in the calculation of the thermodynamic potential of weak solutions ([187], Sec. 85). Let us consider a binary substitutional alloy $A - B$, with a nearly stoichiometric composition, consisting of N atoms and in a state of almost complete order. Let us denote by $\Phi_y = \Phi_y(T, P)$ the thermodynamic potential of a completely ordered solution (of stoichiometric composition), also containing N atoms, which exists at the temperature T (i.e., in a state which does not correspond to equilibrium at this temperature). With the substitution in a completely ordered alloy of atom A (located at the first type of sites) by atom B, the thermodynamic potential changes by some quantity, which we shall denote by $\chi_1 = \chi_1(T, P)$. Similarly the change of thermodynamic potential with a substitution of atom B by atom A on the second type of site will be denoted by $\chi_2 = \chi_2(T, P)$. When a relatively small number $N_B^{(1)} = N\nu p_B^{(1)}$ of B atoms is introduced by such substitutions into the first type of sites, and $N_A^{(2)} = N(1-\nu)\, p_A^{(2)}$ of atoms A on a second type of site, then, assuming that the atoms are localized at definite sites of the lattice (rearrangements are impossible), the thermodynamic potential will change by

$$N\nu p_B^{(1)}\chi_1 + N(1-\nu)\, p_A^{(2)}\chi_2. \tag{12.1}$$

Here we have neglected very improbable states (for the smallest concentration of atoms on the "foreign" sites), when two or more atoms on the "foreign" locations are situated close to one another. In a real crystal, however, atoms are not placed in position and in calculating the thermodynamic potential one should take into account the possibility of their rearrangement. We should, therefore, consider only those different arrangements of atoms which correspond to a given degree of long-range order, i.e., rearrangements inside each sublattice. Since the magnitude of (12.1) remains constant for each such rearrangement in an almost completely ordered solution, then, taking into account the possibility of rearrangement changes of the thermodynamic potential by $-kT\ln W$, where W is the number of rearrangements of atoms, we have

$$-kT\ln W = -kT\ln\left(\frac{N^{(1)}!}{N_A^{(1)}!\,N_B^{(1)}!}\,\frac{N\ \ !}{N_A^{(2)}!\,N_B^{(2)}!}\right) = \tag{12.2}$$
$$= kTN\left[\nu\left(p_A^{(1)}\ln p_A^{(1)} + p_B^{(1)}\ln p_B^{(1)}\right) + (1-\nu)\left(p_A^{(2)}\ln p_A^{(2)} + p_B^{(2)}\ln p_B^{(2)}\right)\right].$$

Here, we used Stirling's formula

$$\ln X! \approx X(\ln X - 1) \qquad (X \gg 1). \tag{12.3}$$

With a further increase of the number of "foreign" atoms $N_B^{(1)}$ and $N_A^{(2)}$ the variation of the thermodynamic potential no longer reduces to the sum of Eqs. (12.1) and (12.2), but will also contain square terms in $p_B^{(1)}$ and $p_A^{(2)}$, and terms of higher order. Hence, we shall limit ourselves to the square terms in the thermodynamic potential and shall denote their sum by

$$N\left(\lambda_{11}p_B^{(1)^2}+2\lambda_{12}p_B^{(1)}p_A^{(2)}+\lambda_{22}p_A^{(2)^2}\right).$$

As a result, the following expression is obtained for the thermodynamic potential [169]:

$$\begin{aligned}\Phi=\Phi_y+N\nu\chi_1p_B^{(1)}+N(1-\nu)\chi_2p_A^{(2)}+\\+kTN\left[\nu\left(p_A^{(1)}\ln p_A^{(1)}+p_B^{(1)}\ln p_B^{(1)}\right)+(1-\nu)\left(p_A^{(2)}\ln p_A^{(2)}+p_B^{(2)}\ln p_B^{(2)}\right)\right]+\\+N\left(\lambda_{11}p_B^{(1)^2}+2\lambda_{12}p_B^{(1)}p_A^{(2)}+\lambda_{22}p_A^{(2)^2}\right).\end{aligned}\qquad(12.4)$$

In the following it is essential that we define the quantities Φ_y, χ_1, χ_2 and λ_{ij} $(i, j=1, 2)$ in such a manner that they depend only on temperature and pressure, and are independent of the composition and of the degree of long-range order.

The temperature dependence of χ_1 and χ_2 at low temperature may be investigated, using Nernst's theorem. As $T\rightarrow 0$ the quantities χ_1 and χ_2 tend to the finite limiting values χ_1 and χ_2. Nernst's theorem can be applied not only to solid solutions in a state of equilibrium, but also to solutions where the atoms at foreign sites are either "frozen" or "localized," i.e., rearrangements among them are forbidden, and such distribution of atoms can be regarded as independent of temperature. In the expression for the thermodynamic potential of such a system, as we have seen, the first three terms of Eq. (12.4) remain, while the fourth term vanishes, and the small last term changes. We shall choose such small values of $p_B^{(1)}$ and $p_A^{(2)}$, in the alloy, that the last term can be neglected. Then, the entropy S_f of the "frozen" solution under consideration (in which $p_B^{(1)}$ and $p_A^{(2)}$ are independent of temperature) will have the following form:

$$S_f=S_y-N\nu\frac{\partial\chi_1}{\partial T}p_B^{(1)}-N(1-\nu)\frac{\partial\chi_2}{\partial T}p_A^{(2)},\qquad(12.5)$$

where S_y is the entropy of the completely ordered solution At temperature T. Since both the entropy of the solution with atoms "localized" on "foreign" sites S_f, and the entropy

S_y tend to zero, at $T \to 0$ the derivatives $\frac{\partial \chi_1}{\partial T}$ and $\frac{\partial \chi_2}{\partial T}$ at absolute zero must also become zero. Consequently, no linear terms will occur in the expansion of χ_1 and χ_2 in powers of T.

The equilibrium value of the degree of long-range order η is found from the condition of minimum thermodynamic potential, which can be determined from Eq (12.4). Within the limits of applicability of this approximation one may neglect the small last term in (12.4). Then, condition $\frac{\partial \Phi}{\partial \eta} = 0$, considering (1.5), leads to the following equation for the degree of long-range order at low temperatures (substantially below the critical temperature):

$$\frac{(c_A - \nu\eta)\,[1 - c_A - (1 - \nu)\,\eta]}{[c_A + (1 - \nu)\,\eta]\,(1 - c_A + \nu\eta)} = e^{-\frac{\chi}{kT}}, \tag{12.6}$$

where $\chi = \chi_1 + \chi_2$: This equation determines the temperature and concentration dependence of the degree of long-range order in an almost completely ordered solid solution. At sufficiently low temperatures (when the quadratic terms with respect to T may be neglected in the expansion of χ in T) χ in (12.6) may be replaced by $\chi^0 = \chi_1^0 + \chi_2^0$. The absence of linear terms in the expansions of χ_1 and χ_2 in powers of T leads to the fact that the preexponential factor occuring on the right-hand side of Eq. (12.6) does not differ from unity.

It follows from (12.6) that in case of a solid solution of stoichiometric composition $(c_A = \nu)$ the degree of long-range order at low temperature tends to unity according to the exponential equation

$$1 - \eta = \frac{1}{\sqrt{\nu(1 - \nu)}}\, e^{-\frac{\chi^0}{2kT}}. \tag{12.7}$$

Substituting (1.5) and (12.7) into (12.4) at $c_A = \nu$, and discarding the small last term in (12.4), we obtain explicitly the dependence of the equilibrium thermodynamic potential of stoichiometric alloy on temperature

$$\Phi = \Phi_y - 2NkT\sqrt{\nu(1 - \nu)} \cdot e^{-\frac{\chi^0}{2kT}}. \tag{12.8}$$

Hence, it follows that the entropy of the system is

$$S = S_y + kN\sqrt{\nu(1 - \nu)}\left(2 + \frac{\chi^0}{kT}\right)e^{-\frac{\chi^0}{2kT}}. \tag{12.9}$$

For the heat capacity $C_P = T\left(\frac{\partial S}{\partial T}\right)_P$, we obtain

$$C_P = C_{Py} + \frac{1}{2} kN \sqrt{\nu(1-\nu)} \left(\frac{\chi^0}{kT}\right)^2 e^{-\frac{\chi^0}{2kT}}. \qquad (12.10)$$

The second terms in Eqs. (12.8)-(12.10) express the configurational parts of the corresponding thermodynamic quantities which become zero as $T \to 0$. The configurational parts of the other thermodynamic quantities are also proportional to the factor $e^{-\chi^0/2kT}$. Since the terms Φ_y, S_y and C_{Py} (representing the thermodynamic potential, entropy and the specific heat of a completely ordered solution) change with temperature according to a power law, the configurational parts of the thermodynamic quantities at sufficiently low temperatures (at which the exponential terms are small compared with the power terms) play a small role in thermal effects. The exponential temperature dependence will appear in those properties of the alloy which are mainly determined by the arrangement of different types of atoms on the sites of the crystal lattice (for example, in investigating the residual electrical resistivity of an alloy, quenched at different temperatures well below the critical temperature). A similar situation exists in the thermodynamic quantities in almost completely ordered alloys, whose composition differs from but is close to stoichoimetric. By considering that $|c_A - \nu| \ll 1$ and $1 - \eta \ll 1$, we find from (12.6) that the temperature and concentration dependence of the degree of long-range order takes the form

$$1 - \eta = \frac{2\nu - 1}{2\nu(1-\nu)} \delta c_A + \frac{1}{2\nu(1-\nu)} \sqrt{(\delta c_A)^2 + 4\nu(1-\nu)\exp\left(-\frac{\chi^0}{kT}\right)}, \qquad (12.11)$$

where $\delta c_A = c_A - \nu$.

The treatment presented here for almost completely ordered solution was developed by Landau and Lifshits,* who considered the concentration dependence of the degree of long-range order in terms of the law of mass action.

It is apparent from Eq. (12.11) that for alloys in which the stoichiometric composition corresponds to the formula

*See Eq. (63.13) in the first edition of [167], which will coincide with our Eq. (12.11), if we transform the degree of long-range order introduced in [167] to η as defined by Eq. (1.4), and set $\lambda_0^2 = \nu(1-\nu) e^{-\frac{\chi^0}{kT}}$ in (63.13).

$AB\left(\nu=\frac{1}{2}\right)$, the degree of long-range order for a given temperature should be a maximum at $\delta c_A=0$. For other values of ν the maximum of η is shifted somewhat with respect to the point, corresponding to stoichiometric composition $c_A=\nu$, in the direction of the point $c_A=\frac{1}{2}$. In fact, the displacement of the maximum of δc_A^0 corresponding to a given temperature, determined by means of (12.11) from the condition $\frac{\partial\eta}{\partial c_A}=0$, is

$$\delta c_A^0=(1-2\nu)\,e^{-\frac{\chi^0}{2kT}}=(1-2\nu)\sqrt{\nu(1-\nu)}\,(1-\eta_{sto}). \tag{12.12}$$

Here, Eq. (12.7) is taken into account and the degree of long-range order η_{sto} corresponds to an alloy with stoichiometric composition at a given temperature. It follows from (12.12) that the shift of the maximum of the degree of long-range order tends to zero exponentially as the temperature decreases. The relation between δc_A^0 and η_{sto} determines all the theoretical parameters except ν, which permits a comparison of this relation with experimental data.

At $\delta c_A=0$ (12.11) will yield Eq. (12.7) for the degree of long-range order in a stoichiometric alloy. Equation (12.7) remains approximately valid also for alloys in which the departure $|\delta c_A|$ from the stoichiometric composition is appreciably less than $1-\eta$. If the inverse inequality is fulfilled $|\delta c_A|\gg 2\sqrt{\nu(1-\nu)}\exp\left(-\frac{\chi^0}{2kT}\right)$, the expression for the degree of long-range order also acquires a simple form. It follows from (12.11) that

$$1-\eta=\frac{(2\nu-1)\,\delta c_A+|\delta c_A|}{2\nu(1-\nu)}+\frac{1}{|\delta c_A|}\exp\left(-\frac{\chi^0}{kT}\right). \tag{12.13}$$

The last term in (12.13) gives the deviation $\delta\eta$ of the degree of long-range order at a given temperature from the maximum possible value in an alloy of given composition. Considering that $2\sqrt{\nu(1-\nu)}\sim 1$, the criterion of applicability of Eq. (12.13) may be given as $\delta\eta\ll|\delta c_A|$. Thus, in this limiting case the degree of long-range order again changes exponentially with temperature, but with an exponent twice as large (at the same temperature) and with a substantially larger pre-exponential factor. The variation of the configurational parts of the various thermodynamic quantities at low temperatures in the case under consideration are proportional to $\left(-\frac{\chi^0}{kT}\right)$.

If the critical temperature is very large, the solid solution may be found to be in an almost completely ordered state, even at comparatively high temperatures (near or above the Debye temperature), when it is impossible to use Nernst's theorem. Thus, in the formula cited above the constant χ^0 is replaced by the function χ, which is slightly dependent on temperature (in the conventional approximation of the statistical theory of ordering, as will be shown in Chap. III, this function generally is independent of T). The expansion of χ in powers of T converges slowly at such high temperatures. For investigation of the temperature dependence of η over a narrow temperature range near any temperature T_1, χ should be expanded in powers of $T-T_1$. Thus, the derivative $\frac{\partial\chi}{\partial T}$ at the point $T=T_1$ is in general different from zero, and a linear term appears in the expansion.

Owing to the presence of a linear (in $T-T_1$) term in the numerator of the exponent in Eqs. (12.7) and (12.13), the experimentally determined preexponential factors may be significantly different from $\frac{1}{\sqrt{\nu(1-\nu)}}$ and $\frac{1}{|\delta c_A|}$. This will be true only when Nernst's theorem is inapplicable, and when the inequalities $\delta\eta \gg |\delta c_A|$ or $\delta\eta \ll |\delta c_A|$ are fulfilled.

To explain the dependence of the degree of long-range order on pressure in an almost completely ordered alloy, we may use the fact that in the expansion of χ in powers of P for moderate pressures (at constant T) only the linear terms of the following expansion can be retained:

$$\chi(P)=\chi(0)+\chi' P. \tag{12.14}$$

In an almost completely ordered alloy, when the last term in (12.4) is negligible, the volume of the alloy is

$$V=\frac{\partial\Phi}{\partial P}=V_y+N\left[\nu p_B^{(1)}\frac{\partial\chi_1}{\partial P}+(1-\nu)\,p_A^{(2)}\frac{\partial\chi_2}{\partial P}\right], \tag{12.15}$$

where V_y is the value of a completely ordered alloy at the same temperature. Using (1.5), this gives a simple formula for χ'

$$\chi'=\frac{\partial\chi}{\partial P}=\frac{\partial(\chi_1+\chi_2)}{\partial P}=-\left(\frac{\partial V}{\partial\eta}\right)_{T,P,c_A}\frac{1}{N\nu(1-\nu)}, \tag{12.16}$$

enabling us to express χ' in terms of experimentally measurable quantities. Substituting (12.14) into (12.7) (writing χ instead of χ^0), we find that $1-\eta$ in almost completely ordered alloys of the stoichiometric composition depends exponentially on pressure:

$$1-\eta=\frac{1}{\sqrt{\nu(1-\nu)}}e^{-\frac{\chi(0)+\chi' P}{2kT}}, \tag{12.17}$$

the coefficient of P in the exponent being inversely proportional to temperature. In the limiting case in which $|\delta c_A| \gg \delta\eta$, in a similar manner we find from (12.13)

$$1-\eta=\frac{(2\nu-1)\,\delta c_A+|\delta c_A|}{2\nu(1-\nu)}+\frac{1}{|\delta c_A|}e^{-\frac{\chi(0)+\chi' P}{kT}}, \tag{12.18}$$

i.e., $\delta\eta$ also depends exponentially on P, but with an exponent twice as large.

It should be emphasized that for all nonstoichiometric alloys this treatment becomes invalid not only at high temperatures, but also at very low temperatures. In fact, in the calculation we have omitted quadratic and higher order terms in the expansion (12.4). In order to be able to neglect these terms in formulating the equation for the degree of long-range order, it is necessary that the products $p_B^{(1)}\lambda_{1i}$ and $p_A^{(2)}\lambda_{i2}$ $(i=1,\ 2)$ be much less than unity. However, for nonstoichiometric alloys when $T\to 0$, one of the probabilities $p_B^{(1)}$ or $p_A^{(2)}$ tends to a nonzero limit, while (as will be shown in the next chapter*) λ_{ij} may increase. The region of applicability of this approximation, therefore, has a limit not only in the direction of high temperatures, but also in the direction of low temperatures, which decreases with a decrease in $|\delta c_A|$.

It has been emphasized in [167] that nonstoichiometric alloys cannot exist in the equilibrium state at sufficiently low temperatures and must become disordered. In fact, the entropy of such an alloy, according to (12.4) (neglecting the last term of this equation), when $T\to 0$ may be written in the form

*See Eq. (19.5), in which the role of λ_{ij} is played by the sums of terms of the form $-\frac{kT}{n}\left(e^{\frac{w^{ij}}{kT}}-1\right)$ and $-\frac{kT}{n}\left(e^{-\frac{w^{ij}}{kT}}-1\right)$.

$$S = -kN\,[\nu\,(p_A^{(1)}\ln p_A^{(1)} + p_B^{(1)}\ln p_B^{(1)}) + (1-\nu)\,(p_A^{(2)}\ln p_A^{(2)} + p_B^{(2)}\ln p_B^{(2)})]. \tag{12.19}$$

Since for a nonstoichiometric alloy one of the quantities $p_B^{(1)}$ or $p_A^{(2)}$ at $T \to 0$ does not tend to zero, but to a limit different from zero, the entropy will also not tend to zero. This contradicts the Nernst theorem, and consequently, at a sufficiently low temperature, the alloys should decompose into an ordered phase whose composition is closer to stoichiometric than the composition of the original alloy, and a disordered phase in which the concentration c_A is close to either $c_A = 0$ or $c_A = 1$ (or an ordered phase with a different superlattice). The principle of such a decomposition in the case of second-order phase transitions was investigated in the previous section. Thus, only completely ordered alloys and pure metals may exist in equilibrium at absolute zero.

13. Fluctuations in the Degree of Long-Range and Composition in Alloys

Thermal motion of atoms in small volume elements of solid solutions gives rise to departures of the composition (the concentration of the components) and of the degree of long-range order from their equilibrium values. Such fluctuations exert a substantial influence on some properties of solid solutions, for instance, scattering of different types of waves (x-rays, neutrons, electrons and others) by the crystal lattices of the alloy. Of particular interest is the investigation of these fluctuations near the second-order phase transition point, where, as Landau has shown [170], especially large fluctuations of the degree of the long-range order arise.

The probabilities of fluctuations of composition and of the degree of long-range order may be calculated from the general theory of thermodynamic fluctuations (see, for example, [167], Chap. XII). The fluctuation of the degree of long-range order has been calculated in [170] in connection with the scattering of x-rays near a second-order phase transition point. The same authors also considered crystals, consisting of one type of molecule, in which only fluctuations of the degree of long-range order (for example, with orientational ordering) occur. In solid solutions, particularly in alloys, fluctuations of the degree of long-range order are accompanied by fluctuations of composition. Since, as will be shown below, fluctuations of the degree of long-range order and composition are not statistically independent,* the probability of fluctuations must be determined by considering simultaneously the departure of the composition and of the degree of long-range order from their equilibrium values. Such a calculation has been performed in [171].

As follows from the theory of thermodynamic fluctuations, the fluctuation probability is proportional to

$$w \sim e^{-\frac{R}{kT}}, \tag{13.1}$$

where R is the minimum work required to create reversibly a given fluctuation in some small part of the crystal of volume V, containing a constant total number of atoms N; and T is the temperature of the external medium (equal to the mean temperature of the body). Fluctuations of the composition and of the degree of long-range order, on one hand, and fluctuations of temperature and pressure, on the other, may be shown to be statistically independent. The expression for the minimum work in the case of the binary alloy A—B may therefore be represented as (see [167], Sec. 94)

$$R = \Delta\Phi - \mu_A \Delta N_A - \mu_B \Delta N_B. \tag{13.2}$$

Here

$$\Delta\Phi = \Phi(\bar{c}_A + \Delta c_A, \bar{\eta} + \Delta\eta) - \Phi(\bar{c}_A, \bar{\eta})$$

denotes the change in the thermodynamic potential when the atomic concentration c_A in the precipitated portion of the alloy for a given total number of atoms $N = N_A + N_B$ changes from the mean value $\bar{c}_A$ to $\bar{c}_A + \Delta c_A$, and the degree of long-range order changes from the mean (equilibrium) value $\bar{\eta}$ to $\bar{\eta} + \Delta\eta$ (for constant T and P)$\cdot$ ΔN_A and $\Delta N_B = -\Delta N_A$ are the changes in the number of atoms A and B of this part of the alloy, while μ_A and μ_B are the equilibrium values of the chemical potentials of A and B. Using Eq. (10.12), the minimum work takes the form

$$R = \Delta\Phi - \frac{\partial\Phi}{\partial c_A} \Delta c_A. \tag{13.3}$$

*The fluctuations of the various quantities are called statistically independent if the probability of their simultaneous departure from the mean value decomposes into the product of factors, each depending only on the fluctuations of the corresponding quantities. The mean value of the product of such fluctuations equals zero.

Then, expanding $\Delta\Phi$ in powers of Δc_A and $\Delta\eta$, taking into account the fact that equilibrium $\frac{\partial\Phi}{\partial\eta}=0$, and restricting ourselves to quadratic terms of the expansion, we obtain the following expression for R:

$$R=\frac{1}{2}\left[\frac{\partial^2\varphi}{\partial c_A^2}(\Delta c_A)^2+2\frac{\partial^2\varphi}{\partial c_A\,\partial\eta}\Delta c_A\,\Delta\eta+\frac{\partial^2\varphi}{\partial\eta^2}(\Delta\eta)^2\right]V. \tag{13.4}$$

Here φ is the specific thermodynamic potential (per unit volume of the crystal) and the derivatives are calculated at constant N, T and P.*

Using Eqs. (13.1) and (13.4), we may find the mean values of the square of the fluctuations of composition and of the degree of long-range order, and also the mean value of the product $\overline{\Delta c_A \Delta\eta}$ in a volume V, containing N particles

$$\overline{(\Delta c_A)^2}=\frac{\int\int(\Delta c_A)^2 w\,dc_A\,d\eta}{\int\int w\,dc_A\,d\eta}=\frac{kT}{V}\frac{\frac{\partial^2\varphi}{\partial\eta^2}}{\frac{\partial^2\varphi}{\partial c_A^2}\frac{\partial^2\varphi}{\partial\eta^2}-\left(\frac{\partial^2\varphi}{\partial c_A\,\partial\eta}\right)^2} \tag{13.5}$$

and similarly

$$\overline{(\Delta\eta)^2}=\frac{kT}{V}\frac{\frac{\partial^2\varphi}{\partial c_A^2}}{\frac{\partial^2\varphi}{\partial c_A^2}\frac{\partial^2\varphi}{\partial\eta^2}-\left(\frac{\partial^2\varphi}{\partial c_A\,\partial\eta}\right)^2}, \tag{13.6}$$

$$\overline{\Delta c_A\,\Delta\eta}=-\frac{kT}{V}\frac{\frac{\partial^2\varphi}{\partial c_A\,\partial\eta}}{\frac{\partial^2\varphi}{\partial c_A^2}\frac{\partial^2\varphi}{\partial\eta^2}-\left(\frac{\partial^2\varphi}{\partial c_A\,\partial\eta}\right)^2}. \tag{13.7}$$

In Eqs. (13.5) and (13.7) the derivatives with respect to c_A and η are taken at the equilibrium values of the degree of long-range order and composition. Sometimes it is convenient to express

---

*Here, the partial derivatives with respect to c_A are calculated for constant η in contrast with Sec. 10, where it was assumed that the degree of long-range order changes simultaneously with composition in accordance with the condition of equilibrium. The latter type of derivatives will be denoted by $\frac{d}{dc_A}$ in this section.

$\overline{(\Delta c_A)^2}$ in terms of second derivative $\frac{d^2\varphi}{dc_A^2}$, which can be calculated at values of the degree of long-range order, which corresponds to variations (fluctuations) of the composition (at the same temperature). Using the equilibrium condition $\frac{\partial\varphi}{\partial\eta}=0$, Eq. (13.5) is easily represented as

$$\overline{(\Delta c_A)^2}=\frac{kT}{V\frac{d^2\varphi}{dc_A^2}}. \tag{13.8}$$

Near the second-order phase transition point, the second derivative, occurring in Eqs. (13.5)–(13.7), may be calculated using the expansion given above for φ in powers of η:

$$\varphi=\varphi_0+\frac{A_2''}{2}\eta^2+\frac{A_4''}{4}\eta^4, \tag{13.9}$$

where $A_2''=a''(T-T_0)$. Taking into account the equilibrium condition $(A_2''+A_4''\bar{\eta}^2)\bar{\eta}=0$, for $\frac{\partial^2\varphi}{\partial\eta^2}$, we obtain the expression

$$\frac{\partial^2\varphi}{\partial\eta^2}=\begin{cases} A_2''=a''(T-T_0) & \text{for } T>T_0, \\ A_2''+3A_4''\bar{\eta}^2=2a''(T_0-T) & \text{for } T<T_0. \end{cases} \tag{13.10}$$

In a calculation of $\frac{\partial^2\varphi}{\partial c_A\,\partial\eta}$ one may use the relation

$$\frac{\partial A_2''}{\partial c_A}=-\frac{\partial A_2''}{\partial T}\frac{dT_0}{dc_A}=-a''\frac{dT_0}{ac_A},$$

and also neglect the last term in the expansion (13.9) (which is not essential for calculation of the derivative with respect to c_A). Then

$$\left(\frac{\partial^2\varphi}{\partial c_A\,\partial\eta}\right)^2=\left(\frac{\partial A_2''}{\partial c_A}\right)^2\bar{\eta}^2=\begin{cases} 0 & \text{for } T>T_0, \\ \frac{a''^3(T_0-T)}{A_4''}\left(\frac{dT_0}{dc_A}\right)^2 & \text{for } T<T_0. \end{cases} \tag{13.11}$$

The second derivation $\frac{\partial^2\varphi}{\partial c_A^2}$ near the transition point is approximately equal to $\frac{\partial^2\varphi_0}{\partial c_A^2}$. Since for $T<T_0$ $\frac{\partial^2\varphi}{\partial c_A\,\partial\eta}$, and consequently,

$\overline{\Delta c_A \Delta\eta}$ also differ from zero, the fluctuations of composition and of the degree of long-range order in an ordered phase are actually not statistically independent.

Substituting Eqs. (13.10) and (13.11) into Eq. (13.6), we show that as the transition temperature is approached, the fluctuations of the degree of long-range order become extremely large both in the ordered and in the disordered phases:

$$\left.\begin{aligned} \overline{(\Delta\eta)^2} &= \frac{kT}{Va''(T-T_0)} \quad \text{for } T > T_0, \\ \overline{(\Delta\eta)^2} &= \frac{kT}{2Va''(T_0-T)\left[1 - \dfrac{a''^3}{2A''_4}\dfrac{\left(\dfrac{dT_0}{dc_A}\right)^2}{\dfrac{\partial^2\varphi_0}{\partial c_A^2}}\right]} \quad \text{for } T < T_0. \end{aligned}\right\} \quad (13.12)$$

The mean square fluctuation of composition according to (13.5) remains finite at the transition point.

At the second-order phase transition point Eq. (13.12) gives an infinitely large value $\overline{(\Delta\eta)^2}$. This result is due to the fact that some terms were neglected in setting up an expression for the minimum work. To obtain the correct final value of the mean square fluctuation η, it is necessary to take into account the inhomogeneity of composition and of degree of long-range order, related to fluctuations of these quantities. The presence of such inhomogeneities should lead to the appearance of terms in the expansion of the thermodynamic potential which contain various derivatives of c_A and η. The linear terms with respect to the first derivatives of c_A and η will be absent in this expansion. In fact, if in any volume element we replace, for example, $\frac{\partial\eta}{\partial x}$ by $-\frac{\partial\eta}{\partial x}$ when the other derivatives with respect to η and c_A are equal to zero, obviously, the thermodynamic potential of this volume element $d\Phi$ should not change. Since the term proportional to $\frac{\partial\eta}{\partial x}$ in the expansion changes sign, the coefficient occurring in the corresponding term should identically be equal to zero. Therefore, the coefficients for the other first derivatives become zero. Next, we shall consider the case of cubic crystals. In this case the replacement of the fluctuations

$$\frac{\partial\eta}{\partial x} = \text{u}, \ \frac{\partial\eta}{\partial y} = \frac{\partial\eta}{\partial z} = \frac{\partial c_A}{\partial x} = \frac{\partial c_A}{\partial y} = \frac{\partial c_A}{\partial z} = 0$$

by the fluctuations

$$\frac{\partial\eta}{\partial y}=\mathrm{u},\ \frac{\partial\eta}{\partial x}=\frac{\partial\eta}{\partial z}=\frac{\partial c_A}{\partial x}=\frac{\partial c_A}{\partial y}=\frac{\partial c_A}{\partial z}=0$$

does not change $d\Phi$. Therefore, the expansion will contain identical coefficients for $\left(\frac{\partial\eta}{\partial x}\right)^2$ and $\left(\frac{\partial\eta}{\partial y}\right)^2$ (and also for $\left(\frac{\partial\eta}{\partial z}\right)^2$). Finally, in the expansion of $d\Phi$ there will occur no terms of the type $\frac{\partial\eta}{\partial x}\frac{\partial\eta}{\partial y}$ for $\frac{\partial\eta}{\partial x}\frac{\partial c_A}{\partial y}$, $\frac{\partial c_A}{\partial x}\frac{\partial c_A}{\partial y}$, because the replacement of $\frac{\partial\eta}{\partial x}$ by $-\frac{\partial\eta}{\partial x}$, for constant $\frac{\partial\eta}{\partial y}$, should not change the value of $d\Phi$. Thus, restricting ourselves to quadratic terms in the expansion of the thermodynamic potential of the volume element $d\tau$, one may write a term containing the first derivatives, in the following form:

$$\left\{\frac{1}{2}\alpha\left[\left(\frac{\partial\eta}{\partial x}\right)^2+\left(\frac{\partial\eta}{\partial y}\right)^2+\left(\frac{\partial\eta}{\partial z}\right)^2\right]+\frac{1}{2}\beta\left[\left(\frac{\partial c_A}{\partial x}\right)^2+\left(\frac{\partial c_A}{\partial y}\right)^2+\left(\frac{\partial c_A}{\partial z}\right)^2\right]+\right.$$
$$\left.+\gamma\left[\frac{\partial\eta}{\partial x}\frac{\partial c_A}{\partial x}+\frac{\partial\eta}{\partial y}\frac{\partial c_A}{\partial y}+\frac{\partial\eta}{\partial z}\frac{\partial c_A}{\partial z}\right]\right\}d\tau= \qquad (13.13)$$
$$=\left[\frac{\alpha}{2}(\nabla\eta)^2+\frac{\beta}{2}(\nabla c_A)^2+\gamma\nabla\eta\nabla c_A\right]d\tau.$$

Terms containing second derivatives of c_A and η of the type $\Delta\eta\frac{\partial^2\eta}{\partial x^2}$, $\Delta\eta\frac{\partial^2 c_A}{\partial x^2}$, $\Delta c_A\frac{\partial^2\eta}{\partial x^2}$, $\Delta c_A\frac{\partial^2 c_A}{\partial x^2}$ in integration over the entire volume V of the precipitated part of the crystal (neglecting surface effects) lead to the same types of expression as were obtained in the integration of (13.13). Assuming the fluctuations to be sufficiently smooth, we neglect terms in the expansion containing higher derivatives of c_A and η. Thus, when the gradients of concentration and of degree of long-range order are taken into account the expression for the minimum work (13.14) is replaced by

$$R=\frac{1}{2}\int\left[\frac{\partial^2\varphi}{\partial c_A^2}(\Delta c_A)^2+2\frac{\partial^2\varphi}{\partial c_A\partial\eta}\Delta c_A\Delta\eta+\frac{\partial^2\varphi}{\partial\eta^2}(\Delta\eta)^2+\right. \qquad (13.14)$$
$$\left.+\alpha(\nabla\eta)^2+2\gamma\nabla\eta\nabla c_A+\beta(\nabla c_A)^2\right]d\tau.$$

For many applications of the theory of fluctuations, it is not necessary to know the mean square fluctuations in any volume element, but instead the mean values of square of the Fourier components of these fluctuations must be given. The Fourier series expansions of $\Delta\eta$ and Δc_A may be written in the following manner:

$$\left.\begin{aligned}\Delta\eta &= \sum_{\varkappa} (\eta_{\varkappa} e^{i\varkappa r} + \overset{*}{\eta}_{\varkappa} e^{-i\varkappa r}),\\ \Delta c_A &= \sum_{\varkappa} (c_{A\varkappa} e^{i\varkappa r} + \overset{*}{c}_{A\varkappa} e^{-i\varkappa r}),\end{aligned}\right\} \tag{13.15}$$

where the components of the vector $\varkappa$ acquire values corresponding to definite conditions at the boundary of the volume under consideration, when $\varkappa_z > 0$.* Using Eqs. (13.14) and (13.15) one may express the minimum work in terms of coefficients of Fourier fluctuations of c_A and η:

$$R = V\Big[\sum_{\varkappa}\Big(\frac{\partial^2\varphi}{\partial\eta^2} + \alpha\varkappa^2\Big)|\eta_{\varkappa}|^2 + \Big(\frac{\partial^2\varphi}{\partial c_A^2} + \beta\varkappa^2\Big)|c_{A\varkappa}|^2 + \\ + \Big(\frac{\partial^2\varphi}{\partial c_A\,\partial\eta} + \gamma\varkappa^2\Big)\big(\eta_{\varkappa} c^*_{A\varkappa} + \eta^*_{\varkappa} c_{A\varkappa}\big)\Big]. \tag{13.16}$$

Writing the complex quantities $\eta_{\varkappa}$ and $c_{A\varkappa}$ in the form

$$\eta_{\varkappa} = \eta'_{\varkappa} + i\eta''_{\varkappa}, \quad c_{A\varkappa} = c'_{A\varkappa} + ic''_{A\varkappa} \tag{13.17}$$

and substituting the obtained expression for R into (13.1), we find that the probability of the fluctuation under consideration is proportional to

$$w \sim \exp\Big\{-\frac{V}{kT}\sum_{\varkappa}\Big[\Big(\frac{\partial^2\varphi}{\partial\eta^2} + \alpha\varkappa^2\Big)\big(\eta'^2_{\varkappa} + \eta''^2_{\varkappa}\big) + \\ + \Big(\frac{\partial^2\varphi}{\partial c_A^2} + \beta\varkappa^2\Big)\big(c'^2_{A\varkappa} + c''^2_{A\varkappa}\big) + 2\Big(\frac{\partial^2\varphi}{\partial c_A\,\partial\eta} + \gamma\varkappa^2\Big)\big(\eta'_{\varkappa} c'_{A\varkappa} + \eta''_{\varkappa} c''_{A\varkappa}\big)\Big]\Big\}. \tag{13.18}$$

It is apparent from Eq. (13.18) that each term in the sum over $\varkappa$ depends only on those $\eta'_{\varkappa}$ and $c'_{A\varkappa}$ (or $\eta''_{\varkappa}$ and $c''_{A\varkappa}$) which correspond to a given $\varkappa$. Actually the fluctuations of the Fourier

---

*The condition $\varkappa_z > 0$ is imposed so that in (13.15) we do not encounter sums with the same exponential factors. For $\varkappa_z = 0$, similar conditions are applied to $\varkappa_x$ or $\varkappa_y$

components corresponding to different $\varkappa$ are statistically independent (but the fluctuations of the quantities $\eta'_\varkappa$ and $c'_{A\varkappa}$ or $\eta''_\varkappa$ and $c''_{A\varkappa}$ with the same $\varkappa$ are statistically dependent). It is easy therefore to find the mean square of these fluctuations. In such a calculation we limit ourselves to the case in which $\varkappa$ is rather small (compared to the reciprocal of the lattice constant). This assumption also justifies the omission of higher derivatives of c_A and η in Eq. (13.14). Then, one may neglect terms proportional to $\varkappa^4$, and retain terms containing $\varkappa^2$, only in a series with small terms, independent of $\varkappa$, which vanish at the critical point. By calculating in this manner the average values of $|\eta'_\varkappa|^2$ and $|\eta''_\varkappa|^2$, and adding the results, we obtain:

$$\overline{|\eta_\varkappa|^2} = \frac{kT}{V} \frac{\dfrac{\partial^2\varphi}{\partial c_A^2}}{\dfrac{\partial^2\varphi}{\partial\eta^2}\dfrac{\partial^2\varphi}{\partial c_A^2} - \left(\dfrac{\partial^2\varphi}{\partial c_A\,\partial\eta}\right)^2 + \left(\alpha\dfrac{\partial^2\varphi}{\partial c_A^2} + \beta\dfrac{\partial^2\varphi}{\partial\eta^2} - 2\gamma\dfrac{\partial^2\varphi}{\partial c_A\,\partial\eta}\right)\varkappa^2}. \tag{13.19}$$

Thus, we find $\overline{|c_{A\varkappa}|^2}$

$$\overline{|c_{A\varkappa}|^2} = \frac{kT}{V} \frac{\dfrac{\partial^2\varphi}{\partial\eta^2} + \alpha\varkappa^2}{\dfrac{\partial^2\varphi}{\partial\eta^2}\dfrac{\partial^2\varphi}{\partial c_A^2} - \left(\dfrac{\partial^2\varphi}{\partial c_A\partial\eta}\right)^2 + \left(\alpha\dfrac{\partial^2\varphi}{\partial c_A^2} + \beta\dfrac{\partial^2\varphi}{\partial\eta^2} - 2\gamma\dfrac{\partial^2\varphi}{\partial c_A\partial\eta}\right)\varkappa^2}. \tag{13.20}$$

The quantity $\overline{|c_{A\varkappa}|^2}$ may also be expressed in terms of derivatives of the thermodynamic potential, calculated for such degrees of long-range order, that correspond to an increment in composition ($c_A = \overline{c}_A + \Delta c_A$), as was done in the derivation of Eq. (13.8). Then,

$$\overline{|c_{A\varkappa}|^2} = \frac{kT}{V} \frac{1}{\dfrac{d^2\varphi}{dc_A^2} + \beta'\varkappa^2}, \tag{13.20a}$$

where $\frac{1}{2}\beta'$ is the coefficient of $(\nabla c_A)^2$ in the expansion of φ with respect to ∇c_A, made for the indicated values of η.

Substituting Eqs. (13.10) and (13.11) for the second derivative of the thermodynamic potential (13.9), we find near the transition temperature [where one neglects $\beta\dfrac{\partial^2\varphi}{\partial\eta^2}\varkappa^2$ and $\gamma\dfrac{\partial^2\varphi}{\partial c_A\,\partial\eta}\varkappa^2$ in the denominator of Eq. (13.19)] that for $T > T_0$

$$\overline{|\eta_\varkappa|^2} = \frac{kT}{V} \frac{1}{a''(T - T_0) + \alpha\varkappa^2} \qquad (T > T_0), \tag{13.21}$$

and at temperatures somewhat below the critical temperature,

$$\overline{|\eta_\varkappa|^2} = \frac{kT}{V} \frac{1}{2a''\left[1 - \dfrac{a''^2}{2A_4''\dfrac{\partial^2\varphi_0}{\partial c_A^2}}\left(\dfrac{dT_0}{dc_A}\right)^2\right](T_0 - T) + \alpha\varkappa^2}. \tag{13.22}$$

It is apparent from Eqs. (13.21) and (13.22) that for sufficiently small $\varkappa$ the mean square Fourier components of fluctuations of the degree of long-range order $\overline{|\eta_\varkappa|^2}$ become exceedingly large near the transition point. This magnitude, however, does not become infinite for any finite $\varkappa$. The quantity $\overline{|\eta_\varkappa|^2}$ (corresponding to a given $\varkappa$) reaches its largest value at $T = T_0$, and then decreases linearly both with increase and decrease of temperature. The Fourier components of fluctuations of composition, as is apparent from Eq. (13.20), do not take anomalously large values.

Fluctuations of η and c_A have interesting properties near the critical point, at which the curve of the second order phase transition points changes into a decomposition curve (point K in Fig. 89). The expression in the square brackets of the denominator of Eq. (13.22), on the basis of (11.12), can be represented in the form

$$1 - \frac{a''^2}{2A_4''\dfrac{\partial^2\varphi_0}{\partial c_A^2}}\left(\frac{dT_0}{dc_A}\right)^2 = 1 - \frac{\Delta C_P''\left(\dfrac{dT_0}{dc_A}\right)^2}{T_0\dfrac{\partial^2\varphi_0}{\partial c_A^2}}. \tag{13.23}$$

At the critical point, as follows from (11.30), this expression becomes zero. Near the critical point, Eq. (13.23) considered as a function of T and c_A, can be expanded in a series with respect to $T - T_K$ and $c_A - c_{AK}$. Taking $c_A - c_{AK}$ to be proportional to $T_0 - T_K$, this expansion may be written in the form

$$1 - \frac{a''^2\left(\dfrac{dT_0}{dc_A}\right)^2}{2A_4''\dfrac{\partial^2\varphi_0}{\partial c_A^2}} = g_1(T - T_K) + g_2(T_0 - T_K), \tag{13.24}$$

where g_1 and g_2 are coefficients that are independent of T.

Substituting this expression into (13.22), we find that for an ordered alloy $\overline{|\eta_\varkappa|^2}$ is, in this case, given by

$$\overline{|\eta_\varkappa|^2} = \frac{kT}{V} \frac{1}{2a''[g_1(T-T_K)+g_2(T_0-T_K)](T_0-T)+\alpha\varkappa^2}. \quad (13.25)$$

Thus, as apparent from (13.21) and (13.25), near the critical point as we depart from the temperature T_0 into the ordered region, $\overline{|\eta_\varkappa|^2}$ increases substantially more slowly than in the disordered region. Equation (13.20) for $\overline{|c_{A\varkappa}|^2}$ for the case when the alloy is in the ordered state near the critical state may, by using Eqs. (13.10), (13.11) and (13.24), be represented in the form

$$\overline{|c_{A\varkappa}|^2} = \frac{kT}{V} \frac{2a''(T_0-T)+\alpha\varkappa^2}{2a''\frac{\partial^2\varphi_0}{\partial c_A^2}(T_0-T)[g_1(T-T_K)+g_2(T_0-T_K)]+\alpha\frac{\partial^2\varphi_0}{\partial c_A^2}\varkappa^2}, \quad (13.26)$$

where the small terms $\left(\beta\frac{\partial^2\varphi}{\partial\eta^2}-2\gamma\frac{\partial^2\varphi}{\partial c_A\,\partial\eta}\right)\varkappa^2$ are discarded. It follows from (13.26) that at very small $\varkappa$, where $\alpha\varkappa^2 \ll 2a''(T_0-T)$, the Fourier components of fluctuations of composition of the alloy, like the Fourier components of fluctuation of the degree of long-range order, become anomalously large near the critical point for $T < T_0$. Around the temperature T_0 (for $2a''(T_0-T) \ll \alpha\varkappa^2$) anomalies of this type should not occur.

With the aid of Eqs. (13.19) and (13.20) one may also determine $\overline{|\eta_\varkappa|^2}$ and $\overline{|c_{A\varkappa}|^2}$ for an almost completely ordered alloy (whose composition is nearly stoichiometric). In addition one may use Eq. (12.4) for the thermodynamic potential, having discarded the last small term. The second derivatives of thermodynamic potential per unit volume of the alloy with respect to c_A and η in this case are not small but, to the contrary, increase without limit as the temperature is lowered, so that terms proportional to $\varkappa^2$ can be omitted. Differentiating the expression(12.4), using Eqs. (1.5) and substituting the results into (13.19) and (13.20), we obtain:

$$\overline{|\eta_\varkappa|^2} = \frac{\Omega^0}{V}[c_Ac_B+(1-2\nu)(1-2c_A)\eta-(1-3\nu+3\nu^2)\eta^2]\frac{1}{\nu(1-\nu)}, \quad (13.27)$$

$$\overline{|c_{A\varkappa}|^2} = \frac{\Omega^0}{V}[c_Ac_B-\nu(1-\nu)\eta^2], \quad (13.28)$$

where Ω^0 is the volume occupied by a single atom.

Eqs. (13.27) and (13.28) also determine the mean square fluctuations $\overline{(\Delta\eta)^2}$ and $\overline{(\Delta c_A)^2}$ in the volume V. For $c_A = \nu$ and $\eta = 1$, expressions (13.27) and (13.28) become zero. This means that in an almost completely ordered alloy, fluctuations of composition and degree of long-range order become very small. For a stoichiometric alloy, the mean square fluctuations are proportional to $1-\eta$, i.e., according to Eq. (12.7) at low temperatures $\overline{(\Delta\eta)^2}$ and $\overline{(\Delta c_A)^2}$ decrease exponentially as the temperature decreases.

Expressions for the mean square fluctuation may be obtained within the framework of the thermodynamic theory in other cases, also, when the thermodynamic expression for Φ is known, in particular, in this manner we can analyze weak as well as concentrated ideal solid solutions (occurring in the disordered state). The thermodynamic potential per unit volume of a weak solution ($c_A \ll 1$), neglecting quadratic terms in c_A, has the form (see, for instance, [167]

$$\varphi = \varphi_B + nc_A kT \ln c_A + c_A G(P, T). \qquad (13.29)$$

Here φ_B is the thermodynamic potential per unit volume of a pure component B, n is the number of atoms per unit volume, and $G(P, T)$ is independent of c_A.

The mean square fluctuations $\overline{(\Delta c_A)^2}$ and $\overline{|c_{A\varkappa}|^2}$ for a disordered alloy are more conveniently calculated using Eqs. (13.8) and (13.28). The derivative of the thermodynamic potential (13.29) $\frac{d^2\varphi}{dc_A^2} = \frac{nkT}{c_A}$ at sufficiently small c_A substantially exceeds the second term in the denominator of Eq. (13.20a). Therefore,

$$\overline{|c_{A\varkappa}|^2} = \overline{(\Delta c_A)^2} = \frac{\Omega_0}{V} c_A. \qquad (13.30)$$

Hence, in weak solutions, the mean square fluctuation of composition is proportional to the concentration c_A.

In the case of ideal solutions (i.e., solutions in which the atoms are randomly distributed on the crystal lattice sites) the expression for the thermodynamic potential has a simple form

$$\varphi = \varphi^0 + kTn\left[c_A \ln c_A + (1 - c_A)\ln(1 - c_A)\right], \qquad (13.31)$$

where φ^0 is independent of composition. Since in the ideal solution, approximation of the thermodynamic potential of a nonequilibrium alloy is independent of the gradients of c_A (i.e., the coefficient β' in its expansion in powers of ∇c_A equals zero), then Eqs. (13.8), (13.20a) and (13.31)show that the mean square fluctuations are

$$\overline{(\Delta c_A)^2} = \overline{|c_{A\varkappa}|^2} = \frac{\Omega_0}{V} c_A (1 - c_A). \tag{13.32}$$

The maximum value of $\overline{(\Delta c_A)^2}$ occurs at $c_A = \frac{1}{2}$.

Chapter III

Statistical Theory of Order-Disorder Phenomena

14. Statement of the Problem

In the preceding chapter the thermodynamic theory of the order-disorder transition in alloys was expounded. This type of theory can be developed in a number of cases, when it is possible to select a suitable small parameter and expand the thermodynamic potential in powers of this parameter. As a result, the dependence of the degree of long-range order on temperature, pressure and composition of the alloy was obtained for temperatures close to the critical temperature (for a second-order phase transition), and also in the case of an almost completely ordered alloy. The possibility of a second-order phase transition in alloys of different structures was considered; the features of fluctuations in the degree of long-range order and composition near the second-order phase transition point were also presented. Owing to the fact that the thermodynamic calculations were carried out without the use of a concrete model of the alloy, the results are quite general.

However, these results are not correct over the entire range of the variation of the expansion parameter of the thermodynamic potential, but only over the range of values in which only the first terms of these expansions can be used (for instance, when the degree of long-range order η is small or close to unity). In particular, this method can not be used to investigate the order-disorder transition, which is a first-order phase transition, because upon passage to the ordered state the value of η at once assumes a large value. Moreover, the expansion coefficients occurring in the thermodynamic theory are not related to the physical quantities which characterize the atomic structure of the alloy, such as the energy of inter-atomic interactions, etc. These expansion coefficients, which are derived at temperatures near the critical temperature and at low temperatures, have no correlation within the framework of such a theory. Furthermore, the possibility of a power series expansion of certain quantities near the second-order phase transition point must be proven.

As distinguished from the thermodynamic theory, in the statistical theory a concrete atomic model of the alloy is used explicitly. On the basis of this model we determine the free energy and find the equilibrium properties of the system. In addition, the degree of long-range order is not assumed to be small or close to unity, and results applicable for the entire region of variation of the degree of long-range order and of composition can be obtained. The parameters occurring in the statistical theory may be related to the interaction energies of atoms and to quantities describing the structure of the alloy. In the statistical theory the short-range order can be investigated in detail by determining the temperature and concentration dependence of the correlation parameters in the different coordination spheres. In principle, it is also possible, within the framework of this model, to clarify the nature of the singularities in the thermodynamic potential of an alloy at the second-order phase transition point.

As already pointed out, in constructing a statistical theory one must use a simplified model of the alloy. In this theory we explicitly take into account the atomic structure of the alloy, consider the different arrangements of atoms on the lattice sites and calculate the energies of such arrangements, using certain assumptions concerning the nature of the interatomic interaction. The usual model of an alloy is based on the assumption that the energy of a crystal can be represented as a sum of the interaction energies of pairs of atoms, which depend on the arrangement of atoms (configurational part of the energy), and of a constant term, which is the same for all configurations of atoms on crystal lattice sites.

The interaction energies of individual pairs are assumed to be independent of alloy composition, degree of long-range order, temperature and the type of atoms surrounding a given pair. In this model we neglect the geometrical distortions of the crystal lattice, and also the effect of the variation of the lattice constant with composition, order and temperature on the interaction energy of the pairs of atoms. Since the energy of the alloy is reduced to the energies of interaction of pairs of atoms (the magnitudes of the interaction energies are regarded as negative), it is assumed that the dependence of the energy of the conduction of electrons (or that fraction thereof which cannot be included in the interaction energy of pairs), the vibrational energy of atoms in the lattice, etc., on the arrangement of atoms can be neglected. In the statistical theory, usually interactions between all but nearest neighbor atoms are neglected, but sometimes the interactions of more distant atoms is also considered. The selected model does not in all cases correspond well to a real alloy.* Nevertheless, such a model makes possible an explanation of a number of phenomena occurring in ordered alloys.

To determine the equilibrium properties of an alloy at a given temperature T, volume V and concentrations of the components, one must calculate the partition function

$$Z=\sum_n e^{-\frac{E_n}{kT}}. \tag{14.1}$$

Here n is the number of states of the system, determined by the distribution of atoms on the sites (configuration) and by the quantum numbers characterizing the thermal vibrations of atoms, the state of conduction electrons, and also the states associated with other degrees of freedom in the given configuration. The set of these quantum numbers for a given configuration is denoted by m. In the framework of the model the energy of the alloy E_n may be represented as the sum

$$E_n=E_m+E_i, \tag{14.2}$$

where E_i is the configurational energy of the crystal, determined by the assignment of the type of atom to each lattice site (the ground state energy for the given i configuration); and E_m is the fraction of the energy that is independent of the number of configuration i and is determined by the quantum number m. Then the partition function Z may be written as a product of

where

$$\left.\begin{array}{c} Z=Z_0 Z_k, \\ Z_0=\sum_m e^{-\frac{E_m}{kT}}, \quad Z_k=\sum_i e^{-\frac{E_i}{kT}}, \end{array}\right\} \tag{14.3}$$

*The model under consideration is applicable with a great degree of accuracy to alloys of nontransition metals. If, however, transition metals occur in the alloy, the three-dimensional redistribution of the electron density can have a significant influence on the interaction energy of atoms in the crystal. This redistribution may lead to a change in the short-range order, which in turn leads to a change in some properties of the alloys. Such effects have been examined by Borovskiy and Gurov [172].

and the free energy of the alloy $F=-kT\ln Z$ takes the form

$$\left.\begin{aligned} F=F_0+F_k, \\ F_0=-kT\ln Z_0, \quad F_k=-kT\ln Z_k. \end{aligned}\right\} \qquad (14.4)$$

F_k represents the configurational part of the free energy.

The configurational energy E_i (corresponding to the ith configuration) may be expressed in terms of the interaction energy of pairs of different ions located at various separations. In the particular case of a binary alloy $A-B$ this energy is

$$E_i=-\sum_{l=1}^{\infty}\left[N_{AA}^{(l)}v_{AA}(\rho_l)+N_{BB}^{(l)}v_{BB}(\rho_l)+N_{AB}^{(l)}v_{AB}(\rho_l)\right], \qquad (14.5)$$

where $v_{AA}(\rho_l)$, $v_{BB}(\rho_l)$, $v_{AB}(\rho_l)$ are taken with the opposite sign to the interaction energy of the pairs of atoms $A-A$, $B-B$ and $A-B$ in the crystal at distances ρ_l, that are equal to the radius of the lth coordination sphere, and $N_{AA}^{(l)}$, $N_{BB}^{(l)}$, $N_{AB}^{(l)}$ are the number of pairs of atoms $A-A$, $B-B$ and $A-B$, located at distances ρ_l. These numbers evidently depend on the number of the configuration i. In the nearest-neighbor approximation E_i takes the form

$$E_i=-(N_{AA}v_{AA}+N_{BB}v_{BB}+N_{AB}v_{AB}).$$

Here the numbers of pairs N_{AA}, N_{BB}, N_{AB} and the interaction energies $-v_{AA}$, $-v_{BB}$, $-v_{AB}$ must be taken for neighboring atoms. The number of pairs N_{AA} may be expressed in terms of the number N_A of A atoms in the alloy and the number of pairs N_{AB}. The total number of atoms adjacent to the A atoms is equal to zN_A (z is the coordination number for the first coordination sphere). By subtracting the number of pairs for different atoms N_{AB} from the total number of atoms, we obviously obtain double the number of pairs N_{AA}

$$N_{AA}=\frac{1}{2}(zN_A-N_{AB}). \qquad (14.7)$$

Similarly

$$N_{BB}=\frac{1}{2}(zN_B-N_{AB}). \qquad (14.8)$$

Substituting (14.7) and (14.8) into (14.6), we obtain another expression for the configurational energy

$$E_i = -\frac{1}{2}\left[wN_{AB} + z\left(N_A v_{AA} + N_B v_{BB}\right)\right], \tag{14.9}$$

where

$$w = 2v_{AB} - v_{AA} - v_{BB}. \tag{14.10}$$

The quantity w is called the ordering energy. The second term in (14.9) is independent of the configuration of atoms, i.e., of the degree of long-range and short-range orders, and depends only on the composition of the alloy (this term could be referred to the part of the energy E_m which is independent of configuration). The first term in (14.9) is a function of the configuration.

To determine the equilibrium properties of alloys, associated with the arrangement of the atoms over the lattice sites, one must next use the condition that the configurational part of free energy $F_{\text{к}}$ is a minimum. However, in spite of the simplification, calculation of the partition function is an extremely difficult mathematical problem, which has been solved only for one-dimensional and two-dimensional lattices. To obtain a solution in the three-dimensional case, one must resort to various approximate methods of calculation.

15. Matrix Methods

An exact calculation of the partition function under the simplifying assumptions formulated in the preceeding section for two-dimensional crystal lattices has been performed by means of algebraic methods, employing matrix and spinor calculus. Here we shall not present in detail all the calculations that must be performed to determine the partition function. We shall merely indicate the main stages of calculation and discuss the results.* In addition we examine ordered alloys in the approximation which takes into account only the interaction of nearest atoms.

For applications of this method it is convenient to use the grand partition function Ξ of the alloy $A-B$ instead of the partition function Z_k:

---

*A detailed discussion of the subjects under consideration may be found in the reviews [173–175], which cite original literature.

$$\Xi = \sum_{\substack{N'_A N'_B \\ (N'_A + N'_B = N)}} \sum_i \exp \frac{\mu_A N'_A + \mu_B N'_B - E_i}{kT}, \tag{15.1}$$

where μ_A and μ_B are the chemical potentials of the components A and B. The summation in (15.1) extends over all values of the number of atoms A and B, N'_A and N'_B for the constant sum $N'_A + N'_B = N$ and over all configurations i which may be realized for given values of N'_A and N'_B. Thus, we consider ensembles of alloys with different compositions (numbers N'_A and N'_B) for a given total number of atoms N. The chemical potentials μ_A and μ_B are regarded as given in this manner, so that the mean values $\overline{N'_A}$ and $\overline{N'_B}$ over the ensemble equal the numbers N_A and N_B of atoms A and B in the alloys under consideration:

$$N_A = \overline{N'_A}, \quad N_B = \overline{N'_B}. \tag{15.2}$$

We introduce the quantities σ_l, equal to $+1$ if site number l is occupied by atom A, and -1 if it is occupied by atom B. Then the configuration of atoms on the crystal lattice sites may be given by N discrete variables σ_l. Clearly, the numbers of atoms A and B in the crystal N'_A and N'_B may be expressed in terms of the variables of σ_l by the formula

$$\left.\begin{aligned} N'_A &= \frac{1}{2} \sum_{l=1}^{N} (1 + \sigma_l), \\ N'_B &= \frac{1}{2} \sum_{l=1}^{N} (1 - \sigma_l). \end{aligned}\right\} \tag{15.3}$$

Let us introduce further the quantities $a_{ll'}$, which equal zero if site number l is not a nearest neighbor of site number l' and equal unity in the case when these sites are nearest neighbors. For sites l and l', occupied by atoms of different species, $\sigma_l \sigma_{l'} = -1$. If, however, these sites are occupied by atoms of the same species, $\sigma_l \sigma_{l'} = +1$. The number of pairs of neighboring atoms type AB for an alloy of a given configuration, characterized by the numbers $\sigma_1, \sigma_2, \ldots, \sigma_N$, therefore, equals

$$N'_{AB} = \frac{1}{4} \sum_{l, l'=1}^{N} (1 - \sigma_l \sigma_{l'}) a_{ll'} = \frac{1}{4} Nz - \frac{1}{4} \sum_{l, l'=1}^{N} \sigma_l \sigma_{l'} a_{ll'}. \quad (15.4)$$

Rewriting (14.9) for the specimen of the ensemble with numbers of atoms A and B, equal to N'_A and N'_B and number of pairs $A-B$ equal N'_{AB},

$$E_i = -\frac{1}{2} \left[wN'_{AB} + z (N'_A v_{AA} + N'_B v_{BB}) \right] \quad (15.5)$$

and substituting (15.5), (15.3) and (15.4) into (15.1), we obtain,

$$\Xi = L^N \sum_{\sigma_1=-1}^{+1} \sum_{\sigma_2=-1}^{+1} \cdots \sum_{\sigma_N=-1}^{+1} \exp \left(m \sum_{l=1}^{N} \sigma_l - \frac{w}{8kT} \sum_{l, l'=1}^{N} \sigma_l \sigma_{l'} a_{ll'} \right) \quad (15.6)$$

Here

$$L = \exp \left(\frac{\mu_A + \mu_B}{2kT} + \frac{z}{8} \frac{2v_{AB} + v_{AA} + v_{BB}}{kT} \right), \quad (15.7)$$

$$m = \frac{\mu_A - \mu_B}{2kT} + \frac{z}{4} \frac{v_{AA} - v_{BB}}{kT} \quad (15.8)$$

and the summation over N'_A, N'_B and i is replaced by an equivalent summation over all values of σ_1, σ_2, ..., σ_N. We shall show that when the chemical potentials μ_A and μ_B are chosen so that $m=0$, the alloy has the same number of atoms A and B, i.e., $N_A = N_B$. Using (15.3) we obtain

$$N_A = \overline{N'_A} = \frac{N}{2} + \frac{1}{2} \frac{\sum_{\sigma_1=-1}^{+1} \sum_{\sigma_2=-1}^{+1} \cdots \sum_{\sigma_N=-1}^{+1} \sum_{l=1}^{N} \sigma_l \exp \left(m \sum_{l=1}^{N} \sigma_l - \frac{w}{8kT} \sum_{l, l'=1}^{N} \sigma_l \sigma_{l'} a_{ll'} \right)}{\sum_{\sigma_1=-1}^{+1} \sum_{\sigma_2=-1}^{+1} \cdots \sum_{\sigma_N=-1}^{+1} \exp \left(m \sum_{l=1}^{N} \sigma_l - \frac{w}{8kT} \sum_{l, l'=1}^{N} \sigma_l \sigma_{l'} a_{ll'} \right)}, \quad (15.9)$$

$$N_B = N - N_A.$$

Since the quantities σ_l assume the values -1 and $+1$, replacement of all σ_l by σ_l does not alter the sums occurring in (15.9). On the other hand, for $m=0$ after such a substitution the exponential function in (15.9) remains unaltered, and the σ_l occurring in front of the function in the numerator change sign, i.e., the second term in the formula for N_A changes sign. This term therefore equals zero, and for $m=0$ we obtain

$$N_A = N_B = \frac{N}{2}.$$

Let us now represent the partition function Ξ in matrix form [176-178]. For this purpose we subdivide the crystal lattice into s different layers in such a manner that the atoms of each jth layer interact with one another and with those of the j+1th and j-1th neighboring layers. For example, in the nearest neighbor approximation in a three-dimensional crystal such layers may be exhibited by atomic planes, in a two-dimensional lattice by linear chains of atoms and in a one-dimensional lattice by individual atoms. The configurational energy of the crystal E_i may be written as

$$E_i = \sum_{j=1}^{s} V(\nu_j, \nu_{j+1}). \tag{15.10}$$

Here ν_j denotes the set of quantities σ_l for the jth layer, i.e., ν_j completely determines the configuration of atoms A and B in the jth layer. $V(\nu_j, \nu_{j+1})$ denotes the interaction energy of atoms of the $j+1$th layer among themselves and with atoms of the jth layer. The magnitude of $V(\nu_j, \nu_{j+1})$ for the first layer $(j=1)$ should not contain the energy of interaction with the previous layer. However, this interaction will not be taken into account, because a periodicity condition will be imposed on the crystal, according to which the layer s is considered as neighboring on the first layer. The error made in this assumption is attributed to the surface effect and is vanishingly small for microscopic crystals. Since N'_A and N'_B, according to (15.3), may also be decomposed into a sum of terms referring to each layer, the expression $\mu_A N'_A + \mu_B N'_B$ may be written as

$$\mu_A N'_A + \mu_B N'_B = \sum_{j=1}^{s} M(\nu_j, \nu_{j+1}). \tag{15.11}$$

Here

$$M(\nu_j, \nu_{j+1}) = q\frac{\mu_A + \mu_B}{2} + \frac{\mu_A - \mu_B}{4}\left(\sum_{l_j=1}^{q} \sigma_{l_j} + \sum_{l_{j+1}=1}^{q} \sigma_{l_{j+1}}\right) \quad (15.12)$$

where q is the number of atoms in the layer and l_j is the number of sites of the crystal lattice in the jth layer.

Introducing the notation

$$\exp\frac{M(\nu_j, \nu_{j+1}) - V(\nu_j, \nu_{j+1})}{kT} = P_{\nu_j \nu_{j+1}}, \quad (15.13)$$

the partition function (15.1) can be represented in the form

$$\Xi = \sum_{\nu_1}\sum_{\nu_2} \ldots \sum_{\nu_s} P_{\nu_1\nu_2}P_{\nu_2\nu_3} \ldots P_{\nu_{s-1}\nu_s}P_{\nu_s\nu_1}. \quad (15.14)$$

When the layer is monatomic, (15.14) may be obtained from (15.6). The quantities $P_{\nu_j\nu_{j+1}}$ may be considered as matrix elements of some matrix **P** of the order 2^q. Using the formula for multiplication of matrices, we shall write Eq. (15.14) in the form

$$\Xi = \mathrm{Sp}\,\mathbf{P}^s. \quad (15.15)$$

If the eigenvalues of matrix **P** equal α_k, as is well known, the eigenvalues of the matrix $\mathbf{P}^s$ form a set of quantities α_k^s. Therefore

$$\Xi = \sum_{k=1}^{2^q} \alpha_k^s. \quad (15.16)$$

In the limit of an infinitely large crystal, when $s \to \infty$, in a calculation of $\ln \Xi$ only the maximum eigenvalue α_{max} is significant. In fact, labelling the eigenvalues in such a matter that $\alpha_1 = \alpha_{max}$, we represent $\frac{1}{s}\ln \Xi$ in the form

$$\frac{1}{s}\ln \Xi = \ln \alpha_{max} + \frac{1}{s}\ln\left[1 + \sum_{k=2}^{2^q}\left(\frac{\alpha_k}{\alpha_{max}}\right)^s\right]. \quad (15.17)$$

Since $\frac{\alpha_k}{\alpha_{max}} < 1$, the second term vanishes when $s \to \infty$, so that in this case*

$$\ln \Xi = s \ln \alpha_{max}. \tag{15.18}$$

As an example of application of the matrix method let us consider a one-dimensional model of the alloy $A-B$. Since the layers in this case are the individual atoms, the variables ν_j, characterizing the configuration in the jth layer, coincide with the variables σ_j (where j is the number of the atom). The quantities $V(\nu_j, \nu_{j+1})$ will be equal to the interaction energy of the pairs of neighboring atoms, which can be expressed in terms of the variables σ_j, σ_{j+1} as follows

$$V(\nu_j, \nu_{j+1}) = V(\sigma_j, \sigma_{j+1}) = \\ = \frac{1}{4}\left[-(v_{AA} + v_{BB} + 2v_{AB}) + (v_{BB} - v_{AA})(\sigma_j + \sigma_{j+1}) + w\sigma_j\sigma_{j+1}\right]. \tag{15.19}$$

The quantity $M(\nu_j)$, which can be determined from Eq. (15.12) (in which one must put $q = 1$), is

$$M(\nu_j, \nu_{j+1}) = M(\sigma_j, \sigma_{j+1}) = \frac{\mu_A + \mu_B}{2} + \frac{\mu_A - \mu_B}{4}(\sigma_j + \sigma_{j+1}). \tag{15.20}$$

The matrix $\mathbf{P}$ with matrix elements (15.13) in the case under consideration is a two-by-two matrix. With aid of the notation of (15.5) and (15.8) (where the coordination number z is set equal to two) its matrix elements may be reduced to the form

$$P_{\nu_j \nu_{j+1}} = L \exp\left[\frac{m}{2}(\sigma_j + \sigma_{j+1}) - \frac{w}{4kT}\sigma_j\sigma_{j+1}\right], \tag{15.21}$$

i.e.,

$$\mathbf{P} = \begin{Vmatrix} Le^{m - \frac{w}{4kT}} & Le^{\frac{w}{4kT}} \\ Le^{\frac{w}{4kT}} & Le^{-m - \frac{w}{4kT}} \end{Vmatrix}. \tag{15.22}$$

---

*The result (15.18) is obvious for finite q, when the crystal is infinite only in one direction. When $s \to \infty$ and $q \to \infty$ in (15.17), it is impossible to neglect the second term if, for example, the maximum eigenvalue α_{max} asymptotically becomes degenerate with an exponentially increasing degree of degeneracy for $q \to \infty$, because the sum occurring in (15.17) contains unity an infinite number of times. However, in all the investigated applications of the theory, the second term in (15.17) may also be neglected in the case when $s \to \infty$ and $q \to \infty$.

The eigenvalues α_1 and α_2 of the matrix P are determined from the quadratic equations

$$\begin{vmatrix} Le^{m-\frac{w}{4kT}}-\alpha & Le^{\frac{w}{4kT}} \\ Le^{\frac{w}{4kT}} & Le^{-m-\frac{w}{4kT}}-\alpha \end{vmatrix}=0, \qquad (15.23)$$

whence

$$\alpha_{1,\,2}=L\left(e^{-\frac{w}{4kT}}\cosh m \pm \sqrt{e^{-\frac{w}{2kT}}\cosh^2 m+2\sinh\frac{w}{2kT}}\right), \qquad (15.24)$$

where the largest eigenvalue α_1 corresponds to the + sign. The grand partition function Ξ for a one-dimensional crystal containing $A=N$ atoms, according to (15.18) and (15.24), for $N\to\infty$ may be written in the form

$$\Xi=L^N\left(e^{-\frac{w}{4kT}}\cosh m+\sqrt{e^{-\frac{w}{2kT}}\cosh^2 m+2\sinh\frac{w}{2kT}}\right)^N. \qquad (15.25)$$

For an alloy with an equal number of atoms A and B, $m=0$ and

$$\Xi=\left(2L\cosh\frac{w}{4kT}\right)^N. \qquad (15.26)$$

The functions Ξ, which are given by Eqs. (15.25) and (15.26), are analytic functions of T over the entire temperature range $0<T<\infty$. This means that in a one-dimensional chain of atoms all derivatives of Ξ with respect to T, and, hence, also all derivatives of thermodynamic potential with respect to temperature are continuous, namely, order-disorder phase transitions are impossible. Using formula (15.26) for the partition function, we can find the temperature dependence of the specific heat of a linear chain of atoms in the case $N_A=N_B$; a plot of this relationship is shown in Fig. 90. The specific heat in this model is related to the temperature variation of the short-range order. The curve given in Fig. 90 shows that as the temperature is lowered from high temperatures, the degree of short-range order increases first slowly, then more and more rapidly, and at a certain temperature its rate of change reaches a maximum. With a further decrease of temperature, the degree of short-range order almost reaches its maximum value, its rate of change again becomes small, causing a reduction of the specific heat, which tends to zero at $T \to 0$, in conformity with Nernst's theorem.

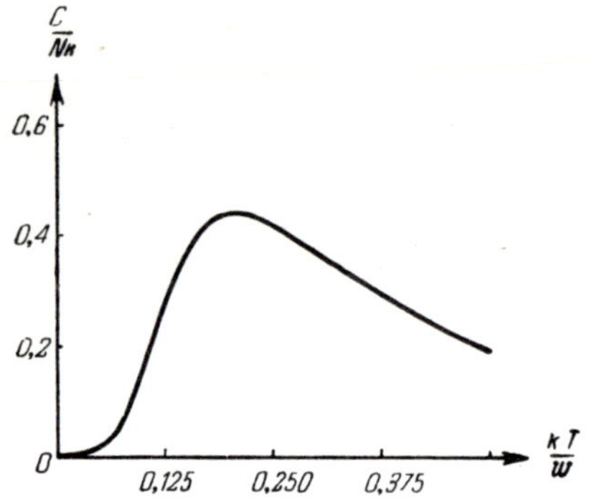

Fig. 90. Temperature dependence of the specific heat of a linear model of a solid solution having an equal number of atoms of different types.

We shall now consider possibility of the emergence of long-range order in the crystal model under consideration. As already mentioned in Sec. 1, long-range order occurs in a crystal lattice if there is correlation between the filling of infinitely remote sites by atoms. Dividing the crystal into layers as before and giving the configuration of atoms on sites of the first layer, we may determine whether or not long-range order will exist in the alloy by considering the probability configuration of atoms in the sth layer when $s \to \infty$. If this probability depends on the choice of configuration in the first layer, long-range order exists in the crystal. If, however, there is no such dependence, only short-range order exists. Suppose we are given the probabilities $p_1(\nu_1)$ of realization of various configurations in the first layer, which can be characterized by the different values of the variable ν_1. In the special case when a definite configuration is given for the first layer, the probability p_1 differs from zero only for a definite ν_1. The probability $p(\nu_1\nu_2 \ldots \nu_s)$ that the first layer has the configuration ν_2, the second layer—the configuration ν_1, etc. and lastly, the sth layer—the configuration ν_s, is given by the formula for the probability of state in a Gibbs ensemble with a variable number of particles

$p(\nu_1\nu_2\ldots\nu_s)=$

$$\frac{p_1(\nu_1)\exp\left\{\frac{1}{kT}\left[\mu_A\sum_{j=2}^{s}N_A'^{(j)}+\mu_B\sum_{j=2}^{s}N_B'^{(j)}-\sum_{j=1}^{s-1}V(\nu_j\nu_{j+1})\right]\right\}}{\sum\limits_{\nu_1\ldots\nu_s}p_1(\nu_1)\exp\left\{\frac{1}{kT}\left[\mu_A\sum_{j=2}^{s}N_A'^{(j)}+\mu_B\sum_{j=2}^{s}N_B'^{(j)}-\sum_{j=1}^{s-1}V(\nu_j\nu_{j+1})\right]\right\}}. \qquad (15.27)$$

Here $N_A'^{(j)}$ and $N_B'^{(j)}$ are the numbers of atoms A and B in the jth layer. The periodicity condition is not imposed in this case.

In order to obtain the probability $p_s(\nu_s)$ of the configuration ν_s in the sth layer, one must sum expression (15.27) over all configurations $\nu_1\ldots\nu_{s-1}$ in all the layers except the sth. Introducing the notation

$$R_{\nu_j\nu_{j+1}}=\exp\frac{\mu_A N_A'^{(j+1)}+\mu_B N_B'^{(j+1)}-V(\nu_j,\nu_{j+1})}{kT}, \qquad (15.28)$$

we obtain

$$p_s(\nu_s)=\frac{\sum\limits_{\nu_1\ldots\nu_{s-1}}\prod\limits_{j=1}^{s-1}R_{\nu_j\nu_{j+1}}p_1(\nu_1)}{\sum\limits_{\nu_1\ldots\nu_s}\prod\limits_{j=1}^{s-1}R_{\nu_j\nu_{j+1}}p_1(\nu_1)}. \qquad (15.29)$$

The quantities $R_{\nu_j\nu_{j+1}}$ may be considered as matrix elements of the matrix $\mathbf{R}$ and the quantities $p_1(\nu_1)$ and $p_s(\nu_s)$ as components of the vectors $\mathbf{p}_1$ and $\mathbf{p}_s$. Then Eq. (15.29) becomes

$$\mathbf{p}_s=\frac{\mathbf{R}^{s-1}\mathbf{p}_1}{\sum\limits_{\nu_s}(\mathbf{R}^{s-1}\mathbf{p}_1)_{\nu_s}}=\frac{\mathbf{R}^{s-1}\mathbf{p}_1}{\mathbf{I}\mathbf{R}^{s-1}\mathbf{p}_1}, \qquad (15.30)$$

where $\mathbf{I}$ is a vector with components equal to unity for each ν. Vector $\mathbf{p}_1$ may be expanded in eigenvectors ψ_k of a matrix $\mathbf{R}$

$$\mathbf{p}_1=\sum_k C_k\psi_k. \qquad (15.31)$$

The vectors ψ_k are determined from the equation $\mathbf{R}\psi_k=\alpha_k\psi_k$. Substituting the expansion (15.31) into (15.30) we obtain

$$\mathbf{p}_s=\frac{\sum\limits_k C_k\alpha_k^{s-1}\psi_k}{\sum\limits_k C_k\alpha_k^{s-1}\mathbf{I}\psi_k}. \qquad (15.32)$$

If we divide the numerator and denominator of (15.32) by the maximum eigenvalue of the matrix R to the power $s-1$, it is easy to show that in the limit when $s \to \infty$ only terms corresponding to this maximum value α_{max} may be retained in the sums in (15.32). If α_{max} is a nondegenerate eigenvalue,

$$\lim_{s \to \infty} \mathbf{p}_s = \frac{\psi_{max}}{\mathbf{I}\psi_{max}} = \frac{\psi_{max}}{\sum_{\nu} (\psi_{max})_{\nu}}, \tag{15.33}$$

where ψ_{max} is the eigenvector of matrix R, corresponding to the eigenvalue α_{max}. If, however, α_{max} is degenerate (or asymptotically degenerate as $s \to \infty$, when the number of sites in the layer also tends to infinity), several terms remain in the sums occurring in (15.32), and

$$\lim_{s \to \infty} \mathbf{p}_s = \frac{\sum_{i=1}^{r} C_i \psi_{max\, i}}{\sum_{i=1}^{r} C_i \mathbf{I} \psi_{max\, i}}, \tag{15.34}$$

where C_i are coefficients in the expansion (15.31) for vectors $\psi_{max\, i}$, corresponding to α_{max}, and r is the degree of the degeneracy. According to Eq. (15.31) the probabilities of different configurations in the first layer (which are components of the vector $\mathbf{p}_1$) are given by the coefficients C_k. It is seen from (15.33) that in the case when the maximum eigenvalue α_{max} is not degenerate, when $s \to \infty$ the quantities C_k, in general, come from an expression for the limiting value of $\mathbf{p}_s$. Thus, the probabilities of different configurations in the sth layer are found to be independent of the configurations at the first layer, which is infinitely far from it. This means that there is no correlation between the replacement of the sites by atoms in these layers, i.e., the degree of long-range order in the alloy equals zero, and there is only short-range order. If, however, the maximum eigenvalue α_{max} is degenerate, according to (15.34) even when $s \to \infty$ the probabilities of different configurations in the sth layer still depend on C_i, namely, the probability of the configurations in the first layer. This means that in this case correlation exists at infinity, i.e., the alloy has long-range as well as short-range order. This relationship between the degeneracy of the maximum eigenvalue of matrix R and the presence of long-range order in the alloy has been established [179].

In the case in which the crystal lattice is infinite only in one direction, i.e., when the number of sites q in the layer remains finite as $s \to \infty$, the matrix R has an infinite number of rows and columns. Since it follows from the definition of this matrix (15.28) that its matrix elements $R_{\nu_j \nu_{j+1}}$ are always greater than zero, using the Frobenius theorem one may show that the maximum eigenvalue of matrix R in this case is not degenerate. Consequently, a state with long-range order is impossible in such lattices. Therefore, according to the foregoing, long-range order can not occur in a linear chain of atoms (where $q=1$).

If, however, the crystal is infinite in two or three dimensions, i.e., if the number of atoms q in this layer simultaneously tends to infinity when the number of layers s tends to infinity, the number of rows and columns of the matrix R also tends to infinity. The Frobenius theorem is inapplicable in this case, and the maximum eigenvalue α_{max} may turn out to be asymptotically degenerate (for $q \to \infty$). In such alloys, it can be shown that the degeneracy of α_{max} occurs only at sufficiently low temperatures, when $T < T_0$. Thus, correlation exists in the alloy at infinity, i.e., there exists long-range order which disappears with an increase of temperature at the point $T = T_0$.

In order to explain why a transition to the ordered state is impossible in a one-dimensional case, we note that the existence of even a small number of energetically unfavorable pairs of atoms (for ordered alloys, pairs of atoms of a single type) may lead to the appearance of arbitrarily large antiphase domains. Fig. 91 depicts, for instance, two states of a one-dimensional chain of the alloy AB. The first state is perfectly ordered and has no neighboring AA or BB pairs, while the second state differs from the first by the presence of only two such pairs, as a consequence of which there emerges an arbitrarily long antiphase domain, lying between the dashed lines. Therefore, the appearance of fluctuations due to the creation of a negligibly small numbers of pairs AA or AB leads to the destruction of long-range order. By contrast, in two-dimensional and three-dimensional lattices the formation of a sufficiently large antiphase domain requires the presence of a large numbers of pairs AA and BB, located on the boundary of this domain. Therefore, the destruction of long-range order is made difficult,

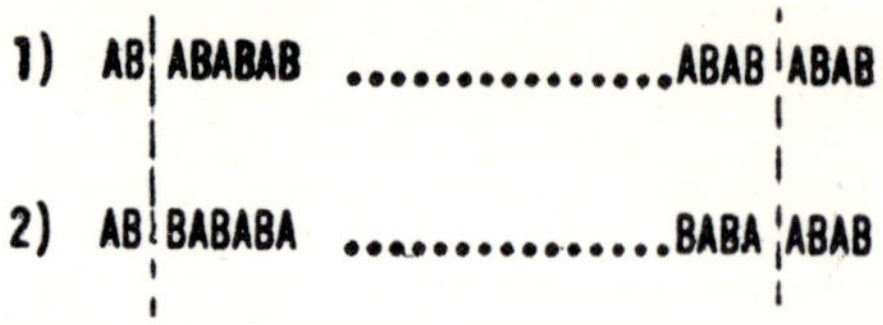

Fig. 91. Antiphase domains in a linear chain.

and, as we have indicated, occurs only at the finite temperature $T = T_0$. The ratio of the number of atoms at the boundary of the domain to their number inside is different in two-dimensional and three-dimensional crystals, which may lead to the existence of different features of the ordering in these cases.

Ordering of a two-dimensional lattice was analyzed exactly by Onsager [180] in the model under consideration. Using the matrix method, he successfully analyzed the case in which the quantity m occurring in Eq. (15.6) for the partition function is equal to zero (which corresponds to an alloy with an equal number of atoms A and B). This problem has been solved by Kaufman [181], who used a simpler method based on spinor analysis. Another type of calculation of the partition function was proposed by Kac and Ward [182], who used the combinatorial method, based on a consideration of the different closed graphs drawn by connecting lattice sites.

Since calculation of the partition function by these methods for a two-dimensional lattice is cumbersome, we present only the final results. The grand partition function Ξ for a square two-dimensional lattice of an alloy, infinite in two dimensions (in which the total number of atoms N tends to infinity) with the same number of atoms A and B, is defined by the formula

$$\lim_{N\to\infty} \frac{1}{N} \ln \Xi = \ln\left(2\cosh\frac{w}{2kT}\right) + \frac{1}{2\pi^2}\int_0^\pi\int_0^\pi \ln\left[1 + \frac{\varkappa}{2}(\cos x + \cos y)\right] dx\, dy, \tag{15.35}$$

where

$$\varkappa = \frac{2\sinh\frac{w}{2kT}}{\cosh^2\frac{w}{2kT}}. \tag{15.36}$$

Having determined the mean thermodynamic energy of the

crystal per atom, $\frac{E}{N}$, from the formula $E = kT^2 \frac{\partial \ln \Xi}{\partial T}$, we may find

$$\frac{E}{N} = -\frac{w}{4}\left[1 + \frac{2}{\pi}\left(2\tanh^2\frac{w}{2kT} - 1\right)K(\varkappa)\right]\coth\frac{w}{2kT}, \quad (15.37)$$

where $K(\varkappa)$ is a complete elliptic integral of the first kind

$$K(\varkappa) = \int_0^{\pi/2} \frac{d\varphi}{\sqrt{1 - \varkappa^2 \sin^2\varphi}}. \quad (15.38)$$

This elliptic integral has a logarithmic singularity at $\varkappa = 1$. Therefore, E and, hence, also the thermodynamic potential Φ have singularities at a certain temperature T_0, for which $\varkappa$, defined by Eq. (15.36), becomes unity, and thus Φ no longer is an analytic function of temperature. Below this temperature, the maximum eigenvalue of matrix **R** is asymptotically degenerate, and the crystal in equilibrium is in an ordered state. When $T > T_0$, the degeneracy is removed and ordering disappears. Thus, the temperature T_0 is the order-disorder phase transition temperature. Setting (15.36) equal to unity, we obtain an equation for determination of the transition temperature T_0

$$\sinh\frac{w}{2kT_0} = 1, \quad (15.39)$$

whence

$$\frac{w}{kT_0} = 1.763, \quad (15.40)$$

i.e., T_0 is proportional to ordering energy w.

It follows from (15.39) that the coefficient of $K(\varkappa)$ in parentheses in (15.37) approaches zero at $T = T_0$, as the first power of the difference $T - T_0$. Therefore, the corresponding term in the energy $\frac{E}{N}$ near T_0 is proportional to $|T - T_0| \ln |T - T_0|$. Thus, the energy at the transition point changes continuously, and no heat is liberated or absorbed.

This term introduces in the expression for the specific heat $C = \frac{\partial E}{\partial T}$ a term proportional to $\ln|T - T_0|$, i.e., as we approach the point T_0 both from high and low temperatures the specific

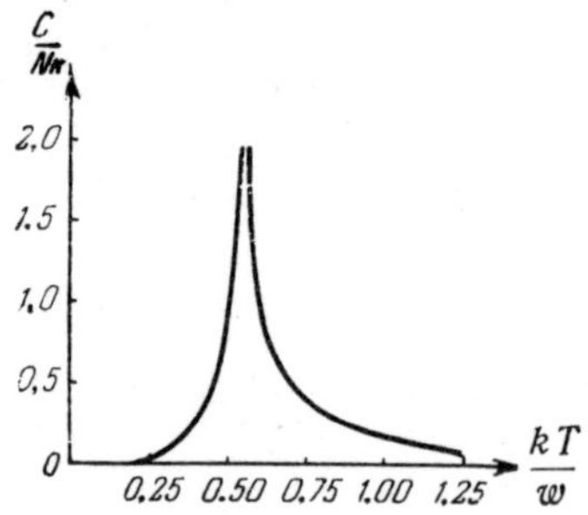

Fig. 92. Temperature dependence of specific heat of a two-dimensional square lattice.

heat tends to infinity logarithmically. The temperature dependence of specific heat, calculated on the basis of (15.37), is illustrated in Fig. 92.

Yang's calculations [183] show that the temperature dependence of the degree of long-range order η in a two-dimensional square lattice of the alloy with an equal number of atoms A and B is given by the formula

$$\eta = \frac{\left(1 + e^{-\frac{w}{kT}}\right)^{1/4}\left(1 - 6e^{-\frac{w}{kT}} + e^{-\frac{2w}{kT}}\right)^{1/8}}{\left(1 - e^{-\frac{w}{kT}}\right)^{1/2}}. \qquad (15.41)$$

It is apparent from (15.41) that for $T = 0$ $\eta = 1$ and for $T = T_0$, on the basis of (15.39), η becomes zero. Near the critical point $T = T_0$ (when $T < T_0$) the degree of long-range order varies with temperature according to the law

$$\eta \sim (T_0 - T)^{1/8}. \qquad (15.42)$$

The results obtained for the temperature dependence of the degree of long-range order and of the specific heat near the point $T = T_0$ differ essentially from those obtained in the thermodynamic theory. As already mentioned, the results obtained for two-dimensional lattices may differ qualitatively from those found for three-dimensional lattices. It has been shown in Sec. 11 that experiment gives for the dependence of η on T near the transition point T_0 the formula $\eta \sim (T_0 - T)^{1/2}$ (see Fig. 88) and corroborates the thermodynamic theory for three-dimensional crystals.

The formalism developed in this section may be applied not only to the problem of ordering of alloys, but also to the problem of thermomagnetism in an analogous simplified model proposed by Ising [184]. In this model the magnetization of the crystal plays the role of the degree of long-range order, and instead of the substitution of atoms A and B along the sites of the lattice we consider the different configurations of "right" and "left" spins.

16. Theory of Ordering, Neglecting Correlation in the Alloy (theory of Gorsky, Bragg and Williams)

As already mentioned, for three-dimensional crystals no one has succeeded up to the present time in accurately calculating the partition function in the simplified model of the alloy described in Sec. 14. Such calculations can be successfully performed by means of different approximate methods. In this section we present the simplest theory of ordering, which was first developed by Gorsky [185] and improved by Bragg and Williams [84, 186, 187].

The solution of the problem of ordering reduces to the calculation of the configuration factor of the partition function Z_k from Eqs. (14.3) and (14.9) for an alloy with given numbers N_A and N_B of atoms A and B. The expression for Z_k may be represented in the form

$$Z_k = \sum_{\eta} Z_{\eta}, \tag{16.1}$$

where

$$Z_{\eta} = \sum_{i}^{(\eta)} e^{-\frac{E_i}{kT}} \tag{16.2}$$

denotes a sum over all configurations of atoms on the lattice sites for a given value of the degree of long-range order η, and the summation in (16.1) extends over all values of η. It follows from the general principles of statistical mechanics, that in the limit of an infinitely large crystal, a vanishingly small error is made, upon replacement of $\ln Z_k$ by $\ln Z_{\bar{\eta}}$, where $\bar{\eta}$ – is the value of the degree of long-range order, which makes Z_{η} a maximum at a given temperature. Thus $\bar{\eta}$ is the equilibrium value of the degree of long-range order. Therefore, in the following, we shall calculate Z_η for different values of η, and then determine the η for which this value will be a maximum. Since the configurational free energy is related to Z_k by $F_k = -kT \ln Z_k$, the problem of finding the equilibrium value $\eta = \bar{\eta}$ reduces to the determination of the value of η from the condition that the free energy be a minimum, $\frac{\partial F_k}{\partial \eta} = 0$. Since in the following we shall investigate only the configurational parts of thermodynamic quantities (in this approximation they depend only on η, and $\frac{\partial F_0}{\partial \eta} = 0$, i.e., $\frac{\partial F}{\partial \eta} = \frac{\partial F_k}{\partial \eta}$), hereafter the subscript in F_k will be omitted.

The approximation adopted in this section consists of the fact that the energies E_i of different configurations, corresponding to a given value of the degree of long-range order, are considered to be equal and are calculated on the assumption of random distribution of atoms on sites of each type. This means that the total number N_{AB} of pairs AB, which according to (14.9) determine E_i, is found by assuming that the probabilities of replacement of a given site by atoms A and B are independent of the configuration of the atoms on the neighboring sites and are equal to the a priori probabilities (1.5), i.e., they are the same for all sites of a given type. Thus, in this approximation, we neglect correlation in the alloy.

Let us consider a binary alloy A—B, having the same number $\frac{N}{2}$ of sites of the first and second type, where each site of a given type is surrounded only by sites of the other type. Such structures occur, for example, in β-brass type alloys (coordination number $z=8$), NaCl type crystals ($z=6$), a two-dimensional square lattice ($z=4$), and Fe_3Al type alloys ($z=6$), if we consider only the sublattices b and d (see Fig. 8), in which ordering grows strongly. The quantity N_{AB}, which is equal to the total number of pairs of the atoms in which atom A is at the first type of site and atom B at the second type of site, for these alloys is in this approximation equal to

$$N_{AB} = \frac{zN}{2}\left(p_A^{(1)} p_B^{(2)} + p_A^{(2)} p_B^{(1)}\right). \tag{16.3}$$

Substituting this expression into (14.9) and using (15.1) (where one must put $\nu = \frac{1}{2}$), we find the energy E_i in the nearest-neighbor approximation

$$E_i = -\frac{zN}{2}\left[c_A v_{AA} + c_B v_{BB} + w\left(c_A c_B + \frac{\eta^2}{4}\right)\right] \tag{16.4}$$

where $c_B = 1 - c_A$. In Eq. (16.2), all terms $e^{-\frac{E_i}{kT}}$, corresponding to different configurations of atoms on the lattice sites, are equal in the nearest-neighbor approximation. The total number of terms equals the number of different permutations W of atoms A and B on the first and second types of sites for a given value of the degree of long-range order in an alloy of the given composition. To determine W it is necessary to compute the number of such permutations on sites of each type for given values of the numbers of atoms N_A, N_B (which are determined by assignment of the composition and of the degree of long-range order) and then multiply together the results

$$W = \frac{\left(\frac{N}{2}\right)!}{N_A^{(1)}!\, N_B^{(1)}!}\ \frac{\left(\frac{N}{2}\right)!}{N_A^{(2)}!\, N_B^{(2)}!}. \tag{16.5}$$

Thus, the free energy in this case is equal to

$$F = -kT \ln Z_k = E_i - kT \ln W. \tag{16.6}$$

Substituting Eqs. (16.4) and (16.5) into (16.6), using Stirling's formula $X! = X(\ln X - 1)$ (valid for large values of X) and taking into account (1.1) and (1.5), we obtain the following expression for the free energy:

$$\begin{aligned} F = &-\frac{zN}{2}\left[c_A v_{AA} + c_B v_{BB} + w\left(c_A c_B + \frac{\eta^2}{4}\right)\right] + \\ &+\frac{N}{2} kT\left[\left(c_A + \frac{1}{2}\eta\right)\ln\left(c_A + \frac{1}{2}\eta\right) + \left(c_A - \frac{1}{2}\eta\right)\ln\left(c_A - \frac{1}{2}\eta\right) + \right. \\ &\left. + \left(c_B - \frac{1}{2}\eta\right)\ln\left(c_B - \frac{1}{2}\eta\right) + \left(c_B + \frac{1}{2}\eta\right)\ln\left(c_B + \frac{1}{2}\eta\right)\right]. \end{aligned} \tag{16.7}$$

Differentiating this expression with respect to η, and using the condition $\frac{\partial F}{\partial \eta} = 0$, we find an equation for the equilibrium value of the degree of long-range order ($\bar{\eta}$ is denoted by η):

$$\ln \frac{\left(c_A + \frac{1}{2}\eta\right)\left(c_B + \frac{1}{2}\eta\right)}{\left(c_A - \frac{1}{2}\eta\right)\left(c_B - \frac{1}{2}\eta\right)} = z\,\frac{w}{kT}\,\eta. \tag{16.8}$$

In particular, for the stoichiometric alloy AB, when $c_A = c_B = \frac{1}{2}$

$$\ln\frac{1+\eta}{1-\eta}=\frac{z}{2}\frac{w}{kT}\eta. \tag{16.9}$$

To determine the order-disorder transition temperature, we expand the left-hand side of (16.8) in powers of η. Then this equation for small η becomes

$$\frac{1}{c_Ac_B}\eta+\frac{1}{12}\left(\frac{1}{c_A^3}+\frac{1}{c_B^3}\right)\eta^3=z\frac{w}{kT}\eta. \tag{16.10}$$

As the critical point is approached, the term proportional to η^3 becomes less and less significant, and a necessary condition for the fulfillment of (16.10) is that the critical temperature T_0 equals

$$T_0(c_A)=zc_A(1-c_A)\frac{w}{k}. \tag{16.11}$$

It follows from (16.7) and (16.11) that for $T=T_0$ and $\eta=0$ not only does $\frac{\partial F}{\partial\eta}=0$, but the second derivative $\frac{\partial^2 F}{\partial\eta^2}$ also vanishes, as should be the case at a second-order phase transition point (see Sec. 11). It is apparent from (16.8) and (16.11) that η is a function only of c_A and of the ratio $\frac{T}{T_0}$, determined by the equation

$$\ln\frac{\left(c_A+\frac{1}{2}\eta\right)\left(c_B+\frac{1}{2}\eta\right)}{\left(c_A-\frac{1}{2}\eta\right)\left(c_B-\frac{1}{2}\eta\right)}=\frac{\eta}{c_Ac_B}\frac{T_0}{T}, \tag{16.12}$$

from which w and z have been eliminated. To illustrate the influence of alloy composition on the temperature dependence of the degree of long-range order, Fig. 93 gives graphs of the dependences of η on the ratio of the absolute temperatures $\frac{T}{T_0\left(\frac{1}{2}\right)}$ for different c_A, plotted from (16.12). Figure 94 (curve a) illustrates the dependence of T_0 on c_A in accordance with (16.11).

The mean thermodynamic energy of an alloy, defined by the Gibbs-Helmholtz relation

$$E=F-T\frac{\partial F}{\partial T}, \tag{16.13}$$

in this approximation, evidently, agrees with the energy E_i

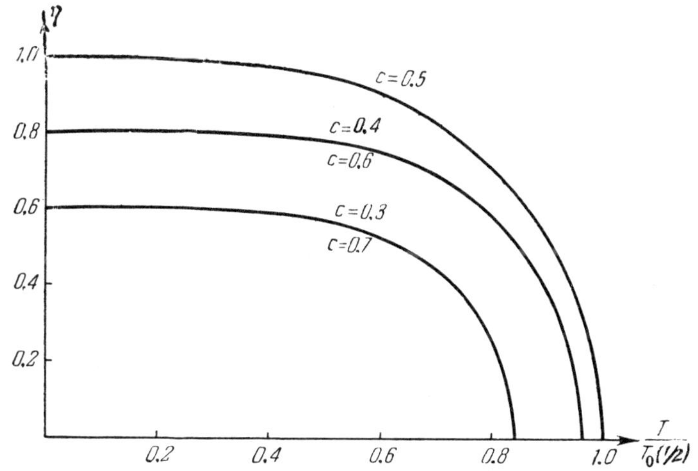

Fig. 93. The temperature dependence of the degree of long-range order of alloys of different compositions $\left(\text{for } \nu = \frac{1}{2}\right)$ (Gorsky-Bragg-Williams theory).

that is obtained upon substituting the equilibrium value η into (16.4). According to this formula, the energy of a completely ordered alloy is found to be less than the energy of a completely

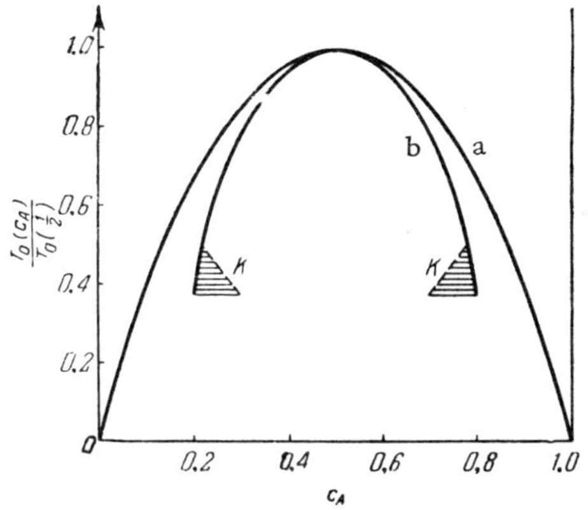

Fig. 94. Dependence of the critical temperature on alloy composition $\left(\text{for } \nu = \frac{1}{2}\right)$. a-Gorsky-Bragg-Williams theory, b-Kirkwood theory, including terms with $\left(\frac{w}{kT}\right)^2$ (for $z=8$), K-critical points.

disordered alloy by the amount

$$\delta E = \frac{zN}{8} w. \tag{16.14}$$

The configurational part of the entropy of the alloy $S = -\frac{\partial F}{\partial T}$ is equal to

$$S = -\frac{Nk}{2}\Big[\Big(c_A + \frac{1}{2}\eta\Big)\ln\Big(c_A + \frac{1}{2}\eta\Big) + \Big(c_A - \frac{1}{2}\eta\Big)\ln\Big(c_A - \frac{1}{2}\eta\Big) + \\ + \Big(c_B - \frac{1}{2}\eta\Big)\ln\Big(c_B - \frac{1}{2}\eta\Big) + \Big(c_B + \frac{1}{2}\eta\Big)\ln\Big(c_B + \frac{1}{2}\eta\Big)\Big]. \tag{16.15}$$

Here when differentiating the free energy with respect to temperature, we have neglected dependence of η on T, as $\frac{\partial F}{\partial \eta} = 0$.

The configurational part of the specific heat of the alloy, according to (16.4), is

$$C = -\frac{zN}{8} w \frac{d\eta^2}{dT}. \tag{16.16}$$

Near the critical point T_0 for $T < T_0$ the derivative $\frac{d\eta^2}{dT}$ may be determined with the aid of Eqs. (16.10) and (16.11)

$$\frac{d\eta^2}{dT}\Big|_{T=T_0-0} = -\frac{12zc_A^3 c_B^3}{c_A^3 + c_B^3}\,\frac{w}{kT_0^2} = -\frac{12c_A^2 c_B^2}{c_A^3 + c_B^3}\,\frac{1}{T_0}. \tag{16.17}$$

Above the temperature T_0, the derivative $\frac{d\eta^2}{dT}$ evidently equals zero. It follows therefore from (16.16), (16.17) and (16.11) that, during the transition into the ordered state, the specific heat at the critical point increases discontinuously by the amount

$$\Delta C = C|_{T_0-0} - C|_{T_0+0} = \frac{3}{2} zN \frac{w}{T_0} \frac{c_A^2 c_B^2}{{}^3_A + c_B^3} = \frac{3}{2} kN \frac{c_A c_B}{c_A^3 + c_B^3}. \tag{16.18}$$

In particular, for a stoichiometric alloy

$$\Delta C = \frac{3}{2} kN, \quad \Big(c_A = \frac{1}{2}\Big). \tag{16.19}$$

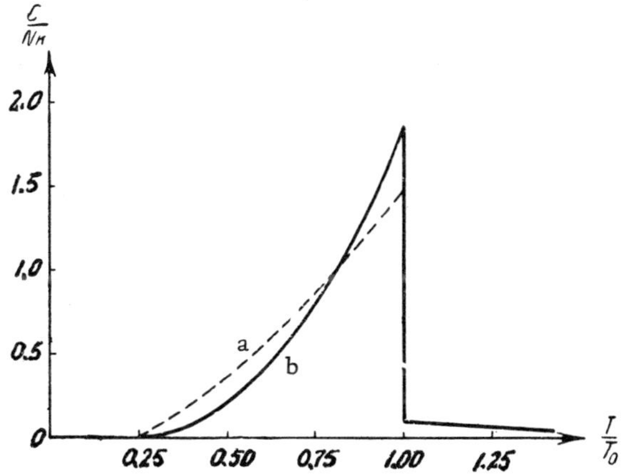

Fig. 95. Temperature dependence of the configurational part of the specific heat of alloys AB. a-Gorsky-Bragg-Williams theory, b-quasi-chemical method (for $z=6$).

A graph of the temperature dependence of specific heat for stoichiometric alloys is shown in Fig. 95 (curve a). Above the transition temperature, the configurational part of the specific heat is in this approximation equal to zero.

It is apparent from these results that in the Gorsky-Bragg-Williams theory, for the structures listed above, the entropy changes continuously at the transition point [Eqs. (16.15) and (16.18)], i.e., there is no heat of transition, but the specific heat undergoes a finite jump. Thus, the order-disorder transition occurring here is a second-order phase transition.

To illustrate the solution of the problem of ordering of alloys in which the number of sites of the first and second type are unlike, let us consider a type $AuCu_3$ face-centered cubic lattice. In this lattice the number of sites of the first type is one-third as large as those of the second type $\left(\nu=\frac{1}{4}\right)$. Each site of the first type is surrounded by the twelve neighboring sites of the second type, and a site of the second type has among its nearest neighbors four sites of the first type and eight of the second type. The number of pairs AB in this case is

$$N_{AB}=3N\left(p_A^{(1)}p_B^{(2)}+p_A^{(2)}p_B^{(1)}+2p_A^{(2)}p_B^{(2)}\right). \qquad (16.20)$$

Taking into account (14.9) and (1.5), the energy E_i becomes

$$E_i=-6N\left[c_A v_{AA}+c_B v_{BB}+w\left(c_A c_B+\frac{1}{16}\eta^2\right)\right]. \qquad (16.21)$$

The number of different permutations of atoms on the lattice sites for a given η in alloys with a face-centered cubic lattice is equal to

$$W = \frac{\left(\frac{N}{4}\right)!}{N_A^{(1)}!\,N_B^{(1)}!}\;\frac{\left(\frac{3N}{4}\right)!}{N_A^{(2)}!\,N_B^{(2)}!}. \tag{16.22}$$

The free energy $F = E_i - kT\ln W$ therefore takes the form

$$F = -6N\left[c_A v_{AA} + c_B v_{BB} + w\left(c_A c_B + \frac{1}{16}\eta^2\right)\right] + \\ + \frac{kTN}{4}\left[\left(c_A + \frac{3}{4}\eta\right)\ln\left(c_A + \frac{3}{4}\eta\right) + \left(c_B - \frac{3}{4}\eta\right)\ln\left(c_B - \frac{3}{4}\eta\right) + \\ + 3\left(c_A - \frac{1}{4}\eta\right)\ln\left(c_A - \frac{1}{4}\eta\right) + 3\left(c_B + \frac{1}{4}\eta\right)\ln\left(c_B + \frac{1}{4}\eta\right)\right]. \tag{16.23}$$

Hence from the condition $\frac{\partial F}{\partial \eta} = 0$ we obtain an equation for determination of the degree of long-range order [188]

$$\ln\frac{\left(c_A + \frac{3}{4}\eta\right)\left(c_B + \frac{1}{4}\eta\right)}{\left(c_A - \frac{1}{4}\eta\right)\left(c_B - \frac{3}{4}\eta\right)} = 4\,\frac{w}{kT}\,\eta. \tag{16.24}$$

In the special case of a stoichiometric alloy, when $c_A = \frac{1}{4}$, Eq. (16.24) takes the form

$$\ln\frac{(1+3\eta)(3+\eta)}{3(1-\eta)^2} = 4\,\frac{w}{kT}\,\eta. \tag{16.25}$$

A graph of the dependence of the solution of this equation η on $\frac{kT}{w}$ is shown in Fig. 96. This equation has a root $\eta = 0$ for any temperature. At sufficiently low temperatures, in addition to this root, there appear nonvanishing roots of the equation. In order to find the solution that corresponds to the equilibrium value of the degree of long-range order at a given temperature, it is necessary to investigate the dependence of the free energy on the degree of long-range order at this temperature. According to Eqs. (15.23) and (16.7), the family of curves $F(\eta)$ at different temperatures, plotted for alloys with a face-centered

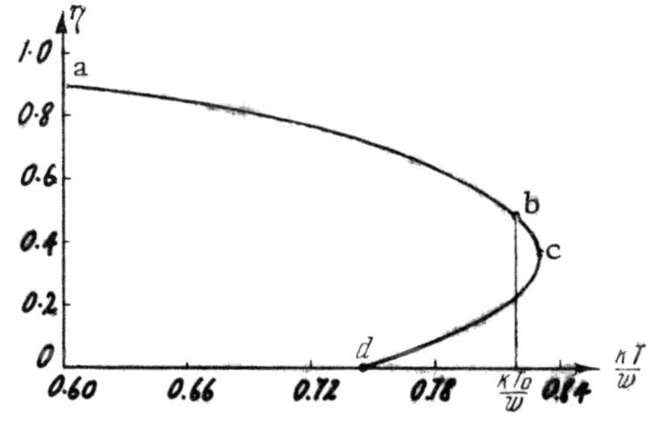

Fig. 96. Temperature dependence of the degree of long-range order in the alloy AB_3 with a face-centered cubic lattice (Gorsky-Bragg-Williams theory).

cubic lattice and for alloys with the structures considered at the beginning of the section, differ appreciably. At high temperatures in both cases the curves $F(\eta)$ have a single minimum at $\eta = 0$ (Figs. 97a and b [4, 189]). As the temperature is decreased below some temperature T_0 in the second case (see Fig. 97a) the minimum of the curve $F(\eta)$ is shifted continuously to larger values of η, and the point $\eta = 0$ becomes a maximum

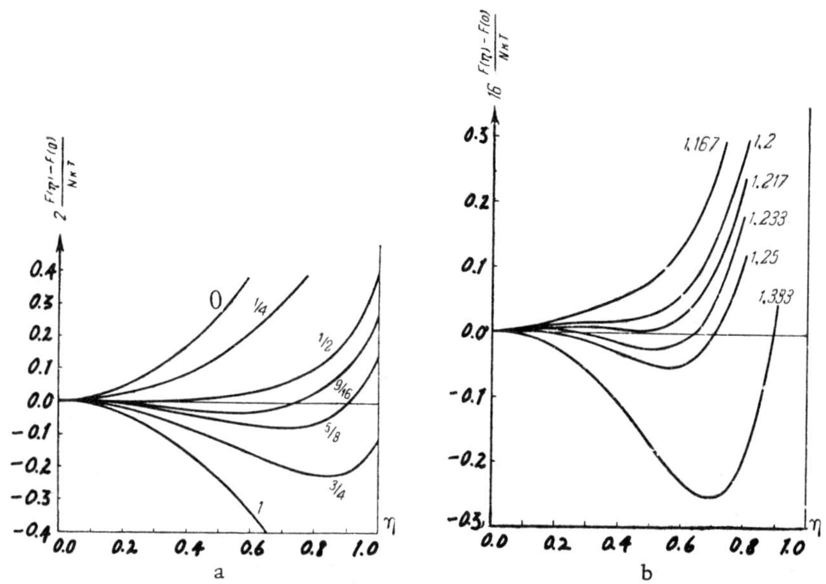

Fig. 97. Dependence of the configurational part of the free energy on the degree of long-range order (Gorsky-Bragg-Williams theory).
a–β-brass type alloys, b–AuCu3 type alloys. The numbers on the curve indicate the ratio $\frac{w}{kT}$.

point of F. Such a pattern corresponds to a continuous increase of η from zero with a decrease of temperature, i.e. a second-order phase transition. Conversely, in alloys with a face-centered cubic lattice, as may be seen from Fig. 97b, as the temperature is lowered a minimum appears at a nonzero value of η, the point $\eta=0$ as before is a minimum, and the maximum lies between these two minima. Thus, in the case of equilibrium there may exist both an ordered and a disordered phase, but one of them (corresponding to a large value of F at the minimum point) is found to be in a metastable state. A second minimum appears at once at an η which differs from zero by a finite magnitude. This behavior describes a first-order phase transition. This transition is realized when both minima are at the same level.

The temperature of the first-order phase transition is determined from the condition of equilibrium of the free energy $F(0)$ at $\eta=0$ and a $F(\eta_0)$ corresponding to a nonzero value of the degree of long-range order η_0, which is formed in the alloy directly below the transition temperature (and also from the condition $\frac{\partial F}{\partial \eta}=0$). A numerical solution of the system of equations

$$F(0)=F(\eta_0), \quad \left.\frac{\partial F}{\partial \eta}\right|_{\eta=\eta_0}=0 \tag{16.26}$$

gives

$$kT_0=0.82w, \quad \eta_0=0.46. \tag{16.27}$$

It follows from (16.21) and (16.27) that the energy of the alloy at the transition point changes discontinuously by the amount

$$E(0)-E(\eta_0)=0.098\,NkT_0, \tag{16.28}$$

i.e. there is a heat of transition. In the curve $\eta(T)$ (Fig. 96) the section a-b corresponds to the equilibrium values of η, which are realized when $T<T_0$. Section b-c corresponds [4] to metastable states, and the values η over the range c-d correspond to the maximum of F and can not be realized.

It follows from (16.21) that the configurational part of the specific heat of an alloy with a $AuCu_3$ type crystal lattice (for $c_A=\frac{1}{4}$) is

$$C=\frac{dE}{dT}=-\frac{3}{8}Nw\frac{d\eta^2}{dT}. \tag{16.29}$$

For $T = T_0 - 0$, as follows from (16.25) and (16.27), $\frac{d\eta^2}{dT} = -\frac{5.4}{T_0}$ and its specific heat is equal to

$$C\,|_{T_0 - 0} \approx 2.5Nk. \tag{16.30}$$

In this approximation $C\,|_{T_0+0} = 0$ and the magnitude (16.30) also equals the change in specific heat at the critical point.

Although the theory of ordering which neglects correlation in the alloy correctly predicts many of the previously mentioned features of an order-disorder transition, in some cases it gives qualitatively incorrect results. Thus, for alloys having the same number of sites of the first and second type, the phase diagram has the form illustrated in Fig. 94 (curve a). If we use for the free energy the equation (16.7) obtained in this approximation, there is no value of c_A that will satisfy Eq. (11.29), which determines the point on the phase diagram at which the curve of the transition points intersects the decomposition curve. Consequently, in this approximation, an alloy of any composition will not be decomposed at absolute zero. According to Eq. (16.15) the entropy of a nonstoichiometric alloy does not tend to zero when $T \to 0$, which contradicts Nernst's theorem.

The free energy of alloys with an $AuCu_3$ type crystal lattice in this approximation, which neglects the slight tetragonal character of this lattice, has the form

$$F = -6N\,(c_A v_{AA} + c_B v_{BB} + c_A c_B w) - \frac{1}{2} N w \eta^2 +$$
$$+ \frac{N}{2} kT \left[\left(c_A + \frac{1}{2}\eta\right) \ln\left(c_A + \frac{1}{2}\eta\right) + \left(c_A - \frac{1}{2}\eta\right) \ln\left(c_A - \frac{1}{2}\eta\right) \right.$$
$$\left. + \left(c_B - \frac{1}{2}\eta\right) \ln\left(c_B - \frac{1}{2}\eta\right) + \left(c_B + \frac{1}{2}\eta\right) \ln\left(c_B + \frac{1}{2}\eta\right) \right]. \tag{16.31}$$

This expression is analogous in form to (16.7). It cannot result in a first-order transition, and necessarily leads to a second-order phase transition. But, according to the thermodynamic theory (and in agreement with experiment), only first-order phase transitions are possible in alloys with a AuCu type crystal lattice (see Sec. 11). In this case, the Gorsky-Bragg-Williams approximation which neglects correlation also gives a qualitatively erroneous result. The Gorsky-Bragg-Williams approximation leads also to a vanishing configurational part of the specific heat above the critical temperature, i.e. it does not enable one to allow for the specific heat due to a change in the short-range order in the alloy.

More accurate results, which are in qualitative agreement with experiment, may be obtained by statistical theories, which in some approximations consider the correlation in an alloy. Such theories will be considered in the next section.

17. Kirkwood's Method Based on the Expansion of the Free Energy as a Power Series in $\frac{w}{kT}$

One of the methods for determination of the free energy, which accounts for correlation in an alloy, is the evaluation of F as a power series in the ratio of the ordering energy w to kT. This method was proposed by Kirkwood [190]. Only a few of the first terms of the expansion can be calculated with the aid of this method. Hence the expression obtained for F is sufficiently accurate at high temperatures, but at low temperatures it is a poorer approximation to the exact expression for free energy.

To find the expansion of the partition function of the binary alloy A—B in powers of $\frac{w}{kT}$, we write the expression for Z_η with the aid of Eqs. (16.2), (14.9) and (14.7) in the form

$$Z_\eta = \exp\left\{\frac{z}{2kT}\left[N_A(2v_{AB} - v_{BB}) + N_B v_{BB}\right]\right\} \sum_i^{(\eta)} e^{-\frac{w}{kT}N_{AA}}. \qquad (17.1)$$

The total number of terms in the sum (17.1) is equal to the number of different permutations W of A and B atoms on the sites of the first and second types for a given value of η. Expanding this sum in powers of $\frac{w}{kT}$ and replacing the sums of the various powers of the exponent $-\frac{w}{kT}N_{AA}$ by mean values multiplied by the total number of terms W, we obtain

$$Z_\eta = \exp\left\{\frac{z}{2kT}\left[N_A(2v_{AB} - v_{BB}) + N_B v_{BB}\right]\right\} W \times$$
$$\times \left[1 - \frac{w}{kT}\tilde{N}_{AA} + \frac{1}{2!}\left(\frac{w}{kT}\right)^2 \widetilde{N_{AA}^2} - \frac{1}{3!}\left(\frac{w}{kT}\right)^3 \widetilde{N_{AA}^3} + \ldots\right]. \qquad (17.2)$$

Here the averaging extends over all configurations that are

possible for a given η, where each configuration corresponds to the same statistical weight, i.e.

$$\widetilde{N_{AA}^{n}} = \frac{\sum_{i}^{(\eta)} N_{AA}^{n}}{W}. \tag{17.3}$$

The partition function is related to the free energy by the expression

$$Z_{\eta} = e^{-\frac{F}{kT}}. \tag{17.4}$$

The free energy may also be expanded in a power series in $\frac{w}{kT}$,

$$F = -\frac{zN}{2}\left[c_A\left(2v_{AB} - v_{BB}\right) + c_B v_{BB}\right] - kT\ln W + kT\left[\lambda_0 + \lambda_1\frac{w}{kT} + \frac{\lambda_2}{2!}\left(\frac{w}{kT}\right)^2 + \frac{\lambda_3}{3!}\left(\frac{w}{kT}\right)^3 + \dots\right], \tag{17.5}$$

where the first two terms are written out for comparison with (17.2). Substituting (17.5) into (17.4), expanding the resulting expression in powers of $\frac{w}{kT}$, and comparing the coefficients of the different powers of $\frac{w}{kT}$ with the corresponding coefficients in (17.2), we find in succession the expressions for the magnitude of λ_n in terms of the mean values of the different powers of the number of pairs N_{AA}:

$$\lambda_0 = 0, \tag{17.6}$$

$$\lambda_1 = \widetilde{N}_{AA}, \tag{17.7}$$

$$\lambda_2 = -\widetilde{N_{AA}^2} + (\widetilde{N}_{AA})^2, \tag{17.8}$$

$$\lambda_3 = \widetilde{N_{AA}^3} - 3\widetilde{N_{AA}^2 N_{AA}} + 2(\widetilde{N}_{AA})^3, \tag{17.9}$$

$$\lambda_4 = -\widetilde{N_{AA}^4} + 4\widetilde{N_{AA}^3}\widetilde{N}_{AA} + 3(\widetilde{N_{AA}^2})^2 - 12\widetilde{N_{AA}^2}(\widetilde{N}_{AA})^2 + 6(\widetilde{N}_{AA})^4 \tag{17.10}$$

Thus, the problem of calculating the various expansion coefficients of F reduces to a determination of the mean values of the various powers of N_{AA}. We shall first calculate these mean values for alloys with the same number of sites of the first and second types, where each site is surrounded only by sites of the other type [190–194]. On the basis of

Eq. (16.5) and Stirling's formula the magnitude of $\ln W$ in this case equals

$$\ln W = -\frac{N}{2}\left[p_A^{(1)} \ln p_A^{(1)} + p_B^{(1)} \ln p_B^{(1)} + p_A^{(2)} \ln p_A^{(2)} + p_B^{(2)} \ln p_B^{(2)}\right]. \quad (17.11)$$

On the basis of (15.4) and (14.7) the number of AA pairs may be written in the form

$$N_{AA} = \sum_{l=1}^{N/2} \sum_{m=1}^{N/2} a_{lm}\gamma_l\gamma_m. \quad (17.12)$$

Here we replace the quantities occurring in (15.4) by the quantities $\gamma_l = \frac{1+\sigma_l}{2}$, which equal unity if atom A is at the lth site and equal zero if atom B is on this site; the subscript l enumerates the first type of site, and m —the second type of site.

From the definition of the quantities a_{lm} it follows that (see page 146):

$$\sum_{m=1}^{N/2} a_{lm} = z. \quad (17.13)$$

The expression for $\tilde{N}_{AA}$ may therefore be written in the following form:

$$\tilde{N}_{AA} = \frac{Nz}{2}\,\widetilde{\gamma_l\gamma_m}. \quad (17.14)$$

For this averaging (under the assumption of identical probabilities of distinguishable configurations for given numbers of A and B atoms at sites of the first and second type) the quantities γ_l and γ_m, which refer to the sites of a different type, are statistically independent. Therefore, $\widetilde{\gamma_l\gamma_m}$ may be replaced by the product of the mean values of $\tilde{\gamma}_l$ and $\tilde{\gamma}_m$, equal to $p_A^{(1)}$ and $p_A^{(2)}$ respectively. As a result the quantity λ_1, which, according to (17.7), is equal to $\tilde{N}_{AA}$, becomes

$$\lambda_1 = \tilde{N}_{AA} = \frac{Nz}{2}\,p_A^{(1)}\,p_A^{(2)}. \quad (17.15)$$

The quantities γ_l, γ_m and a_{lm} assume only values equal to zero and unity, so that $\gamma_l^2 = \gamma_l$, $\gamma_m^2 = \gamma_m$, $a_{lm}^2 = a_{lm}$. The expression for N_{AA}^2 may therefore be written in the form

$$N_{AA}^2 = \sum_{l=1}^{N/2} \sum_{m=1}^{N/2} a_{lm}\gamma_l\gamma_m + \sum_{l=1}^{N/2} \sum_{m=1}^{N/2} \sum_{\substack{m'=1 \\ (m' \neq m)}}^{N/2} a_{lm}a_{lm'}\gamma_l\gamma_m\gamma_{m'} + \\ + \sum_{l=1}^{N/2} \sum_{\substack{l'=1 \\ (l' \neq l)}}^{N/2} \sum_{m=1}^{N/2} a_{lm}a_{l'm}\gamma_l\gamma_{l'}\gamma_m + \\ + \sum_{l=1}^{N/2} \sum_{\substack{l'=1 \\ (l' \neq l)}}^{N/2} \sum_{m=1}^{N/2} \cdot \sum_{\substack{m'=1 \\ (m' \neq m)}}^{N/2} a_{lm}a_{l'm'}\gamma_l\gamma_{l'}\gamma_m\gamma_{m'}. \tag{17.16}$$

Using Eq. (17.13) and the obvious equalities

$$\sum_{l=1}^{N/2} \sum_{\substack{m, m'=1 \\ (m' \neq m)}}^{N/2} a_{lm}a_{lm'} = \sum_{\substack{l, l'=1 \\ (l' \neq l)}}^{N/2} \sum_{m=1}^{N/2} a_{lm}a_{l'm} = z(z-1)\frac{N}{2}, \tag{17.17}$$

$$\sum_{\substack{l, l'=1 \\ (l' \neq l)}}^{N/2} \sum_{\substack{m, m'=1 \\ (m' \neq m)}}^{N/2} a_{lm}a_{l'm'} = \sum_{\substack{l, l'=1 \\ (l' \neq l)}}^{N/2} \sum_{m, m'=1}^{N/2} a_{lm}a_{l'm'} - \sum_{\substack{l, l'=1 \\ (l' \neq l)}}^{N/2} \sum_{m=1}^{N/2} a_{lm}a_{l'm} = \\ = \frac{N}{2}\left(\frac{N}{2}-1\right)z^2 - \frac{N}{2}z(z-1) = \left(\frac{N}{2}\right)^2 z^2 - \frac{N}{2}z(2z-1), \tag{17.18}$$

we find

$$\widetilde{N_{AA}^2} = \frac{Nz}{2}\tilde{\gamma}_l\tilde{\gamma}_m + \frac{N}{2}z(z-1)\left[\tilde{\gamma}_l\widetilde{\gamma_m\gamma_{m'}} + \widetilde{\gamma_l\gamma_{l'}}\tilde{\gamma}_m\right] + \\ + \left[\left(\frac{N}{2}\right)^2 z^2 - \frac{N}{2}z(2z-1)\right]\widetilde{\gamma_l\gamma_{l'}}\,\widetilde{\gamma_m\gamma_{m'}}. \tag{17.19}$$

The products $\gamma_l\gamma_{l'}$ (and also $\gamma_m\gamma_{m'}$) occurring in (17.19) refer to different sites $(l \neq l')$ of the same type. When calculating the averages of these products it is impossible to regard γ_l and $\gamma_{l'}$ as independent variables, since, if, for instance, atom A is at the site l, the probability of replacement of the other site by atom A will no longer be $\frac{N_A^{(1)}}{\frac{N}{2}}$, but $\frac{N_A^{(1)}-1}{\frac{N}{2}-1}$. Therefore the average of the products $\widetilde{\gamma_l\gamma_{l'}}$ (and $\widetilde{\gamma_m\gamma_{m'}}$) cannot be replaced by the product of the averages. This remark does not refer to the averaging of the product of the quantities γ, referring to sites of different types, because the averaging is carried out for given numbers of A and B atoms at sites of the first and second type and the assignment of a configuration of atoms to one type of site does not influence the number of A atoms on the other type of site. To determine $\widetilde{\gamma_l \gamma_{l'}}$ we shall find the square of the sum of the quantities γ_l over all sites of the first type

$$\left(\sum_{l=1}^{N/2} \gamma_l\right)^2 = \sum^{N/2} \gamma_l + \sum_{\substack{l,\, l'=1 \\ (l' \neq l)}}^{N/2} \gamma_l \gamma_{l'} = \frac{N}{2} p_A^{(1)} + \frac{N}{2}\left(\frac{N}{2} - 1\right) \widetilde{\gamma_l \gamma_{l'}}. \tag{17.20}$$

On the other hand, this expression obviously equals $\left(\frac{N}{2} p_A^{(1)}\right)^2$. Therefore

$$\widetilde{\gamma_l \gamma_{l'}} = \frac{\frac{N}{2} p_A^{(1)} \left(\frac{N}{2} p_A^{(1)} - 1\right)}{\frac{N}{2}\left(\frac{N}{2} - 1\right)}. \tag{17.21}$$

Similarly

$$\widetilde{\gamma_m \gamma_{m'}} = \frac{\frac{N}{2} p_A^{(2)} \left(\frac{N}{2} p_A^{(2)} - 1\right)}{\frac{N}{2}\left(\frac{N}{2} - 1\right)}. \tag{17.22}$$

Substituting (17.21) and (17.22) into (17.19), one may find N_{AA}^2 and then with the aid of (17.8) and (17.15) we find the quantity λ_2:

$$\lambda_2 = -\frac{Nz}{2} p_A^{(1)} p_A^{(2)} (1 - p_A^{(1)})(1 - p_A^{(2)}) = -\frac{Nz}{2} p_A^{(1)} p_A^{(2)} p_B^{(1)} p_B^{(2)}. \tag{17.23}$$

In this manner one may calculate the mean values of higher powers of $\widetilde{N_{AA}^2}$ and, with the aid of equations of the type (17.7)–(17.10), find the other coefficients λ_n in the expansion (17.5) of the free energy in powers of $\frac{w}{kT}$. Chang [194] calculated the quantities $\lambda_1, \lambda_2, \ldots, \lambda_6$ for the binary alloy $A-B$ (arbitrary composition with the same number of sites of first and second types surrounding one another). He found the following expressions for λ_3, λ_4, λ_5 and λ_6:

$$\lambda_3 = \frac{Nz}{2} p_A^{(1)} p_A^{(2)} (1 - p_A^{(1)}) (1 - p_A^{(2)}) (1 - 2p_A^{(1)}) (1 - 2p_A^{(2)}), \tag{17.24}$$

$$\lambda_4 = -\frac{Nz}{2} p_A^{(1)} p_A^{(2)} (1 - p_A^{(1)}) (1 - p_A^{(2)}) (1 - 6p_A^{(1)} + 6p_A^{(1)^2}) (1 - 6p_A^{(2)} + 6p_A^{(2)^2}) - \\ - 3N (y_1 - z) p_A^{(1)^2} p_A^{(2)^2} (1 - p_A^{(1)})^2 (1 - p_A^{(2)})^2, \tag{17.25}$$

$$\lambda_5 = \frac{Nz}{2} p_A^{(1)} p_A^{(2)} (1 - p_A^{(1)}) (1 - p_A^{(2)}) (1 - 2p_A^{(1)}) (1 - 2p_A^{(2)}) \times \\ \times (1 - 12p_A^{(1)} + 12p^{(1)^2}) (1 - 12p_A^{(2)} + 12p_A^{(2)^2}) + \\ 30N (y_1 - z) p_A^{(1)^2} p_A^{(2)^2} (1 - p_A^{(1)})^2 (1 - p_A^{(2)})^2 (1 - 2p_A^{(1)}) (1 - 2p_A^{(2)}), \tag{17.26}$$

$$\lambda_6 = -\frac{Nz}{2} p_A^{(1)} p_A^{(2)} (1 - p_A^{(1)}) (1 - p_A^{(2)}) [1 - 30p_A^{(1)} (1 - p_A^{(1)}) (1 - 2p_A^{(1)})^2 - \\ 30p_A^{(2)} (1 - p_A^{(2)}) (1 - 2p_A^{(2)})^2] - 15N (13y_1 + 17z) p_A^{(1)^2} p_A^{(2)^2} (1 - p_A^{(1)})^2 \times \\ (1 - p_A^{(2)})^2 - 30N (y_2 - 30y_1 - 28z) p_A^{(1)^2} p_A^{(2)^2} (1 - p_A^{(1)})^2 (1 - p_A^{(2)})^2 \times \\ [p_A^{(1)} (1 - p_A^{(1)}) + p_A^{(2)} (1 - p_A^{(2)})] - 60N (y_3 - 4y_2 + 66y_1 + 42z) \times \\ \times p_A^{(1)^3} p_A^{(2)^3} (1 - p_A^{(1)})^3 (1 - p_A^{(2)})^3. \tag{17.27}$$

Here in the expression for λ_n, beginning with λ_4, there occur along with the coordination number z more complicated coordination numbers y_1, y_2 and y_3, which are related to the numbers of atoms adjacent to not one, but to two or three sites:

$$\left. \begin{aligned} y_1 &= \sum_{l'} z_{ll'}^2 - z(z-1), \\ y_2 &= \sum_{l'} z_{ll'}^2 - 3\sum_{l'} z_{ll'}^2 + 2z(z-1), \\ y_3 &= \sum_{l'l''} z_{ll'} z_{l'l''} z_{l''l} - 3(z-2) \sum_{l'} z_{ll'}^2 + 2z(z-1)(z-2), \end{aligned} \right\} \tag{17.28}$$

where $z_{ll'}$ is the number of sites, that are simultaneously nearest neighbors for the sites l and l'. For alloys with a β-brass body-centered cubic lattice

$$z = 8, \quad y_1 = 96, \quad y_2 = 144, \quad y_3 = 1776; \tag{17.29}$$

for crystals with a NaCl cubic lattice

$$z = 6, \quad y_1 = 24, \quad y_2 = 0, \quad y_3 = 264. \tag{17.30}$$

In the case of stoichiometric alloys AB the expressions for the magnitudes $\lambda_1, \lambda_2, \ldots, \lambda_6$, taking into account (1.5), take the following form:

$$\lambda_1 = \frac{Nz}{8}(1-\eta^2), \tag{17.31}$$

$$\lambda_2 = -\frac{Nz}{32}(1-\eta^2)^2, \tag{17.32}$$

$$\lambda_3 = -\frac{Nz}{32}\eta^2(1-\eta^2)^2, \tag{17.33}$$

$$\lambda_4 = -\frac{Nz}{128}(1-\eta^2)^2(1-3\eta^2)^2 - \frac{3}{256}N(y_1-z)(1-\eta^2)^4, \tag{17.34}$$

$$\lambda_5 = -\frac{Nz}{8}\eta^2(1-\eta^2)^2\left(1-\frac{3}{2}\eta^2\right)^2 - \frac{15}{128}N(y_1-z)\eta^2(1-\eta^2)^4, \tag{17.35}$$

$$\lambda_6 = -\frac{Nz}{32}(1-\eta^2)^2(1-15\eta^2+15\eta^4) - \frac{15}{256}N(13y_1+17z)(1-\eta^2)^4 - \\ -\frac{15}{256}N(y_2-30y_1-28z)(1-\eta^2)^5 - \\ -\frac{15}{1024}N(y_3-4y_2+66y_1+42z)(1-\eta^2)^6. \tag{17.36}$$

To investigate the region near the critical temperature, using Eqs. (17.5), (17.11) and (17.31)-(17.36) we find the coefficients A_2 and A_4 in the expansion (11.2) $F = F_0 + \frac{A_2}{2}\eta^2 + \frac{A_4}{4}\eta^4$ in powers of η in the form of a power series in $\frac{w}{kT}$:

$$A_2 = NkT\left\{1 - \frac{z}{4}\left[\frac{w}{kT} - \frac{1}{2\cdot 2!}\left(\frac{w}{kT}\right)^2 + \frac{1}{4\cdot 3!}\left(\frac{w}{kT}\right)^3 - \right.\right. \\ -\frac{1}{8\cdot 4!}\left(1+3\frac{y_1}{z}\right)\left(\frac{w}{kT}\right)^4 + \frac{1}{16\cdot 5!}\left(1+15\frac{y_1}{z}\right)\left(\frac{w}{kT}\right)^5 - \\ \left.\left. -\frac{1}{32\cdot 6!}\left(1+15\frac{y_1}{z}-15\frac{y_2}{z}+\frac{45}{2}\frac{y_3}{z}\right)\left(\frac{w}{kT}\right)^6 + \ldots\right]\right\}, \tag{17.37}$$

$$A_4 = NkT\left\{\frac{1}{3} - \frac{z}{4}\left[\frac{1}{2\cdot 2!}\left(\frac{w}{kT}\right)^2 - \frac{1}{3!}\left(\frac{w}{kT}\right)^3 + \right.\right. \\ +\frac{1}{8\cdot 4!}\left(13+9\frac{y_1}{z}\right)\left(\frac{w}{kT}\right)^4 - \frac{1}{2\cdot 5!}\left(5+15\frac{y_1}{z}\right)\left(\frac{w}{kT}\right)^5 + \\ \left.\left. +\frac{1}{32\cdot 6!}\left(121+765\frac{y_1}{z}-150\frac{y_2}{z}+\frac{225}{2}\frac{y_3}{z}\right)\left(\frac{w}{kT}\right)^6 + \ldots\right]\right\}. \tag{17.38}$$

According to Eq. (11.6) the critical temperature is determined from the condition $A_2(T_0) = 0$. Equating Eq. (17.37) to zero and calculating $\frac{w}{kT_0}$ by the method of successive approximations, we found:

$$\frac{w}{kT_0} = \frac{4}{z}\left[1 + \frac{1}{z} + \frac{4}{3}\frac{1}{z^2} + \left(\frac{y_1}{z}+2\right)\frac{1}{z^3} + \left(4\frac{y_1}{z}+\frac{16}{5}\right)\frac{1}{z^4} + \right. \\ \left. + \left(\frac{y_3}{z} - \frac{2}{3}\frac{y_2}{z} + 10\frac{y_1}{z} + \frac{16}{3}\right)\frac{1}{z^5} + \ldots\right]. \tag{17.39}$$

Next using (17.37), we may find the magnitude of a, which occurs in Eq. (11.8) and is equal to $\frac{\partial A_2}{\partial T}$

$$a = Nk\left\{1 - \frac{z}{16}\left[\left(\frac{w}{kT}\right)^2 - \frac{1}{3}\left(\frac{w}{kT}\right)^3 + \right.\right.$$
$$+ \frac{1}{16}\left(1 + 3\frac{y_1}{z}\right)\left(\frac{w}{kT}\right)^4 - \frac{1}{120}\left(15\frac{y_1}{z} + 1\right)\left(\frac{w}{kT}\right)^5 + \tag{17.40}$$
$$\left.\left. + \frac{1}{1152}\left(\frac{45}{2}\frac{y_3}{z} - 15\frac{y_2}{z} + 15\frac{y_1}{z} + 1\right)\left(\frac{w}{kT}\right)^6 + \ldots\right]\right\}.$$

Knowing the magnitudes of a and A_4, and using (11.12) we can find the jump in the specific heat of the stoichiometric alloy at the critical point

$$\Delta C = \frac{3}{2}Nk\left[1 + \frac{1}{z} + \frac{2}{3z^2} + 3\frac{y_1}{z}\frac{1}{z^3} + \left(6\frac{y_1}{z} - \frac{52}{45}\right)\frac{1}{z^4} + \right.$$
$$\left. + \left(5\frac{y_3}{z} - \frac{40}{3}\frac{y_2}{z} - \frac{26}{3}\frac{y_1}{z} - \frac{28}{9}\right)\frac{1}{z^5} + \ldots\right]. \tag{17.41}$$

In particular, for alloys with a body-centered cubic lattice, $a = 0.67Nk$, $A_4 = 0.054NkT_0$, and $\frac{w}{kT_0}$ [according to (17.39)] at the critical point, ΔC [according to (17.41)] and the ratio $\frac{a}{A_4}$ [which, according to (11.9), gives the temperature dependence of the degree of long-range order near the critical point] are equal to

$$\frac{w}{kT_0} = 0{,}60, \quad \Delta C = 1{,}87Nk, \quad \frac{a}{A_4} = \frac{12}{T_0}. \tag{17.42}$$

To study the dependence of the degree of long-range order for any values of temperature and concentration, in the expansion of the free energy (17.5) we shall consider only terms proportional to $\frac{w}{kT}$ and $\left(\frac{w}{kT}\right)^2$. Substitution of Eqs. (17.11), (17.6), (17.15) and (17.23) into (17.5) gives in the Kirkwood method [192]:

$$F = -\frac{zN}{2}(c_A v_{AA} + c_B v_{BB} + c_A c_B w) - \frac{Nz}{8}w\eta^2 -$$
$$\frac{Nz}{4}w\left(c_A^2 - \frac{1}{4}\eta^2\right)\left(c_B^2 - \frac{1}{4}\eta^2\right)\frac{w}{kT} +$$
$$\frac{1}{2}NkT\left[\left(c_A + \frac{1}{2}\eta\right)\ln\left(c_A + \frac{1}{2}\eta\right) + \left(c_A - \frac{1}{2}\eta\right)\ln\left(c_A - \frac{1}{2}\eta\right) + \right.$$
$$\left.\left(c_B - \frac{1}{2}\eta\right)\ln\left(c_B - \frac{1}{2}\eta\right) + \left(c_B + \frac{1}{2}\eta\right)\ln\left(c_B + \frac{1}{2}\eta\right)\right]. \tag{17.43}$$

Hence, using the equilibrium condition $\frac{\partial F}{\partial \eta}=0$, we find an equation for the degree of long-range order

$$\ln \frac{\left(c_A+\frac{1}{2}\eta\right)\left(c_B+\frac{1}{2}\eta\right)}{\left(c_A-\frac{1}{2}\eta\right)\left(c_B-\frac{1}{2}\eta\right)}=z\frac{w}{kT}\eta-\frac{z}{2}\left(\frac{w}{kT}\right)^2\eta\left(1-2c_Ac_B-\frac{1}{2}\eta^2\right). \tag{17.44}$$

The graph of corresponding function $\eta(T)$ for stoichiometric alloy composition with a body-centered cubic lattice almost coincides with the unbroken curve illustrated in Fig. 13.

The quantities A_2, A_4 and $\left.\frac{\partial^2 F}{\partial c_A^2}\right|_{\eta=0}$ in this approximation for the alloys of an arbitrary composition equal

$$A_2=\frac{1}{4}NkT\left[\frac{1}{c_Ac_B}-z\frac{w}{kT}+\frac{z}{2}(1-2c_Ac_B)\left(\frac{w}{kT}\right)^2\right], \tag{17.45}$$

$$A_4=\frac{NkT}{48}\left[\frac{1}{c_A^3}+\frac{1}{c_B^3}-3z\left(\frac{w}{kT}\right)^2\right], \tag{17.46}$$

$$\left.\frac{\partial^2 F}{\partial c_A^2}\right|_{\eta=0}=NkT\left[\frac{1}{c_Ac_B}+z\frac{w}{kT}-\frac{z}{2}\left(\frac{w}{kT}\right)^2(1-6c_A+6c_A^2)\right]. \tag{17.47}$$

From the condition $A_2(T_0)=0$ and Eq. (17.45), we derive a formula for the order-disorder transition temperature for different compositions:

$$\frac{w}{kT_0}=\frac{1-\sqrt{1+\frac{2}{z}\left(2-\frac{1}{c_Ac_B}\right)}}{1-2c_Ac_B}. \tag{17.48}$$

The graph of the dependence of T_0 on c_A for the case $z=8$ is illustrated in Fig. 94 (curve b).

As seen in the preceding section, at some point on the phase diagram of the alloy, the curve of the second-order phase transition points changes into a decomposition curve. The statistical theory of ordering enables us to determine at what concentration $c_A=c_{A\text{K}}$ this critical point lies. Substituting Eqs. (17.45)-(17.47) into (11.29), which gives the position of the critical point, and using (17.48), we find that in the Kirkwood approximation for $z=8$, $c_{A\text{K}}\approx 0.22$, hence, for concentrations $c_A<0.22$ and $c_A>0.78$ the alloy may not be found in an ordered state.

Using the expression for free energy of an alloy, written in the form of a series expansion in $\frac{w}{kT}$, one may obtain not only the critical temperature, the change in specific heat, the equilibrium value of the degree of long-range order etc., but also the equilibrium numbers of pairs of neighboring atoms N_{AA}, N_{AB}, N_{BB} and the associated correlation parameters for the first coordination sphere. Indeed, from the equation (17.1) for Z_η, it follows that

$$N_{AA} = -\frac{\partial \ln Z_\eta}{\partial \frac{w}{kT}} = \frac{\partial \frac{F}{kT}}{\partial \frac{w}{kT}}. \tag{17.49}$$

Substituting the expansion (17.5) in place of F and taking into account Eqs. (1.11) and (7.15), we find that the correlation parameter ε^{12}_{AA} for the first coordination sphere equals

$$\varepsilon^{12}_{AA} = p^{12}_{AA} - p^{(1)}_A p^{(2)}_A = \left(\frac{N_{AA}}{\frac{1}{2} Nz} - p^{(1)}_A p^{(2)}_A \right) = \\ = \frac{2}{zN} \left[\lambda_2 \frac{w}{kT} + \frac{\lambda_3}{2!} \left(\frac{w}{kT} \right)^2 + \frac{\lambda_4}{3!} \left(\frac{w}{kT} \right)^3 + \ldots \right]. \tag{17.50}$$

In particular, for a disordered stoichiometric alloy

$$\varepsilon^{12}_{AA} = \varepsilon_{AA} = -\frac{1}{16} \left[1 + \frac{3 \frac{y_1}{z} - 1}{48} \left(\frac{w}{kT} \right)^2 + \frac{15 \frac{y_3}{z} - 30 \frac{y_1}{z} + 2}{3840} \left(\frac{w}{kT} \right)^4 + \ldots \right] \frac{w}{kT}. \tag{17.51}$$

Restricting ourselves to the first term in expansion (17.50), we obtain for alloys of arbitrary composition

$$\varepsilon^{12}_{AA} = -\left(c^2_A - \frac{\eta^2}{4} \right) \left(c^2_B - \frac{\eta^2}{4} \right) \frac{w}{kT}. \tag{17.52}$$

Simple relations exist between the various correlation parameters for the first coordination sphere. In order to derive them, we consider that in a multicomponent alloy of an arbitrary structure, the sums of the probabilities $p^{LL'}_{\alpha\alpha'}$ of substitution of sites of the Lth and L'th types of atoms α and α', respectively, for all types of atoms α and α' must be equal to the a priori probabilities $p^{(L')}_{\alpha'}$ or $p^{(L)}_\alpha$:

$$\sum_\alpha p^{LL'}_{\alpha\alpha'} = p^{(L')}_{\alpha'}, \qquad \sum_{\alpha'} p^{LL'}_{\alpha\alpha'} = p^{(L)}_\alpha. \tag{17.53}$$

Using the relation $\sum_{\alpha} p_{\alpha}^{(L)} = 1$ and the definition of the correlation parameters $\varepsilon_{\alpha\alpha'}^{LL'} = p_{\alpha\alpha'}^{LL'} - p_{\alpha}^{(L)} p_{\alpha'}^{(L')}$, by means of Eq. (17.53) we find

$$\sum_{\alpha} \varepsilon_{\alpha\alpha'}^{LL'} = \sum_{\alpha'} \varepsilon_{\alpha\alpha'}^{LL'} = 0. \tag{17.54}$$

From Eq. (17.54) and the obvious equality $\varepsilon_{\alpha\alpha'}^{LL'} = \varepsilon_{\alpha'\alpha}^{L'L}$ in the special case of a binary alloy with two types of sites we obtain

$$\varepsilon_{AB}^{12} = \varepsilon_{BA}^{12} = \varepsilon_{AB}^{21} = \varepsilon_{BA}^{21} = -\varepsilon_{AA}^{12} = -\varepsilon_{BB}^{12}. \tag{17.55}$$

Thus, in the case of binary alloys, if one correlation parameter is known, the remaining parameters may be computed by means of Eq. (17.55). For example, discarding terms proportional to $\left(\frac{w}{kT}\right)^2$, we obtain for ε_{AB}^{12}

$$\varepsilon_{AB}^{12} = \left(c_A^2 - \frac{\eta^2}{4}\right)\left(c_B^2 - \frac{\eta^2}{4}\right)\frac{w}{kT}. \tag{17.56}$$

In the Kirkwood approximation the correlation parameters are non-zero not only for an ordered, but also for a disordered alloy. The temperature variation of these parameters and the numbers of pairs N_{AA}, N_{BB} or N_{AB}, which are related to them, should lead to the existence of a configurational part of the specific heat above the transition temperature. Differentiating the expression for the mean thermodynamic energy

$$E = -\frac{zN}{2}\left[c_A(2v_{AB} - v_{BB}) + c_B v_{BB}\right] + wN_{AA} \tag{17.57}$$

(where one must substitute the equilibrium value N_{AA}) with respect to temperature and taking into account (17.49) and (17.5), we find that above the order–disorder transition temperature the specific heat of the alloy is

$$\frac{C}{k} = -\lambda_2\left(\frac{w}{kT}\right)^2 - \frac{2\lambda_3}{2!}\left(\frac{w}{kT}\right)^3 - \frac{3\lambda_4}{3!}\left(\frac{w}{kT}\right)^4 - \dots \tag{17.58}$$

Thus, in the formulas for λ_2, λ_3, etc. in the case of a disordered alloy the probabilities $p_A^{(1)}$ and $p_A^{(2)}$ must be replaced by the concentration c_A. It follows from (17.58) and (17.23) that for sufficiently high temperatures, the specific heat is inversely proportional to T^2 and equals

$$C = Nk \frac{z}{2} c_A^2 c_B^2 \left(\frac{w}{kT}\right)^2. \tag{17.59}$$

For a given temperature this quantity is a maximum for stoichiometric alloys for $c_A = c_B = \frac{1}{2}$, when $C = Nk\frac{z}{32}\left(\frac{w}{kT}\right)^2$.

We shall now give an expression for the various expansion coefficients of the free energy in powers of $\frac{w}{kT}$ in the case of alloys with a face-centered cubic structure. In this lattice any type of site may be surrounded not only by the other type of site, but also by the same type of site. To avoid difficulties arising from this fact, one may subdivide the crystal into four geometrically identical sublattices, each of which is a simple cubic lattice resulting from a translation of any basis atom of the original lattice by the edge length of the unit cell. Thus, each atom has among its nearest neighbors only atoms of the other sublattices. We denote by p_{A1}, p_{A2}, p_{A3}, p_{A4} the probabilities of replacement of the sites of different sublattices by the A atoms. In a disordered alloy all p_{Ai} $(i = 1, 2, 3, 4)$ are equal to c_A; in different ordered structures some of these probabilities are equal to each other. Thus, in alloys with a $AuCu_3$ type crystal lattice $p_{A2} = p_{A3} = p_{A4}$, in alloys with a AuCu type crystal lattice

$$p_{A1} = p_{A2}, \quad p_{A3} = p_{A4}.$$

Performing the same calculation of the free energy for alloys with a face-centered cubic structure, as in the case considered above, we find that the quantity $\ln W$ and the coefficients in the expansion (17.5) of F in powers of $\frac{w}{kT}$ equal [169]

$$\ln W = -\frac{N}{4} \sum_{i=1}^{4} [p_{Ai} \ln p_{Ai} + (1 - p_{Ai}) \ln (1 - p_{Ai})], \tag{17.60}$$

$$\lambda_1 = N \sum_{\substack{i,j=1 \\ (i<j)}}^{4} p_{Ai} p_{Aj}, \tag{17.61}$$

$$\lambda_2 = -N \sum_{\substack{i,j=1 \\ (i<j)}}^{4} p_{Ai} p_{Aj} (1 - p_{Ai})(1 - p_{Aj}), \tag{17.62}$$

$$\lambda_3 = N\left[\sum_{\substack{i,j=1\\(i<j)}}^{4} p_{Ai}p_{Aj}(1-p_{Ai})(1-p_{Aj})(1-2p_{Ai})(1-2p_{Aj})+\right.$$

$$\left.+12\sum_{\substack{i,j,k=1\\(i<j<k)}}^{4} p_{Ai}p_{Aj}p_{Ak}(1-p_{Ai})(1-p_{Aj})(1-p_{Ak})\right], \quad (17.63)$$

$$\lambda_4 = -N\left\{\sum_{\substack{i,j=1\\(i<j)}}^{4}[p_{Ai}p_{Aj}(1-p_{Ai})(1-p_{Aj})(1-6p_{Ai}+6p_{Ai}^2)\times\right.$$

$$\times(1-6p_{Aj}+6p_{Aj}^2)+6p_{Ai}^2p_{Aj}^2(1-p_{Ai})^2(1-p_{Aj})^2]+$$

$$+24\sum_{\substack{i,j,k=1\\(i<j<k)}}^{4} p_{Ai}p_{Aj}p_{Ak}(1-p_{Ai})(1-p_{Aj})(1-p_{Ak})\times$$

$$\times[(1-2p_{Ai})(1-2p_{Aj})+(1-2p_{Ai})(1-2p_{Ak})+$$

$$+(1-2p_{Aj})(1-2p_{Ak})]+48\sum_{\substack{i,j,k=1\\(i<j<k)}}^{4} p_{Ai}p_{Aj}p_{Ak}(1-p_{Ai})\times$$

$$\times(1-p_{Aj})(1-p_{Ak})[p_{Ai}(1-p_{Ai})+p_{Aj}(1-p_{Aj})+p_{Ak}(1-p_{Ak})]+$$

$$\left.+144p_{A1}p_{A2}p_{A3}p_{A4}(1-p_{A1})(1-p_{A2})(1-p_{A3})(1-p_{A4})\right\}. \quad (17.64)$$

The Kirkwood method for the calculation of the free energy takes into account correlation in an alloy and is more accurate than the Gorsky-Bragg-Williams method, of the previous section, in which correlation was neglected. The expressions for the free energy, obtained in Sec. 16, result from (17.5), if we set the quantities λ_2, λ_3, λ_4 etc. in this formula equal to zero. The expansion terms of the free energy which have been determined here were calculated exactly. Therefore, the agreement (or slight discrepancy) of the first terms of the expansion, in any approximate theory, with the terms given here may serve as an indication of the adequate accuracy of the approximate theory. The expansion of free energy in powers of $\frac{w}{kT}$, obviously converges rapidly at high temperatures (above the critical temperature), when $\frac{w}{kT}$ is sufficiently small. This means that in this case there is already sufficient accuracy in the expressions for free energy, containing several of the first terms of the expansion. Thus, the Kirkwood method enables us to determine accurately (within the framework of the assumed model for an alloy) the various equilibrium properties of an alloy (associated with the existence of short-range order) at temperatures appreciably above the critical temperature. For example, the correlation parameter at high temperatures in disordered alloys with a face-centered cubic structure, according to (17.50), (17.54) and (17.62) (where the terms with λ_2 are retained), is given from the formula

$$\varepsilon_{AB} = c_A^2 c_B^2 \frac{w}{kT}. \qquad (17.65)$$

At temperatures near the critical temperature, when $\frac{w}{kT} \sim 1$, this method, as distinguished from the method that neglects correlation, makes possible a qualitative explanation of a number of phenomena in alloys (such as the existence of short-range order above the critical temperature T_0), an investigation of the transition of the ordering curve to the decomposition curve, etc. However, since $\frac{w}{kT}$ is small for $T \sim T_0$, in this region the expansion of the various physical quantities converges slowly. Thus, for instance, in Eq. (17.41) in the case of a body-centered cubic lattice the third, fifth and sixth terms in the square brackets equal 0.010, 0.017 and 0.023, respectively. The series should be sufficiently exact if one includes an even greater number of terms in the expansion (higher than the term proportional to $\left(\frac{w}{kT}\right)^6$). This requires immense labor. The power series in $\frac{w}{kT}$ at temperatures near the transition temperature converge particularly slowly in the case of alloys with a face-centered structure, for which $\frac{w}{kT_0} > 1$. Since the results concerning a phase transition of this structure are less reliable, they will not be given here.

18. Quasi-Chemical Method

The Kirkwood method (see Sec. 17) is rather cumbersome and is not always suitable for practical calculations. Therefore, we shall present another method of treatment of order-disorder transitions in alloys, the so-called quasi-chemical method [195, 196, 4]. Since correlation is accounted for in this method, one must find the free energy of the given alloy as a function of temperature, of degree of long-range order and of N_{AB}—the number of pairs of neighboring atoms AB. This free energy characterizes correlation in an alloy. The configurational factor of the partition function Z_K may be represented as the sum

$$Z_k = \sum_{\eta,\, N_{AB}} Z_{\eta,\, N_{AB}}, \tag{18.1}$$

where

$$Z_{\eta,\, N_{AB}} = \sum_i^{(\eta,\, N_{AB})} e^{-\frac{E_i}{kT}} \tag{18.2}$$

denotes the sum over all configurations of atoms for given values η and N_{AB}. By analogy with Sec. 16, the sum (18.1) in the expression for $\ln Z_k$ may be replaced by the maximum term, and the equilibrium values η and N_{AB} may be found from the condition of maximum $Z_{\eta,\, N_{AB}}$. Since the energies for different configurations are the same for a given number of pairs N_{AB}, the sum (18.2) equals a constant factor $e^{-\frac{E_i}{kT}}$, multiplied by the number of terms $W(\eta,\, N_{AB})$, which is equal to the number of different permutations of atoms of the alloy for given η and N_{AB}, i.e.

$$F = -kT \ln Z_{\eta,\, N_{AB}} = E_i - kT \ln W(\eta,\, N_{AB}). \tag{18.3}$$

Let us consider a binary ordered alloy $A—B$ of arbitrary composition, in which the crystal lattice has the same number of sites of the first and second types and the sites of each type are surrounded only by sites of the other type. The energy of such an alloy in the nearest-neighbor approximation may, according to Eq. (14.6), be written in the form

$$E_i = -N_{AA}v_{AA} - N_{BB}v_{BB} - (N_{AB}^{12} + N_{BA}^{12})\, v_{AB}, \tag{18.4}$$

where $N_{\alpha\alpha'}^{12}$ is the number of pairs, in which atom α is located at a site of the first type and atom α' on a neighboring site of the second type ($\alpha,\, \alpha' = A,\, B$). The quasi-chemical method is based on the assumption that separate pairs of neighboring atoms may be considered as independent "molecules," the binding energy of each "molecule" being equal to $-v_{AA}$, $-v_{AB}$ and $-v_{BB}$ respectively. The number $W(\eta,\, N_{AB})$ of distinguishable configurations of the system is therefore assumed to be proportional to the number of ways in which one may decompose the total number $\frac{zN}{2}$ of pairs of atoms into four groups containing the pairs AA, BB, AB and BA*:

$$W(\eta, N_{AB}) = h(\eta)\frac{\left(\frac{zN}{2}\right)!}{N_{AA}!N_{BB}!N_{AB}^{12}!N_{BA}^{12}!}. \tag{18.5}$$

Here the factor $h(\eta)$ is regarded as independent of the number of pairs of atoms, but depends on η, i.e., on the number of A and B atoms on sites of the first and second types. To determine $h(\eta)$ one may consider the value $W(\eta, N_{AB})$, which corresponds to a random distribution of atoms on each type of site (i.e., the value corresponding to the case when no correlation exists between the filling of the neighboring sites of the lattice by atoms of the alloy). In this case the numbers of pairs N^0_{AA}, N^0_{BB}, N^0_{AB}, N^0_{BA} are given by the formula

$$N^0_{\alpha\alpha'} = 2z\frac{N^{(1)}_{\alpha}N^{(2)}_{\alpha'}}{N} \qquad (\alpha, \alpha' = A, B). \tag{18.6}$$

When the numbers of pairs equals $N^0_{\alpha\alpha'}$, the magnitude of $W(\eta, N_{AB})$ for a given η obviously is a maximum and in the limit of an infinite crystal $\ln W(\eta, N_{AB})$ may be replaced by the logarithm of the sum of all $W(\eta, N_{AB})$ corresponding to different numbers of pairs for a given η.

The last sum evidently equals the number of ways in which one may distribute the given numbers of atoms $N^{(1)}_A$, $N^{(2)}_A$, $N^{(1)}_B$ and $N^{(2)}_B$ on the lattice sites, i.e.,

$$\frac{\left(\frac{N}{2}\right)!}{N^{(1)}_A!N^{(1)}_B!}\frac{\left(\frac{N}{2}\right)!}{N^{(2)}_A!N^{(2)}_B}. \tag{18.7}$$

Equating $W(\eta, N^{(0)}_{AB})$ given by Eq. (18.5) and Eq. (18.7) we obtain

$$h(\eta) = \frac{\left[\left(\frac{N}{2}\right)!\right]^2 N^0_{AA}!N^0_{BB}!N^0_{AB}!N^0_{BA}!}{N^{(1)}_A!N^{(1)}_B!N^{(2)}_A!N^{(2)}_B!\left(\frac{zN}{2}\right)!}. \tag{18.8}$$

---

*In the pairs AB the A atom is at a site of the first type and the B atom is at a site of the second type, while in pairs BA the converse is true.

It follows from (18.3), (18.4), (18.5) and (18.8) that the free energy, expressed as a function of the numbers of A and B atoms on the sites of the first and second types and the numbers of pairs of neighboring atoms, is, in the quasi-chemical method, equal to

$$F = -N_{AA}v_{AA} - N_{BB}v_{BB} - (N_{AB}^{12} + N_{BA}^{12})\, v_{AB} -$$
$$- kT \Big\{ N\left(\ln \frac{N}{2} - 1\right) - N_A^{(1)}(\ln N_A^{(1)} - 1) - N_B^{(1)}(\ln N_B^{(1)} - 1) -$$
$$- N_A^{(2)}(\ln N_A^{(2)} - 1) - N_B^{(2)}(\ln N_B^{(2)} - 1) +$$
$$+ \sum_{\alpha,\, \alpha' = A,\, B} \left[N_{\alpha\alpha'}^{0}(\ln N_{\alpha\alpha'}^{0} - 1) - N_{\alpha\alpha'}(\ln N_{\alpha\alpha'} - 1)\right]\Big\}. \quad (18.9)$$

To determine the equilibrium properties of alloys, it is necessary to find the minimum of the free energy (18.9) with respect to the number of atoms $N_A^{(1)}, N_B^{(1)}, \; N_A^{(2)}, \; N_B^{(2)}$ and the number of pairs of atoms $N_{\alpha\alpha'}^{12}$ $(\alpha, \; \alpha' = A, \, B)$. These variables are not independent, but are related by the obvious conditions

$$N_A^{(1)} + N_B^{(1)} - N_A^{(2)} - N_B^{(2)} = 0, \quad (18.10)$$
$$N_A^{(1)} + N_A^{(2)} - N_A = 0, \quad (18.11)$$
$$N_B^{(1)} + N_B^{(2)} - N_B = 0, \quad (18.12)$$
$$N_{AA} + N_{AB}^{12} - zN_A^{(1)} = 0, \quad (18.13)$$
$$N_{BB} + N_{BA}^{12} - zN_B^{(1)} = 0, \quad (18.14)$$
$$N_{AA} + N_{BA}^{12} - zN_A^{(2)} = 0, \quad (18.15)$$
$$N_{BB} + N_{AB}^{12} - zN_B^{(2)} = 0. \quad (18.16)$$

When finding the minimum of F these additional conditions may be taken into account using the method of Lagrangian multipliers, by comparing Eqs. (18.10)-(18.16) with the factors $a, a_A, a_B, b_A, b_B, d_A, d_B$, respectively. Then the conditions for minimization take the following form

$$kT\left[\ln N_A^{(1)} - 2z\frac{N_A^{(2)}}{N}\ln N_{AA}^{0} - 2z\frac{N_B^{(2)}}{N}\ln N_{AB}^{0}\right] + a + a_A - zb_A = 0, \quad (18.17)$$

$$kT\left[\ln N_B^{(1)} - 2z\frac{N_A^{(2)}}{N}\ln N_{BA}^{0} - 2z\frac{N_B^{(2)}}{N}\ln N_{BB}^{0}\right] + a + a_B - zb_B = 0, \quad (18.18)$$

$$kT\left[\ln N_A^{(2)} - 2z\frac{N_A^{(1)}}{N}\ln N_{AA}^{(0)} - 2z\frac{N_B^{(1)}}{N}\ln N_{BA}^{0}\right] - a + a_A - zd_A = 0, \quad (18.19)$$

$$kT\left[\ln N_B^{(2)} - 2z\frac{N_A^{(1)}}{N}\ln N_{AB}^0 - 2z\frac{N_B^{(1)}}{N}\ln N_{BB}^0\right] - a + a_B - zd_B = 0, \tag{18.20}$$

$$-v_{AA} + kT\ln N_{AA} + b_A + d_A = 0, \tag{18.21}$$

$$-v_{AB} + kT\ln N_{AB}^{12} + b_A + d_B = 0, \tag{18.22}$$

$$-v_{AB} + kT\ln N_{BA}^{12} + b_B + d_A = 0, \tag{18.23}$$

$$-v_{BB} + kT\ln N_{BB} + b_B + d_B = 0. \tag{18.24}$$

With the aid of Eqs. (18.10)-(18.24), one may determine the long-range order and the correlation parameter for the first coordination sphere. Adding Eqs. (18.22) and (18.23) and subtracting (18.21) and (18.24), we find

$$\frac{N_{AB}^{12}N_{BA}^{12}}{N_{AA}N_{BB}} = e^{\frac{w}{kT}}. \tag{18.25}$$

Defining the numbers of pairs $N_{\alpha\alpha'}^{12}$ in terms of the probabilities $p_{\alpha\alpha'}^{12}$ so that atom α is on the first type of site, and atom α' on the neighboring site of the second type: $N_{\alpha\alpha'}^{12} = \frac{1}{2}Nzp_{\alpha\alpha'}^{12}$, and expressing these probabilities in terms of $\varepsilon_{AB}^{12} = \varepsilon_{BA}^{12} = \varepsilon_{AB}$

$$p_{AA}^{12} = p_A^{(1)}p_A^{(2)} - \varepsilon_{AB}, \quad p_{AB}^{12} = p_A^{(1)}p_B^{(2)} + \varepsilon_{AB}, \quad p_{BA}^{12} = p_B^{(1)}p_A^{(2)} + \varepsilon_{AB},$$
$$p_{BB}^{12} = p_B^{(1)}p_B^{(2)} - \varepsilon_{AB}, \tag{18.26}$$

we find from Eq. (18.25) a formula for ε_{AB}

$$\varepsilon_{AB} = \\ = \frac{1 - 2c_Ac_B\left(1 - e^{-\frac{w}{kT}}\right) - \frac{1}{2}\eta^2\left(1 - e^{-\frac{w}{kT}}\right) - \sqrt{(1-2c_A)^2 + 4c_Ac_Be^{-\frac{w}{kT}} - \eta^2\left(e^{-\frac{w}{kT}} - e^{-2\frac{w}{kT}}\right)}}{2\left(1 - e^{-\frac{w}{kT}}\right)}. \tag{18.27}$$

Equation (18.27) enables us to determine the correlation parameter as a function of the degree of long-range order and temperature. For a stoichiometric alloy the expression for ε_{AB} takes a simple form above the critical temperature

$$\varepsilon_{AB} = \frac{1}{4}\tanh\frac{w}{4kT}. \tag{18.28}$$

To obtain an equation specifying the degree of long-range order, we add together Eqs. (18.17) and (18.20), then we subtract (18.18) and (18.19) and take into account (18.22) and (18.23). Then, taking note of (1.1), (1.5) and (18.6), we obtain

$$(z-1)\ln\frac{\left(c_A+\frac{\eta}{2}\right)\left(c_B+\frac{\eta}{2}\right)}{\left(c_A-\frac{\eta}{2}\right)\left(c_B-\frac{\eta}{2}\right)}=z\ln\frac{c_Ac_B+\frac{\eta}{2}+\frac{\eta^2}{4}+\varepsilon_{AB}}{c_Ac_B-\frac{\eta}{2}+\frac{\eta^2}{4}+\varepsilon_{AB}}. \quad (18.29)$$

Equation (18.29) in conjunction with (18.27) gives the dependence of the degree of long-range order on temperature and alloy composition. In particular, with stoichiometric alloys

$$(z-1)\ln\frac{1+\eta}{1-\eta}=\frac{z}{2}\ln\frac{(1+\eta)^2+4\varepsilon_{AB}}{(1-\eta)^2+4\varepsilon_{AB}}, \quad (18.30)$$

where according to (18.27)

$$\varepsilon_{AB}=\frac{1}{4}\coth\frac{w}{2kT}-\frac{1}{4}\eta^2-\frac{\sqrt{e^{\frac{w}{kT}}-\eta^2\left(e^{\frac{w}{kT}}-1\right)}}{2\left(e^{\frac{w}{kT}}-1\right)}. \quad (18.31)$$

Expanding the right-and left-hand sides of Eq. (18.29) in powers of η and noting that near the transition temperature, when $\eta\ll 1$, the coefficients of the linear terms of the expansion of the right and left-hand sides should be identical, after simple transformations we find an expression for the dependence of the order-disorder transition temperature on the composition of the alloy

$$\frac{w}{kT_0}=-\ln\left(1-\frac{z-1}{z^2c_Ac_B}\right). \quad (18.32)$$

In particular, for stoichiometric alloys

$$\frac{w}{kT_0}=-2\ln\left(1-\frac{2}{z}\right). \quad (18.33)$$

In the case of an alloy with a β-brass crystal lattice ($z=8$) $\frac{w}{kT_0}\approx$ 0.58; for crystals with a NaCl type lattice ($z=6$) $\frac{w}{kT_0}\approx 0.81$.

The energy of an alloy, which on the basis of (14.9), the relation $N_{AB}=\frac{Nz}{2}\left(p_{AB}^{12}+p_{BA}^{12}\right)$ and (18.26), equals

$$E=-\frac{zN}{2}\left[c_Av_{AA}+c_Bv_{BB}+c_Ac_Bw+\frac{\eta^2}{4}w+\varepsilon_{AB}w\right], \quad (18.34)$$

is determined with the aid of Eq. (18.26). It follows from (18.34) and (18.28) that, for example, for a stoichiometric alloy, the specific heat of the alloy above the critical temperature is equal to

$$C=\frac{z}{32}Nk\frac{\left(\frac{w}{kT}\right)^2}{\cosh^2\frac{w}{4kT}}. \tag{18.35}$$

Next, by using Eqs. (18.29) and (18.27) to determine the value of the derivative $\frac{d\eta^2}{dT}$ at the transition point, one may find the increase in the specific heat Nk. For a stoichiometric alloy this increase equals

$$\Delta C=\frac{3}{8}kN\frac{z-2}{z-1}\left(z\ln\frac{z}{z-2}\right)^2. \tag{18.36}$$

In particular, for alloys with a β-brass crystal lattice $\Delta C = 1.70\,\Delta C$.

The accuracy of the quasi-chemical method may be estimated by expanding the free energy deduced thereby in powers of $\frac{w}{kT}$ and comparing the expansion coefficients with the exact coefficients found by the Kirkwood method (in the framework of the model under consideration). Such a comparison shows that these expansions coincide, provided in the coefficients of (17.15), (17.23)-(17.27) for different powers of $\frac{w}{kT}$, we set the magnitudes y_1, y_2, y_3 and similar quantities equal to zero, which can be accounted for in the next approximation. Since the coefficients of $\frac{w}{kT}$, $\left(\frac{w}{kT}\right)^2$ and $\left(\frac{w}{kT}\right)^3$ in the Kirkwood expansion do not contain these quantities, the expansion of the free energy derived by the quasi-chemical method is accurate up to and including terms proportional to $\left(\frac{w}{kT}\right)^3$.

We have presented only the simplest version of the quasi-chemical method, based on the treatment of individual pairs of neighboring atoms. In more complicated versions of this method, atoms constituting a triangle, a tetrahedron, etc. are taken as the independent "molecules." An increase in the number of atoms in a single "molecule" obviously improves the accuracy of the method.

For a face-centered cubic lattice the version of the quasi-chemical method which considers only pairs of nearest neighbors leads to qualitatively incorrect results (the ordered state is not stable at any temperature [197]). Qualitatively correct results are obtained if we choose as independent "molecules" the tetrahedrons formed by four atoms, each of which belongs only to its sublattice [190, 199]. In such a theory a first-order phase transition is obtained for alloys with an $AuCu_3$ crystal lattice. In the case of the stoichiometric alloy AB_3 the critical temperature and the magnitude of the increase of the degree of long-range order at the critical point are determined from the formulas

$$\frac{w}{kT_0}=2.43, \quad \eta_0=0.956 \tag{18.37}$$

A graph of the temperature dependence of the degree of long-range order in the quasi-chemical method is illustrated in Fig. 15 (curve a).

For alloys with a AuCu crystal lattice in contrast with the Gorsky-Bragg-Williams approximation, which neglects correlation, the quasi-chemical method, based on a consideration of tetrahedrons, leads to a first-order phase transisition also in agreement with the thermodynamic theory (see Sec. 11) and with experiment [200, 189]. The critical temperature of the stoichiometric alloy AB and the magnitude of the jump at the critical point are determined by the following relations:

$$\frac{w}{kT_0}=2.73, \quad \eta_0=0.983 \tag{18.38}$$

A somewhat different version of the quasi-chemical method has been applied to the problem of order-disorder phenomena in the alloy AB [201].

Correlation is taken into account approximately also in the method developed by Bethe for the alloys with an equal number of sites of the first and second types [202]. In this method a partition function is constructed for a certain group of atoms, consisting of a central atom and neighboring atoms. The influence of the remaining atoms on this particular group is considered in the Gorsky-Bragg-Williams approximation, which neglects correlation in the alloy, and the probability of replacement of any site of the neighborhood of the central atom depends only on what atom is located at the center (this also gives an account of correlation in the alloy), and is independent of the replacement of other sites of this neighborhood. This method has been applied by Peierls [203] to alloys with a face-centered cubic lattice of such a type. It has been shown [204] that for β-brass alloys the Bethe method leads to results that are just as accurate as the quasi-chemical method, but the calculations are more complicated. In view of this we shall not present the calculations of thermodynamic quantities of alloys according to the Bethe method.

19. Statistical Theory of Almost Completely Ordered Alloys

In Sec. 17 the free energy of an alloy F was calculated using a method based on the expansion of F in powers of the ratio $\frac{w}{kT}$, which may be considered as a small parameter. This approximation, as already mentioned, is sufficiently accurate at high temperatures, but does not give reliable results at low temperatures. In the latter case another method can be developed for calculation of the free energy, in which, as in the thermodynamic calculation of Sec. 12, we allow for the fact that at low temperatures (below the critical temperature) the alloy is in an almost completely ordered state. At low temperatures atoms located on "foreign" sites form a weak imperfect gas of unusually low-mobility "quasi-particles." The small parameter in this case is the departure of the degree of long-range order from unity, and the calculation may be carried out by analogy with the treatment of ordinary imperfect gases. This type of calculation has been performed for the case of stoichiometric composition within the nearest-neighbor approximation [205, 206].

Let us consider a binary, almost completely ordered alloy A—B, whose composition is close to stoichiometric corresponding to a ratio of the sites of the first and second types $\frac{\nu}{1-\nu}$ [169]. A statistical calculation of the free energy of such an alloy may be performed using the pair interaction approximation. In this approximation, we will take into account the interaction energy of a given atom not only with atoms lying in the first coordinations sphere, but also with the more remote atoms, which still contribute appreciably to the interaction energies. In the crystal lattice we single out a polyhedron (which does not pass through the lattice sites) containing a rather large number of unit cells, so that we may neglect the interaction of the basis atoms of the unit cell of the polyhedron with atoms lying outside its boundaries. The crystal lattice of an alloy may be subdivided into geometrically identical sublattices, the number of which equals the number of sites inside the specified polyhedron, in such a manner that each site belongs to its sublattice. In this partitioning each atom interacts only with a single atom of another sublattice (some of these interaction energies may equal zero) and does not interact with atoms of its sublattice. The isolated polyhedron is comprised of n_1 sites of the first type, at which the A atoms are arranged in a state of complete order, and n_2 sites of the second type, correct for the B atoms. Let N_{Ai}, N_{Bi} denote the numbers of the A and B atoms on the sites of the ith sublattice. For almost completely ordered solutions the number of atoms on the "foreign" sublattice is substantially less than the total number of atoms in the sublattice, N_0. The numbers of pairs of neighboring atoms located on "foreign" sites in the ith and jth sublattices are second-order small quantities. The case in which there is interaction between three or more atoms located on foreign sites is encountered very rarely in an almost completely ordered alloy, hence this type of configuration will not be considered in what follows.

Let v_{AA}^{ij}, v_{AB}^{ij} and v_{BB}^{ij} denote the interaction energies of the pairs AA, AB and BB, one of the atoms being located on a site of the ith sublattice, and the other atom on its neighboring site in the jth sublattice. In a determination of the configurational energy E_i (hereafter we shall omit the index i in E_i) of an almost completely ordered alloy, we must remember that the interaction energy between an atom on a site of the ith sublattice, correct for the A atoms, and other atoms of the crystal changes when an A atom is replaced by a B atom from the value $-\sum_{\substack{j=1 \\ (j\neq i)}}^{n_1} v_{AA}^{ij} - \sum_{j=n_1+1}^{n} v_{AB}^{ij}$ to the value $-\sum_{\substack{j=1 \\ (j\neq i)}}^{n_1} v_{AB}^{ij} - \sum_{j=n_1+1}^{n} v_{BB}^{ij}$, where $n = n_1 + n_2$. The difference between these quantities equals the change of crystal energy when a B atom appears at a foreign site if there are no other atoms at foreign sites in the sphere of "molecular action." If, however, two B atoms are located at "foreign" sites at a distance less than the radius of the sphere of "molecular action," these atoms must be regarded as a single pair in this theory. With the appearance of such a pair the change in the interaction energy of the ions at the neighboring sites of the ith and jth sublattices $-2(v_{AB}^{ij} - v_{AA}^{ij})$, due to the formation of two isolated B atoms at foreign sites, is replaced by the change of energy $-v_{BB}^{ij} + v_{AA}^{ij}$. Consequently, each pair BB on neighboring sites of the ith and jth sublattices corresponds to the energy $w^{ij}=2v^{ij}_{AB}-v^{ij}_{AA}-v^{ij}_{BB}$. Thus the energy of an almost completely ordered alloy is equal to

$$E=E^0-\sum_{i=1}^{n_1} N_{Bi}\left[\sum_{\substack{j=1\\(j\neq i)}}^{n_1}(v^{ij}_{AB}-v^{ij}_{AA})+\sum_{j=n_1+1}^{n}(v^{ij}_{BB}-v^{ij}_{AB})\right]-$$
$$-\sum_{i=n_1+1}^{n} N_{Ai}\left[\sum_{j=1}^{n_1}(v^{ij}_{AA}-v^{ij}_{AB})+\sum_{\substack{j=n_1+1\\(j\neq i)}}^{n}(v^{ij}_{AB}-v^{ij}_{BB})\right]+$$
$$+\sum_{\substack{i,\,j=1\\(i<j)}}^{n_1} N^{ij}_{BB}w^{ij}+\sum_{\substack{i,\,j=n_1+1\\(i<j)}}^{n} N^{ij}_{AA}w^{ij}-\sum_{i=1}^{n_1}\sum_{j=n_1+1}^{n} N^{ij}_{BA}w^{ij}, \tag{19.1}$$

where E^0 is the energy of an almost completely ordered stoichiometric alloy containing the same total number of atoms as in the alloy being considered, N^{ij}_{AA}, N^{ij}_{BA} and N^{ij}_{BB} are the number of pairs of atoms located on neighboring sites of the ith and jth sublattices $(i<j)$.

Furthermore, to calculate the free energy one must find the total number of different permutations W of atoms on the sites of the crystal lattice for the given numbers N_{Ai}, N_{Bi}, N^{ij}_{AA}, N^{ij}_{BA} and N^{ij}_{BB}. In this approximation when we neglect configurations in which more than two neighboring atoms in the ith and jth sublattices are located at "foreign" sites, it is necessary to consider the possibility of the permutations of the atoms located at "foreign" sites, and also pairs of atoms at "foreign" sites. Thus, on N_0 sites of the first sublattice it is necessary to rearrange N^{1j}_{BB} B atoms occurring in the pair of "foreign" BB atoms at sites of the first and jth sublattices $(1\leqslant j\leqslant n_1)$ (by simultaneously rearranging also the second atoms of the pair, i.e., by rearranging the pair), N^{1j}_{BA} B atoms occurring in the pairs of "foreign" BA atoms at the sites of the first and jth sublattices and $N_{B1}-\sum_{j=2}^{n_1}N^{1j}_{BB}-\sum_{j=n_1+1}^{n}N^{1j}_{BA}$ "isolated" B atoms at the sites of the first sublattice, which do not interact with the other atoms at the "foreign" sites. In a permutation of the "foreign" B atoms at the sites of the second sublattice, one must note that these atoms must not be in the sphere of the "molecular action" of B atoms at the sites of the first sublattice, since the number of pairs

N_{BB}^{12} would be changed, and this number is regarded as given in the calculation of W. By examining the remaining sublattices in the same manner, we obtain

$$
W = \frac{N_0!}{\prod_{j=2}^{n_1} N_{BB}^{1j}! \prod_{j=n_1+1}^{n} N_{BA}^{1j}! \left(N_{B1} - \sum_{j=2}^{n_1} N_{BB}^{1j} - \sum_{j=n_1+1}^{n} N_{BA}^{1j}\right)! \, (N_0 - N_{B1})!} \times
$$

$$
\times \frac{(N_0 - N_{B1})!}{\prod_{j=3}^{n_1} N_{BB}^{2j}! \prod_{j=n_1+1}^{n} N_{BA}^{2j}! \left(N_{B2} - \sum_{\substack{j=1 \\ j\neq 2}}^{n_1} N_{BB}^{2j} - \sum_{j=n_1+1}^{n} N_{BA}^{2j}\right)! \, (N_0 - N_{B1} - N_{B2} + N_{BB}^{12})!} \times
$$

$$
\times \cdots \times \frac{\left(N_0 - \sum_{j=1}^{n_1-1} N_{Bj}\right)!}{\prod_{j=n_1+1}^{n} N_{BA}^{n_1 j}! \left(N_{Bn_1} - \sum_{j=1}^{n_1-1} N_{BB}^{n_1 j} - \sum_{j=n_1+1}^{n} N_{BA}^{n_1 j}\right)! \left(N_0 - \sum_{j=1}^{n_1} N_{Bj} + \sum_{j=1}^{n_1-1} N_{BB}^{j n_1}\right)} \times
$$

$$
\times \frac{\left(N_0 - \sum_{j=1}^{n_1} N_{Bj}\right)!}{\prod_{j=n_1+2}^{n} N_{AA}^{n_1+1\, j}! \left(N_{An_1+1} - \sum_{j=1}^{n_1} N_{BA}^{j n_1+1} - \sum_{j=n_1+2}^{n} N_{AA}^{n_1+1\, j}\right)! \left(N_0 - \sum_{j=1}^{n_1} N_{Bj} - N_{An_1+1} + \sum_{j=1}^{n_1} N_{BA}^{j n_1+1}\right)!} \cdot
$$

$$
\times \cdots \times \frac{\left(N_0 - \sum_{j=1}^{n_1} N_{Bj} - \sum_{j=n_1+1}^{n} N_{Aj}\right)!}{\left(N_{An} - \sum_{j=1}^{n_1} N_{BA}^{jn} - \sum_{j=n_1+1}^{n} N_{AA}^{jn}\right)! \left(N_0 - \sum_{j=1}^{n_1} N_{Bj} - \sum_{j=n_1+1}^{n} N_{Aj} + \sum_{j=1}^{n_1} N_{BA}^{jn} + \sum_{j=n_1+1}^{n} N_{AA}^{jn}\right)!} \times \tag{19.2}
$$

Omitting terms that are cubic in N_{Bi} $(1 \leqslant i \leqslant n_1)$ and N_{Ai} $(n_1 < i \leqslant n)$ and quadratic in N_{AA}^{ij}, N_{BB}^{ij} and N_{BA}^{ij} terms, one may represent the expression for W in a simpler form

$$
W = \frac{(N_0!)^n}{\prod_{i=1}^{n} [N_{Ai}! \, N_{Bi}!]} \prod_{j=1}^{n_1} \left(\frac{N_{Bj}}{N_0}\right)^{\left(\sum_{\substack{i=1 \\ (i\neq j)}}^{n_1} N_{BB}^{ij} + \sum_{i=n_1+1}^{n} N_{BA}^{ij}\right)} \times
$$

$$
\times \prod_{j=n_1+1}^{n} \left(\frac{N_{Aj}}{N_0}\right)^{\left(\sum_{i=1}^{n_1} N_{BA}^{ij} + \sum_{\substack{i=n_1+1 \\ (i\neq j)}}^{n} N_{AA}^{ij}\right)} \frac{N_0^{\left(\sum_{\substack{i,j=1 \\ (i<j)}}^{n_1} N_{BB}^{ij} + \sum_{\substack{i,j=n_1+1 \\ (i<j)}}^{n} N_{AA}^{ij} + \sum_{i=1}^{n_1} \sum_{j=n_1+1}^{n} N_{BA}^{ij}\right)}}{\prod_{\substack{i,j=1 \\ (i<j)}}^{n_1} N_{BB}^{ij}! \prod_{\substack{i,j=n_1+1 \\ (i<j)}}^{n} N_{AA}^{ij}! \prod_{i=1}^{n_1} \prod_{j=n_1+1}^{n} N_{BA}^{ij}!} \times
$$

$$\times \exp\left[-\frac{1}{N_0}\left(\sum_{\substack{i,j=1\\(i<j)}}^{n_1} N_{Bi}N_{Bj}+\sum_{\substack{i,j=n_1+1\\(i<j)}}^{n} N_{Ai}N_{Aj}+\sum_{i=1}^{n_1}\sum_{j=n_1+1}^{n} N_{Bi}N_{Aj}\right)\right]. \tag{19.3}$$

With aid of Eqs. (19.1) and (19.3), the free energy of an alloy $F=E-kT\ln W$ is specified as a function of the variables T, N_{Ai}, N_{AA}^{ij}, N_{BA}^{ij}, N_{BB}^{ij}. From the extremum condition on the expression for the free energy with respect to a variation of the number of pairs of atoms at foreign sites (for given N_{Ai}), we may find the equilibrium numbers of such pairs:

$$\left.\begin{aligned} N_{BB}^{ij}&=\frac{N_{Bi}N_{Bj}}{N_0}e^{-\frac{w^{ij}}{kT}} \qquad (i\leqslant n_1,\quad j\leqslant n_1),\\ N_{AA}^{ij}&=\frac{N_{Ai}N_{Aj}}{N_0}e^{-\frac{w^{ij}}{kT}} \qquad (i>n_1,\quad j>n_1),\\ N_{BA}^{ij}&=\frac{N_{Bi}N_{Aj}}{N_0}e^{\frac{w^{ij}}{kT}} \qquad (i\leqslant n_1,\quad j>n_1),\end{aligned}\right\} \tag{19.4}$$

Substituting these expressions into Eqs. (19.1) and (19.3) for E and W, after simple transformations we obtain an expression for the free energy as a function of temperature and the number N_{Ai} of A atoms on the different sublattices

$$\begin{aligned} F = E^0 &- N_0\sum_{i=1}^{n} p_{Bi}\left[\sum_{\substack{j=1\\(j\neq i)}}^{n_1}(v_{AB}^{ij}-v_{AA}^{ij})+\sum_{j=n_1+1}^{n}(v_{BB}^{ij}-v_{AB}^{ij})\right]-\\ &- N_0\sum_{i=n_1+1}^{n} p_{Ai}\left[\sum_{j=1}^{n_1}(v_{AA}^{ij}-v_{AB}^{ij})+\sum_{\substack{j=n_1+1\\(j\neq i)}}^{n}(v_{AB}^{ij}-v_{BB}^{ij})\right]+\\ &+kTN_0\sum_{i=1}^{n}(p_{Ai}\ln p_{Ai}+p_{Bi}\ln p_{Bi})-kTN_0\left[\sum_{\substack{i,j=1\\(i<j)}}^{n_1} p_{Bi}p_{Bj}\left(e^{-\frac{w^{ij}}{kT}}-1\right)+\right.\\ &\left.+\sum_{\substack{i,j=n_1+1\\(i<j)}}^{n} p_{Ai}p_{Aj}\left(e^{-\frac{w^{ij}}{kT}}-1\right)+\sum_{i=1}^{n_1}\sum_{j=n_1+1}^{n} p_{Bi}p_{Aj}\left(e^{\frac{w^{ij}}{kT}}-1\right)\right], \end{aligned} \tag{19.5}$$

where

$$p_{Ai}=\frac{N_{Ai}}{N_0}, \quad p_{Bi}=\frac{N_{Bi}}{N_0}. \tag{19.6}$$

According to (1.5) the probabilities p_{Ai} and p_{Bi} of replacement of different sublattices by A and B atoms are related to c_A and

η as follows

$$\left.\begin{array}{lll} p_{Ai}=c_A+(1-\nu)\eta, & p_{Bi}=1-c_A-(1-\nu)\eta & (1\leqslant i\leqslant n), \\ p_{Bi}=c_A-\nu\eta, & p_{Bi}=1-c_A+\nu\eta & (n_1<i\leqslant n). \end{array}\right\} \quad (19.7)$$

To find an equation for the equilibrium value of the degree of long-range order, we must set the derivative of expression (19.5) with respect to η equal to zero. Noting that $\sum\limits_{\substack{j=1 \\ (j\neq i)}}^{n} v_{\alpha\alpha'}^{ij}$, ($\alpha\alpha'=A, B$) does not depend on i*, we find that the temperature and concentration dependence of the degree of long-range order is given by

$$\begin{aligned} \ln\frac{[c_A+(1-\nu)\eta][1-c_A+\nu\eta]}{[c_A-\nu\eta][1-c_A-(1-\nu)\eta]} &= \\ =\frac{\beta}{kT}-\frac{2}{n_1}[1-c_A-(1-\nu)\eta] &\sum_{\substack{i,\,j=1 \\ (i<j)}}^{n_1}\left(e^{-\frac{w^{ij}}{kT}}-1\right)- \\ -\frac{2}{n_2}(c_A-\nu\eta) &\sum_{\substack{i,\,j=n_1+1 \\ (i<j)}}^{n}\left(e^{-\frac{w^{ij}}{kT}}-1\right)- \\ -\frac{1}{n}\left(\frac{c_A}{\nu}+\frac{c_B}{1-\nu}-2\eta\right) &\sum_{i=1}^{n_1}\sum_{j=n_1+1}^{n}\left(e^{\frac{w^{ij}}{kT}}-1\right), \end{aligned} \quad (19.8)$$

where

$$\beta=\frac{n}{n_1n_2}\sum_{i=1}^{n_1}\sum_{j=n_1+1}^{n}w^{ij}-\sum_{\substack{j=1 \\ (j\neq i)}}^{n}w^{ij}. \quad (19.9)$$

This approximation is correct, if, first, the concentration of atoms at "foreign" sites is sufficiently small, i.e.,

$$p_{Bi}\ll 1 \quad \text{for} \quad 1\leqslant i\leqslant n_1, \quad p_{Ai}\ll 1 \quad \text{for} \quad n_1<i\leqslant n \quad (19.10)$$

and, secondly, if the difference between the numbers of atoms at

---

*Here we consider only alloys whose disordered state has a Bravais type crystal lattice.

the "foreign" sites, in the sphere of "molecular action" of a given atom also located at a foreign site, and the corresponding number of completely randomly distributed atoms is substantially less than unity. Then the numbers of pairs of "foreign" atoms will be second-order quantities, the numbers of triads will be third-order quantities, etc. Using (19.4) and (19.6), the last condition may be written in the form

$$\left.\begin{aligned}&\sum_{\substack{j=1\\(j\neq i)}}^{n_1} p_{Bj}\left(e^{-\frac{w^{ij}}{kT}}-1\right)+\sum_{j=n_1+1}^{n} p_{Aj}\left(e^{\frac{w^{ij}}{kT}}-1\right)\ll 1 \quad (1\leqslant i\leqslant n_1),\\&\sum_{j=1}^{n_1} p_{Bj}\left(e^{\frac{w^{ij}}{kT}}-1\right)+\sum_{\substack{j=n_1+1\\ \neq i)}}^{n} p_{Aj}\left(e^{-\frac{w^{ij}}{kT}}-1\right)\ll 1 \quad (n_1<i\leqslant n).\end{aligned}\right\} \qquad (19.11)$$

Therefore, only the first term may be retained on the right-hand side of Eq. (19.8), while remaining terms give a small correction. Then Eq. (19.8) takes the form

$$\ln\frac{[c_A+(1-\nu)\eta][1-c_A+\nu\eta]}{[c_A-\nu\eta][1-c_A-(1-\nu)\eta]}=\frac{\beta}{kT}, \qquad (19.12)$$

which agrees with (12.6), if we put $\beta=\chi$*. Consequently, in stoichiometric alloys when $c_A=\nu$, at sufficiently low temperatures the degree of long-range order tends to unity according to the exponential law

$$\eta=1-\frac{1}{\sqrt{\nu(1-\nu)}}e^{-\frac{\beta}{2kT}}. \qquad (19.13)$$

This formula agrees with Eq. (12.7), if in the latter we put $\chi^0=\beta$. Thus the statistical calculation enabled us to give physical meaning to the quantity χ^0. Obviously the various expressions for different thermodynamic quantities (F, E, C, etc.) in this approximation of the statistical theory are the same as in the thermodynamic theory, if in the thermodynamic formula (see Sec. 12) χ^0 is replaced by β. Thus the configurational part of the free energy of a stoichiometric alloy at low temperature is

$$F=E^0-2NkT\sqrt{\nu(1-\nu)}\,e^{-\frac{\beta}{2kT}}.$$

*We note that in this statistical model of an alloy the quantity χ is independent of temperature.

Similarly, the dependence of η on c_A for alloys of nearly stoichiometric composition is of the same type as in thermodynamic theory.

With the aid of Eqs. (19.4) and (17.54) the correlation parameters in an almost completely ordered alloy may also be determined for the various coordination spheres

$$\left.\begin{aligned} \varepsilon_{AB}^{11}(\rho_l) &= (p_B^{(1)})^2\left(1 - e^{-\frac{w^{ij}}{kT}}\right), \\ \varepsilon_{AB}^{22}(\rho_l) &= (p_A^{(2)})^2\left(1 - e^{-\frac{w^{ij}}{kT}}\right), \\ \varepsilon_{AB}^{12}(\rho_l) &= p_B^{(1)} p_A^{(2)}\left(e^{\frac{w^{ij}}{kT}} - 1\right). \end{aligned}\right\} \qquad (19.14)$$

From (19.14) and (19.11) it follows that at low temperatures the correlation parameters are small.

We shall now apply the derived general formulas to specific crystal structures of alloys. For simplicity let us restrict ourselves to the nearest-neighbor approximation. In alloys with a β-brass type body-centered cubic lattice the polyhedron, which includes atoms interacting with atoms of the unit cell, contains sixteen sites confined to two cubic cells displaced by one half the diagonal of the cube with respect to each other. The numbers of sites of the first and second types in the polyhedron are the same ($n_1 = n_2 = 8$). The set of ordering energies w^{ij} in the nearest-neighbor approximation reduces to a single ordering energy w. The general expression (19.5) for the free energy of an alloy in this case becomes

$$\begin{aligned} F = E^0 &- \frac{Nz}{2} p_B^{(1)}(v_{BB} - v_{AB}) - \frac{Nz}{2} p_A^{(2)}(v_{AA} - v_{AB}) + \\ &+ kT\frac{N}{2}\left[p_A^{(1)} \ln p_A^{(1)} + p_B^{(1)} \ln p_B^{(1)} + p_A^{(2)} \ln p_A^{(2)} + p_B^{(2)} \ln p_B^{(2)}\right] - \\ &- kT\frac{Nz}{2} p_A^{(2)} p_B^{(1)}\left(e^{\frac{w}{kT}} - 1\right). \end{aligned} \qquad (19.15)$$

Thus the coordination number z for β-brass type alloys is eight. Eq. (19.15) for the free energy of an alloy in the nearest-neighbor approximation is valid also for other structures in which the numbers of sites of the first and second type are the same and each type of site is surrounded only by sites of the other type (for example, in the case of NaCl type crystals).

If we discard the last term in (19.15), the resulting expression will agree accurately with the expression for F, obtained by neglecting correlation (see Sec. 16.). The last term in (19.15) is a correction to the free energy of alloy. This correction takes correlation into account. Since $w>0$ for an ordered alloy, this correction is always negative. Differentiating Eq. (19.15) with respect to η and setting $\frac{\partial F}{\partial \eta}$ equal to zero, we find an equation for the equilibrium degree of long-range order in the case under consideration

$$\ln \frac{\left(c_A+\frac{1}{2}\eta\right)\left(c_B+\frac{1}{2}\eta\right)}{\left(c_A-\frac{1}{2}\eta\right)\left(c_B-\frac{1}{2}\eta\right)}=z\frac{w}{kT}-z(1-\eta)\left(e^{\frac{w}{kT}}-1\right). \quad (19.16)$$

In stoichiometric alloys as $T\to 0$ in accordance with Eq. (19.13) $1-\eta\approx 2e^{-\frac{z}{2}\frac{w}{kT}}$. The second term on the right-hand side of (19.16) therefore decreases rapidly as the temperature is lowered $\left[\text{as } 2z\exp\left(-\frac{z-2}{2}\frac{w}{kT}\right)\right]$ According to (16.8) in the method which neglects correlation in an alloy, the expression $z\frac{w}{kT}\eta = z\frac{w}{kT}-z(1-\eta)\frac{w}{kT}$ should occur on the right-hand side of the equation for η. The difference between this type of expression and the right-hand side of Eq. (19.16) decreases exponentially as the temperature is lowered. This means that the role of correlation in an alloy becomes small not only at high temperatures, when $\frac{w}{kT}\ll 1$, but also at low temperatures, when $1-\eta\ll 1$. Thus, the Gorsky-Bragg-Williams theory, which neglects correlation, is applicable at low temperatures for alloys of almost stoichiometric composition. As is apparent from Eq. (19.16), when correlation is taken into account in the case under consideration there is a decrease in the value of the degree of long-range order corresponding to the given temperature.

In alloys with a $AuCu_3$ type face-centered cubic lattice the polyhedron which surrounds the atoms of the central unit cell contains 28 sites — seven of them being sites of first type, while 21 are sites of the second type. In contrast to the previously considered structures, sites of the second type now have among their nearest-neighbors sites both of the first and of the second types. Eq. (19.5) gives the free energy in this case as

$$F = E^0 - 3Np_B^{(1)}(v_{BB} - v_{AB}) - 3Np_A^{(2)}(v_{AA} + v_{AB} - 2v_{BB}) + \\ + \frac{}{4} kT[p_A^{(1)} \ln p_A^{(1)} + p_B^{(1)} \ln p_B^{(1)} + 3p_A^{(2)} \ln p_A^{(2)} + 3p_B^{(2)} \ln p_B^{(2)}] - \\ - 3NkT\left[p_B^{(1)}p_A^{(2)}\left(e^{\frac{w}{kT}} - 1\right) + (p_A^{(2)})^2\left(e^{-\frac{w}{kT}} - 1\right)\right]. \quad (19.17)$$

The equation for the degree of long-range order in this case is

$$\ln\frac{\left(c_A + \frac{3}{4}\eta\right)\left(c_B + \frac{1}{4}\eta\right)}{\left(c_A - \frac{1}{4}\eta\right)\left(c_B - \frac{3}{4}\eta\right)} = 4\frac{w}{kT} - 4\left(2c_A + 1 - \frac{3}{2}\eta\right)\left(e^{\frac{w}{kT}} - 1\right) - \\ - 8\left(c_A - \frac{1}{4}\eta\right)\left(e^{-\frac{w}{kT}} - 1\right). \quad (19.18)$$

For alloys with a AuCu type crystal lattice in a similar manner the following expression can be obtained for the free energy, if the influence of the usual small tetragonal character of the lattice on the energy of the interatomic actions is neglected

$$F = E^0 - 2Np_B^{(1)}(2v_{BB} - v_{AA} - v_{AB}) - 2Np_A^{(2)}(2v_{AA} - v_{BB} - v_{AB}) + \\ + kT\frac{N}{2}[p_A^{(1)} \ln p_A^{(1)} + p_B^{(1)} \ln p_B^{(1)} + p_A^{(2)} \ln p_A^{(2)} + p_B^{(2)} \ln p_B^{(2)}] - \\ - kTN[(p_B^{(1)})^2 + (p_A^{(2)})^2]\left(e^{-\frac{w}{kT}} - 1\right) + 4kTNp_B^{(1)}p_A^{(2)}\left(e^{\frac{w}{kT}} - 1\right).$$

The equilibrium degree of long-range order is given by the equation

$$\ln\frac{\left(c_A + \frac{1}{2}\eta\right)\left(c_B + \frac{1}{2}\eta\right)}{\left(c_A - \frac{1}{2}\eta\right)\left(c_B - \frac{1}{2}\eta\right)} = \\ = 4\frac{w}{kT} - 4(1 - \eta)\left(e^{-\frac{w}{kT}} - 1\right) - 8(1 - \eta)\left(e^{\frac{w}{kT}} - 1\right). \quad (19.20)$$

It is seen from (19.18) and (19.20) that because of the existence of interactions between atoms located on the sites of one type, in alloy types $AuCu_3$ and AuCu the degree of long-range order is less than in β-brass type alloys for the same $\frac{w}{kT}$. A comparison of (19.18) and (19.20) with (19.16) also shows that correlation in alloys in these cases is more significant than in the case of β-brass type alloys (for the same $\frac{w}{kT}$). The Gorsky-Bragg-Williams approximation, which neglects correlation (see Sec. 16), is also found to be applicable in the case of $AuCu_3$ and AuCu type alloys of almost stoichiometric compositions at temperatures substantially below the critical temperatures, just as in the case of β-brass type alloys.

20. Influence of a Third Element on Ordering of Alloys

The addition of a third element to a binary alloy may, as is well known, strongly influence the order-disorder transition. In this section we shall consider the statistical theory of the influence of atoms of a third element C, located both at interstitial sites, and at sites of the crystal lattice of a substitutional alloy $A—B$, on the ordering of such a ternary alloy.

We first consider the case [207] in which the C atoms are introduced into the octahedral interstitial sites of the substitutional alloy $A—B$, having a β-brass type body-centered cubic lattice. We shall confine our treatment to the framework of the Gorsky-Bragg-Williams theory of Sec. 16, which neglects correlation, for the case of binary alloys. We assume that the atomic radius of the added C atoms is small enough that we may neglect the distortions of the crystal lattice which result from their insertion. Other phases of the alloy are also neglected.

In a body-centered cubic lattice of the alloy $A—B$, the C atoms may be placed at centers of the cubic cell faces (first type of interstitial site) and at the centers of the edges (second type of site). The first of these have as neighbors four sites of the first type at a distance $\frac{a}{\sqrt{2}}$ and two sites of the second type at a distance $\frac{a}{2}$ (a is the lattice constant); the second have four sites of the second type at a distance $\frac{a}{\sqrt{2}}$ and two sites of the first type at a distance $\frac{a}{2}$. The total number of octahedral type interstitial sites in an alloy having N sites is $3N$ ($\frac{3N}{2}$ interstitial sites of the first and an equal number of sites of the second type). N_C is the total number of C atoms in the alloy and $N_C^{(1)}$, $N_C^{(2)}$ their numbers at the first and second types of sites. Next we introduce the concentration c_C of C atoms in the alloy

$$c_C = \frac{N_C}{N} \tag{20.1}$$

and their concentrations at the first and second types of interstitial sites

$$\left.\begin{aligned} c_C^{(1)} = \frac{N_C^{(1)}}{N}, \quad c_C^{(2)} = \frac{N_C^{(2)}}{N}, \\ c_C^{(1)} + c_C^{(2)} = c_C. \end{aligned}\right\} \tag{20.2}$$

The negative of the interaction energies of the pairs AC, BC and CC at the distances $\frac{a}{\sqrt{2}}$ and $\frac{a}{2}$ will be denoted by

$$\left.\begin{aligned} v_{AC}\left(\frac{a}{2}\right) = v_{AC}, \quad v_{BC}\left(\frac{a}{2}\right) = v_{BC}, \quad v_{CC}\left(\frac{a}{2}\right) = v_{CC}, \\ v_{AC}\left(\frac{a}{\sqrt{2}}\right) = v'_{AC}, \quad v_{BC}\left(\frac{a}{\sqrt{2}}\right) = v'_{BC}. \end{aligned}\right\} \tag{20.3}$$

Then the configurational part of the energy of the alloy is equal to

$$\begin{aligned} E = -4N\left(v_{AA}c_A^2 + v_{BB}c_B^2 + 2v_{AB}c_Ac_B + \frac{1}{4}w\eta^2\right) + \\ + Nw'\left(\frac{1}{2}c_C - c_C^{(1)}\right)\eta - 4Nc_C\Big[c_Av'_{AC} + c_Bv'_{BC} + \\ + \frac{1}{2}(c_Av_{AC} + c_Bv_{BC})\Big] - \frac{8}{3}Nv_{CC}c_C^{(1)}(c_C - c_C^{(1)}). \end{aligned} \tag{20.4}$$

Here

$$w' = 4(v'_{AC} - v'_{BC}) - 2(v_{AC} - v_{BC}) \tag{20.5}$$

and equals the difference between the potential energies of the C atom at the first and second types of interstitial sites in the case of complete order in a stoichiometric alloy AB.

The number W of permutations of A and B atoms at the sites and of C atoms at the interstices for given numbers $N_A^{(1)}$, $N_A^{(2)}$, $N_B^{(1)}$, $N_B^{(2)}$, $N_C^{(1)}$ and N_C, obviously, is equal to

$$W = \frac{\left(\frac{N}{2}\right)!}{N_A^{(1)}!\,N_B^{(1)}!}\,\frac{\left(\frac{N}{2}\right)!}{N_A^{(2)}!\,N_B^{(2)}!}\,\frac{\left(\frac{3}{2}N\right)!}{N_C^{(1)}!\left(\frac{3}{2}N - N_C^{(1)}\right)!}\,\frac{\left(\frac{3}{2}N\right)!}{N_C^{(2)}!\left(\frac{3}{2}N - N_C^{(2)}\right)!}. \tag{20.6}$$

Using Eqs. (1.1), (1.5), (20.2), (20.4) and (20.6), we find from the formula $F = E - kT \ln W$ an expression for the configurational part of the free energy of the alloy

$$
\begin{aligned}
F = & -4N\left(c_A^2 v_{AA} + c_B^2 v_{BB} + 2c_A c_B v_{AB} + \frac{1}{4}\eta^2 w\right) + \\
& + Nw'\left(\frac{1}{2}c_C - c_C^{(1)}\right)\eta - 4Nc_C\left[c_A v'_{AC} + c_B v'_{BC} + \right. \\
& \left. + \frac{1}{2}(c_A v_{AC} + c_B v_{BC})\right] - \frac{8}{3} N v_{CC} c_C^{(1)} (c_C - c_C^{(1)}) + \\
& + \frac{1}{2} NkT\left[\left(c_A + \frac{1}{2}\eta\right)\ln\left(c_A + \frac{1}{2}\eta\right) + \left(c_A - \frac{1}{2}\eta\right)\ln\left(c_A - \frac{1}{2}\eta\right) + \right. \\
& \left. + \left(c_B + \frac{1}{2}\eta\right)\ln\left(c_B + \frac{1}{2}\eta\right) + \left(c_B - \frac{1}{2}\eta\right)\ln\left(c_B - \frac{1}{2}\eta\right)\right] + \\
& + 2c_C^{(1)}\ln c_C^{(1)} + (3 - 2c_C^{(1)})\ln(3 - 2c_C^{(1)}) + 2(c_C - c_C^{(1)})\ln(c_C - c_C^{(1)}) + \\
& + (3 - 2c_C + 2c_C^{(1)})\ln(3 - 2c_C + 2c_C^{(1)}) + 2c_C \ln 2 - 6\ln 3. \qquad (20.7)
\end{aligned}
$$

Regarding F as a function of η and $c_C^{(1)}$ for given N, c_A, c_B, c_C and T (i.e., for given N_A, N_B, N_C and T), we may obtain the following equilibrium conditions for the determination of the equilibrium values of η and $c_C^{(1)}$:

$$
\frac{\partial F}{\partial \eta} = -2Nw\eta + \frac{1}{4}kTN \ln \frac{\left(c_A + \frac{1}{2}\eta\right)\left(c_B + \frac{1}{2}\eta\right)}{\left(c_A - \frac{1}{2}\eta\right)\left(c_B - \frac{1}{2}\eta\right)} + Nw'\left(\frac{1}{2}c_C - c_C^{(1)}\right) = 0, \qquad (20.8)
$$

$$
\frac{\partial F}{\partial c_C^{(1)}} = -Nw'\eta + kTN \ln \frac{c_C^{(1)}(3 - 2c_C + 2c_C^{(1)})}{(c_C - c_C^{(1)})(3 - 2c_C^{(1)})} - \frac{16}{3} N v_{CC}\left(\frac{1}{2}c_C - c_C^{(1)}\right) = 0. \qquad (20.9)
$$

To determine the order-disorder transition temperature we note that the equations of equilibrium (20.8) and (20.9) for all values T are satisfied by the values $\eta = 0$ and $c_C^{(1)} = \frac{1}{2}c_C$, corresponding to a disordered state of the alloy and to the equilibrium distribution of C atoms over the first and second types of interstitial sites. However, in order for this state to correspond to stable equilibrium, it is necessary that the free energy $F = F(\eta, c_C)$ have a minimum at the point $\eta = 0$ and $c_C^{(1)} = \frac{1}{2}c_C$, i.e. that the following two conditions be simultaneously fulfilled:

$$\left.\begin{aligned}&\frac{\partial^2 F}{\partial \eta^2} > 0,\\ &\frac{\partial^2 F}{\partial \eta^2}\frac{\partial^2 F}{\partial c_C^{(1)^2}} - \left(\frac{\partial^2 F}{\partial \eta\, \partial c_C^{(1)}}\right)^2 > 0\end{aligned}\right\} \qquad (20.10)$$

at $\eta = 0$ and $c_C^{(1)} = \frac{1}{2} c_C$. Calculating the derivatives from (20.7), we obtain the conditions

$$\frac{kT}{c_A c_B} - 8w > 0,$$

$$\left(\frac{kT}{c_A c_B} - 8w\right)\left[\frac{3kT}{c_C(3 - c_C)} + \frac{4}{3} v_{CC}\right] - w'^2 > 0.$$

It is apparent from (20.11) and (20.12) that these conditions are not fulfilled for all values of T. The temperature at which one of these inequalities ceases to be fulfilled is the critical temperature of the ternary alloy T_{0C}, since then the disordered state no longer corresponds to a minimum of F. At sufficiently high temperatures both conditions (20.11) and (20.12) are fulfilled. As the temperature is reduced the second of these conditions ceases to be fulfilled sooner. This happens at a temperature T_{0C} equal to the large root of the quadratic equation, which is obtained by setting the left-hand side of condition (20.12) equal to zero. As a result we obtain

$$kT_{0C} = \frac{2}{9}\Big\{18 w c_A c_B - v_{CC} c_C (3 - c_C) + \\ + \sqrt{[18 w c_A c_B + v_{CC} c_C (3 - c_C)]^2 + \frac{27}{4} w'^2 c_A c_B c_C (3 - c_C)}\Big\}. \qquad (20.13)$$

Comparing this expression for the critical temperature T_{0C} of the ternary alloy with Eq. (16.11) for $z = 8$) for the critical temperature T_0 of the binary alloy [condition (20.11) ceases to be fulfilled at this temperature], it is easy to show that

$$T_{0C} \geqslant T_0. \qquad (20.14)$$

Thus, in this approximation the presence of interstitial atoms of any type elevates the critical temperature of the alloy.

We note that c_C and the magnitude $3 - c_C$, which equals the concentration of unoccupied interstitial sites $\frac{3N - N_C}{N}$, occur symmetrically in Eq. (20.13). Therefore in the model adopted for an alloy, the presence of a certain number of C atoms at interstitial sites leads to the same critical temperature as in the presence of the same number of unoccupied interstitial sites*. In the special case in which C atoms interact weakly among themselves and terms containing the v_{CC} may also be neglected, it follows from (20.13) that

$$kT_{0C}=4wc_Ac_B+\sqrt{16w^2c_A^2c_B^2+\frac{1}{3}w'^2c_Ac_Bc_C(3-c_C)}. \qquad (20.15)$$

If the concentrations c_C of interstitial C atoms is so small that we may restrict ourselves to linear terms in the expansion of (20.13) in powers of c_C, then

$$kT_{0C}=8wc_Ac_B+\frac{w'^2}{8w}c_C=kT_0'+\frac{w'^2}{8w}c_C. \qquad (20.16)$$

If $c_C\approx 3$, i.e. almost all the interstitial sites are occupied by C atoms, to obtain T_{0C} one must replace c_C in Eq. (20.16) by $3-c_C$. According to (20.16) the change in the critical temperature, caused by the C atoms, is independent of the composition of the alloy $A-B$. This conclusion evidently is no longer valid when wc_Ac_B is so small that the expansion of (20.13) becomes incorrect. For example, for $w=0$, when the binary alloy $A-B$ is disordered, Eq. (20.13) gives an expression for the critical temperature of the ternary alloy

$$kT_{0C}=-\frac{2}{9}v_{CC}c_C(3-c_C)+ +\frac{2}{9}\sqrt{[c_C(3-c_C)v_{CC}]^2+\frac{27}{4}c_Ac_Bc_C(3-c_C)w'^2}, \qquad (20.17)$$

i.e., T_{0C} generally speaking is not to equal zero. Thus, the presence of added atoms causes the alloys to become ordered. In the special case in which $w=0$ and c_C is small, from (20.17) we obtain

$$kT_{0C}=|w'|\sqrt{c_Ac_Bc_C}, \qquad (20.18)$$

i.e. T_{0C} is proportional to $\sqrt{c_C}$. Thus, the ordering caused by

*This conclusion, of course, ceases to be valid if the interaction energy of pairs of atoms depends on composition, which is to be expected, for instance, in alloys containing transition metals.

C atoms is due to the difference in the interaction energies of C atoms with A and B atoms ($w' \neq 0$). C atoms are preferentially surrounded by atoms of a definite type, wherein the presence of short-range ordering for $T < T_{0C}$ leads to the appearance of long-range order in the sites.

When $w = 0$ and $w' = 0$, and also in the case when C atoms are injected into the interstitial sites of a pure metal (c_A or c_B equal zero), Eq. (20.13) for $v_{CC} < 0$ is

$$kT_{0C} = \frac{4}{9} | v_{CC} | c_C (3 - c_C). \tag{20.19}$$

Consequently, even in this case $T_{0C} \neq 0$, which corresponds to the possibility of the emergence of long-range order in the distribution of C atoms and vacancies at the interstitial sites. Eq. (20.19), as always in the case of an approximation that neglects correlation, is qualitatively correct over a range of mean concentrations c_C which do not differ appreciably from the value $c_C = 1.5$. The case $v_{CC} > 0$ corresponds to the negative energy of "ordering" of C atoms and vacancies at the interstitial sites, i.e., no ordering of C atoms and vacancies occurs; rather the alloy decomposes into phases, one of which is enriched with C atoms in comparison with the other.

We shall consider in greater detail the often encountered case of low concentration of interstitial impurity C atoms. Thus, from the equilibrium condition (20.9) we find approximately

$$c_C^{(1)} = c_C \frac{e^{\frac{w'\eta}{kT}}}{1 + e^{\frac{w'\eta}{kT}}}. \tag{20.20}$$

Substituting this expression for $c_C^{(1)}$ into (20.8) we obtain an equation for determination of the dependence of the equilibrium degree of long-range order η on temperature and composition of the alloy

$$-8w\eta + kT \ln \frac{\left(c_A + \frac{1}{2}\eta\right)\left(c_B + \frac{1}{2}\eta\right)}{\left(c_A - \frac{1}{2}\eta\right)\left(c_B - \frac{1}{2}\eta\right)} = 2w' c_C \tanh \frac{w'\eta}{2kT}. \tag{20.21}$$

When $c_C = 0$ Eq. (20.21) goes over into Eq. (16.8) (written for $z = 8$) for the binary alloy $A - B$. To compare the behavior of the curves $\eta(T)$ for type $A-B$ binary and ternary alloys we shall determine the difference ΔT between the temperatures of the ternary alloy $T+\Delta T$ and the binary alloy T at which η has the same value. Retaining only first-order terms in ΔT and c_C, from Eqs. (20.21) and (16.8) we find

$$\Delta T=\frac{2w'c_C}{k}\,\frac{\tanh\dfrac{w'\eta}{2kT}}{\ln\dfrac{\left(c_A+\frac{1}{2}\eta\right)\left(c_B+\frac{1}{2}\eta\right)}{\left(c_A-\frac{1}{2}\eta\right)\left(c_B-\frac{1}{2}\eta\right)}}. \tag{20.22}$$

It is clear from (20.22) that ΔT is always positive, i.e., regardless of the type of impurity of added atoms, the curve $\eta(T)$ for the alloy with an impurity lies above the corresponding curve for an alloy without an impurity. The approximate difference of the curves $\eta(T)$ is illustrated in Fig. 98, where the dashed curve corresponds to the binary alloy $A-B$, and the solid curve corresponds to the same alloy with an impurity of interstitial atoms. Thus, the presence of added atoms leads (in this model) to an increase of the degree of long-range order of the alloy.

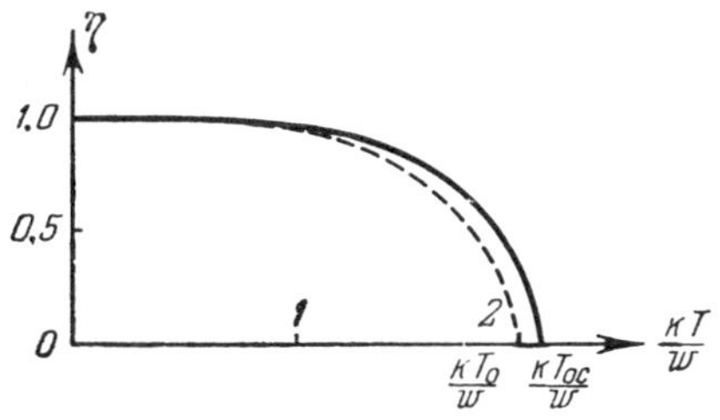

Fig. 98. Influence of impurities of added atoms on the temperature dependence of the degree of long-range order of binary (dashed curve) and of ternary (solid curve) alloys.

From Eqs. (20.20) and (20.2) we obtain the probabilities $p_C^{(1)}$ and $p_C^{(2)}$ of substitution of the interstitial sites of the first and second type by C atoms in the case of a small concentration c_C of C atoms

$$\left.\begin{aligned} p_C^{(1)}&=\frac{N_C^{(1)}}{\frac{3}{2}N}=\frac{2}{3}c_C\frac{1}{1+e^{-\frac{w'\eta}{kT}}},\\ p_C^{(2)}&=\frac{N_C^{(2)}}{\frac{3}{2}N}=\frac{2}{3}c_C\frac{1}{1+e^{\frac{w'\eta}{kT}}}.\end{aligned}\right\} \tag{20.23}$$

It follows from (20.23) that for an ordered alloy (as distinguished from the case of a disordered alloy) $p_C^{(1)} \neq p_C^{(2)}$ and the C atoms are mainly concentrated on the interstitial sites of a definite type. This nonuniformity in the distribution of added atoms on the interstitial sites of different types increases with decreasing temperature.

Let us now consider the influence of a third element, whose atoms are situated at the crystal lattice sites, on the ordering of the alloy. Ordering in a ternary substitutional alloy with two types of sites, in which $\nu=\frac{1}{2}$ and each type of site is surrounded only by sites of the other type, was investigated [208] by means of the quasi-chemical method.

The configurational part of the energy of the ternary alloy $A—B—C$ equals

$$E=-N_{AA}v_{AA}-N_{BB}v_{BB}-N_{CC}v_{CC}-(N_{AB}^{12}+N_{BA}^{12})v_{AB}-\\ -(N_{AC}^{12}+N_{CA}^{12})v_{AC}-(N_{BC}^{12}+N_{CB}^{12})v_{BC}, \quad (20.24)$$

where the notation is similar to that given in Sec. 18. The number of distinguishable configurations of atoms of the ternary alloy in the quasi-chemical method may be determined for the binary alloy in the same manner as in Sec. 18

$$W=\frac{\left(\frac{N}{2}\right)!}{N_A^{(1)}!\,N_B^{(1)}!\,N_C^{(1)}!}\,\frac{\left(\frac{N}{2}\right)!}{N_A^{(2)}!\,N_B^{(2)}!\,N_C^{(2)}!}\,\frac{\prod\limits_{\alpha,\alpha'=A,B,C} N_{\alpha\alpha'}^{0}!}{\prod\limits_{\alpha,\alpha'=A,B,C} N_{\alpha\alpha'}^{12}!}, \quad (20.25)$$

where

$$N_{\alpha\alpha'}^{0}=2z\frac{N_\alpha^{(1)}N_{\alpha'}^{(2)}}{N} \qquad (\alpha,\ \alpha'=A,\ B,\ C),$$

$N_\alpha^{(1)}$ and $N_{\alpha'}^{(2)}$ are the numbers of type α atoms on the first and second types of sites. From (20.24) and (20.25) we obtain the following expression for the configurational part of the free energy

$$F=-\sum_{\alpha,\alpha'=A,B,C} N_{\alpha\alpha'}^{12}v_{\alpha\alpha'}-kT\Big\{N\left(\ln\frac{N}{2}-1\right)-\\ -\sum_{\alpha=A,B,C}\left[N_\alpha^{(1)}\left(\ln N_\alpha^{(1)}-1\right)+N_\alpha^{(2)}\left(\ln N_\alpha^{(2)}-1\right)\right]+ \quad (20.26)\\ +\sum_{\alpha,\alpha'=A,B,C}\left[N_{\alpha\alpha'}^{0}\left(\ln N_{\alpha\alpha'}^{0}-1\right)-N_{\alpha\alpha'}^{12}\left(\ln N_{\alpha\alpha'}^{12}-1\right)\right]\Big\}.$$

In order to determine the equilibrium values of the numbers of A, B and C atoms at the first and second types of sites and the numbers of different pairs of neighboring atoms, we must use the condition that the free energy (20.26) be a minimum with respect to the variables $N_\alpha^{(1)}$, $N_\alpha^{(2)}$, $N_{\alpha\alpha'}^{12}$. These quantities are not independent variables, but are related by the auxiliary conditions

$$N_A^{(1)}+N_B^{(1)}+N_C^{(1)}=N_A^{(2)}+N_B^{(2)}+N_C^{(2)}, \tag{20.27}$$

$$N_\alpha^{(1)}+N_\alpha^{(2)}=N_\alpha, \tag{20.28}$$

$$\left.\begin{aligned} &N_{\alpha A}^{12}+N_{\alpha B}^{12}+N_{\alpha C}^{12}=zN_\alpha^{(1)},\\ &N_{A\alpha}^{12}+N_{B\alpha}^{12}+N_{C\alpha}^{12}=zN_\alpha^{(2)}\\ &(\alpha=A,\ B,\ C). \end{aligned}\right\} \tag{20.29}$$

The auxiliary conditions are taken into account most simply by the method of Lagrangian multipliers, by associating with Eq. (20.27) the factor a, with Eq. (20.28) the factors a_A, a_B, a_C and with Eq. (20.29)— b_A, b_B, b_C, d_A, d_B, d_C. Then the condition for minimization takes the following form:

$$kT\left[\ln N_\alpha^{(1)}-2z\frac{N_A^{(2)}}{N}\ln N_{\alpha A}^0-2z\frac{N_B^{(2)}}{N}\ln N_{\alpha B}^0-\right.$$
$$\left.-2z\frac{N_C^{(2)}}{N}\ln N_{\alpha C}^0\right]+a+a_\alpha-zb_\alpha=0, \tag{20.30}$$

$$kT\left[\ln N_\alpha^{(2)}-2z\frac{N_A^{(1)}}{N}\ln N_{A\alpha}^0-2z\frac{N_B^{(1)}}{N}\ln N_{B\alpha}^0-\right.$$
$$\left.-2z\frac{N_C^{(1)}}{N}\ln N_{C\alpha}^0\right]-a+a_\alpha-zd_\alpha=0, \tag{20.31}$$

$$-v_{\alpha\alpha'}+kT\ln N_{\alpha\alpha'}^{12}+b_\alpha+d_{\alpha'}=0. \tag{20.32}$$

From Eqs. (20.30)-(20.32) we may determine the quantities $N_\alpha^{(1)}, N_\alpha^{(2)}$, $N_{\alpha\alpha'}^{12}$, which characterize the long-range order and the correlation in the alloy. The solution of this system of equations is in general difficult. Therefore in the following we shall consider some special cases.

In the case of a disordered alloy (when the numbers of α atoms on the first and second types of sites are equal) $N_\alpha^{(1)}=N_\alpha^{(2)}$, $N_{\alpha\alpha'}^{12}=N_{\alpha'\alpha}^{12}$ and $b_\alpha=d_\alpha$. We introduce the notation

$$\left.\begin{aligned} X &= \sqrt{\frac{2N_{AA}}{zN}} = \frac{1}{\sqrt{\frac{zN}{2}}}\, e^{\frac{v_{AA}}{2kT}} e^{-\frac{b_1}{kT}}, \\ Y &= \sqrt{\frac{2N_{BB}}{zN}} = \frac{1}{\sqrt{\frac{zN}{2}}}\, e^{\frac{v_{BB}}{2kT}} e^{-\frac{b_2}{kT}}, \\ Z &= \sqrt{\frac{2N_{CC}}{zN}} = \frac{1}{\sqrt{\frac{zN}{2}}}\, e^{\frac{v_{CC}}{2kT}} e^{-\frac{b_3}{kT}} \end{aligned}\right\} \tag{20.33}$$

[where Eqs. (20.32) are used]. With the aid of (20.32) we express N_{AB}^{12}, N_{AC}^{12} and N_{BC}^{12} in terms of the quantities X, Y, Z introduced by (20.33)

$$\left.\begin{aligned} N_{AB}^{12} &= \frac{zN}{2}\, e^{\frac{w_{AB}}{2kT}}\, XY, \\ N_{AC}^{12} &= \frac{zN}{2}\, e^{\frac{w_{AC}}{2kT}}\, XZ, \\ N_{BC}^{12} &= \frac{zN}{2}\, e^{\frac{w_{BC}}{2kT}}\, YZ; \end{aligned}\right\} \tag{20.34}$$

where

$$w_{\alpha\alpha'} = 2v_{\alpha\alpha'} - v_{\alpha\alpha} - v_{\alpha'\alpha'}. \tag{20.35}$$

Substituting $N_{\alpha\alpha'}$, expressed with the aid of Eqs. (20.33) and (20.34) in terms of X, Y and Z, into the condition (20.29) written for a disordered alloy, we obtain a system of equations for the determination of X, Y and Z

$$\left.\begin{aligned} X^2 + e^{\frac{w_{AB}}{2kT}} XY + e^{\frac{w_{AC}}{2kT}} XZ &= c_A, \\ e^{\frac{w_{AB}}{2kT}} XY + Y^2 + e^{\frac{w_{BC}}{2kT}} YZ &= c_B, \\ e^{\frac{w_{AC}}{2kT}} XZ + e^{\frac{w_{BC}}{2kT}} YZ + Z^2 &= c_C. \end{aligned}\right\} \tag{20.36}$$

Solving this system of equations, from Eqs. (20.33) and (20.34) we may find the numbers of pairs $N_{\alpha\alpha'}^{12}$, which determine the short-range order in an alloy. If for a specific alloy the quantities w_{AB}, w_{AC} and w_{BC} are known (which may be determined, for example, from data for the binary alloys $A-B$, $A-C$ and $B-C$ that have the same structure as the ternary alloy), then the system of equations (20.36) is solved numerically. We shall investigate in analytic form two special cases: high temperature, and low concentrations of atoms of one of the components of the alloy.

The high temperature case is realized when the following conditions are fulfilled

$$|w_{AB}| \ll kT, \quad |w_{AC}| \ll kT, \quad |w_{BC}| \ll kT. \tag{20.37}$$

Thus, the system (20.36) may be solved by the method of successive approximations, by choosing as the zeroth approximation the solution $X_0 = c_A$, $Y_0 = c_B$, $Z_0 = c_C$, corresponding to an infinitely high temperature. Determining then the numbers of pairs $N^{12}_{\alpha\alpha'}$, we find the correlation parameters $\varepsilon^{12}_{\alpha\alpha'} = \varepsilon_{\alpha\alpha'} = \frac{2N^{12}_{\alpha\alpha'}}{Nz} - c_\alpha c_{\alpha'}$ (corresponding to the first coordination sphere). Restricting ourselves to terms linear in $\frac{w_{\alpha\alpha'}}{kT}$, we obtain

$$\varepsilon_{AA} = -2c_A^2 c_B (1-c_A)\frac{w_{AB}}{2kT} - 2c_A^2 c_C (1-c_A)\frac{w_{AC}}{2kT} + 2c_A^2 c_B c_C \frac{w_{BC}}{2kT}, \tag{20.38}$$

$$\varepsilon_{AB} = c_A c_B (2c_A c_B + c_C)\frac{w_{AB}}{2kT} + c_A c_B c_C (2c_A - 1)\frac{w_{AC}}{2kT} + c_A c_B c_C (2c_B - 1)\frac{w_{BC}}{2kT}. \tag{20.39}$$

The remaining correlation parameters (ε_{BB}, ε_{CC}, ε_{AC} and ε_{BC}) may be obtained from (20.38) and (20.39) by a cyclic permutation of the indexes A, B, C.

Substituting the values obtained for the numbers of pairs and (20.26), we obtain an expression for the free energy of an alloy at high temperatures

$$\begin{aligned} F = NkT \sum_{\alpha = A, B, C} c_\alpha \ln c_\alpha - \frac{Nz}{2} \sum_{\alpha, \alpha' = A, B, C} c_\alpha c_{\alpha'} v_{\alpha\alpha'} - \\ - \frac{Nz}{8kT}[c_A c_B (2c_A c_B + c_C) w_{AB}^2 + c_A c_C (2c_A c_C + c_B) w_{AC}^2 + \\ + c_B c_C (2c_B c_C + c_A) w_{BC}^2 + 2c_A c_B c_C (2c_A - 1) w_{AB} w_{AC} + \\ + 2c_A c_B c_C (2c_B - 1) w_{AB} w_{BC} + 2c_A c_B c_C (2c_C - 1) w_{AC} w_{BC}]. \end{aligned} \tag{20.40}$$

Hence we may obtain, for instance, a formula for the configurational part of the specific heat of a disordered alloy

$$C=\frac{zN}{4kT^2}[c_Ac_B(2c_Ac_B+c_C)w_{AB}^2+c_Ac_C(2c_Ac_C+c_B)w_{AC}^2+$$
$$+c_Bc_C(2c_Bc_C+c_A)w_{BC}^2+2c_Ac_Bc_C(2c_A-1)w_{AB}w_{AC}+$$
$$+2c_Ac_Bc_C(2c_B-1)w_{AB}w_{BC}+2c_Ac_Bc_C(2c_C-1)w_{AC}w_{BC}]. \qquad (20.41)$$

Thus the specific heat of a ternary alloy, as in the case of a binary alloy, is proportional to $\frac{1}{T^2}$ at high temperatures. In particular, upon adding a small amount of impurity C atoms to the stoichiometric alloy $\left(c_C\ll 1,\ c_A\approx c_B\approx\frac{1}{2}\right)$ the specific heat becomes equal to

$$C=\frac{zN}{32kT^2}\{1+2c_C[(w_{AC}-w_{BC})^2-w_{AB}^2]\}, \qquad (20.42)$$

i.e. the configurational part of the specific heat increases (in comparison with the binary alloy), if $|w_{AC}-w_{BC}|$ is greater than $|w_{AB}|$.

In the case of a disordered alloy with a small concentration of C atoms at any temperature the solution of Eq. (20.36) may be sought also by a method of successive approximations, using as the zeroth approximation the solution for a binary alloy. In particular, with an alloy of almost stoichiometric composition $\left(c_A\approx c_B\approx\frac{1}{2},\ c_C\ll 1\right)$ the correlation parameters ε_{AB}, ε_{AC} and ε_{BC} for the first coordination sphere are

$$\left.\begin{aligned}\varepsilon_{AB}&=\left(\frac{1}{4}-\frac{c_C}{2}\right)\tanh\frac{w_{AB}}{4kT},\\ \varepsilon_{AC}&=\frac{c_C}{2}\tanh\frac{w_{AC}-w_{BC}}{4kT},\\ \varepsilon_{BC}&=\frac{c_C}{2}\tanh\frac{w_{BC}-w_{AC}}{4kT}.\end{aligned}\right\} \qquad (20.43)$$

It is apparent from (20.43) that if $w_{AC}-w_{BC}\gg kT$, then $\varepsilon_{AC}\approx\frac{c_C}{2}$, $\varepsilon_{BC}\approx-\frac{c_C}{2}$ and the C atoms in an alloy are almost completely surrounded by B atoms. If, however, $w_{BC}-w_{AC}\gg kT$, then the C atoms are surrounded by B atoms. As a result around the C atoms are formed sections with a sharply increased concentration of A and B atoms. In the case of disordered decomposed alloys, this may prove to have a substantial influence on the kinetics of decomposition of alloys, facilitating the nucleation of the new phase and leading in this manner to a substantial increase in the rate of decomposition with the addition of an impurity of the third element to a binary alloy.

We shall now determine the critical temperature of a ternary alloy T_{0C}. Eliminating from Eqs. (20.30)-(20.32) the Lagrangian multipliers, we may obtain the following equations:

$$\left.\begin{aligned}(z-1)\ln\frac{N_A^{(1)}N_B^{(2)}}{N_A^{(2)}N_B^{(1)}} &= z\ln\frac{N_{AB}^{12}}{N_{BA}^{12}},\\(z-1)\ln\frac{N_A^{(1)}N_C^{(2)}}{N_A^{(2)}N_C^{(1)}} &= z\ln\frac{N_{AC}^{12}}{N_{CA}^{12}},\\(z-1)\ln\frac{N_B^{(1)}N_C^{(2)}}{N_B^{(2)}N_C^{(1)}} &= z\ln\frac{N_{BC}^{12}}{N_{CB}^{12}},\end{aligned}\right\}\qquad(20.44)$$

which relate the numbers of A, B and C atoms at the first and second types of sites and the numbers of pairs of neighboring atoms of different species. The differences $N_A^{(1)}-N_A^{(2)}$, $N_B^{(1)}-N_B^{(2)}$, $N_C^{(1)}-N_C^{(2)}$ and also $N_{AB}^{12}-N_{BA}^{12}$, $N_{AC}^{12}-N_{CA}^{12}$, $N_{BC}^{12}-N_{CB}^{12}$ are small near the order-disorder transition point. Using the notation

$$N_A^{(1)}-N_A^{(2)}=\eta_A N,\qquad N_B^{(1)}-N_B^{(2)}=\eta_B N,\qquad N_C^{(1)}-N_C^{(2)}=\eta_C N \qquad (20.45)$$

and (20.28) we obtain an approximate expression for the ratios $\frac{N_\alpha^{(1)}}{N_\alpha^{(2)}}$ ($\alpha=A,\ B,\ C$) near the transition point

$$\frac{N_A^{(1)}}{N_A^{(2)}}=1+2\frac{\eta_A}{c_A},\quad \frac{N_B^{(1)}}{N_B^{(2)}}=1+2\frac{\eta_B}{c_B},\quad \frac{N_C^{(1)}}{N_C^{(2)}}=1+\frac{\eta_C}{c_C}. \qquad (20.46)$$

Substituting (20.46) into (20.44) and expanding the left and the right sides of Eqs. (20.44) in a power series in the small differences, we obtain in the same approximation

$$\left.\begin{aligned} N_{AB}^{12}-N_{BA}^{12}&=2N_{AB}^{12}\,\frac{z-1}{z}\left(\frac{\eta_A}{c_A}-\frac{\eta_B}{c_B}\right),\\ N_{AC}^{12}-N_{CA}^{12}&=2N_{AC}^{12}\,\frac{z-1}{z}\left(\frac{\eta_A}{c_A}-\frac{\eta_C}{c_C}\right),\\ N_{BC}^{12}-N_{CB}^{12}&=2N_{BC}^{12}\,\frac{z-1}{z}\left(\frac{\eta_B}{c_B}-\frac{\eta_C}{c_C}\right).\end{aligned}\right\} \tag{20.47}$$

We substitute these expressions into the equations

$$\left.\begin{aligned} N_{AB}^{12}-N_{BA}^{12}+N_{AC}^{12}-N_{CA}^{12}&=z\eta_A N,\\ N_{BA}^{12}-N_{AB}^{12}+N_{BC}^{12}-N_{CB}^{12}&=z\eta_B N,\\ N_{CA}^{12}-N_{AC}^{12}+N_{CB}^{12}-N_{BC}^{12}&=z\eta_C N,\end{aligned}\right\} \tag{20.48}$$

which follow from (20.29) and (20.45). As a result of this substitution we obtain three linear homogeneous equations in η_A, η_B and η_C. Such a system, as is well known, has a nonvanishing solution if its determinant vanishes. This condition is

$$\begin{aligned} 1-\frac{z-1}{z}\,\frac{2}{zN}\left(\frac{N_{AB}^{12}+N_{AC}^{12}}{c_A}+\frac{N_{AB}^{12}+N_{BC}^{12}}{c_B}+\frac{N_{AC}^{12}+N_{BC}^{12}}{c_C}\right)+\\ +\left(\frac{z-1}{z}\right)^2\left(\frac{2}{zN}\right)^2\left(N_{AB}^{12}N_{AC}^{12}+N_{AB}^{12}N_{BC}^{12}+N_{AC}^{12}N_{BC}^{12}\right)=0.\end{aligned} \tag{20.49}$$

Obviously, above the critical temperature the system of equations for η_A, η_B and η_C has only the solution zero. Below the critical temperature these equations are only approximately linear. Therefore the condition for existence of a nonvanishing solution, written in the form (20.49), is also fulfilled approximately; it is fulfilled more exactly the closer the temperature of the alloy approaches the critical temperature. Condition (20.49) is fulfilled exactly only at the critical point itself. Hence, one may determine the critical temperature T_{0C} from Eq. (20.49). For this purpose, by solving Eq. (20.36) (in the general case, these equations can be solved numerically since the energies w_{AB}, w_{AC} and w_{BC} are known for a given alloy), one must determine the numbers of pairs as a function of temperature and substitute them into (20.49). The temperature at which the left-hand side of Eq. (20.49) becomes zero is the critical temperature.

Since Eqs. (20.36) are not solved analytically in the general case, we shall find the transition temperature when one of the alloy components has a low concentration ($c_C \ll 1$). In the special case of a binary alloy formula (18.32) for the critical temperature T_0 follows from (20.49), (20.36) and (20.34). Solving Eqs. (20.36) and (20.49) for small c_C by the method of successive approximations and retaining only linear terms in c_C, we find that the change in critical temperature $\Delta T = T_{0C} - T_0$ when a small number of atoms of a third element are added to the binary alloy is given by the formula

$$\frac{\Delta T}{T_0} = \frac{kT_0}{w_{AB}} \frac{z(z-1)}{z^2 c_A c_B - z + 1} \left\{ -(z-4)\, c_C + \frac{1}{z}\left(\frac{c_A - c'_A}{c_A} + \frac{c_B - c'_B}{c_B}\right) + \right.$$

$$+ (z-3)\, c_C \left[\frac{\exp \dfrac{w_{AC} - w_{BC} - w_{AB}}{2kT_0}}{1 + \dfrac{z}{z-1}\, c_A \left(\exp \dfrac{w_{AC} - w_{BC} - w_{AB}}{2kT_0} - 1\right)} + \right.$$

$$\left. + \frac{\exp \dfrac{w_{BC} - w_{AC} - w_{AB}}{2kT_0}}{1 + \dfrac{z}{z-1}\, c_B \left(\exp \dfrac{w_{BC} - w_{AC} - w_{AB}}{2kT_0} - 1\right)} \right] - (z-2)\, c_C \times$$

$$\left. \times \frac{1 - \dfrac{z-1}{z^2} \dfrac{1}{c_A c_B}}{\left[1 + \dfrac{z}{z-1}\, c_A \left(\exp \dfrac{w_{AC} - w_{BC} - w_{AB}}{2kT_0} - 1\right)\right]\left[1 + \dfrac{z}{z-1}\, c_B \left(\exp \dfrac{w_{BC} - w_{AC} - w_{AB}}{2kT_0} - 1\right)\right]} \right\}. \tag{20.50}$$

Here c'_A and c'_B are the concentrations of A and B atoms in a binary alloy, which is obtained from the ternary if the C atoms are removed from it ($c'_A + c'_B = 1$).

For specific structures, for example, type β-brass ($z = 8$) or NaCl and Fe_3Al ($z = 6$) alloys, Eq. (20.50) takes a simpler form. In particular, for β-brass type alloys of nearly stoichiometric composition $\left(z = 8,\ c_A \approx c_B \approx \frac{1}{2},\ \frac{w_{AB}}{kT_0} \approx 0{,}575\right)$

$$\frac{\Delta T}{T_0} \approx 49 \left(1 - \frac{49}{48} \frac{1}{\cosh^2 \dfrac{w_{AC} - w_{BC}}{4kT_0}}\right) c_C. \tag{20.51}$$

The ratio $\frac{\Delta T}{T_0}$ in this case varies over a range from $49\, c_C$ for $\frac{w_{AC} - w_{BC}}{kT_0} \to \pm\infty$ to $-1{,}02 c_C$ for $\frac{w_{AC} - w_{BC}}{kT_0} = 0$.

Thus, the relative variation of the critical temperature is particularly large if the difference of the ordering energies in binary alloys $A-C$ and $B-C$ (having the same structure as a ternary alloy) is much greater in absolute value than kT_0. In this case, the critical temperature increases with the addition of C atoms. This temperature may also decrease, however, if the energies w_{AC} and w_{BC} are almost equal.

From Eqs. (20.30)-(20.32) one may also find the distribution of impurity C atoms (for $c_C \ll 1$) on the first and second types of sites. Omitting simple calculations, we give only the final results

$$\left.\begin{aligned} \frac{N_C^{(1)}}{N} &= c_C \frac{1}{1+K}, \\ \frac{N_C^{(2)}}{N} &= c_C \frac{K}{1+K}, \end{aligned}\right\} \tag{20.52}$$

where

$$K = \left(\frac{c_A + \frac{\eta}{2}}{c_A - \frac{\eta}{2}} \right)^{z-1} \times$$
$$\times \left\{ \frac{\left(c_A^2 - \frac{\eta^2}{4} - \varepsilon_{AB} \right) + \left[\left(c_A - \frac{\eta}{2} \right)\left(c_B - \frac{\eta}{2} \right) + \varepsilon_{AB} \right] e^{\frac{v_{BC} - v_{AC} + v_{AA} - v_{AB}}{kT}}}{\left(c_A^2 - \frac{\eta^2}{4} - \varepsilon_{AB} \right) + \left[\left(c_A + \frac{\eta}{2} \right)\left(c_B + \frac{\eta}{2} \right) + \varepsilon_{AB} \right] e^{\frac{v_{BC} - v_{AC} + v_{AA} - v_{AB}}{kT}}} \right\}^{z}.$$

Here the correlation parameter ε_{AB} and the degree of long-range order η refer to a binary alloy $A-B$ and are determined from Eqs. (18.27) and (18.29).

21. Discussion of Results of the Statistical Theory and Comparison with Experimental Data

In addition to the approximate methods presented above there have also been developed other approximate methods for the statistical theory of alloys. These include, for instance, the variational method [209–211]. In this method we start with an expression for the partition function expressed in terms of the maximum eigenvalue of a certain matrix [see (15.18)]. It is well known that the maximum eigenvalue of a matrix may be calculated by means of the variational method. In this method, in the formula for the maximum eigenvalue we substitute in place of the eigenvalue of the matrix a vector containing certain parameters and determine them from the condition that the eigenvalue be a maximum. If we successfully choose the vectors with the aid of which the approximation is carried out, we may obtain a sufficiently accurate approximate expression for the partition function. To determine the partition function of ordered alloys another variational method has also been used, in which the variational parameters are introduced not in the eigenvector, but directly into the expression for the energy of the system [212]. It should be emphasized, however, that if the exact partition function has a singularity at a certain temperature, then near this point the approximate solution obtained by means of the variational method (as in other approximate methods) may give substantially different results. This happens because even when the exact and approximate solutions are almost coincident, the nature of the singularity of the approximate solution may be entirely different.

Other approximate methods for solution of the problem of ordering have also been developed by Zernike [213], Cowley [214], Kikuchi [215], Hijmans and De Boer [216].

The interaction of atoms with atoms of the second coordination sphere has been considered by Chang [217]. Chang has shown that if the ordering energy for the second coordination sphere (just as for the first coordination sphere) is positive, taking account of the second coordination sphere leads to a decrease of the critical temperature T_0, an increase of the degree of short-range order at the temperature T_0+0 and to a small increase in the jump in specific heat ΔC at $T=T_0$ for type AB alloys, in which sites of one type are surrounded by sites of the other type. Interaction with atoms lying in the next coordination sphere has also been taken into account by Pines [218], who considered the phase diagram, in particular, the decomposition of ordered solid solutions taking this interaction into account.

Lesnik [219] took into account ionic bonding, which appears when electrons fill the vacant places in the d-states of the transition metal atoms in alloys between transition metals and non-transition metals.

The influence of thermal vibrations on the ordering of an alloy with a body-centered cubic lattice has been investigated by Stepanov [220], who used for this purpose the theory of lattice vibrations due to Einstein, and Chang [221]. The influence of thermal vibrations, and also volume change, on the order-disorder transition of a $AuCu_3$ alloy was considered by Kholodenko [222], who reached the conclusion that taking these effects into account does not lead to a great improvement of the theory.

As mentioned in the preceding sections, an exact solution of the problem of ordering in the framework of our simplified statistical model for an alloy can be found only for the one-dimensional and two-dimensional cases. For three-dimensional crystals there has been found only an approximate solution, valid at high or low temperatures. In the temperature range of greatest interest, near the order-disorder transition temperature, these approximate methods gave less accurate results. In order to study the accuracy of the various statistical methods, we present a comparison of the results with those of exact theory (for two-dimensional lattices), the thermodynamic theory, and also with experimental data.

The approximate methods make possible a reasonably accurate determination of the critical temperature of an alloy of stoichiometric composition (corresponding to a given value of the ordering energy), if the numbers of sites of the first and second types are the same and each site of a given type is surrounded only by sites of the other type. The values of the ratios $\frac{w}{kT_0}$, obtained in the different approximate theories for structures with the coordination numbers $z=4$ (two-dimensional square lattice), $z=6$ (NaCl type lattice) and $z=8$ (β-brass type body-centered cubic lattice) are given in Table 12. This Table also gives (in the last row) the exact value of $\frac{w}{kT_0}$ for a two-dimensional lattice and values for NaCl and β-brass type lattices, obtained [206] by extrapolation of the expansion of expressions for reciprocal specific heat, which are valid at high and low temperatures. These last values, of course, entail a certain error. The cited values of $\frac{w}{kT_0}$ obtained by the Kirkwood method were not determined from Eq. (17.39), but by solving the equation $A_2=0$, where A_2 is given by (17.37). It is clear from the Table that the accuracy of the different approximate methods increases as the coordination number z increases. The different methods, which to some approximation take into account correlation in the alloy (quasi-chemical method, Kirkwood method, variational method, Vernike method [213]), in three-dimensional cases gives almost equal values of $\frac{w}{kT_0}$ which, however, are approximately 15-20% smaller than the exact value.

The dependence of the critical temperature on alloy composition is essentially different in the Gorsky-Bragg-Williams

Table 12

Values $\frac{w}{kT_0}$ in the various approximations of the statistical theory of ordering

Approximations	for z=4	for z=6	for z=8
Bragg-Williams	1.00	0.67	0.50
Quasi-chemical method	1.39	0.81	0.58
Kirkwood's method up to $\left(\frac{w}{kT}\right)^2$	2.00	0.85	0.59
Kirkwood's method up to $\left(\frac{w}{kT}\right)^3$	1.36	0.81	0.58
Kirkwood's method up to $\left(\frac{w}{kT}\right)^4$	1.35	0.85	0.63
Vernike method	1.29	0.79	0.57
Variational method	1.65	—	—
Exact solution	1.76	1.00	0.71

theory and in the different approximations which take into account correlation. It is clear from Fig. 94 that the curve of T_0 versus c_A obtained from the Kirkwood method that includes F terms proportional to $\left(\frac{w}{kT}\right)^2$, unlike the curve obtained when correlation is neglected, changes into a decomposition curve at some values of concentration. The range of existence of the ordered phase when correlation is taken into account does not encompass the entire range of variation of composition, but a narrow range of concentration about that corresponding to the stoichiometric composition.

The values of the discontinuity in the specific heat for β-brass type stoichiometric alloys obtained by the Bragg-Williams theory, which neglects correlation, by the quasi-chemical method and by the theory which uses an expansion of F in powers of $\frac{w}{kT}$ up to $\left(\frac{w}{kT}\right)^6$), are equal respectively to 1.50 kN, 1.70 N and 1.87 kN. In other words they are almost equal. However, series of the type (17.41) for the determination of ΔC converge extremely slowly, so that one may suppose that the cited values of the specific heat discontinuity differ noticeably from the exact value.

The dependence of the degree of long-range order on the ratio $\frac{T}{T_0}$, computed for stoichiometric alloys with a β-brass type crystal lattice by the quasi-chemical method, and according to Kirkwood's method (taking into account

$\left(\frac{w}{kT}\right)^2$), in practice is identical (see the solid curve in Fig. 13). The curve $\eta(T)$ obtained in the Gorsky-Bragg-Williams theory differs somewhat from the curve found by taking into account correlation in the central region, where for identical T_0 (i.e., different w) according to the first theory lower values of the degree of long-range order are obtained.

For alloys with a face-centered cubic lattice the values of the different quantities (the change in the degree of long-range order at the transition point, the critical temperature, etc.), obtained in the Gorsky-Bragg-Williams approximation and in the quasi-chemical method, differ substantially (by a factor of 1.5-2.5). In these alloys, sites of a given type may be partially surrounded by sites of the same type and there is less tendency towards ordering. The ordered state must therefore set in at larger values of the ratio $\frac{w}{kT_0}$ than, for example, in β-brass type alloys [(16.27), (18.37) and (18.38)]. In the theories already discussed, in which the results may be presented as a power series in $\frac{w}{kT}$, the first terms of which coincide with the exact expansion of Kirkwood, more reliable results may be expected for structures with small values of $\frac{w}{kT_0}$. Hence, for alloys with a face-centered cubic lattice we obtain less reliable results than for alloys in which sites of a given type are surrounded only by sites of the other type.

The results obtained by means of the methods of the statistical theory, which take into account correlation in the alloy, agree with the results of the thermodynamic theory. In β-brass and Fe_3Al type structures, the statistical theory leads to a second-order phase transition, in agreement with the thermodynamic theory. In $AuCu_3$ and AuCu type alloys, where according to the thermodynamic theory, second-order phase transitions are impossible, the statistical theory, which takes into account correlation, gives first-order phase transitions. The expansion form of the thermodynamic potential near the second-order phase transition point in these approximations of the statistical theory are the same as in the thermodynamic theory. Consequently, in these approximations, the statistical theory for three-dimensional crystals also leads to a finite specific heat discontinuity and to a linear dependence of η^2 on $T_0 - T$ for small η. The temperature dependence of η at low temperatures given in Sec. 19 also agrees with thermodynamics.

Lastly, the statistical theory, like the thermodynamic theory, leads to the decomposition of a nonstoichiometric alloy into an ordered and a disordered phase at sufficiently low temperatures. The statistical theory enables us to express the various expansion coefficients of thermodynamic potential at small η and in an almost perfectly ordered alloy in terms of the energy of an interatomic interaction [see, for example, (17.37), (17.38), (17.40)].

We now compare the results obtained in the statistical theory with experimental data. In Fig. 13, one may see the agreement between the temperature dependence of the degree of long-range order of β-brass, calculated from the formulas of Gorsky-Bragg-Williams, Kirkwood and the quasi-chemical approximations and the experimental data [8]. It should be expected that the curve plotted according to the theory that neglects correlation (Gorsky-Bragg-Williams) as in poorer agreement with experiment than the curves (which virtually coincide) constructed with consideration of correlation. Using the value for $\frac{w}{kT_0}$ obtained by the Kirkwood method [taking into account $\left(\frac{w}{kT}\right)^6$; see (17.42)], $\frac{w}{kT_0}=0.60$ and with the experimental value of the critical temperature of β-brass ($T=742°$ K), we find the ordering energy of this alloy: $w=0.038$ ev.

The dependence of the critical temperature of β-brass on composition, found with the aid of the Gorsky-Bragg-Williams theory and the quasi-chemical approximation, as well as the experimental values of the critical temperature for different compositions are given in Fig. 4. A similar dependence for the Fe_3Al alloy is given in Fig. 9. It is seen from these figures that both experiment and theory give a decrease of the critical temperature as the composition departs from stoichiometric.

From Eq. (17.42) one may determine the coefficient $\frac{a}{A_4}$ in the temperature dependence of η^2 at small η: $\eta^2=\frac{a}{A_4}(T_0-T)$. For β-brass Eq. (17.42) gives $\frac{a}{A_4}=1.6\cdot 10^{-2}$ deg.$^{-1}$, whereas the experimental value $\frac{a}{A_4}=1.1\cdot 10^{-2}$ deg.$^{-1}$. The computed value of the specific heat jump is in considerably poorer agreement with experiment. According to (17.42), the Kirkwood method [taking into account $\left(\frac{w}{kT}\right)^6$] gives $\Delta C=1.87\,kN$, whereas experiment gives $\Delta C\approx 4kN$.

The experimental values of the degree of long-range order of the alloy $AuCu_3$ at different temperatures are given in Fig. 15, which also shows the theoretical curves obtained in the Gorsky-Bragg-Williams approximation which neglects correlation, and in the quasi-chemical method. It is seen from this figure that the Gorsky-Bragg-Williams theory which neglects correlation gives reduced values of the discontinuity of the degree of long-range order, but the quasi-chemical method gives enhanced values. The ordered phase is found not only in alloys close to the composition $AuCu_3$, but also in alloys corresponding to the stoichiometric composition Au_3Cu. The critical temperature of the alloy $AuCu_3$ is 665 °K, and $T_0 = 516$ °K for the alloy Au_3Cu. The statistical theory (in which the dependence of the ordering energy on composition is neglected) gives the same value for the phase transition of those alloys, regardless of the method of calculation.

The experimental and theoretical temperature dependences of the energy of ordering of the alloy $AuCu_3$ are given in Fig. 32. The experimental values of the energies are greater than the theoretical values obtained neglecting correlation.

An x-ray investigation of the correlation parameters has been made for the concentrations $c_{Au} = 0.5$ and $c_{Au} = 0.25$ for the alloy Au-Ag at the temperature $T = 573$ °K (see Sec. 3). From the value obtained for the correlation parameter for the first coordination sphere $\varepsilon_{AgAu} = 0.02$ for $c_{Au} = 0.05$, one may find with the aid of Eq. (17.65) that $\frac{w}{kT} = 0.32$ (Eq. (17.65) is applicable since $\frac{w}{kT}$ turns out to be small). From this it follows that for a Au-Ag alloy, having a face-centered cubic lattice (in which $\frac{w}{kT_0} \sim 1.5$–2.5), the critical temperature should be of the order 70–100° K. Since the ratio $\frac{w}{kT} = 0.32$ for $T = 573$° K, the ordering energy of the alloy Au-Ag is $w = 0.016$ ev. According to (17.65) this value of w leads to a value of $\varepsilon_{AB} = 0.009$ for the correlation parameter for the first coordination sphere when $c_{Au} = 0.25$ and $T = 573$ °K. The experimental value of ε_{AB} is 0.01 at this concentration.

The influence of gold impurity on the critical temperature of β-phase of the alloy Ag-Zn has been investigated in [14] (Sec. 3). In this alloy a β-brass type structure is formed during a transition into the ordered state, which takes place at the temperature $T \approx 550$° K. Since the composition of the alloys under investigation was almost stoichiometric

$\left(c_{Ag} \approx c_{Zn} \approx \frac{1}{2}\right)$, one may use Eq. (20.51), derived by the quasi-chemical method, to calculate the change in the transition temperature with the addition of impurity. In order to apply this formula, one must know the ordering energy of the alloys Au-Zn and Ag-Au. The alloy AuZn has a β-brass type crystal structure and remains ordered up to the melting point, 725 °C. From the data of [15] the degree of long-range order at this temperature equals 0.7. The ordering energy of the alloy AuZn in the framework of the quasi-chemical method is $w_{AuZn} = 0.058$ ev. Using also the previously cited value $w_{AgAu} = 0.016$ ev, we find from Eq. (20.51) that the computed ratio $\frac{\Delta T}{T_0} = \left(\frac{\Delta T}{T_0}\right)_{calc} \approx 1.5\, c_{Au}$. The value of this ratio measured in the range $c_{Au} < 3.5$ atomic percent equals $\left(\frac{\Delta T}{T_0}\right)_{meas} \approx 2.0\ c_{Au}$

This comparison between the results of the statistical theory and experimental data shows that the statistical theory explains the main features of the order-disorder transformation. Nevertheless, in a number of cases there is a great discrepancy between theory and experiment. As already mentioned, this discrepancy may be attributed to two causes. First, the various approximate methods of calculation of the free energy of a three-dimensional crystal are particularly inaccurate near the critical temperature T_0, and that most of the experimental data was taken in this very range. Second, the model of an alloy adopted in the statistical theory is very crude, since we neglect several features (or treat them only approximately), which may influence the order-disorder transformation (conduction electrons, thermal vibrations of the lattice, lattice distortions, the change of energy of interatomic interaction with composition, with order and with temperature; dependence of the energy of electrons responsible for ferromagnetism or anti-ferromagnetism on the configuration of the atoms, etc.).

The foremost problem for further development of the theory is the use of a sufficiently accurate model of alloy, that will take into account the previously neglected features. In addition, we must develop methods for calculation of the partition function of a partially ordered crystal, which will make possible an explanation of the exact nature of the features of thermodynamic quantities at the phase transition point for the three-dimensional case.

Chapter IV

Theory of Diffusion in Ordered Alloys*

22. General Discussion of Diffusion in Alloys

The mobility of atoms and, consequently, the diffusion coefficients in alloys are strong functions of concentration of the components. These quantities should also depend on the arrangement of atoms in the crystal lattices of the alloy, i.e., on the degree of long-range and short-range order.

Three main mechanisms of migration of atoms in a crystal lattice have been proposed to explain diffusion in solids. The first of them corresponds to a direct interchange of sites of the atoms. However, such a mechanism cannot explain the electrical conductivity of ionic crystals. In this connection there has been proposed an interstitial diffusion mechanism involving the migration of atoms through interstitial sites, and a vacancy mechanism involving movement of atoms into vacant lattice sites. These mechanisms were proposed by Frenkel [224] in his fundamental development of theory of diffusion. On the basis of these three simple types of migration of atoms, more complicated models of diffusion in solids were also constructed. Here we refer, for example, to the simultaneous interchange of sites by groups of three or more atoms migrating around a closed ring; the migration of groups of atoms, near an excess atom, displaced from their equilibrium positions along the direction of the neighboring lattice sites, etc. We shall restrict ourselves hereafter to only the aforementioned simplest diffusion mechanisms.

An interchange of positions by neighboring atoms is very unlikely, because it requires a high activation energy. Therefore, in the majority of crystals such a mechanism plays a minor role in diffusion. This conclusion has been supported

---

*The main contents of this chapter have been presented in [223], which has been rewritten and expanded here.

for some metallic alloys by the rather conclusive experiments of Smigelskas and Kirkendall [225]. In their work a rectangular brass rod was coated with a layer of copper. Thin molybdenum wires were placed between the brass and the copper. After a continuous 56-day annealing at 785 °C, during which diffusion across the brass-copper layer took place, the separation between the wires lying on the opposite faces of the rods decreased. This effect has been investigated for a number of different alloys [226] (Cu-Ni, Cu-Au, Ag-Au, Ag-Pd, Ni-Co, Ni-Au, Fe-Ni, Sn-Cu, Al-Cu). This result excludes the direct interchange of sites (two or several atoms), because if the diffusion were due to this mechanism, the number of atoms inside the boundary formed by the wires must have remained unchanged. Actually, the decrease of the lattice constant due to the change in composition of the alloy yields an effect approximately one-tenth the observed effect. On the other hand, these experiments may be simply explained both by the vacancy diffusion mechanism, and interstitial diffusion.

In both cases zinc diffuses from brass more quickly than copper into brass, and the vacancies remaining in the brass may partially disappear as a result of their aggregation to form holes and the subsequent collapse of the holes by plastic deformation. These experiments were explained in the following manner. The model of direct interchange of sites was also shown to be unsuitable to explain diffusion phenomena in alloys, according to the experiments of Pines and Geguzin [227, 228]. These authors found that the rates of diffusion of nickel and copper from their contact surface are not identical. However, for substitutional alloys, there exists no direct experimental proof in favor of either one of the two possible mechanisms. One can only state with assurance that the diffusion of interstitial atoms is realized via movement between the interstitial sites. In substitutional alloys (and in pure metals), vacancy diffusion evidently plays a leading role.

In the study of diffusion in alloys one should distinguish between the case in which no concentration gradient of any substance exists in the alloy (self-diffusion), and the case in which such a gradient does exist (chemical diffusion). Self-diffusion may be studied by the investigation of the motion of tracer atoms, when only a concentration gradient of traced atoms is present, but no concentration gradient of chemical elements exists. A more complicated case of diffusion is heterophase diffusion, accompanied by the formation of a compound, through which the diffusing atoms will subsequently pass.

It should be noted that the diffusion of atoms is not always directed opposite to the concentration gradient. In other words, it does not always lead to equalization of the inhomogeneities in concentration. In a number of cases, the so-called uphill diffusion is observed, during which atoms of a given type move from a region of lower to a region of higher concentration. Such a process takes place, for example, in the formation of nuclei of a new phase with a composition different from the parent phase and in the presence of inhomogeneous elastic stresses.

A great number of papers have been devoted to the experimental investigation of diffusion in solids. The results of these papers have been reviewed by Barrer [229], Jost [230], Arkharov [231], Bugakov [232], Le Claire [233], and others. Extensive experimental data on self-diffusion have been obtained in a number of papers using tracer atoms [234–235].

At present, two types of diffusion theories in metals and alloys have been developed: the phenomenological and the kinetic microscopic theories. The aim of phenomenological theory of diffusion which will be examined briefly in Sec. 23 is: 1) to derive general macroscopic equations governing the diffusion of substances, 2) to establish a relationship between the chemical diffusion coefficient and the self-diffusion coefficient, 3) to explain the Kirkendall effect, uphill diffusion, etc.

In the kinetic theory of diffusion, the calculations are performed using a definite atomic model of the crystal. Such a theory enables us to explain the temperature and concentration dependence of the diffusion coefficient, as well as its dependence on the degree of long-range order and on the correlation parameters in alloys. A relationship between the diffusion coefficient and energy constants can also be derived.

The kinetic theory of diffusion in pure metals (see, for example, [244]) led to a formula for the variation of the diffusion coefficient D with absolute temperature T

$$D = D_0 e^{-\frac{Q}{RT}}, \tag{22.1}$$

where D_0 is a constant coefficient and Q is the activation energy. This type of relationship is well supported by experimental evidence for pure metals. We note, however, that for alloys there is no basis for expecting a formula of the form (22.1) for the variation of D with T. In fact, diffusing atoms of a given type encounter different conditions and must surmount potential barriers of a different height depending on the nature of the adjacent atoms in the alloy. Different energies are also required for the formation of holes at different sites. Therefore, it is impossible to characterize the diffusion process by a single activation energy. Nevertheless, a plot of the experimentally observed $\ln D$ versus $\frac{1}{T}$ can ordinarily be represented by a straight line. An experimental analysis of the diffusion coefficient as a function of the composition of an alloy (see, for example, [229]) has shown that these relations may have different forms for different alloys. Among them we often encounter the case, where there is a relatively small change of the diffusion coefficient with composition at one end of the phase diagram, and much greater change at the other end.

It is noteworthy that in a number of cases a small amount of impurity may substantially affect the diffusion coefficient. This was observed, for example, by Gertsriken and Dekhtyar [236]. Gruzin, Kornev and Kyrdyumov [237] found that the addition of a small amount (up to 4.5 atomic %) of carbon to γ-iron greatly reduced the activation energy for self-diffusion of iron and decreased the constant coefficient by several orders of magnitude.

The original kinetic theory of diffusion in alloys was developed by using the vacancy mechanism (interchange of position with vacant lattice sites). The case of diffusion of metals across the layer of intermetallic compound by means of the vacancy mechanism has been theoretically treated by Frenkel and Sergeev [238].

In Secs. 24-26, we present the kinetic theory of diffusion in disordered and in ordered alloys, which accounts for the real mechanisms of the movement of atoms between interstitial sites and considers vacancy migration. It should be mentioned that diffusion in alloys is very complicated and cannot be treated rigorously by the modern solid-state theory. To study diffusion we must, therefore, use a greatly simplified model, which will enable us to explain the main qualitative features of the temperature and concentration dependence of the diffusion coefficient. We will be compelled to make a number of simplifying assumptions that are, however, customarily employed in the statistical theory of alloys (see Sec. 14).

23. Principles of the Phenomenological Theory of Diffusion in Alloys

A phenomenological theory of diffusion in alloys should start with the general macroscopic expressions for the flux of the diffusing substances, without considering any specific atomic model. In the initial stages of development of the phenomenological theory of diffusion, the diffusion flux J_α of an alloy component α was assumed to be proportional to the concentration gradient and directed in the opposite direction. In the case where the concentration gradients of all ζ components of the alloy are directed along the x-axis, the expressions for the flux J_α (equal to the numbers of atoms of species α, diffusing across 1 cm^2 of a plane perpendicular to the x-axis per unit time) take the form

$$J_\alpha = -D_\alpha \frac{\partial n_\alpha}{\partial x} \qquad (\alpha = 1, 2, \ldots, \zeta), \tag{23.1}$$

where D_α is the diffusion coefficient and n_α is the number of atoms of species α per unit volume (in this theory D_α is always greater than zero).

According to Eq. (23.1) the diffusion flux is directed from the sites with higher concentration into a region of lower concentration of the diffusing material. Hence, uphill diffusion cannot be explained by this theory, because such diffusion proceeds from sites with lower concentration into a region of higher concentration and does not lead to equalization, but to an increase in the concentration imbalance in the alloy. In order to construct a more general phenomenological theory of diffusion, one should consider the fact that the equilibrium chemical potentials μ_α of all components must have the same value at all points of the alloy. Therefore, if the alloy is not in equilibrium, and diffusion tends to return the alloy to the equilibrium state, the diffusion fluxes must increase with an increase in the chemical potential gradient. In the case of low chemical potential gradients μ_α (this case is usually encountered in experimental investigation of diffusion) we may restrict ourselves to linear terms in $\frac{\partial \mu_\alpha}{\partial x}$ and write the expression for the diffusion fluxes J_α in the following general form:

$$J_\alpha = -\sum_{\alpha'=1}^{\zeta} L_{\alpha\alpha'} \frac{\partial \mu_{\alpha'}}{\partial x}, \tag{23.2}$$

where $L_{\alpha\alpha'}$ are coefficients, that may depend on temperature, pressure and composition of the alloy, and the gradients of chemical potentials are directed along the x axis. As follows from the thermodynamics of irreversible processes, the following relations apply between the coefficients $L_{\alpha\alpha'}$ [239] in the absence of a magnetic field:

$$L_{\alpha\alpha'} = L_{\alpha'\alpha}. \tag{23.3}$$

The fluxes J_α obviously depend on whether a stationary or a moving coordinate system is selected. It is convenient to choose a coordinate system in which the net flux of all components is equal to zero, i.e.,

$$\sum_{\alpha=1}^{\zeta} J_\alpha = 0. \tag{23.4}$$

In the special case when the off-diagonal terms in (23.2) may be neglected (i.e., $L_{\alpha\alpha'}$ are small when $\alpha \neq \alpha'$), we obtain a simpler formula for the diffusion flux J_α

$$J_\alpha = -L_{\alpha\alpha}\frac{\partial \mu_\alpha}{\partial x}. \tag{23.5}$$

If diffusion is realized via a vacancy mechanism, then, in setting up a system of equations (23.2), one of the alloy components may be considered as vacancies, and this type of equation can also be used for determining the diffusion flux of vacancies.

In the case when Eq. (23.5) applies, we may use it to establish a relation between the self-diffusion coefficient (the diffusion coefficient of tracer atoms with vanishing concentration gradients of the chemical substances) and the coefficient of chemical diffusion (when concentration gradients of the chemical substances exist). For this purpose we first apply Eq. (23.5) to the binary alloy $A-B$, where the concentrations n_A and n_B of atoms A and B are constant over the entire crystal, but the concentration gradient of the radioactive isotope of A is directed along the x axis. It is also assumed that no external fields and elastic stresses exist. The thermodynamic potential of an alloy, consisting of N_A atoms A and N_B atoms B, whose isotopic composition is determined by the numbers N_{Ak} of atoms of the different isotopes of element A (k is the number of the isotope) and by the numbers N_{Bl} of atoms of the different isotopes (number l) of element B, is

$$\Phi = \Phi^0(T, P, N_A, N_B) - kT\ln\left[\frac{N_A!}{\prod\limits_k N_{Ak}!}\,\frac{N_B!}{\prod\limits_l N_{Bl}!}\right]. \tag{23.6}$$

Here $\Phi^0(T, P, N_A, N_B)$ is that part of the thermodynamic potential which is independent of isotopic composition, and which corresponds to the case when elements A and B in the alloy have only one isotope each.*

Using Stirling's formula and differentiating Eq. (23.6) with respect to the number of atoms of the kth isotope of element A, we find the chemical potential μ_{Ak} of this isotope

$$\mu_{Ak} = \frac{\partial \Phi}{\partial N_{Ak}} = \mu_A^0 + kT \ln \frac{N_{Ak}}{N_A}, \tag{23.7}$$

where $\mu_A^0 = \frac{\partial \Phi^0}{\partial N_{Ak}} = \frac{\partial \Phi^0}{\partial N_A}$ is independent of the isotopic composition and is equal to the chemical potential of element A when the latter consists of a single (kth) isotope. In the case when the concentration gradients of elements A and B are zero, the quantity μ_A^0 is independent of the coordinates. Since, however, there does exist a concentration gradient of the radioactive isotope, whose number of atoms per unit volume is $n_{Ak} = \frac{N_{Ak}}{V}$ (V is the volume of the crystal) and will be denoted by n_A^*, μ_{Ak} by μ_A^*, the derivative $\frac{\partial \mu_{Ak}}{\partial x} = \frac{\partial \mu_A^*}{\partial x}$ will be different from zero $\frac{\partial \mu_A^*}{\partial x} = \frac{kT}{n_A^*} \frac{\partial n_A^*}{\partial x}$. Substituting this expression for the flux J_A^* of the tracer atoms A into (23.5), we obtain:

$$J_A^* = -\frac{L_{AA}^* kT}{n_A^*} \frac{\partial n_A^*}{\partial x}, \tag{23.8}$$

where L_{AA}^* is the coefficient of proportionality in Eq. (23.5), written for the flux of tracer A atoms. The coefficient of proportionality with $-\frac{\partial n_A^*}{\partial x}$ in Eq. (23.8) is the diffusion coefficient D_A^* of the tracer atoms, or the self-diffusion coefficient of atoms A

$$D_A^* = \frac{L_{AA}^* kT}{n_A^*}. \tag{23.9}$$

---

*In the majority of processes, the isotopic composition remains constant and the second term in (23.6) is usually included in an immaterial function (which is linearly dependent on temperature). In this section the dependence of the thermodynamic potential on the concentration of isotopes will be considered as essential.

We shall now consider the case when a concentration gradient of atoms of an element exists, and the isotopic composition of this element is constant. The quantity μ_A^0, equal to the chemical potential of the A atoms, neglecting the presence of isotopes, may be presented in the form

$$\mu_A^0 = M(P, T) + kT\ln\left(\frac{n_A}{n}\gamma_A\right), \tag{23.10}$$

where $M(P, T)$ is the chemical potential of pure metal A, n is the total number of sites per unit volume, and γ_A is the activity coefficient. Substituting (23.10) into (23.7) and noting the relation $\frac{N_{Ak}}{N_A} = \frac{n_{Ak}}{n_A}$ we find that the chemical potential of the kth isotope of atoms A is

$$\mu_{Ak} = M(P, T) + kT\ln\left(\frac{n_{Ak}}{n}\gamma_A\right). \tag{23.11}$$

In this case, not only the concentrations of isotopes n_{Ak}, but also the activity coefficient γ_A depend on the coordinate x, because the activity coefficient is a function of the concentration n_A, and the latter, in turn, is a variable. Noting this fact, we find* that the derivative $\frac{\partial\mu_{Ak}}{\partial x}$ is equal to $kT\left(\frac{1}{n_{Ak}}\frac{\partial n_{Ak}}{\partial x} + \frac{1}{\gamma_A}\frac{\partial\gamma_A}{\partial n_A}\frac{\partial n_A}{\partial x}\right)$. We substitute this expression into (23.5). Then the flux of the tracer atoms A in the presence of a concentration gradient of A will equal

$$J_A^* = -\frac{L_{AA}^* kT}{n_A^*}\left(1 + \frac{\partial\ln\gamma_A}{\partial\ln n_A}\right)\frac{\partial n_A^*}{\partial x}. \tag{23.12}$$

By determining the coefficient of chemical diffusion D_A as a proportionality factor at $-\frac{\partial n_A^*}{\partial x}$ in Eq. (23.2), we find that D_A is equal to

$$D_A = \frac{L_{AA}^* kT}{n_A^*}\left(1 + \frac{\partial\ln\gamma_A}{\partial\ln n_A}\right). \tag{23.13}$$

---

*Thus, we neglect the usually small change of volume during diffusion, i.e., we assume n to be independent of x.

Comparing (23.9) and (23.13), we find a relation between D_A and D_A^*

$$D_A = D_A^* \left(1 + \frac{\partial \ln \gamma_A}{\partial \ln n_A}\right) = D_A^* \frac{\partial \frac{\mu_A^0}{kT}}{\partial \ln n_A}. \quad (23.14)$$

The relation (23.14) between the coefficient of chemical diffusion and the coefficient of self-diffusion was obtained by Darken [240] (see also reviews [233] and [241]). It must be emphasized that this relation is not a rigorous one since it was obtained by neglecting the off-diagonal terms ($\alpha \neq \alpha'$) in Eq. (23.2). An estimate of the role of the off-diagonal terms may be carried out by means of the kinetic theory of diffusion, based on a specific model of an alloy [242, 243]. However, this subject is still not sufficiently clear at present.

Using the coefficients of chemical diffusion D_A and D_B in the binary alloy $A - B$, we can define the diffusion coefficient D_{AB}, and the velocity of the tracers in Kirkendall experiments. We shall consider the diffusion between two samples of the alloy $A - B$ of a different composition. The thickness of the samples is sufficiently great so that the concentrations of components on the outer ends of the samples may be considered constant during the diffusion. If, for example, the diffusion coefficient of A atoms is greater than that of B atoms, in a sample enriched with atoms A, vacancies will appear and then disappear because of plastic deformation. Therefore, in addition to the oppositely directed diffusion fluxes of atoms A and B relative to the atomic planes of the crystal lattice, one should observe a motion of the atomic planes of the crystal lattice associated with plastic flow. The rate of this flow v, agrees with the velocity of the macroscopic inclusions in the lattice (for example, of molybdenum in the experiments of Kirkendall).

We choose a coordinate system which is stationary relative to the outer end of the sample in the vicinity of which no diffusion occurs, letting the direction of the x axis coincide with the direction of the concentration gradient. The flux of A atoms at a point x across a plane perpendicular to the x axis, is given in terms of both the diffusion and the lattice flow by

$$-D_A \frac{\partial n_A}{\partial x} + n_A v. \quad (23.15)$$

A similar flux across the plane $x + dx$ is given by

$$-D_A \frac{\partial n_A}{\partial x} + n_A v + \frac{\partial}{\partial x}\left(-D_A \frac{\partial n_A}{\partial x} + n_A v\right) dx. \quad (23.16)$$

The difference between these two fluxes must be equal to the change of the number of atoms A per unit time in a layer between two planes drawn perpendicular to the x axis at points x and $x+dx$. The area of these planes is equal to unity. Such a change in the number of atoms is obviously equal to $\frac{\partial n_A}{\partial t}dx$. Therefore,

$$\frac{\partial n_A}{\partial t}=\frac{\partial}{\partial x}\left(D_A\frac{\partial n_A}{\partial x}-vn_A\right). \tag{23.17}$$

Considering the B atoms in the same manner, we obtain

$$\frac{\partial n_B}{\partial t}=\frac{\partial}{\partial x}\left(D_B\frac{\partial n_B}{\partial x}-vn_B\right). \tag{23.18}$$

Since it is assumed that the vacancies formed during diffusion disappear as a result of plastic deformation (and that the change in the specific volume with composition is neglected), the total concentration n_A+n_B may be regarded as constant. Therefore, $\frac{\partial(n_A+n_B)}{\partial t}=0$, and adding (23.17) and (23.18), we obtain

$$\frac{\partial}{\partial x}\left[D_A\frac{\partial n_A}{\partial x}+D_B\frac{\partial n_B}{\partial x}-v\left(n_A+n_B\right)\right]=0. \tag{23.19}$$

Choosing the origin of the coordinates at the outer end, where no diffusion takes place, we obtain $\frac{\partial n_A}{\partial x}=0$, $\frac{\partial n_B}{\partial x}=0$ at $x=0$ and $v=0$ at all times. The expression in square brackets in Eq. (23.19) should not be a function of x, and since at $x=0$ (and for all values of t) it is equal to zero, then for all values of x and t

$$D_A\frac{\partial n_A}{\partial x}+D_B\frac{\partial n_B}{\partial x}-v(n_A+n_B)=0. \tag{23.20}$$

From Eq. (23.20) we may determine the velocity v. Recalling that $\frac{\partial n_B}{\partial x}=-\frac{\partial n_A}{\partial x}$, we obtain

$$v=\frac{1}{n_A+n_B}\left(D_A-D_B\right)\frac{\partial n_A}{\partial x}. \tag{23.21}$$

It is apparent from (23.21) that the velocity of the inclusions v is proportional to the difference of the coefficients of chemical diffusion of A and B atoms and to the concentration gradient $\frac{\partial n_A}{\partial x}$.

The velocity v in (23.17) can be eliminated with the aid of Eq. (23.21)

$$\frac{\partial n_A}{\partial t} = \frac{\partial}{\partial x}\left[\left(\frac{n_A}{n_A+n_B} D_B + \frac{n_B}{n_A+n_B} D_A\right)\frac{\partial n_A}{\partial x}\right]. \tag{23.22}$$

Noting that $\frac{n_A}{n_A+n_B} = c_A$ and $\frac{n_B}{n_A+n_B} = c_B$, we rewrite (23.22) in the form

$$\frac{\partial c_A}{\partial t} = \frac{\partial}{\partial x}\left[(c_A D_B + c_B D_A)\frac{\partial c_A}{\partial x}\right]. \tag{23.23}$$

Eq. (23.23) has the form of the ordinary diffusion equation

$$\frac{\partial c_A}{\partial t} = \frac{\partial}{\partial x}\left(D_{AB}\frac{\partial c_A}{\partial x}\right) \tag{23.24}$$

with the diffusion coefficient

$$D_{AB} = c_A D_B + c_B D_A. \tag{23.25}$$

If after the diffusion annealing of the two adjoining samples of alloys of different compositions, the distribution of concentration c_A of atoms A (or the concentration $c_B = 1 - c_A$ of atoms B) is studied, and then this distribution is used to determine the diffusion coefficient, we will obtain the quantity D_{AB}, which is called the mutual diffusion coefficient. Thus, such types of experiments measured the quantity D_{AB}, and not the diffusion coefficients D_A and D_B of the individual components.

Measuring simultaneously the mutual diffusion coefficient D_{AB} and the velocity of the inclusions v, Eqs. (23.25) and (23.21) can be used to determine the coefficients of chemical diffusion D_A and D_B.

Let us note that if the concentration of one of the components of the alloy tends to zero ($c_A \ll 1$), as is apparent from (23.25), the coefficient of mutual diffusion agrees approximately with the diffusion coefficient D_A of that component which is considered as an impurity. D_{AB} coincides with D_A and with D_B only when $D_A = D_B$. Thus, according to (23.21) no plastic flow takes place. Usually, such behavior takes place in the case of mutual diffusion of isotopes.

The coefficient of mutual diffusion and the velocity of the inclusion may be expressed by Eqs. (23.25), (23.21) and (23.14) in terms of the self-diffusion coefficients D_A^* and D_B^*. Taking into account the Gibbs-Duhem relation

$$\frac{\partial \ln \gamma_A}{\partial \ln n_A} = \frac{\partial \ln \gamma_B}{\partial \ln n_B}, \tag{23.26}$$

we obtain

$$D_{AB} = (c_A D_B^* + c_B D_A^*)\left(1 + \frac{\partial \ln \gamma_A}{\partial \ln c_A}\right), \tag{23.27}$$

$$v = (D_A^* - D_B^*)\left(1 + \frac{\partial \ln \gamma_A}{\partial \ln c_A}\right)\frac{\partial c_A}{\partial x}. \tag{23.28}$$

Eq. (23.27) is in good agreement with the experiments of Johnson [244] concerning diffusion in Ag-Au alloys. In these experiments measurements of the diffusion coefficients of tagged Ag and Au atoms were made in a chemically homogeneous sample of an alloy with a 50% composition. Also, the coefficient of mutual diffusion between the samples, whose composition differed only slightly from the 50% composition, were determined. Using also the experimental values of activity [245] for this alloy, Darken [240] obtained satisfactory agreement between the values of D_{AB} calculated from Eq. (23.27) and measured experimentally. Satisfactory agreement has also been obtained in the comparison of the experimentally found velocity of inclusions in the Au-Ag alloy with the value of v calculated from Eq. (23.28) [246].

As follows from Eqs. (23.14), (23.27) when

$$1 + \frac{\partial \ln \gamma_A}{\partial \ln n_A} = \frac{1}{kT}\frac{\partial \mu_A^0}{\partial \ln n_A} = \frac{n_A}{kT}\frac{\partial \mu_A^0}{\partial n_A} = \frac{n_A}{kT}\frac{\partial^2 \varphi}{\partial n_A^2} < 0 \tag{23.29}$$

(where φ is the thermodynamic potential per unit volume of the alloy), the diffusion coefficients D_A, D_B and D_{AB} are negative. Thus, the diffusion fluxes are directed along the gradients of the corresponding concentrations, i.e., the diffusion of atoms of a given species occurs from sites with a lower concentration into a site with a higher concentration. Consequently, uphill diffusion occurs here. An interesting case of uphill diffusion has been considered by Konodevskiy [247], who considered the effect of an inhomogeneous elastic deformation on diffusion in alloys. Lyubov and Fastov [248] studied this problem in the framework of the phenomenological theory.

24. Theory of Diffusion of Interstitial Atoms through Interstitial Sites

In contrast to the phenomenological theory of diffusion expounded in the preceding section, we shall now consider the theory of diffusion of atoms in alloys, based on the application of a specific atomic model of an alloy.*

1. *Diffusion of interstitial atoms in alloys with a body-centered cubic lattice [249-251]*

Of the two most probable diffusion mechanisms indicated above, we shall first investigate the simplest diffusion mechanism of atoms of a third element through the interstitial sites of the crystal lattice of a binary alloy.

We shall consider a substitutional type alloy of metals A and B, having a β-brass body-centered cubic lattice, which may be found either in an ordered or a disordered state. Let atoms of any third element C be introduced at an interstitial lattice site of the alloy A–B. We assume that the C atoms have positions of stable equilibrium at the centers of the faces and at the middle of the edges of the cubic unit cells.** The first position which is called (in Sec. 20) the first type of interstitial site will be denoted by O_1, and the second type by O_2 (Fig. 99). As indicated in Sec. 20, the interstitial sites O_1 have four neighboring sites of the first type at a distance $\frac{a}{\sqrt{2}}$ and two neighboring sites of the second type at the distance $a/2$ (a is the length of the edge of a cubic cell). The interstitial sites O_2 have four neighboring sites of the second type at a distance $\frac{a}{\sqrt{2}}$ and two neighboring sites of the first type at a distance $a/2$. Here, we shall note not only the two nearest neighbor atoms, which would give results that are too inaccurate, but also the next nearest four atoms.

We shall denote the position of the interstitial C atoms, corresponding to the tops of the potential barriers for a transition from one interstitial site to a neighboring one, by the letter P (see Fig. 99). Since substitutional alloys consist of atoms A and B that do not differ greatly in atomic radii, one

---

*Just as in the statistical theory of ordering, the interaction energies of atoms will be considered as constants that are independent of composition and of degree of order.

**For example, in the case of a solution of carbon in α-iron this approximation is supported by x-ray studies [252].

may regard the position P as located in the middle of a straight line joining the middle of the sites O_1 and O_2. Then, the positions P have four neighboring sites at a distance $\frac{\sqrt{5}}{4}a$. Two of them will be sites of the first type, and two of the second type.

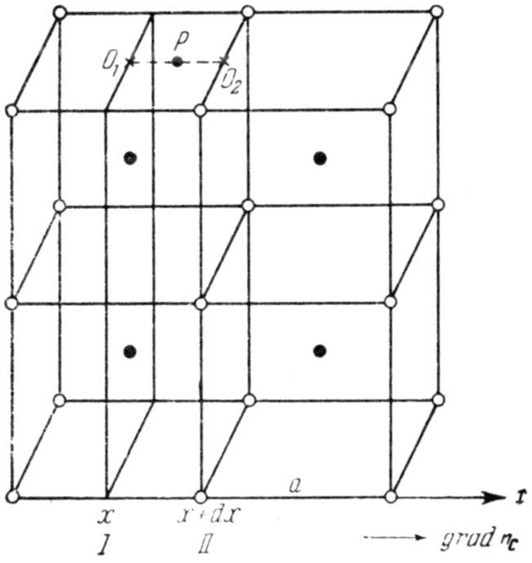

Fig. 99. Equilibrium positions for atoms, introduced at the interstitial sites of a body-centered cubic lattice.

In addition to the notation v_{AC}, v_{BC}, v'_{AC} and v'_{BC} introduced in accordance with (20.3) for the negative energy of interaction of the pairs of atoms CA and CB at distances $a/2$ and $\frac{a}{\sqrt{2}}$, we also introduce the notation

$$v_{AC}\left(\frac{\sqrt{5}}{4}a\right)=v''_{AC},\quad v_{BC}\left(\frac{\sqrt{5}}{4}a\right)=v''_{BC}. \tag{24.1}$$

We shall choose the direction of the x axis to be along an edge of the cubic cell and assume that a concentration gradient of the interstitial C atoms is produced in this direction. Such a choice of direction of the gradient does not limit the generality of the discussion, because diffusion is isotropic in cubic crystals. We shall consider two neighboring atomic planes I and II, perpendicular to the x axis. Let plane I pass through the centers of the cubes, and plane II through their corners. The number of C atoms per cm³ of the alloy is denoted by n_C. We assume n_C to be small compared with the total number n of atoms A and B at the sites. Thus, $n_C = n_{C1} + n_{C2}$, where n_{C1} is the concentration of atoms C at the interstitial sites O_1 and n_{C2} is the same concentration at O_2. It is easy to show that 1 cm² of plane I contains $\frac{2}{3}an_{C1}(x)$ of C atoms at interstitial sites of type O_1, and 1 cm² of plane II contains $\frac{2}{3}an_{C2}\left(x+\frac{a}{2}\right)$ atoms C at type O_2 interstitial sites.

The diffusing atoms C, which occupy different interstitial sites O_1 (or O_2), occur under different energy conditions, because they are surrounded by different numbers of atoms A and B at the six sites neighboring the interstitial sites. Furthermore, we shall explicitly take into account all possible configurations of atoms A and B at these sites. We shall divide all n_{C1} atoms C into groups, each of which contains a C atom with the same configuration of A and B atoms at the neighboring sites around the interstitial site O_1. The number of C atoms in such groups will be noted by n^m_{C1} (m is the number of the configuration). We shall limit ourselves to the usual case in which the mobility of atoms C in the alloy is substantially greater than the mobility of atoms A and B. Then, with a change of the state of ordering of alloy $A-B$, atoms C are able to reach the equilibrium state. In the case under consideration, when the number of interstitial atoms is small compared with the number of interstitial sites of the crystal, the number n^m_{C1} may be determined from the Boltzmann distribution

$$n^m_{C1} = \Lambda W^m_{O_1} \exp\left(\frac{U^m_O}{kT}\right). \tag{24.2}$$

Here Λ is a normalization constant, $W^m_{O_1}$ the probability of the attainment of the mth configuration around the interstitial site O_1, and U^m_O is the negative of the potential energy of atom C at this interstitial site (Fig. 100). Similarly, the number n^m_{C2} of C atoms, surrounded by the neighboring atoms with the configurations m around the interstitial site O, is equal to

$$n^m_{C2} = \Lambda W^m_{O_2} \exp\left(\frac{U^m_O}{kT}\right), \tag{24.3}$$

where $W^m_{O_2}$ is the probability of the mth configuration around O_2.

The C atom can make a transition into a neighboring interstitial site by surmounting a barrier with a height $U^m_O - U^m_P$, where U^m_P is the negative of the potential energy of atom C at position P. Thus, for a given structure the neighbors of point P are simultaneously neighbors of O_1, so that both energies U^m_O and U^m_P are uniquely determined by assignment of the mth configuration. The transition $\mathfrak{W}^m$ of an atom C from a given

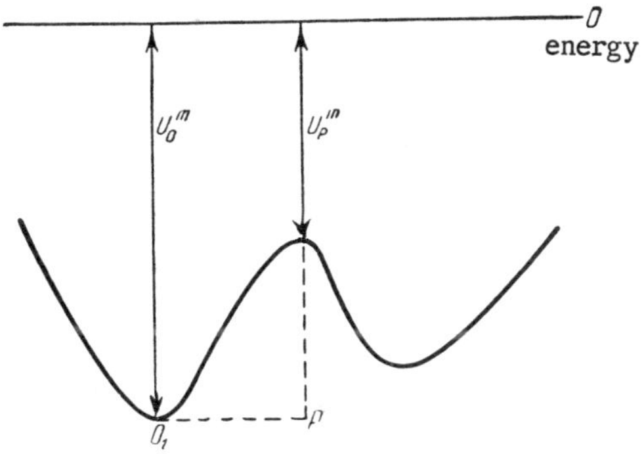

Fig. 100. Potential energy of the diffusing atom.

interstitial site with an mth configuration of the atoms surrounding it, into a neighboring interstitial site is determined by the known formula [224]

$$\mathfrak{W}^m = \frac{1}{\tau_0} \exp\left(-\frac{U_O^m - U_P^m}{kT}\right), \tag{24.4}$$

where τ_0 is a constant, having the dimensions of time and equal in order of magnitude (according to [224]) to 10^{-13} sec. The magnitude of τ_0 will be regarded as approximately the same for all configurations.*

The number of C atoms, having the mth configuration of neighbors and passing per unit time per cm² of plane I into plane II, may be then written in the following form (transitions from the O_2 sites, located in the plane I are obviously impossible):

$$J^m_{I\to II} = \frac{2}{3} a n^m_{C1} \mathfrak{W}^m = \frac{2}{3} \frac{a}{\tau_0} \Lambda(x) W^m_{O1} \exp\left(\frac{U_P^m}{kT}\right). \tag{24.5}$$

The total number of C atoms, passing per unit time per cm² from plane I to plane II, will equal

$$J_{I\to II} = \frac{2}{3} \frac{a}{\tau_0} \Lambda(x) \sigma_1, \tag{24.6}$$

where

$$\sigma_1 = \sum_m W^m_{O_1} \exp\left(\frac{U_P^m}{kT}\right)$$

*Strictly speaking, this assumption means that we limit ourselves to the cases where τ_{0A} for the diffusion of C atoms into a pure metal A differs little from τ_{0B} for the diffusion into a pure metal B. There are grounds, however, for assuming that in substitutional alloys $A-B$, formed of atoms A and B with nearly equal atomic radii and having the same type of fields of forces, τ_0 will not depend critically on the number of the configuration.

and the summation extends over all 64 configurations. Similarly, an expression may be found for the inverse flux of C atoms from plane II to I

$$J_{II \to I} = \frac{2}{3} \frac{a}{\tau_0} \Lambda \left(x + \frac{a}{2} \right) \sigma_1. \tag{24.7}$$

The normalization constant Λ is determined from the condition

$$\sum_m n_{C1}^m + \sum_m n_{C2}^m = n_C.$$

With the aid of (24.2) and (24.3) we obtain

$$\Lambda = \frac{n_C(x)}{\sigma_2},$$

where

$$\sigma_2 = \sum_m \left(W_{O_1}^m + W_{O_2}^m \right) \exp\left(\frac{U_O^m}{kT} \right). \tag{24.8}$$

From (24.6), (24.7), and (24.8) it follows that the net diffusion flux of C atoms is

$$J = J_{I \to II} - J_{II \to I} = -\frac{1}{3} \frac{a^2}{\tau_0} \frac{\sigma_1}{\sigma_2} \frac{dn_C}{dx}. \tag{24.9}$$

Comparing (24.9) with (23.1), which defines the diffusion coefficient, and introducing the notation

$$D_0 = \frac{1}{3} \frac{a^2}{\tau_0}, \tag{24.10}$$

we obtain the following expression for the diffusion coefficient of C atoms:

$$D = D_0 \frac{\sigma_1}{\sigma_2} = D_0 \frac{\sum_m W_{O_1}^m \exp\left(\frac{U_P^m}{kT} \right)}{\sum_m \left(W_{O_1}^m + W_{O_2}^m \right) \exp\left(\frac{U_O^m}{kT} \right)}. \tag{24.11}$$

To determine the dependence of the diffusion coefficient on the alloy composition, temperature and degree of order, we should obtain explicit expressions for the probabilities $W_{O_1}^m$ and $W_{O_2}^m$. If we neglect correlation between the substitution of sites in the alloy, the quantities $W_{O_1}^m$ and $W_{O_2}^m$ can be expressed as a product of six a priori probabilities of substituion of neighboring sites with interstitial lattice sites by the given atoms. These probabilities are related to the concentrations of atoms c_A and c_B, and also to the degree long-range order η by formulas (1.5), in which we must set $\nu = \frac{1}{2}$. Using the fact that the quantities U_O^m and U_P^m are sums of the negative interaction energies of atom C with the neighboring atoms of the alloy $A-B$, we can readily calculate the sums occurring in (24.11). Using Eqs. (24.1) and (1.5), we obtain

$$D = D_0 \frac{K_+''^2 K_-''^2}{K_+'^4 K_-^2 + K_-'^4 K_+^2}, \qquad (24.12)$$

where

$$\left.\begin{aligned}
K_+ &= \left(c_A + \frac{1}{2}\eta\right) e^{\frac{v_{AC}}{kT}} + \left(c_B - \frac{1}{2}\eta\right) e^{\frac{v_{BC}}{kT}},\\
K_- &= \left(c_A - \frac{1}{2}\eta\right) e^{\frac{v_{AC}}{kT}} + \left(c_B + \frac{1}{2}\eta\right) e^{\frac{v_{BC}}{kT}},\\
K'_+ &= \left(c_A + \frac{1}{2}\eta\right) e^{\frac{v'_{AC}}{kT}} + \left(c_B - \frac{1}{2}\eta\right) e^{\frac{v'_{BC}}{kT}},\\
K'_- &= \left(c_A - \frac{1}{2}\eta\right) e^{\frac{v'_{AC}}{kT}} + \left(c_B + \frac{1}{2}\eta\right) e^{\frac{v'_{BC}}{kT}},\\
K''_+ &= \left(c_B + \frac{1}{2}\eta\right) e^{\frac{v''_{AC}}{kT}} + \left(c_B - \frac{1}{2}\eta\right) e^{\frac{v''_{BC}}{kT}},\\
K''_- &= \left(c_A - \frac{1}{2}\eta\right) e^{\frac{v''_{AC}}{kT}} + \left(c_B + \frac{1}{2}\eta\right) e^{\frac{v''_{BC}}{kT}}.
\end{aligned}\right\} \qquad (24.13)$$

In particular, for a disordered alloy, when $\eta = 0$, from (24.12) and (24.13) we obtain

$$D = D_{\text{dis}} = \frac{D_0}{2} \frac{\left(c_A e^{\frac{v''_{AC}}{kT}} + c_B e^{\frac{v''_{BC}}{kT}}\right)^4}{\left(c_A e^{\frac{v'_{AC}}{kT}} + c_B e^{\frac{v'_{BC}}{kT}}\right)^4 \left(c_A e^{\frac{v_{AC}}{kT}} + c_B e^{\frac{v_{BC}}{kT}}\right)^2}. \qquad (24.14)$$

To investigate the temperature and concentration dependence of the diffusion coefficient, we must allow for the fact that the degree of long-range order η in (24.12), for the equilibrium state of alloy $A-B$, is also a function of temperature and composition. For an approximate evaluation of this dependence in alloys of a given structure, in which the order-disorder transition is a second-order phase transition, we can use Eq. (16.8). The latter was obtained in the statistical theory of ordering by neglecting correlation (for alloys of a given structure $z=8$).

We shall consider now the case of a disordered alloy. As indicated in Sec. 24, there is no basis for expecting a linear dependence of $\ln D$ on $1/T$ for alloys, although such a dependence is usually observed rather accurately in experiments. In fact, inspection of Eq. (24.14) shows that no such dependence occurs. However, when the constants of (20.3) and (24.1) are chosen to be of the same order of magnitude, as is usually the case in real alloys, the deviations from a straight line given by (24.14) are small and are noticeable only over a wide range of temperatures (which are usually included in the experimental investigation). The dependence of D on c_A, according to (24.14), may be of a different type. The following case is apparently rather common: when there is a slight dependence of D on composition at one end of the concentration diagram, and a considerably stronger dependence at the other end.

We shall now consider the case of ordered alloys. In ordered alloys at equilibrium, as shown in Chap. III (see Fig. 13), as the temperature decreases below the critical temperature T_0, the quantity η continuously increases from zero to the maximum value. This increase occurs first rapidly, and then more and more slowly. This leads to a characteristic sharp change in the diffusion coefficient D determined from (24.12) at T which are lower than but close to T_0. The curve of $\ln D$ vs $1/T$ has a break at $T=T_0$, then as T decreases the curve deviates greatly from the almost linear behavior characteristic of disordered alloys. Finally, at sufficiently low temperatures, the curve takes again a form differing little from a straight line. Figure 101 gives a typical curve, illustrating such a relation (for the case where $c_A=1/2$, $v_{AC}=v_{BC}=0.75$ ev, $v'_{AC}=0.3$ ev, $v'_{BC}=0.2$ ev, $v''_{AC}=0.4$ ev, $v''_{BC}=0.25$ ev and the ordering energy $w=0.0352$ ev). Here, the dependence of η on T was determined from (16.9).

For a more detailed investigation of the temperature range near T_0 (for $T\lessdot T_0$), where $\eta\ll 1$, we expand (24.12) for D into a power series in η and limit ourselves to quadratic terms. This gives

$$D=D_{\text{dis}}(1-b\eta^2), \tag{24.15}$$

where D_{dis} is the diffusion coefficient for the disordered alloy, which can be determined from (24.14), and

$$b=\frac{1}{4}\left(\frac{e^{\frac{v_{AC}}{kT}}-e^{\frac{v_{BC}}{kT}}}{c_Ae^{\frac{v_{AC}}{kT}}+c_Be^{\frac{v_{BC}}{kT}}}\right)^2+\frac{3}{2}\left(\frac{e^{\frac{v'_{AC}}{kT}}-e^{\frac{v'_{BC}}{kT}}}{c_Ae^{\frac{v'_{AC}}{kT}}+c_Be^{\frac{v'_{BC}}{kT}}}\right)^2+$$

$$+\frac{1}{2}\left(\frac{e^{\frac{v''_{AC}}{kT}}-e^{\frac{v''_{BC}}{kT}}}{c_Ae^{\frac{v''_{AC}}{kT}}+c_Be^{\frac{v''_{BC}}{kT}}}\right)^2-2\,\frac{\left(e^{\frac{v_{AC}}{kT}}-e^{\frac{v_{BC}}{kT}}\right)\left(e^{\frac{v'_{AC}}{kT}}-e^{\frac{v'_{BC}}{kT}}\right)}{\left(c_Ae^{\frac{v_{AC}}{kT}}+c_Be^{\frac{v_{BC}}{kT}}\right)\left(c_Ae^{\frac{v'_{AC}}{kT}}+c_Be^{\frac{v'_{BC}}{kT}}\right)}. \tag{24.16}$$

It is apparent from (24.16) that the sign of b may be either positive or negative, although the most likely (typical) case is that of a positive b corresponding to the form of the curve $\ln D$ vs $1/T$, illustrated in Fig. 101. For $b<0$ this curve should lie above the dotted curve, which illustrates this relationship for the case when the given alloy remains disordered also at $T<T_0$.

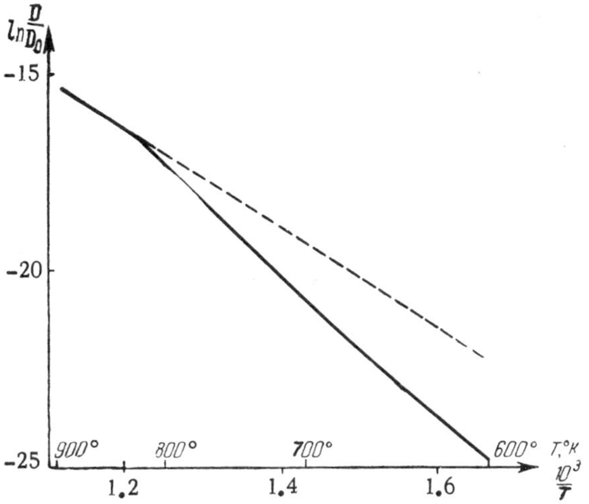

Fig. 101. Approximate correlation between the logarithm of the diffusion coefficient in alloys with a β-brass type crystal lattice and the reciprocal of temperature.

The slope of this curve can be characterized by introducing an effective activation energy defined by

$$Q=-\frac{\partial \ln D}{\partial \frac{1}{kT}}. \tag{24.17}$$

In the case where the dependence of $\ln D$ on $1/T$ is linear, the Q defined in this manner will obviously coincide with the activation energy.

The derivative of η^2 with respect to $1/kT$ for $T = T_0$, changes discontinuously from zero to some finite value, with a decrease in temperature. This derivative, according to (16.17), for an alloy with the structure under consideration ($z = 8$) will be equal to

$$\left.\frac{\partial \eta^2}{\partial \dfrac{1}{kT}}\right|_{T=T_0-0} = 96\, w \frac{c_A^3 c_B^3}{c_A^3 + c_B^3}. \tag{24.18}$$

Consequently, Q has a discontinuity, whose magnitude, according to (24.17), (24.15) and (24.18) is given by

$$\Delta Q = Q|_{T_0-0} - Q|_{T_0+0} = 96\, b(T_0) \frac{c_A^3 c_B^3}{c_A^3 + c_B^3} w. \tag{24.19}$$

To estimate the possible values of the discontinuity ΔQ, we shall consider the case in which atoms C have at all distances such a greater interaction energy with atoms A than with atoms B that

$$e^{\frac{v_{AC}}{kT_0}} \gg e^{\frac{v_{BC}}{kT_0}}, \quad e^{\frac{v'_{AC}}{kT_0}} \gg e^{\frac{v'_{BC}}{kT_0}}, \quad e^{\frac{v''_{AC}}{kT_0}} \gg e^{\frac{v''_{BC}}{kT_0}}. \tag{24.20}$$

Then we may approximately set $b(T_0) = \frac{1}{4c_A^2}$ and from (24.19) obtain

$$\Delta Q = \frac{24 w c_A c_B^3}{c_A^3 + c_B^3}. \tag{24.21}$$

For example, at $c_A = c_B = \frac{1}{2}$ from (24.21) we obtain $\Delta Q = 6w$. Choosing a value for T_0 close to that observed experimentally, for example $T_0 = 810\,°\mathrm{K}$, from (16.11) at $c_A = \frac{1}{2}$ we find $w = 0.035$ ev, which gives $\Delta Q = 0.21$ ev, i.e., a magnitude that is easily measured with present-day experimental accuracy. It is not difficult to show that (24.19) in individual cases may lead to even higher values of ΔQ (for example, for $v_{AC} = v_{BC}$, $e^{\frac{v'_{AC}}{kT_0}} \gg e^{\frac{v'_{BC}}{kT_0}}$, $e^{\frac{v''_{AC}}{kT_0}} \gg e^{\frac{v''_{BC}}{kT_0}}$ we obtain ΔQ, which is eight times greater than in the case just examined).

We shall now investigate the concentration dependence of the diffusion coefficient (at constant temperature). Since the degree of long-range order η (as distinguished from the degree of long-range order η_*, defined by Eq. (1.8)) at $c_A = \frac{1}{2}$ does not have a break, $D(c_A)$ will also lack a break at this point. However, the curves $D(c_A)$ (for different T) exhibit a break at those values of concentration $c_A = c_0$ and $c_A = 1 - c_0$ where the given temperature T is the order-disorder transition (critical) temperature. At these values of concentration, as follows from Eq. (24.15), $\frac{\partial D}{\partial c_A}$ changes by the magnitude

$$\delta = \left(\frac{\partial D}{\partial c_A}\right)_{\text{ord}} - \left(\frac{\partial D}{\partial c_A}\right)_{\text{dis}} = -\left(D_{\text{dis}} b \frac{\partial \eta^2}{\partial c_A}\right)_{\substack{c_A = c_0 \\ c_A = 1 - c_0}} \tag{24.22}$$

As mentioned above, those cases in which b is positive are typical, i.e., the curve $D(c_A)$ at the points $c_A = c_0$ and $c_A = 1 - c_0$ during passage into the region of ordered alloys has a change of slope. This type of curve is illustrated in Fig. 102 (where we set $T = 650\,°\text{K}, v_{AC} = v_{BC} = 0.825$ ev, $v'_{AC} = 0.3$ ev, $v'_{BC} = 0.19$ ev, $v''_{AC} = 0.4$ ev, $v''_{BC} = 0.3$ ev and $w = 0.0352$ ev). The magnitudes of ΔQ (for $c_A = c_0$) and δ [for $T_0(c_0)$] are related to one another. To explain this relation it is convenient to consider the discontinuity not of $\frac{\partial D}{\partial c_A}$, but of the magnitude $\frac{\partial \ln D}{\partial c_A}$:

$$\delta^* = \left(\frac{\partial \ln D}{\partial c_A}\right)_{\text{ord}} - \left(\frac{\partial \ln D}{\partial c_A}\right)_{\text{dis}} = \frac{1}{D_{\text{dis}}}\delta. \tag{24.23}$$

The discontinuity (step change) δ^*, as follows from (24.23), (24.22), (24.19) and (24.18), is related to ΔQ by an expression where all the energy constants, with the exception of w, have been cancelled

$$w\frac{\delta^*}{\partial Q} = -w\frac{\dfrac{\partial \eta^2}{\partial c_A}}{\dfrac{\partial \eta^2}{\partial \dfrac{1}{kT}}} = \frac{d\left(\dfrac{w}{kT_0}\right)}{dc_A}. \tag{24.24}$$

Using Eq. (16.11) we obtain

$$\frac{\delta^*}{\Delta Q} = \left(\frac{c_A - c_B}{8wc_A^2c_B^2}\right)_{c_A = c_0} = \left(\frac{1}{kT_0}\frac{c_A - c_B}{c_Ac_B}\right)_{c_A = c_0}. \tag{24.25}$$

This relation can be checked experimentally.

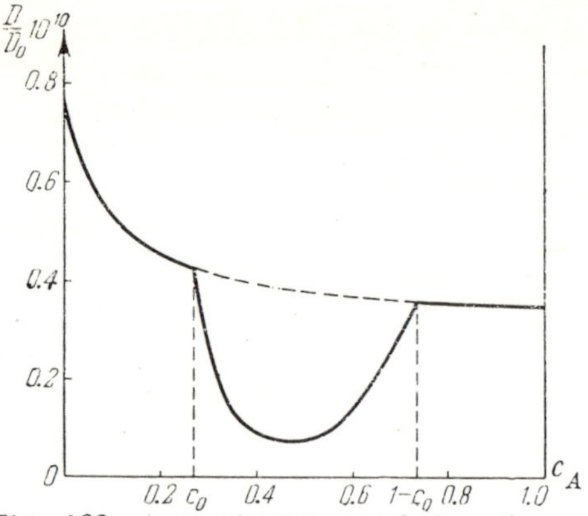

Fig. 102. Approximate correlation of concentration and the diffusion coefficient in alloys with a β-brass type crystal lattice.

Equation (24.12) is applicable also to those cases in which the sequence of the alternation of atoms of the components is practically independent of temperature. This occurs in compounds, and also in ordered alloys existing in a quenched state with a definite temperature-independent degree of order η, but which can be studied at high temperatures so that the diffusion of interstitial atoms C is still marked. Apparently such a case may occur in the diffusion of hydrogen. For this type of substance there is no break on the curve of $\ln D$ vs $1/T$ and no sharp deviations from rectilinear behavior caused by rapid variations of η with temperature will occur.

The subject of the diffusion of interstitial atoms in ordered alloys was first treated [253] by means of an approximate method often used in the theory of alloys, which may be called the method of mean energies. In this method one disregards the various configurations and assumes that all atoms at position O_1, O_2 or P have the same potential energy, equal to the mean potential energy of all positions under consideration. The following expression was obtained by this method for the diffusion coefficient in an alloy considered here:

$$D = \frac{D_0}{2}\exp\left\{\frac{2}{kT}\left[c_A v_{AC} + c_B v_{BC} + 2\left(c_A v'_{AC} + c_B v'_{BC}\right) - \right.\right.$$
$$\left.\left. - 2\left(c_A v''_{AC} + c_B v''_{BC}\right)\right]\right\}\left[\operatorname{ch}\frac{\left(2v'_{AC} - 2v'_{BC} - v_{AC} + v_{BC}\right)\eta}{kT}\right]^{-1}. \qquad (24.26)$$

This formula enables us to obtain qualitatively most of the results mentioned above, with one exception: For disordered alloys the variation of $\ln D$ with $1/T$ resulting from (24.26) is always linear during the passage into an ordered state but has a break in the curves for the temperature and concentration dependence of the diffusion coefficient, which according to (24.6) are always directed towards the x axis.

We shall clarify next the limits of applicability of the method of mean energies. We note first of all, that for $c_A = \frac{1}{2}$ and $\eta = 1$, Eq. (24.6) coincides with (24.12), which is to be expected, because in this case there exists only one, completely definite configuration of atoms around all points O_1, O_2 and P. In the general case of arbitrary c_A, however, approximate agreement of these formulas may be expected only when the following inequality is fulfilled:

$$\frac{|v_{AC} - v_{BC}|}{kT} \ll 1, \quad \frac{|v'_{AC} - v'_{BC}|}{kT} \ll 1, \quad \frac{|v''_{AC} - v''_{BC}|}{kT} \ll 1. \quad (24.27)$$

The power series expansion of the expression for D in these small quantities shows that the above formulas coincide up to the first order terms. These terms, however, do not contain η, and therefore this result can be applied only to disordered alloys ($\eta = 0$). The largest terms containing η are found to be second order terms with respect to the small quantities (24.27). The agreement of these formulas for D results from fulfillment of the additional condition

$$2(v''_{AC} - v''_{BC})^2 = 2(v'_{AC} - v'_{BC})^2 + (v_{AC} - v_{BC})^2. \quad (24.28)$$

Thus the method of mean energies can be applied approximately to the diffusion problem with disordered alloys when (24.27) is fulfilled, and with ordered alloys when the additional condition (24.28) is fulfilled; e.g., in the case where $|v'_{AC} - v'_{BC}| \approx |v''_{AC} - v''_{BC}|$ and $v_{AC} \approx v_{BC}$. However, the basic qualitative features of the temperature and concentration dependence of the diffusion coefficient may be demonstrated by using the method of mean energies.

The expression for the diffusion coefficient also takes a simple form in another limiting case, when atoms C have greatly differing energies of interaction with atoms A and B. Thus, for example, if conditions (24.20) are fulfilled, Eq. (24.12) becomes

$$D = \frac{D_0}{2} \frac{1}{c_A^2 + \frac{1}{4}\eta^2} \exp\left\{\frac{4v''_{AC} - 4v'_{AC} - 2v_{AC}}{kT}\right\}. \quad (24.29)$$

In this case the curves of $\ln D$ vs $1/T$ and D vs c_A, during a transition of the alloy into the ordered state, always bend toward the abscissa axis.

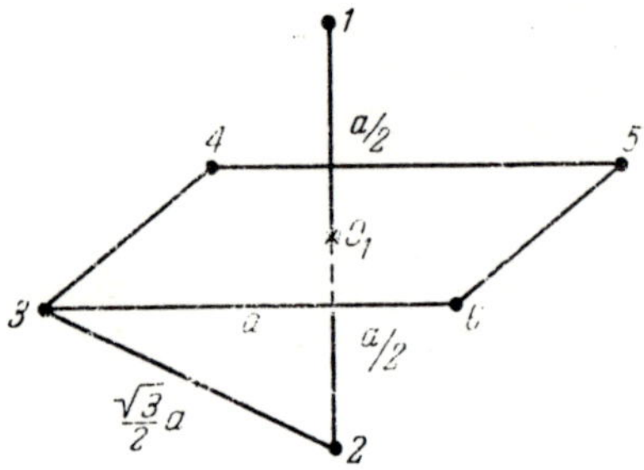

Fig. 103. Arrangement of sites around an interstitial atom in an octahedral interstitial site of an alloy with a body-centered cubic lattice.

We shall now examine the problem of determining the diffusion coefficient under the assumptions previously adopted, but taking into account correlation between the substitution of different lattice sites by atoms A and B. In this case the probabilities $W^m_{O_1}$ and $W^m_{O_2}$, occurring in (24.11), can no longer be calculated as products of six a priori probabilities. To determine $W^m_{O_1}$ we shall fill sites 1-6 in succession (Fig. 103) with atoms A and B, corresponding to the given mth configuration. Strictly speaking, $W^m_{O_1}$ is equal to the product of the a posteriori probabilities of substitution of these sites, calculated in the case of an actual substitution of previously filled sites (see, for example, [192]). Correlation can be approximately taken into account, however, only between the nearest neighbors. It is then possible to neglect correlation in the filling of sites 1 and 2, and also in filling of sites 3, 4, 5 and 6. The magnitude $W^m_{O_1}$ may therefore be determined as a product of two a priori probabilities of substitution of sites 1 and 2 and four a posteriori probabilities of substitution of sites 3, 4, 5, 6 in the case of actual substitution of sites 1 and 2. The probability $W^m_{O_2}$ is calculated in a similar manner. The corresponding a posteriori probabilities may be determined as functions of alloy composition, degree of long-range order and temperature. Thus, in all calculations only linear terms in $\frac{w}{kT}$ are retained ($\frac{w}{kT}$ plays the role of a small parameter). Afterwards the sums occurring in (24.11) can be computed. The following expression for the diffusion coefficient is obtained in this approximation:

$$D = D_0 \frac{K''^2_+ K''^2_-}{K'^4_+ K^2_- + K'^4_- K^2_+} \left[1 + 4 \frac{w}{kT} \left(c_A^2 - \frac{\eta^2}{4} \right) \left(c_B^2 - \frac{\eta^2}{4} \right) \times \right.$$

$$\left. \times \left(- \frac{R''^2}{K''_+ K''_-} + 2 \frac{K'^3_+ K_- + K'^3_- K_+}{K'^4_+ K^2_- + K'^4_- K^2_+} RR' \right) \right]. \quad (24.30)$$

where

$$R = e^{\frac{v_{AC}}{kT}} - e^{\frac{v_{BC}}{kT}}, \quad R' = e^{\frac{v'_{AC}}{kT}} - e^{\frac{v'_{BC}}{kT}}, \quad R'' = e^{\frac{v''_{AC}}{kT}} - e^{\frac{v''_{BC}}{kT}}. \quad (24.31)$$

In Eq. (24.30) the dependence of η on composition and temperature is determined from (17.44) for the degree of long-range order, obtained by the Kirkwood method. The latter takes into account terms proportional to $\left(\frac{w}{kT}\right)^2$ in the free energy term. For disordered alloys the expression for the diffusion coefficient is

$$D_{\text{dis}} = \frac{1}{2} D_0 \frac{K''^4}{K^2 K'^4} \left[1 + 4 c_A^2 c_B^2 \frac{w}{kT} \left(2 \frac{RR'}{KK'} - \frac{R''^2}{K''^2} \right) \right]. \quad (24.32)$$

where

$$K = K_+ \big|_{\eta=0}, \quad K' = K'_+ \big|_{\eta=0}, \quad K'' = K''_+ \big|_{\eta=0}. \quad (24.33)$$

Equations (24.30) and (24.32) become Eqs. (24.12) and (24.14), if the second term in the square brackets is neglected. From (24.32) it follows that the correlation correction tends to zero as the temperature increases, or as the concentration of one of the components of the alloy tends to zero. The correlation correction to the diffusion coefficient is also small, if the diffusing C atoms interact almost equally with A and B atoms, i.e., if the differences R, R' and R'' are small. Finally, as may be seen from Eq. (24.30), the correlation correction to the diffusion coefficient becomes negligible when the degree of long-range order in the alloy approaches a maximum value.

To estimate the magnitude of the change of the diffusion coefficient upon establishment of correlation in a disordered alloy we may assume that atoms C interact much more strongly with atoms A than with atoms B, so that inequalities (24.20) are fulfilled.

For a stoichiometric alloy $\left(c_A = c_B = \frac{1}{2}\right)$ at the order-disorder transition temperature T_0 $\left(\frac{w}{kT_0} \approx 0.59\right)$ the expression in the square bracket of Eq. (24.32) is equal to 1.59 (when correlation is neglected this expression will be equal to unity). Thus, upon establishment of short-range order in a disordered alloy (as a result of annealing at a temperature somewhat above T_0) the diffusion coefficient of interstitial atoms may change markedly. All the qualitative features of the temperature and concentration dependence of the diffusion coefficient indicated in the preceding section remain valid when correlation is taken into account. There are, however, changes in the estimate of the magnitude of the breaks in the curves $D(T)$ for $c_A = \text{const}$ and $D(c_A)$ for $T = \text{const}$ at points corresponding to the order-disorder transition. For small η, the diffusion coefficient has the form (24.15), also when correlation is taken into account. Here, however, D_{dis} is determined from Eq. (24.32), and b is equal to

$$b = \frac{1}{4}\frac{R^2}{K^2} + \frac{3}{2}\frac{R'^2}{K'^2} + \frac{1}{2}\frac{R''^2}{K''^2} - 2\frac{RR'}{KK'} + \frac{w}{kT}(c_A^2 + c_B^2)\left(2\frac{RR'}{KK'} - \frac{R''^2}{K''^2}\right) + \\ + \frac{w}{kT}c_A^2 c_B^2\left[\frac{R''^4}{K''^4} + 2\frac{RR'}{KK'}\left(\frac{R^2}{K^2} - 5\frac{RR'}{KK'} + 3\frac{R'^2}{K'^2}\right)\right]. \quad (24.34)$$

When (17.44) is used instead of (16.18) to determine the degree of long-range order, the magnitude of the discontinuity of the derivative of η^2 with respect to $\frac{1}{kT}$ also changes (increases). From Eq. (17.44), the following expression can be obtained for this discontinuity

$$\left.\frac{d\eta^2}{d\frac{1}{kT}}\right|_{T_0 - 0} = 48c_A^2 c_B^2 \frac{\sqrt{6c_A^2 c_B^2 - c_A c_B}}{1 - 3c_A c_B - 24c_A^3 c_B^3 \left(\frac{w}{kT_0}\right)^2} w, \quad (24.35)$$

where $\frac{w}{kT_0}$ is determined by (17.48) (for $z = 8$). The magnitude of the step change in the effective activation energy is

$$\Delta Q = 48c_A^2 c_B^2 \frac{\sqrt{6c_A^2 c_B^2 - c_A c_B}}{1 - 3c_A c_B - 24c_A^3 c_B^3 \left(\frac{w}{kT_0}\right)^2} wb(T_0). \quad (24.36)$$

To estimate this quantity we may consider the case in which inequalities (24.20) are fulfilled at $T = T_0$. Then $b(T_0) = \frac{1}{4c_A^2} + \frac{w}{kT_0}$ and for $c_A = c_B = \frac{1}{2}$ we have $\Delta Q \approx 14w \approx 8kT_0$. If $T_0 = 800\,°\text{K}$, $\Delta Q \approx 0.6$ ev. Thus, the noting of correlation in this case will lead to a substantial increase of ΔQ (almost threefold).

The relation between the increase of δ^* in the derivative of $\frac{d \ln D}{dc_A}$ during a transition to the ordered state and the increase in ΔQ when correlation is taken into account is

$$w \frac{\delta^*}{\Delta Q} = \frac{d \frac{w}{kT_0}}{dc_A} =$$

$$= \left[\frac{(c_A - c_B)(24c_A^2 c_B^2 - 6c_A c_B + 1 - 8c_A c_B \sqrt{6c_A^2 c_B^2 - c_A c_B}}{4c_A c_B (1 - 2c_A c_B)^2 \sqrt{6c_A^2 c_B^2 - c_A c_B}} \right]_{\substack{c_A = c_0 \\ c_B = 1 - c_0}} . \quad (24.37)$$

A graph of $w \frac{\delta^*}{\Delta Q}$ as a function of $c_A = c_0$ is illustrated in Fig. 104 (curve for $z = 8$).

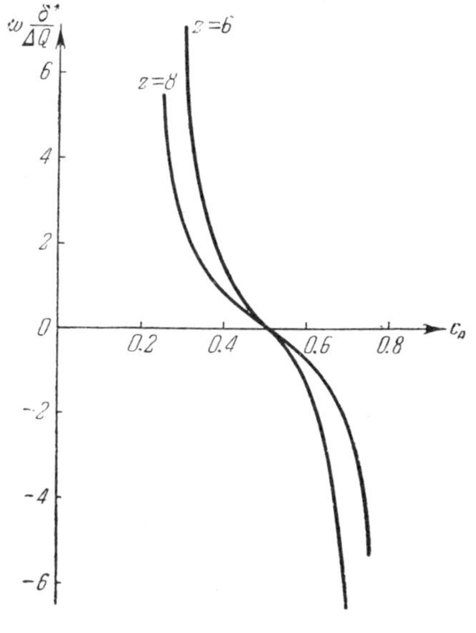

Fig. 104. The ratio $\frac{w\delta^*}{\Delta Q}$ as a function of concentration c_A.

By a similar method we may examine the diffusion of interstitial atoms in Fe_3Al type alloys. An investigation of the resulting formulas shows that the temperature and concentration dependence of the diffusion coefficient in such alloys has the same qualitative features as in β-brass alloys. However, in Fe_3Al alloys the graph of $\ln D$ vs $\frac{1}{T}$ apparently has a break more often away from the abscissa axis. The magnitude of

ΔQ in most cases is much smaller than for alloys with a β-brass type lattice. Correlation in an alloy has a lesser effect on D. This is related to the fact that the order-disorder transformation affects only one-half the sites in a Fe_3Al lattice. Moreover, each of the interstitial sites O_1 and O_2 in this lattice is surrounded by the same number of sites of the first and second type, which also weakens the effect of ordering on the diffusion of interstitial atoms. The graph of the ratio $w\frac{\delta^*}{\Delta Q}$ versus concentration $c_A = c_0$ for Fe_3Al alloys is given in Fig. 104 (curve $z = 6$).

2. Diffusion of interstitial atoms in alloys with a face-centered cubic lattice [254]

We shall now treat the diffusion of interstitial atoms in an alloy having a type $AuCu_3$ face-centered cubic lattice, using the previously adopted simplified model of an alloy. Let us consider a binary ordered substitutional alloy of metals A and B in a state of thermodynamic equilibrium. Let us assume that the number of A atoms is lower than the number of B atoms. Then, as already noted in Sec. 2, the first type of sites (correct for atoms A) is located at the corners of the cubic cells, and the second type of sites (correct for atoms B)—at the centers of their faces. We assume that the interstitial atoms C in the alloy have positions of stable equilibrium at type O_1 interstitial sites, located at the centers of cubic cells, and type O_2 at the centers of their edges (Fig. 105).* The interstitial site O_1 is surrounded by six nearest neighbor sites of the second type spaced at a distance $\frac{a}{2}$. Interstitial site O_2 is surrounded by two sites of the first type, four sites of the second type, also spaced at a distance $\frac{a}{2}$. By choosing the direction of the x axis and separating two neighboring planes I and II which are perpendicular to it as shown in Fig. 105, we can show that the C atom may go over from interstitial site O_1 in plane I into the neighboring interstitial site O_2 in plane II by four different, but geometrically equivalent, paths. The saddle point of the potential barrier for such a transition will be regarded as located at P_1, midway between O_1 and O_2. Atom C, which occupies

*Such an arrangement of interstitial atoms on the interstitial sites for a solution of carbon and nitrogen in γ-iron has been demonstrated by x-ray analysis [255].

interstitial site O_2 in plane II, may go over into intersite O_2 in plane II also by four equivalent paths. Hence, it must pass through the saddle point of the potential barrier P_2, located in the middle between two neighboring interstitial sites O_2. Atom C may go over from interstitial site O_2 in plane II into interstitial site O_1 in plane I by two means, passing through the type P_1 position and into interstitial site O_1 in plane I also by two paths, passing through position type P_2. Position P_1 has two neighboring sites of the second type at a distance $\frac{\sqrt{2}}{4}a$, and also two neighboring sites of the first type and two of the second type at distances $\frac{\sqrt{6}}{4}a$. Position P_2 has one neighboring site of the first type and one of the second type at distances $\frac{\sqrt{2}}{4}a$, and also four neighboring sites of the second type at distances $\frac{\sqrt{6}}{4}a$.

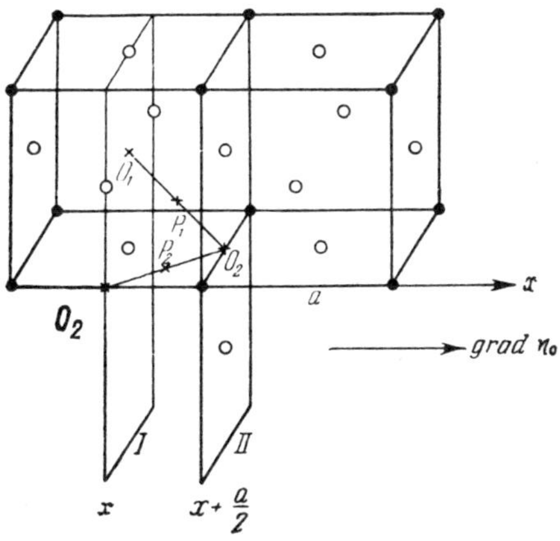

Fig. 105. Equilibrium positions for interstitial atoms of a face centered cubic lattice.

Let us denote the negative interaction energies by v_{AC} and v_{BC}, for atom C with atoms A and B, located at distances $\frac{\sqrt{2}}{4}a$, $\frac{a}{2}$ and $\frac{\sqrt{6}}{4}a$, in the following manner:

$$\left.\begin{aligned} v_{AC}\left(\frac{\sqrt{2}}{4}a\right) &= u_{AC}, & v_{BC}\left(\frac{\sqrt{2}}{4}a\right) &= u_{BC}, \\ v_{AC}\left(\frac{a}{2}\right) &= u'_{AC}, & v_{BC}\left(\frac{a}{2}\right) &= u'_{BC}, \\ v_{AC}\left(\frac{\sqrt{6}}{4}a\right) &= u''_{AC}, & v_{BC}\left(\frac{\sqrt{6}}{4}a\right) &= u''_{BC}. \end{aligned}\right\} \quad (24.38)$$

We shall use the same method as for alloys with β-brass type crystal lattice to calculate the inverse flux of atoms *C* between planes *I* and *II* and their net flux. Then by considering (1.5), at $\nu = \frac{1}{4}$, we obtain (neglecting correlation, but allowing for the existence of different configurations of the *A* and *B* atoms) the following formula for the diffusion coefficient of *C* atoms:

$$D = D_0' \frac{\left(\mathfrak{f} - \frac{1}{4}\eta r\right)\left(\mathfrak{f}'' - \frac{1}{4}\eta r''\right)^2}{\left(\mathfrak{f}' - \frac{1}{4}\eta r'\right)^4} \left[\frac{\left(\mathfrak{f} - \frac{1}{4}\eta r\right)\left(\mathfrak{f}'' + \frac{3}{4}\eta r''\right)^2}{\left(\mathfrak{f}' - \frac{1}{4}\eta r'\right)^2 + 3\left(\mathfrak{f}' + \frac{3}{4}\eta r'\right)^2} + \right.$$
$$\left. + \frac{\left(\mathfrak{f}'' - \frac{1}{4}\eta r''\right)^2\left(\mathfrak{f} + \frac{3}{4}\eta r\right)}{\left(\mathfrak{f}' - \frac{1}{4}\eta r'\right)^2 + 3\left(\mathfrak{f}' + \frac{3}{4}\eta r'\right)^2} \right] \tag{24.39}$$

where $D_0' = 2\frac{a^2}{\tau_0}$,

$$\mathfrak{f} = c_A e^{\frac{u_{AC}}{kT}} + c_B e^{\frac{u_{BC}}{kT}}, \quad r = e^{\frac{u_{AC}}{kT}} - e^{\frac{u_{BC}}{kT}}, \tag{24.40}$$

$$\left.\begin{aligned} \mathfrak{f}' &= c_A e^{\frac{u'_{AC}}{kT}} + c_B e^{\frac{u_{B}^{\prime C}}{kT}}, \quad r' = e^{\frac{u'_{AC}}{kT}} - e^{\frac{n'_{BC}}{kT}}, \\ \mathfrak{f}'' &= c_A e^{\frac{u''_{AC}}{kT}} + c_B e^{\frac{u''_{BC}}{kT}}, \quad r'' = e^{\frac{u''_{AC}}{kT}} - e^{\frac{u''_{BC}}{kT}}. \end{aligned}\right\} \tag{24.41}$$

In the case of a disordered alloy, the diffusion coefficient is

$$D_{\text{dis}} = \frac{D_0'}{2} \frac{\left(c_A e^{\frac{u_{AC}}{kT}} + c_B e^{\frac{u_{BC}}{kT}}\right)^2 \left(c_A e^{\frac{u''_{AC}}{kT}} + c_B e^{\frac{u''_{BC}}{kT}}\right)^4}{\left(c_A e^{\frac{u'_{AC}}{kT}} + c_B e^{\frac{u'_{BC}}{kT}}\right)^6}. \tag{24.42}$$

A study of (24.42) shows that the temperature and concentration dependence of the diffusion coefficient in disordered alloys of a $AuCu_3$ face-centered structure has the same qualitative features as in alloys with a body-centered lattice. However, essentially different results are obtained for these two structures in studying the transition of the alloy through the critical temperature T_0. In an alloy with a $AuCu_3$ type lattice, a first-order phase transition occurs and the degree of long-range order changes discontinuously during ordering from zero to η_0. Therefore, the diffusion coefficient at temperature T_0 also changes discontinuously unlike the β-brass alloys, in which the change is continuous and only the effective activation energy experiences a step-change. A step-change in the effective activation energy also occurs in $AuCu_3$ type alloys.

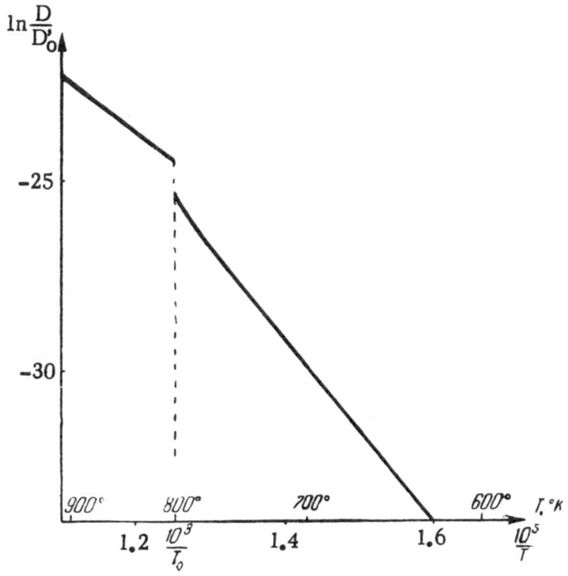

Fig. 106. Approximate correlation between the logarithm of the diffusion coefficient in alloys with AuCu type crystal lattice and the reciprocal of temperature.

For a qualitative explanation of the temperature dependence of the degree of long-range order at temperatures below T_0 we may use the approximate formula (16.24). The latter was obtained by the statistical theory of ordering in which correlation in the alloy was neglected. As the temperature is lowered, the degree of long-range order increases smoothly. Thus, the graph of $\ln D$ vs $\frac{1}{T}$ at first differs markedly from a straight line, and then usually approximates a straight line rather well. As an example, Fig. 106 shows a graph of $\ln \frac{D}{D_0'}$ vs $\frac{1}{T}$ for $c_A = \frac{1}{4}$ and the following parameters: $u_{AC} = 0.95$ ev, $u_{BC} =$

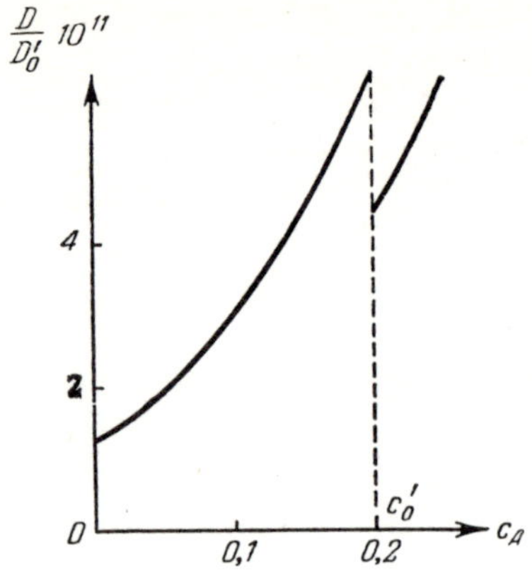

Fig. 107. Approximate correlation of the concentration and the diffusion coefficient in alloys with an $AuCu_3$ type crystal lattice.

0.75 ev, $u'_{AC}=u'_{BC}=0.70$ ev, $u''_{AC}=0.29$ ev, $u''_{BC}=0.09$ ev, $T_0=800\,°K$.

The formula for the diffusion coefficient is considerably simplified in the limiting cases, when the C atoms interact much more strongly with atoms of one component of the alloy than with atoms of the other component. In these limiting cases with ordering (at the point $T=T_0$) D decreases in a step-change by 20–50%. A substantially greater decrease of the diffusion coefficient during ordering is also possible; such a case is shown for example in Fig. 106. Equation (24.39) also allows an increase of the diffusion coefficient during ordering. We note that taking correlation into account would lead to a large value of the step-change in η, and consequently also in D.

There are also step changes in the diffusion coefficient on the curve of the function $D(c_A)$ (for $T=$ const). These step-changes obviously must occur at those values of concentration $c_A=c_0'$ and $c_A=c_0''$, when the alloy is at the critical temperature.

Figure 107 shows the curve $\frac{D(c_A)}{D_0'}$ (in the range $c_A \ll \frac{1}{4}$), plotted for the following parameters: $u_{AC}=0.95$ ev, $u_{BC}=0.866$ ev, $u'_{AC}=u'_{BC}=0.7$ ev, $u''_{AC}=0.29$ ev, $u''_{BC}=0.248$ ev, $w=0.084$ ev and at $T=700\,°K$.

Thus, at temperatures close to T_0, ordering in an $AuCu_3$ type structure exerts a stronger influence on the diffusion of the interstitial atoms than in the case of alloys with a body-centered cubic lattice.

25. Determination of Equilibrium Concentration of Vacancies At Sites of the Crystal Lattice of an Alloy

1. *Alloy with a body-centered cubic lattice* [256, 257]

We shall consider here the second possible mechanism of diffusion, in substitutional alloys, i.e. the movement of atoms into vacancies. To investigate diffusion in alloys occurring by the vacancy mechanism, it is necessary first of all to find the equilibrium number of vacancies. If the alloy is in equilibrium at a given temperature, the concentration of vacancies on each type of site must be known. First, these concentrations will be determined by the method of mean energies, neglecting the effect of enrichment of the neighboring coordination spheres with vacancies by atoms of any species. The latter effect will be taken into account subsequently.

Using our simplified model of an alloy, we shall consider a binary, substitutional type ordered solid solution of metals A and B with a β-brass lattice. We assume that a transition of atoms of the alloy into interstitial sites is impossible. We will denote by $\mathfrak{N}$ the total number of lattice sites in the alloy and by N the total number of atoms (including N_A of atoms A and N_B of atoms B). The number of atoms A and B at sites of the first and second types, just as before, will be denoted by $N_A^{(1)}$, $N_B^{(1)}$, $N_A^{(2)}$ and $N_B^{(2)}$, respectively, and the number of vacancies at sites of the first and second types by $N_{\text{d}}^{(1)}$ and $N_{\text{d}}^{(2)}$. Then,

$$N_A^{(1)}+N_B^{(1)}+N_{\text{д}}^{(1)}=N_A^{(2)}+N_B^{(2)}+N_{\text{d}}^{(2)}=\frac{\mathfrak{N}}{2}. \tag{25.1}$$

The probabilities of substitution of sites of the first and second types by atoms A, B or by a vacancy are, respectively,

$$\left.\begin{aligned}
p_A^{(1)}&=2\frac{N_A^{(1)}}{\mathfrak{N}}=\left(c_A+\frac{\eta'}{2}\right)\left(1+\frac{N_{\text{d}}^{(1)}+N_{\text{d}}^{(2)}}{N}\right)^{-1},\\
p_A^{(2)}&=2\frac{N_A^{(2)}}{\mathfrak{N}}=\left(c_A-\frac{\eta'}{2}\right)\left(1+\frac{N_{\text{d}}^{(1)}+N_{\text{d}}^{(2)}}{N}\right)^{-1},\\
p_B^{(1)}&=2\frac{N_B^{(1)}}{\mathfrak{N}}=\left(c_B-\frac{\eta'}{2}-g\right)\left(1+\frac{N_{\text{d}}^{(1)}+N_{\text{d}}^{(2)}}{N}\right)^{-1},\\
p_B^{(2)}&=2\frac{N_B^{(2)}}{\mathfrak{N}}=\left(c_B+\frac{\eta'}{2}+g\right)\left(1+\frac{N_{\text{d}}^{(1)}+N_{\text{d}}^{(2)}}{N}\right)^{-1},
\end{aligned}\right\} \tag{25.2}$$

$$\left.\begin{aligned} p_{\rm d}^{(1)} &= 2\frac{N_{\rm d}^{(1)}}{\mathfrak{N}} = 2\frac{N_{\rm d}^{(1)}}{N}\left(1+\frac{N_{\rm d}^{(1)}+N_{\rm d}^{(2)}}{N}\right)^{-1}, \\ p_{\rm d}^{(2)} &= 2\frac{N_{\rm d}^{(2)}}{\mathfrak{N}} = 2\frac{N_{\rm d}^{(2)}}{N}\left(1+\frac{N_{\rm d}^{(1)}+N_{\rm d}^{(2)}}{N}\right)^{-1}, \end{aligned}\right\} \quad \begin{array}{c}(25.2)\\ \text{cont'd}\end{array}$$

where

$$c_A = \frac{N_A}{N},\quad c_B = \frac{N_B}{N},\quad \eta' = 4\frac{N_A^{(1)}}{N} - 2c_A \quad \text{and} \quad g = \frac{N_{\rm d}^{(1)} - N_{\rm d}^{(2)}}{N}.$$

The quantity η' obviously changes into the degree of long-range order η, when no vacancies exist in the alloy.

The configurational part of the energy of the alloy in the model assumed can be written as

$$E = -8\left[N_A^{(1)}\left(p_A^{(2)}v_{AA} + p_B^{(2)}v_{AB}\right) + N_B^{(1)}\left(p_A^{(2)}v_{AB} + p_B^{(2)}v_{BB}\right)\right] \quad (25.3)$$

or, taking into account (25.2), as

$$E = -4N\left(1+\frac{N_{\rm d}^{(1)}+N_{\rm d}^{(2)}}{N}\right)^{-1}\left\{\left(c_A^2 - \frac{\eta'^2}{4}\right)v_{AA} + \right.$$
$$\left. + \left[c_B^2 - \left(\frac{\eta'}{2}+g\right)^2\right]v_{BB} + 2\left(c_A c_B + \frac{\eta'^2}{4} + \frac{\eta'}{2}g\right)v_{AB}\right\}. \quad (25.4)$$

The number of different configurations of the crystal W in this approximation is

$$W = \frac{\left(\frac{\mathfrak{N}}{2}\right)!}{N_A^{(1)}!\,N_B^{(1)}!\,N_{\rm d}^{(1)}!}\ \frac{\left(\frac{\mathfrak{N}}{2}\right)!}{N_A^{(2)}!\,N_B^{(2)}!\,N_{\rm d}^{(2)}!}. \quad (25.5)$$

Expressing the configurational part of the free energy of the crystal $F = E - kT\ln W$ as a function of η', $N_{\rm d}^{(1)}$ and $N_{\rm d}^{(2)}$, and using the equilibrium conditions $\frac{\partial F}{\partial \eta'} = 0$, $\frac{\partial F}{\partial N_{\rm d}^{(1)}} = 0$, $\frac{\partial F}{\partial N_{\rm d}^{(2)}} = 0$, we find an expression for the equilibrium values η', $N_{\rm d}^{(1)}$, $N_{\rm d}^{(2)}$ at a given temperature.

The condition $\frac{\partial F}{\partial \eta'} = 0$, neglecting the values of $N_{\rm d}^{(1)}$ and $N_{\rm d}^{(2)}$ which are small compared to N, leads to Eq. (16.8). The conditions $\frac{\partial F}{\partial N_{\rm d}^{(1)}} = 0$ and $\frac{\partial F}{\partial N_{\rm d}^{(2)}} = 0$, neglecting terms of the order $\left(\frac{N_{\rm d}^{(1)}}{N}\right)^2$ and $\left(\frac{N_{\rm d}^{(2)}}{N}\right)^2$ and higher, give the equilibrium values of $N_{\rm d}^{(1)}$ and $N_{\rm d}^{(2)}$ in the form

$$N_d^{(1)} = \frac{N}{2}\sqrt{\frac{c_A + \frac{\eta}{2}}{c_A - \frac{\eta}{2}}}\, e^{-\frac{4(v_{AB} - v_{AA})\eta}{kT}} e^{-\frac{u}{kT}}, \tag{25.6}$$

$$N_d^{(2)} = \frac{N}{2}\sqrt{\frac{c_A - \frac{\eta}{2}}{c_A + \frac{\eta}{2}}}\, e^{\frac{4(v_{AB} - v_{AA})\eta}{kT}} e^{-\frac{u}{kT}}, \tag{25.7}$$

where

$$u = 4\left[\left(c_A^2 - \frac{\eta^2}{4}\right)v_{AA} + \left(c_B^2 - \frac{\eta^2}{4}\right)v_{BB} + 2\left(c_A c_B + \frac{\eta^2}{4}\right)v_{AB}\right] \tag{25.8}$$

and in the given approximation the quantity η' is replaced by the degree of long-range order η.

From Eqs. (25.6) and (25.7) we find for disordered alloys ($\eta = 0$) that

$$N_d^{(1)} = N_d^{(2)} = N_{d0} = \frac{N}{2} e^{-\frac{u_0}{kT}}, \tag{25.9}$$

where

$$u_0 = 4\left[v_{BB} + 2(v_{AB} - v_{BB})c_A - wc_A^2\right]. \tag{25.10}$$

In an ordered alloy the numbers of vacancies $N_d^{(1)}$ and $N_d^{(2)}$ on sites of the first and second types are different.

In a disordered alloy, because the method of mean energies has been used, $\ln N_d^{(1)}$ and $\ln N_d^{(2)}$ are linear functions of $\frac{1}{T}$. The total number of vacancies in the alloy $N_d^{(1)} + N_d^{(2)}$ is an even function of η. Since the derivative of η^2 with respect to $\frac{1}{T}$ at $T = T_0$ experiences a finite step-change, the graph of the function $\ln(N_d^{(1)} + N_d^{(2)})$ will exhibit a break during a transition to the ordered state. On the other hand, $N_d^{(1)}$ and $N_d^{(2)}$ contain linear terms of the expansion in the powers of η. Since $\frac{d\eta}{d\frac{1}{T}}$ tends to infinity as $T \to T_0$ ($T < T_0$), the graphs of $\ln N_d^{(1)}$ and $\ln N_d^{(2)}$ as functions of $\frac{1}{T}$ will have a vertical tangent at the point $T = T_0$. In the immediate vicinity of the point $T = T_0$, during the transition into the ordered condition, one of these curves lies above the point $T = T_0$, while the other, lies below this point.

Turning to the dependence of the vacancy concentration on composition of the disordered alloy, we note that the form of the function u_0 vs c_A, according to (25.10), will be different for ordered alloys at lower temperatures $(w > 0)$ and for disordered alloys $(w < 0)$.

In the first case, the curve $u_0(c_A)$ becomes convex away from the c_A axis, and in the second case convex towards this axis. If in addition $v_{AB} > v_{AA}$ and $v_{AB} > v_{BB}$, the function $u_0(c_A)$ has a maximum, and $N_{d0}(c_A)$ a minimum. If however $v_{AB} < v_{AA}$ and $v_{AB} < v_{BB}$, $u_0(c_A)$ will have a minimum, and $N_{d0}(c_A)$ a maximum. During the transition into the ordered state the curves $N_d^{(1)}$ and $N_d^{(2)}$ vs c_A have breaks at those values of concentration at which the given temperature is equal to the critical temperature T_0.

The formation of vacancies in alloys of the same structure will now be examined by means of the quasichemical approximation, i.e., by taking into account the correlation in the alloy. We shall denote by N_{AA}, N_{BB}, N_{dd}, N_{AB}^{12}, N_{BA}^{12}, N_{Ad}^{12}, N_{dA}^{12}, N_{Bd}^{12}, N_{dB}^{12} the numbers of pairs of atoms A, B and vacancies at neighboring sites of the first and second types. Then, the configurational part of the energy of the alloy is

$$E = -N_{AA}v_{AA} - N_{BB}v_{BB} - (N_{AB}^{12} + N_{AB}^{12})\,v_{AB}. \tag{25.11}$$

The number of configurations W of atoms of the alloy is assumed in the quasichemical approximation to be proportional to the number of ways in which one may divide $4N$ pairs into nine groups containing the following pairs: AA, BB, DD, AB, BA, AD, DA, BD, DB (Sec. 18, Eq. (18.5))

$$W = h\,\frac{(4N)!}{N_{AA}!\,N_{BB}!\,N_{dd}!\,N_{AB}^{12}!\,N_{BA}^{12}!\,N_{Ad}^{12}!\,N_{dA}^{12}!\,N_{Bd}^{12}!\,N_{dB}^{12}!}. \tag{25.12}$$

Here the factor h is independent of the number of pairs, but depends on the distribution of atoms A, B and vacancies on the sites of the first and second kinds. Determining this factor as is usually done in the quasichemical approximation, we obtain

$$h = \frac{(N_A^{(1)} + N_B^{(1)} + N_d^{(1)})!}{N_A^{(1)}!\,N_B^{(1)}!\,N_d^{(1)}!}\,\frac{(N_A^{(2)} + N_B^{(2)} + N_d^{(2)})!}{N_A^{(2)}!\,N_B^{(2)}!\,N_d^{(2)}!}\times$$
$$\times\frac{N_{AA}^0!\,N_{BB}^0!\,N_{dd}^0!\,N_{AB}^0!\,N_{BA}^0!\,N_{Ad}^0!\,N_{dA}^0!\,N_{Bd}^0!\,N_{dB}^0!}{(4N)!}, \tag{25.13}$$

where

$$N_{\alpha\alpha'}^0 = 16\,\frac{N_\alpha^{(1)}N_{\alpha'}^{(2)}}{N} \qquad (\alpha, \alpha' = A, B, d). \tag{25.14}$$

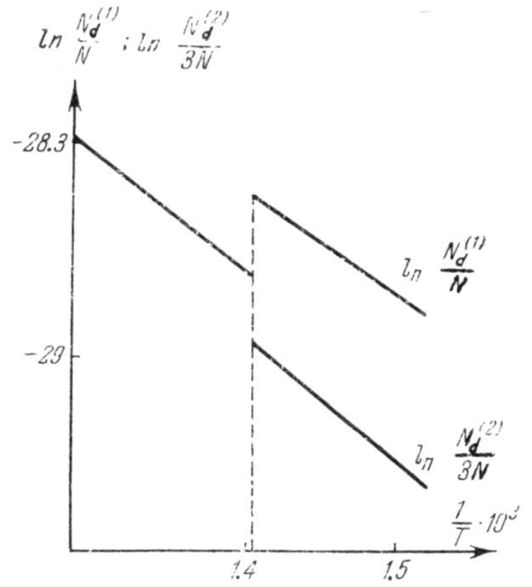

Fig. 108. Approximate correlation of the logarithm of the vacancy concentration at sites of the first and second types in alloys with a $AuCu_3$ type crystal lattice with the reciprocal temperature ($c_A = 0.2, v_{AA} = 0.261$ ev, $v_{BB} = 0.267$ ev, $v_{AB} = 0.306$ ev).

Determining with the aid of (25.11), (25.12), (25.13) and (25.14) the free energy of the alloy $F = E - kT \ln W$ and noting that the free energy is a minimum at thermodynamic equilibrium, we may obtain the following expressions for the number of vacancies of the first and second types:

$$N_d^{(1)} = \frac{N}{2}\left(\frac{c_A - \frac{\eta}{2}}{c_A + \frac{\eta}{2}}\right)^{7/2} (p_{AA}^{12})^4 e^{-\frac{4v_{AA}}{kT}} \left[1 + \frac{p_{AB}^{12}}{p_{AA}^{12}} e^{\frac{v_{AA} - v_{AB}}{kT}}\right]^8, \quad (25.15)$$

$$N_d^{(2)} = \frac{N}{2}\left(\frac{c_A + \frac{\eta}{2}}{c_A - \frac{\eta}{2}}\right)^{7/2} (p_{AA}^{12})^4 e^{-\frac{4v_{AA}}{kT}} \left[1 + \frac{p_{BA}^{12}}{p_{AA}^{12}} e^{\frac{v_{AA} - v_{AB}}{kT}}\right]^8, \quad (25.16)$$

where p_{AA}^{12}, p_{AB}^{12}, p_{BA}^{12} and η in the quasichemical approximation are determined by (18.26), (18.27) and (18.29) (for $z = 8$). For a disordered alloy in which correlation may be neglected (but $\frac{|v_{AA} - v_{AB}|}{kT} > 1$),

$$N_{d0} = N_d^{(1)} = N_d^{(2)} = \frac{N}{2} e^{-\frac{4v_{AA}}{kT}} \left[c_A + c_B e^{\frac{v_{AA} - v_{AB}}{kT}}\right]^8. \quad (25.17)$$

If we use (18.26) and (18.27) for p^{12}_{AA}, p^{12}_{AB}, p^{12}_{BA} and expand $N_{\mathrm{d}}^{(1)}$ in (25.15) and $N_{\mathrm{d}}^{(2)}$ in (25.16) in a power series in $\frac{v_{AB}-v_{AA}}{kT}$ and $\frac{v_{AB}-v_{BB}}{kT}$, and retain only linear terms, these expansions will coincide with the corresponding expansions of (25.6) and (25.7). If, however, the absolute values of the energy differences $|v_{AB}-v_{AA}|$ and $|v_{AB}-v_{BB}|$ are not small in comparison with kT, the variation of $\ln N_{\mathrm{d}0}$ with $\frac{1}{T}$ for a disordered alloy will not, according to (25.17), be exponential. In general, there is no unique relation between the sign of w and the direction of convexity of the concentration curve of the quantity $N_{\mathrm{d}0}$. Thus, for example, if the conditions of applicability of (25.17) are fulfilled, the concentration curve always becomes convex toward the abscissa axis. The dependence of the vacancy concentration on the composition of the alloy according to (25.17) may be very pronounced.

2. *Alloy with a face-centered cubic lattice* [188]

We shall now determine the concentration of vacancies at sites of a $AuCu_3$ type face-centered cubic lattice using the same assumptions as above. Unlike the case of alloys with a body-centered cubic lattice, now the total number $\mathfrak{N}$ of lattice sites contains $\frac{\mathfrak{N}}{4}$ sites of the first type and $\frac{3}{4}\mathfrak{N}$ sites of the second type. The equilibrium number of vacancies $N_{\mathrm{d}}^{(1)}$ and $N_{\mathrm{d}}^{(2)}$ on sites of the first and second types are determined by the method of mean energies, similarly to the previous calculation

$$N_{\mathrm{d}}^{(1)}=\frac{N}{4}\left(\frac{c_B-\frac{3}{4}\eta}{c_B+\frac{\eta}{4}}\right)^{3/4} e^{\frac{3(v_{AB}-v_{BB})\eta}{kT}} e^{-\frac{u'}{kT}}, \qquad (25.18)$$

$$N_{\mathrm{d}}^{(2)}=\frac{3}{4}N\left(\frac{c_B+\frac{\eta}{4}}{c_B-\frac{3}{4}\eta}\right)^{1/4} e^{-\frac{(v_{AB}-v_{BB})\eta}{kT}} e^{-\frac{u'}{kT}}, \qquad (25.19)$$

where

$$u'=6\left[\left(c_A^2-\frac{1}{16}\eta^2\right)v_{AA}+\left(c_B^2-\frac{1}{16}\eta^2\right)v_{BB}+2\left(c_Ac_B+\frac{1}{16}\eta^2\right)v_{AB}\right].$$

Here the degree of long-range order is determined by (16.24), and N denotes the total number of atoms in the alloy.

The temperature and concentration dependence of the concentration of vacancies in a disordered alloy, as follows from (25.18) and (25.19), have the same qualitative features as alloys with a body-centered cubic lattice. In the transition to an ordered state, unlike in the β-brass type alloys, there is a step change not only in the derivatives of vacancy concentration with respect to T and c_A, but also in the concentration itself. Here, a case is possible where the concentration of vacancies increases on sites of one type but decreases on sites of the other type (Fig. 108) (but the vacancy concentration cannot increase on sites of both types).

26. Self-Diffusion in Alloys Via the Vacancy Mechanism

1. *Case of nearly equal interaction energy of atoms* [258]

We shall now consider self-diffusion (diffusion of tracer atoms of one of the components) in ternary and binary disordered alloys with a body-centered cubic lattice attained via the vacancy mechanism. In order to explain merely the qualitative aspects of the phenomenon, we may use the approximate method of mean energies. By analogy with part 1 of Sec. 25 (where the absolute values of the difference in interaction energies of neighboring atoms were regarded as small compared with kT), an expression for the concentration of vacancies N_{d} in ternary disordered alloy $A-B-C$ can be obtained

$$N_{\text{d}} = Ne^{-\frac{U}{kT}}, \tag{26.1}$$

where

$$U = 4\left(c_A^2 v_{AA} + c_B^2 v_{BB} + c_C^2 v_{CC} + 2c_A c_B v_{AB} + 2c_A c_C v_{AC} + 2c_B c_C v_{BC}\right). \tag{26.2}$$

Let us consider an atom A on a site of the crystal lattice and a vacancy on one of its neighboring sites. We shall consider only the most probable type of transitions of atom A, namely, when it replaces a vacancy on a neighboring site. In a body-centered cubic lattice, two neighboring sites have a separation $d = \frac{\sqrt{3}}{2} a \approx 0.86a$. The curve showing the change in the potential energy of atom A when it migrates into a vacancy along the body diagonal of the unit cell, exhibits both a minimum and a maximum. The minimum occurs at the sites, and the maximum will be assumed to be located half way between the sites. Let U_0 denote the negative of the potential energy of atom A on a lattice site and U_P at the point P, corresponding to the saddle of the potential barrier. Then the height of the barrier will be equal to $\Delta U = U_0 - U_P$. Atom A at point P has six nearest neighbors at a distance $\frac{\sqrt{11}}{4} a \approx 0.83a$, which almost coincides with d. In the following we shall therefore assume these distances to be approximately equal. Thus, our approximation gives

$$U_0 = 7\,(c_A v_{AA} + c_B v_{AB} + c_C v_{AC}), \quad U_P = 6\,(c_A v_{AA} + c_B v_{AB} + c_C v_{AC})$$

and consequently,

$$\Delta U = c_A v_{AA} + c_B v_{AB} + c_C v_{AC}. \tag{26.3}$$

Let us assume that a concentration gradient of tagged atoms A is produced in the direction of the axis and parallel to the edge of the cubic cell (under the condition of constancy of the total concentration of all atoms A). As before, we shall consider two neighboring atomic planes I and II perpendicular to the x axis. Each site on one of these planes has four neighboring sites on the other plane. Let $\nu^*(x)$ denote the number of tagged atoms A per cm^2 of the plane with a coordinate x. The probability that a vacancy lies alongside atom A on a neighboring plane is equal in the adopted approximation to $4\frac{N_d}{N}$. Recalling that the probability $\mathfrak{W}$ of a transition of atom A into a neighboring vacancy per unit time is given by the formula $\mathfrak{W} = \frac{1}{\tau_0} e^{-\frac{\Delta U}{kT}}$, we can readily find an expression for the number of tagged atoms A, passing in a unit time per cm^2 of plane I to plane II

$$\overset{*}{J}_{I \to II} = 4\frac{N_d}{N} \nu^*(x) \frac{1}{\tau_0} e^{-\frac{\Delta U}{kT}}. \tag{26.4}$$

Similarly the expression for counterflux is

$$\overset{*}{J}_{II \to I} = 4\frac{N_d}{N} \nu^*\left(x + \frac{a}{2}\right) \frac{1}{\tau_0} e^{-\frac{\Delta U}{kT}}. \tag{26.5}$$

Noting that the number of tagged atoms A per cm^3 of the crystal $\overset{*}{n}_A$ is equal to $\frac{2}{a}\nu^*$, we obtain the following formula for the net flux:

$$J^* = J^*_{I \to II} - J^*_{II \to I} = -\frac{a^2}{\tau_0}\frac{N_d}{N} e^{-\frac{\Delta U}{kT}} \frac{dn^*_A}{dx}. \qquad (26.6)$$

Hence, using (26.1), we may determine the diffusion coefficient for tagged A atoms

$$D^*_A = \frac{a^2}{\tau_0} e^{-\frac{Q}{kT}}, \qquad (26.7)$$

where the activation energy Q is given by

$$Q = U + \Delta U = 4\left(c_A^2 v_{AA} + c_B^2 v_{BB} + c_C^2 v_{CC} + 2c_A c_B v_{AB} + \right.$$
$$\left. + 2c_A c_C v_{AC} + 2c_B c_C v_{BC}\right) + c_A v_{AA} + c_B v_{AB} + c_C v_{AC}. \qquad (26.8)$$

Let us consider some special cases.

Suppose c_A remains constant in a number of alloys, while c_C (and consequently also c_B) takes various values from zero to $1-c_A$. Eliminating c_B from (26.8), we find

$$Q = q_0 + q_1 c_C + q_2 c_C^2, \qquad (26.9)$$

where

$$q_0 = 4(v_{AA} + v_{BB} - 2v_{AB})c_A^2 + (7v_{AB} + v_{AA} - 8v_{BB})c_A + 4v_{BB} + v_{AB},$$
$$q_1 = 8(v_{BB} - v_{AB} + v_{AC} - v_{BC})c_A + 8v_{BC} - 8v_{BB} + v_{AC} - v_{AB},$$
$$q_2 = 4(v_{BB} + v_{CC} - 2v_{BC}) \equiv -4w_{BC}.$$

It follows that the curve $Q(c_C)$ becomes convex away from the c_C axis, when $q_2 < 0$, i.e., $w_{BC} > 0$. This occurs, if the binary alloy of metals B and C (whose diffusion is not being investigated) can be found in an ordered state at lower temperatures.* If, however, $w_{BC} < 0$, i.e., alloy $B-C$ is disordered, the curve $Q(c_C)$ becomes convex towards the c_C axis.

For low concentrations of atoms C, when $c_C \ll 1$, we have $Q \approx q_0 + q_1 c_C$. In this case Q increases (but D^*_A decreases) with an increase in c_C when $q_1 > 0$ (i.e., with the addition of a small admixture of atoms C of the metal replacing atoms B in the alloy $A-B$) and decreases when $q_1 < 0$. When c_A is sufficiently small (which often occurs in the study of diffusion)

*The conclusion drawn here and under similar circumstances elsewhere, will be valid if the binary and ternary alloys have the same structure and almost equal lattice constants.

the first case ($q_1 > 0$) is realized if impurity atoms C have a greater absolute value of energy of interaction with atoms of the alloy $A-B$ than the mutual interaction energies of the atoms of this alloy, i.e., $v_{BC} > v_{BB}, v_{AC} > v_{AB}$. The second case ($q_1 < 0$) is realized (for small c_A), if the inverse inequalities hold.

A similar investigation can be made of the problem of diffusion of tagged atoms A in alloys $A-B-C$ with a constant concentration of atoms C (see [258]).

We shall now consider the case of a binary alloy. Let $c_C = 0$, then $c_B = 1 - c_A$ and (26.7) will give the diffusion coefficient of tagged atoms A in the binary alloy $A-B$. There, the activation energy takes the form

$$Q = q_0' + q_1' c_A + q_2' c_A^2, \tag{26.10}$$

where

$$q_0' = 4v_{BB} + v_{AB},$$
$$q_1' = 7v_{AB} + v_{AA} - 8v_{BB},$$
$$q_2' = 4\left(v_{AA} + v_{BB} - 2v_{AB}\right) \equiv -4w_{AB} = -4w.$$

Just as in the previous case, the curve $Q(c_A)$ becomes convex away from the c_A axis for ordered alloys $A-B$ ($w > 0$), and toward the c_A axis for disordered alloys ($w < 0$).

We shall also clarify the effect of small impurities of metal B in metal A on its self-diffusion. For this purpose we shall consider sections of the curve $Q(c_A)$, where $c_A \approx 1$, and $c_B \ll 1$. There,

$$Q = 5v_{AA} + 9\left(v_{AB} - v_{AA}\right) c_B. \tag{26.11}$$

Thus when $v_{AB} > v_{AA}$, Q increases, but D_A^* decreases as c_B increases, i.e., when metal B is added to metal A.

We note that the obtained variations of Q with composition characteristic for different types of alloys can be used in the selection of those additions to the alloy or the pure metal that cause a change in the diffusion coefficient in the specified direction. For this purpose, however, it is necessary to know the relation between the interaction energies of neighboring atoms.

2. *Case of strongly differing interaction energies of atoms* [257]

In the determination of the self-diffusion coefficient of a binary alloy, we shall take into account the effect of enrichment of the first coordination sphere around the vacancies by atoms A or B, and also the existence of different configurations around the diffusion atom. Here, for simplicity we shall neglect correlation between the filled sites of the alloy, i.e., neglect terms containing the factor $\frac{w}{kT}$ in formulas for the concentration of vacancies. On the other hand, the quantities $\frac{|v_{AB}-v_{AA}|}{kT}$ and $\frac{|v_{AB}-v_{BB}|}{kT}$ are assumed to be greater than unity (hence $v_{AB}-v_{AA}$ and $v_{AB}-v_{BB}$ have different signs). This case is true for a number of alloys, including apparently β-brass. The calculations given below refer to an alloy with a β-brass type body-centered cubic lattice, which may be found either in a disordered or in an ordered state.

In Sec. 25 we cited expressions obtained by means of the quasi-chemical approximation for the concentration of vacancies on sites of the first and second types [see Eqs. (25.15) and (25.16)]. Using the same approximation, we may determine the a posteriori probabilities $p_{dA}^{(L)}$ and $p_{dB}^{(L)}$ ($L=1,\ 2$) that atoms A and B are found alongside the vacancy on an L type site. Then the number of vacancies (per cm³ of the crystal) on sites of the first and second types, surrounded by i atoms A and by $8-i$ atoms B, can be determined from

$$N_{di}^{(1)}=\frac{8!}{i!\,(8-i)!}N_{d}^{(1)}\left[p_{dA}^{(1)}\right]^{i}\left[p_{dB}^{(1)}\right]^{8-i}=$$

$$=\frac{N}{2}\frac{8!}{i!\,(8-i)!}\left(\frac{c_A-\frac{\eta}{2}}{c_A+\frac{\eta}{2}}\right)^{7/2}(p_{AA}^{12})^{4}e^{-\frac{4v_{AA}}{kT}}\left[\frac{p_{AB}^{12}}{p_{AA}^{12}}e^{\frac{v_{AA}-v_{AB}}{kT}}\right]^{8-i}, \qquad (26.12)$$

$$N_{di}^{(2)}=\frac{N}{2}\frac{8!}{i!\,(8-i)!}\left(\frac{c_A+\frac{\eta}{2}}{c_A-\frac{\eta}{2}}\right)^{7/2}(p_{AA}^{12})^{4}e^{-\frac{4v_{AA}}{kT}}\left[\frac{p_{BA}^{12}}{p_{AA}^{12}}e^{\frac{v_{AA}-v_{AB}}{kT}}\right]^{8-i}, \qquad (26.13)$$

where p_{AA}^{12}, p_{AB}^{12} and p_{BA}^{12} are determined from (18.26), (18.27) and (18.29).

If we neglect the correlation between the filled sites of the alloy, we may set $p_{AA}^{12}=c_A^2-\frac{\eta^2}{4}$ in (26.12) and (26.13)

$$p_{AB}^{12}=\left(c_A+\frac{\eta}{2}\right)\left(c_B+\frac{\eta}{2}\right), \quad p_{BA}^{12}=\left(c_A-\frac{\eta}{2}\right)\left(c_B-\frac{\eta}{2}\right).$$

Then,

$$N_{\mathrm{d}i}^{(1)}=\frac{N}{2}\frac{8!}{i!\,(8-i)!}\left(\frac{c_A-\frac{\eta}{2}}{c_A+\frac{\eta}{2}}\right)^{7/2}\left(c_A^2-\frac{\eta^2}{4}\right)^4 e^{-\frac{4v_{AA}}{kT}}\left[\frac{c_B+\frac{\eta}{2}}{c_A-\frac{\eta}{2}}e^{\frac{v_{AA}-v_{AB}}{kT}}\right]^{8-i}, \tag{26.14}$$

$$N_{\mathrm{d}i}^{(2)}=\frac{N}{2}\frac{8!}{i!\,(8-i)!}\left(\frac{c_A+\frac{\eta}{2}}{c_A-\frac{\eta}{2}}\right)^{7/2}\left(c_A^2-\frac{\eta^2}{4}\right)^4 e^{-\frac{4v_{AA}}{kT}}\left[\frac{c_B-\frac{\eta}{2}}{c_A+\frac{\eta}{2}}e^{\frac{v_{AA}-v_{AB}}{kT}}\right]^{8-i}. \tag{26.15}$$

Just as before, we shall consider two atomic planes I and II, passing through the faces of the cubic cells and through their centers respectively. Each cm² of plane I contains $aN_{\mathrm{d}i}^{(1)}$ vacancies surrounded by i atoms A and $(8-i)$ atoms B, while 1 cm² of plane II contains $aN_{\mathrm{d}i}^{(2)}$ vacancies with the indicated environment. The flux $J_{II\to I}$ of atoms A per cm² of plane II into vacancies situated in plane I is given by

$$J_{II\to I}=4a\sum_{i,\,m}N_{\mathrm{d}i}^{(1)}\mathfrak{W}^m W_{mi}. \tag{26.16}$$

Here the index m indicates the configuration of atoms A and B, surrounding the site of second type O_2 at which atom A was located before the transition, and point P at which the potential energy of atom A on the path from O_2 to a neighboring vacancy has a maximum; W_{mi} denotes the probability of the specified configuration of atoms (W_{mi} is assumed to equal zero, if atom B is located at point O_2); $\mathfrak{W}^m$ denotes the probability per unit time of atom A with the mth configuration of neighbor-atoms making a transition from position O_2 to a neighboring vacancy. The factor "four" appears because a transition is possible from any of the four sites adjoining a given vacancy in plane II. The summation in (26.16) is carried out first over m (i.e., all configurations corresponding to a given i), and then over i from $i=0$ to $i=8$.

We shall divide the ten sites constituting the configuration into three groups, the first of which includes four sites adjoining O_2 but not adjoining P, the second includes three sites neighboring both O_2 and P and the third group-three sites from the immediate surroundings of the vacancies adjoining only P. Let the first group contain s atoms A and $4-s$ atoms B, the second group t atoms A and $3-t$ atoms B. The probability W_{st} of attaining configurations where the first group contains

s atoms A, and the second group contains t atoms A, is determined by the formula

$$W_{st}=\frac{4!}{s!\,(4-s)!}\left[p_A^{(1)}\right]^s\left[p_B^{(1)}\right]^{4-s}\frac{3!}{t!\,(3-t)!}\left[p_A^{(1)}\right]^t\left[p_B^{(1)}\right]^{3-t}. \quad (26.17)$$

Next it is necessary to know the probability of realization of some given configuration of atoms A and B in the third group. The probability W_{ri} that four of the eight sites surrounding the vacancy under consideration are occupied by r atoms A and $4-r$ atoms B, while one given site in the indicated quartet (on site O_2), is certainly occupied by atom A, is determined by the following formula:

$$W_{ri}=\frac{4!}{r!\,(4-r)!}\,\frac{4!}{(i-r)!\,[(8-i)-(4-r)]!}\,\frac{i!\,(8-i)!}{8!}\,\frac{r}{4}. \quad (26.18)$$

The probability of realization of an arrangement of atoms on ten sites surrounding points O_2 and P, which is characterized by the numbers s, t, r and i, is equal to the product of W_{st} and W_{ri}.

Having determined the energy of atom A at position O_2 and P, the following expression is obtained for the probability of atom A making a transition (per unit time) from (O_2) to the point P in the neighboring vacancy:

$$\mathfrak{W}^m=\frac{1}{\tau_0}\exp\left[-(s+t)\frac{v_{AA}}{kT}-(7-s-t)\frac{v_{AB}}{kT}+\right.$$
$$\left.+(t+r-1)\frac{\alpha'}{kT}+(6-t-r+1)\frac{\beta'}{kT}\right], \quad (26.19)$$

where α' and β' are the negative interaction energies of atoms A, at a point P, with the neighboring atoms A and B.

We shall calculate the flux (26.16) taking into account (26.14), (26.19), (26.18) and (26.17). Summing over s, t, r and i, we obtain the following expression for $J_{II\to I}$:

$$J_{II\to I}=2\,\frac{a}{\tau_0}N\left(\frac{c_A-\frac{\eta}{2}}{c_A+\frac{\eta}{2}}\right)^{7/2}\left(c_A^2-\frac{\eta^2}{4}\right)^4 e^{-\frac{4v_{AA}}{kT}}\times$$
$$\times\left[\left(c_A+\frac{\eta}{2}\right)e^{-\frac{v_{AA}}{kT}}+\left(c_B-\frac{\eta}{2}\right)e^{-\frac{v_{AB}}{kT}}\right]^4\times$$
$$\times\left[\left(c_A+\frac{\eta}{2}\right)e^{\frac{\alpha'-v_{AA}}{kT}}+\left(c_B-\frac{\eta}{2}\right)e^{\frac{\beta'-v_{AB}}{kT}}\right]^3\times \quad (26.20)$$

$$\times\left[e^{\frac{\alpha}{kT}}+\frac{c_B+\frac{\eta}{2}}{c_A-\frac{\eta}{2}}e^{\frac{\beta'+v_{AA}-v_{AB}}{kT}}\right]^3\left[1+\frac{c_B+\frac{\eta}{2}}{c_A-\frac{\eta}{2}}e^{\frac{v_{AA}-v_{AB}}{kT}}\right]^4. \quad (26.20) \text{ cont'd}$$

Here, in accordance with the statements of the previous section we may set $\alpha'=v_{AA}$, $\beta'=v_{AB}$. The flux $J^*_{II\to I}$ of tagged atoms from plane II into plane I is $\frac{N^*_A\left(x+\frac{a}{2}\right)}{N_A}$ times smaller than the flux $J_{II\to I}$, and the flux $J^*_{I\to II}$ from plane I into plane II is $\frac{N^*_A(x)}{N_A}$ times smaller than the flux $J_{II\to I}$. After determining the net flux of tagged atoms and then, in the same manner as before, the self-diffusion coefficient D^*_A, we obtain

$$D^*_A=D^0e^{-\frac{5v_{AA}}{kT}}\frac{\left(c_A^2-\frac{\eta^2}{4}\right)^{9/2}}{c_A}\left[1+\frac{c_B+\frac{\eta}{2}}{c_A-\frac{\eta}{2}}e^{\frac{v_{AA}-v_{AB}}{kT}}\right]^4\times$$

$$\times\left[1+\frac{c_B-\frac{\eta}{2}}{c_A+\frac{\eta}{2}}e^{\frac{v_{AA}-v_{AB}}{kT}}\right]^4, \quad (26.21)$$

where $D^0=\frac{a^2}{\tau_0}$. For a disordered alloy $\eta=0$ and the self-diffusion coefficient is

$$D^*_A=D^*_{A\,\text{dis}}=D^0e^{-\frac{5v_{AA}}{kT}}\left[c_A+c_Be^{\frac{v_{AA}-v_{AB}}{kT}}\right]^8. \quad (26.22)$$

It is apparent from (26.22) that, just as in the case of diffusion of interstitial atoms, the dependence of $\ln D^*_A$ on $\frac{1}{T}$ is not linear even in a disordered alloy. This relation over small ranges of reciprocal temperatures in which measurements are usually made may, however, be approximated by a linear function. This approximation is particularly accurate when $\frac{|v_{AA}-v_{AB}|}{kT}\gg 1$ and one of the terms in the square brackets of Eq. (26.22) is appreciably smaller than the other term (the last case may also be realized when the concentration of one of the components of the alloy tends to zero).

The variation of the self-diffusion coefficient with composition of the alloy according to formula (26.22) may be very pronounced, if the ratio $\frac{|v_{AA} - v_{AB}|}{kT}$ is sufficiently large. We note that in the case under consideration the plot of D_A^* vs c_A always becomes convex towards the abscissa regardless of the sign of the ordering energy w. If, however, the energies v_{AA}, v_{AB} and v_{BB} are almost equal, as indicated in the preceding section, the direction of the convexity of the curve showing the variation of Q with c_A is determined by the sign of w.

In order to clarify the influence of the addition of a small amount of atoms B to the pure metal A on the self-diffusion coefficient of atoms A, we expand (26.22) in a power series in c_B and retain the linear terms only

$$D_A^* = D^0 e^{-\frac{5v_{AA}}{kT}} \left[1 + 8c_B \left(e^{\frac{v_{AA} - v_{AB}}{kT}} - 1 \right) \right]. \tag{26.23}$$

It follows from (26.23) that the increase of D_A^* with a rise in concentration c_B may be very large, if the quantity $e^{\frac{v_{AA} - v_{AB}}{kT}} - 1$ is large. On the other hand, this quantity can not be less than -1, so that the decrease in the self-diffusion coefficient of atoms A upon the addition of atoms B can not be substantial for homogeneus solid solutions. We note that the above mentioned strong effect of impurities on diffusion can be observed experimentally [236].

In the transition of an alloy into an ordered state, just as in the case of diffusion of interstitial atoms, the characteristic features of the temperature dependence of the diffusion coefficient should be evident. For an investigation of the diffusion coefficient near the critical temperature one should expand the function $D_A^*(T, c_A, \eta)$ in a power series in η.

Restricting ourselves to quadratic terms in the expansion, we obtain

$$D_A^* = D_{A\text{dis}}^*(1 - \eta^2 b'). \tag{26.24}$$

Here $D_{A\text{dis}}^*$ is determined by (26.22), and

$$b' = \frac{9}{8c_A^2} - \frac{2\xi}{c_A^2(c_A + c_B\xi)} + \frac{\xi^2}{c_A^2(c_A + c_B\xi)^2}, \tag{26.25}$$

where

$$\xi = \exp\frac{v_{AA} - v_{AB}}{kT}. \tag{26.26}$$

Since the derivative $\frac{d\eta^2}{d\frac{1}{kT}}$ undergoes a step change at $T=T_0$ [see Eq. (24.18)], the effective activation energy will also undergo a step change at $T=T_0$. This step change is given by

$$\Delta Q = Q|_{T_0-0} - Q|_{T_0+0} = b'\frac{d\eta^2}{d\frac{1}{kT}}, \qquad (26.27)$$

where b' is defined by (26.25), and $\frac{d\eta^2}{d\frac{1}{kT}}$ by (24.18). For alloys with stoichiometric composition

$$\Delta Q = wf(\xi), \qquad (26.28)$$

where

$$f(\xi) = 27 - \frac{96\xi}{(1+\xi)^2}. \qquad (26.29)$$

By investigating the function $f(\xi)$ over the interval $0 < \xi < \infty$, we can prove that this function takes only positive values within the interval

$$3 \leqslant f(\xi) < 27.$$

Hence, the effective activation energy always increases during ordering and the plot of $\ln D_A^*$ vs $\frac{1}{T}$ always exhibits a break in the direction of the abscissa axis. The quantity ΔQ has a maximum at $\xi = 0$ and at $\xi = \infty$, i.e., for $\frac{v_{AA} - v_{AB}}{kT} = \pm\infty$. Thus the experimental detection of the break in the curve of $\ln D_A^*$ vs $\frac{1}{T}$ is most likely for alloys in which the difference between v_{AA} and v_{AB} is large compared with kT_0 and w (hence, v_{AA} and v_{BB} also differ greatly). In such alloys the step change in the effective activation energy is equal in order of magnitude to 13 kT_0, i.e., approximately 1 ev for $T_0 \sim 700-800\,^\circ$K. One should expect a rather pronounced break for self-diffusion in β-brass, because the activation energy for diffusion in pure Zn and Cu differ almost by a factor of three; they are equal to $Q_{Zn} \approx 20$, $Q_{Cu} \approx 57$ kcal/deg, respectively. In fact, the activation energy is determined, to a considerable extent, by the energy of formation of vacancies, and the latter is related to the interaction energies of the atoms.

The concentration curves of the diffusion coefficient should also have a break in the direction towards the x axis, similar to those considered in Sec. 24. The temperature and concentration curves of the self-diffusion coefficients in alloys, where the order-disorder transition is a first-order phase transition (for example, in alloys with a $AuCu_3$ type crystal lattice), exhibit step changes similar to those obtained in diffusion of interstitial atoms (Sec. 24).

The characteristic features of the temperature variation of the diffusion coefficient in ordered alloys, which have been predicted theoretically in the papers cited in this chapter, have also been confirmed experimentally. Plotting the experimental data on the diffusion of tagged copper and zinc atoms in β-brass [157], given in Chapter I, Sec. 9 (see Fig. 86), we observe breaks in the curves of the logarithm of the diffusion coefficient versus the reciprocal of temperature. Thus, in conformity with the theoretical conclusions, the curves have breaks in the direction towards the abscissa, i.e., the effective activation energy increases during ordering. Table 8 and Fig. 86 show that the magnitude of the step-change in activation energy Q at the second-order phase transition point is of the order of 1 ev, i.e., the same order of magnitude as given by the formulas cited. In accordance with the theoretical conclusions expounded in this section, the experimental investigations (see Fig. 86) also reveal a marked departure of the curve of $\lg D$ vs $\frac{1}{T}$ from a straight line in the ordered crystal at temperatures slightly below the critical temperature where the degree of long-range order changes rapidly with temperature. At lower temperatures, when the crystal becomes almost completely ordered, the departures from linearity become less pronounced. In conformity with the theoretical conclusions we also observe small departures from linearity in the graph of $\lg D$ as a function of $\frac{1}{T}$ for a disordered alloy.

3. *Influence of interstitial atoms on self-diffusion of metals* [259]

We shall now consider the influence of a small amount of admixture of element C dissolved in metal A on the self-diffusion of this metal. We assume that metal A has a face-centered cubic lattice, and atoms C are introduced into interstitial positions (centers of the unit cubic cells and the middle of their edges).

Let us denote the ratio of the number of atoms C to the the number of atoms A by $c_C = \frac{N_C}{N_A}$. Furthermore, let E_0 denote the energy of the solid solution, which has no vacancies at the crystal lattice sites. We take into account the fact that in the case of temperatures above zero vacancies will exist in the crystal. These vacancies may be surrounded by a different number i of neighboring atoms C $(0 \leqslant i \leqslant 6)$. Let N_{di} denote the number of vacancies in the crystal, having as neighbors i atoms C. Then the energy of solution of atoms C in metal A in the presence of vacancies will, in the nearest neighbor approximation, be

$$E = E_0(c) + u_A \sum_{i=0}^{6} N_{di} + u'_{AC} \sum_{i=0}^{6} iN_{di}, \tag{26.30}$$

where u_A is the energy necessary for the formation of vacancies in a pure metal A $(u_A > 0)$, u'_{AC} is negative energy of interaction of atoms A and C, separated by a distance $\frac{a}{2}$.

For a determination of the free energy of a crystal we shall find the number of different configurations of atoms A and vacancies on the sites and of atoms C on the interstitial sites for the given values of N_A, N_C and of all N_{di}

$$W = \frac{\left(N_A + \sum_{i=0}^{6} N_{di}\right)!}{N_A! \prod_{i=0}^{6} N_{di}!} \frac{\left(N_A - 5\sum_{i=0}^{6} N_{di}\right)!}{\left(N_C - \sum_{i=0}^{6} iN_{di}\right)! \left[N_A - 5\sum_{i=0}^{6} N_{di} - \left(N_C - \sum_{i=0}^{6} iN_{di}\right)\right]!} \times$$

$$\times \prod_{i=0}^{6} \left[\frac{6!}{i!\,(6-i)!}\right]^{N_{di}}. \tag{26.31}$$

From (26.30) and (26.31) the expression for the free energy of an alloy $F = E - kT \ln W$ can be obtained in the customary manner. The equilibrium concentration $\frac{N_{di}}{N_A}$ of vacancies on the crystal lattice sites can be determined from the equilibrium conditions $\frac{\partial F}{\partial N_{di}} = 0$ $(i = 0, 1, \ldots 6)$. Taking into account the fact that the concentration of vacancies is sufficiently small, we obtain

$$\frac{N_{di}}{N_A} = \frac{1}{1-c_A} e^{-\frac{u_A}{kT}} \frac{6!}{i!\,(6-i)!} \left(c_C e^{-\frac{u'_{AC}}{kT}}\right)^i (1-c_C)^{6-i}. \qquad (26.32)$$

Summing this expression over all values of i, we readily find the total concentration of vacancies $\frac{N_d}{N_A}$ on the crystal lattice sites of the alloy

$$\frac{N_d}{N_A} = \sum_{i=0}^{6} \frac{N_{di}}{N_A} = \frac{1}{1-c_C} e^{-\frac{u_A}{kT}} \left[1 + c_C\left(e^{-\frac{u_{AC}}{kT}} - 1\right)\right]^6. \qquad (26.33)$$

Equation (26.33) shows that even if the energy u'_{AC} is equal to zero, the addition of C atoms leads to an increase in the number of vacancies, because each new vacancy increases the number of interstitial sites in the alloy, and therefore also increases the entropy of the C atoms. This, in turn, leads to a decrease of the free energy of the crystal.

Proceeding to the determination of the diffusion coefficient of atoms A, we shall draw the atomic planes I and II as indicated in Fig. 105. The number of vacancies surrounded by i atoms C per cm² of plane II is equal to $\frac{a}{2} N_{di}$. We find $2aN_{di}$ neighboring atoms A around these vacancies in plane I. The probability $\mathfrak{W}_n$ of a transition of atom A into a neighboring vacancy per a unit time is equal to

$$\mathfrak{W}_n = \frac{1}{\tau_0} e^{-\frac{U_{Pn}-U_{On}}{kT}}. \qquad (26.34)$$

Here U_{On} and U_{Pn} are the energies of atom A at the original position O (at a site) and at the saddle point of the potential barrier (at point P), respectively. The subscript n denotes the number of atoms C adjoining point P ($n = 0, 1, 2$). The height of the potential barrier $U_{Pn} - U_{On}$ is obviously equal to $\Delta U_A + nB$, where ΔU_A is the height of the barrier in pure metal A, and B is the difference of interaction energies of atoms A and C, located at distances $\frac{\sqrt{2}}{4}a$ and $\frac{a}{2}$. The probability W_{ni} that n of the i atoms C surrounding a given vacancy are found at two interstitial sites adjoining point P is determined from the formula

$$W_{ni} = \frac{2!}{n!\,(2-n)!} \frac{4!}{(i-n)!\,(4-i+n)!} \frac{i!\,(6-i)!}{6!}. \qquad (26.35)$$

The flux J_{I-II} of atoms A, passing per unit time across 1 cm from plane I into plane II, is equal to the reverse flux J_{II-I}

$$J_{I-II} = \sum_{i=0}^{6} \sum_{n=0}^{2} 2aN_{\text{d}\,i} W_{ni} \mathfrak{W}_n. \tag{26.36}$$

Let us assume a concentration gradient of tagged atoms A in the direction of the x axis and perpendicular to planes I and II. Calculating the sum in (26.36) and determining D_A^*, we find

$$D_A^* = D_A^0 \frac{1}{1-c_C} \left[1 + c_C \left(e^{-\frac{u'_{AC}}{kT}} - 1\right)\right]^4 \left[1 + c_C \left(e^{-\frac{u'_{AC}+B}{kT}} - 1\right)\right]^2, \tag{26.37}$$

where $D_A^0 = D^0 e^{-\frac{Q_A}{kT}}$ $(Q_A = u_A + \Delta U_A)$ is the self-diffusion coefficient of atoms A in pure metal A, and $D^0 = \frac{a^2}{\tau_0}$. It follows from (26.37) that a smaller amount of admixture of C atoms can greatly increase the value of the diffusion coefficient D_A^*, only if the values of $e^{-\frac{u'_{AC}}{kT}}$ or $e^{-\frac{u'_{AC}+B}{kT}}$ are substantially greater than unity. From this formula it may also be seen that the self-diffusion coefficient is not an exponential function of $\frac{1}{T}$, as in the case of diffusion in pure metals. For alloys, as was indicated above, there is no basis for expecting such a dependence of D_A^* on $\frac{1}{T}$. If, however, the measurement of the diffusion coefficient in alloys is conducted over a small temperature range (200–400°) near some temperature T_1, then the graph of $\ln D_A^*$ vs $\frac{1}{T}$ may be approximated quite accurately by a straight line. Figure 109 gives an example such a graph plotted for the parameters $u'_{AC} = -0.45$ ev, $u'_{AC} + B = 0$, $c_C = 0.05$. Let us introduce the effective activation energy Q at temperature T_1

$$Q = -\left.\frac{\partial \ln D_A^*}{\partial \frac{1}{kT}}\right|_{T=T_1},$$

and also the effective factor D', which is determined by

$$D' = D_A^* e^{\frac{Q}{kT_1}}. \tag{26.38}$$

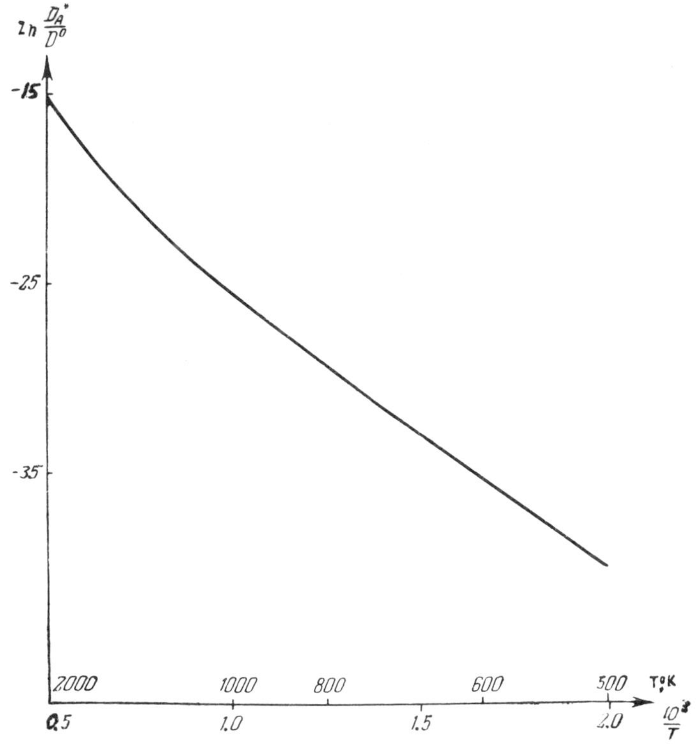

Fig. 109. Approximate correlation of the logarithm of self-diffusion coefficient of a metal in an interstitial alloy and the reciprocal of temperature.

Obviously the magnitude of Q in this case is

$$Q(T_1) = Q_A + 2\left[\frac{2e^{-\frac{u'_{AC}}{kT_1}}u'_{AC}}{1+c_C\left(e^{-\frac{u'_{AC}}{kT_1}}-1\right)} + \frac{e^{-\frac{u'_{AC}+B}{kT_1}}\left(u'_{AC}+B\right)}{1+c_C\left(e^{-\frac{u'_{AC}+B}{kT_1}}-1\right)}\right]c_C. \quad (26.39)$$

At low concentrations of atoms C, when

$$c_C\left(e^{-\frac{u'_{AC}}{kT_1}}-1\right) \ll 1, \quad c_C\left(e^{-\frac{u'_{AC}+B}{kT_1}}-1\right) \ll 1, \quad (26.40)$$

the effective activation energy is a linear function of c_C

$$Q = Q_A - Q'c_C = Q_A + 2\left[2e^{-\frac{u'_{AC}}{kT_1}}u'_{AC} + e^{-\frac{u'_{AC}+B}{kT_1}}\left(u'_{AC}+B\right)\right]c_C. \quad (26.41)$$

If one of the quantities $-\frac{u'_{AC}}{kT_1}$ or $-\frac{u'_{AC}+B}{kT_1}$ exceeds unity, a small quantity of impurity atoms C may cause a substantial decrease in the value of Q. On the other hand a sharp increase in Q in the assumed model with the addition of atoms C is impossible, because the increase of Q occurs with positive energies u'_{AC} and $u'_{AC}+B$ (one of these energies may be negative but small), when $e^{-\frac{u'_{AC}}{kT}}$ and $e^{-\frac{u'_{AC}+B}{kT}}$ are also small.

With an increase of concentration of atoms C, entailing a possible violation of one of conditions (26.40) provided $e^{-\frac{u'_{AC}}{kT}}$ or $e^{-\frac{u'_{AC}+B}{kT}}$ are much greater than unity, the dependence of Q on c_C begins to deviate from linearity and tends toward saturation. Figure 110 gives a typical curve of the correlation between Q and c_C, plotted for the parameters $Q_A = 3$ev; $u'_{AC} = -0.45$ ev, $u'_{AC} + B = 0$, $T = 1400\,°$K.

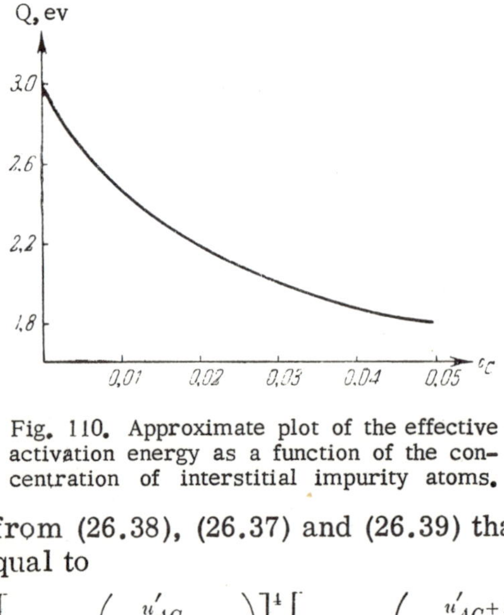

Fig. 110. Approximate plot of the effective activation energy as a function of the concentration of interstitial impurity atoms.

It follows from (26.38), (26.37) and (26.39) that the effective factor D' is equal to

$$D' = \frac{D^0}{1-c_C}\left[1+c_C\left(e^{-\frac{u'_{AC}}{kT_1}}-1\right)\right]^4\left[1+c_C\left(e^{-\frac{u'_{AC}+B}{kT_1}}-1\right)\right]^2 \times$$

$$\times \exp\left\{2\left[\frac{2e^{-\frac{u'_{AC}}{kT_1}}\frac{u'_{AC}}{kT_1}}{1+c_C\left(e^{-\frac{u'_{AC}}{kT_1}}-1\right)}+\frac{e^{-\frac{u'_{AC}+B}{kT_1}}\frac{u'_{AC}+B}{kT_1}}{1+c_C\left(e^{-\frac{u'_{AC}+B}{kT_1}}-1\right)}\right]c_C\cdot\right\} \qquad (26.42)$$

If the concentration of atoms C is sufficiently small $(c_C \ll 1)$, we may discard in Eq. (26.42) the quadratic terms in c_C, and also the linear terms in c_C, that do not contain the factors $e^{-\frac{u'_{AC}}{kT_1}}$ or $e^{-\frac{u'_{AC}+B}{kT_1}}$. The latter can be substantially greater than unity. As a result, D' may be approximately represented by the exponential function

$$D' \approx D^0 e^{-\xi'(T_1)c_C},$$

where

$$\xi'(T_1) = -2\left[2e^{-\frac{u'_{AC}}{kT_1}}\left(\frac{u'_{AC}}{kT_1}+1\right)+e^{-\frac{u'_{AC}+B}{kT_1}}\left(\frac{u'_{AC}+B}{kT_1}+1\right)\right]. \qquad (26.43)$$

Here the quantity $\xi'(T_1)c_C$ may be rather large, if the quantities $-\frac{u'_{AC}}{kT_1}$ or $-\frac{u'_{AC}+B}{kT_1}$ are substantially greater than unity.

Thus the formulas obtained enable us to explain qualitatively the linear decrease of the effective activation energy and the exponential decrease of the effective factor in the self-diffusion coefficient of iron when a small amount of carbon is added to γ-iron [237]. The correct orders of magnitudes of $Q'(T_1)$ and $\xi'(T_1)$ are obtained, if we choose, for example, the following values of the parameters: $u'_{AC} = -0.45$ ev, $u'_{AC}+B = 0$. It should be noted that for these values of the parameters with $c_C \sim$ 2-5 atomic %, the condition (26.40) ceases to be fulfilled and deviations from the linear relation $Q(c_C)$ should be observed. The curve $Q(c_C)$ may however be approximated over this interval by a straight line, so that the deviations from linearity lie within the limits of experimental error. Furthermore, in accord with experiment, the effective factor decreases sharply when carbon is added to iron. For example, for $c_C = 0.02$, its value decreases to $D' \approx 10^{-2}D^0$. Thus in the initial section of the curve of D' versus c_C over the range $c_C = 0$ to $c_C = 0.025$ and with the chosen values of the constants, we obtained an approximate relation for the effective factor D' vs c_C in the form $D' = 10^{-c_C}D^0$ (in this formula c_C is expressed in %). This relationship agrees with the experimental data of [237].

However, at larger values of c_C the value of D' decreases more slowly with an increase of c_C. Nevertheless, in the specified small range of concentration, when c_C increases from zero to 0.025, D' decreases approximately one hundredfold. Thus the theory examined here affords the possibility of explaining the basic experimental results obtained by [237].

The theory of diffusion of atoms in alloys expounded in this chapter leads, as was indicated above, to a number of qualitative conclusions.

Contrary to the popular viewpoint according to which the diffusion coefficient D in alloys is an exponential function of the reciprocal of temperature, we see that a more complicated dependence of D on T should hold even for a disordered alloy. Nevertheless, examination of the formulas cited above shows that the graph of $\ln D$ versus $\frac{1}{T}$ for disordered alloys may be sufficiently accurately approximated by a straight line over small ranges of temperatures, where the measurements are usually carried out. However, in the study of diffusion over a sufficiently wide temperature range, deviations from the indicated linearity may be observed. In this case diffusion in an alloy can not be described by a single constant activation energy.

Upon transition into the ordered state, characteristic features should appear on the curves of the temperature variation of the diffusion coefficient. If the ordering is a second-order phase transition, the diffusion coefficient changes continuously, but the curve of $\ln D$ versus $\frac{1}{T}$ has a break at the critical temperature T_0. If however the transition to the ordered state is a first-order phase transition, then at $T = T_0$ not only the effective activation energy, but also the diffusion coefficient will undergo a step change. In the ordered state near the critical temperature T_0, departures of the graph from this straight line relation should be especially marked. On the concentration curve of the diffusion coefficient we should also encounter step changes and breaks at those compositions where the alloy changes into the ordered state (in the case when $T = \text{const}$).

An investigation of the concentration dependence of D in the case of small concentration of one of the components of the alloy affords an explanation of the strong influence of the impurity on the diffusion coefficient and on the diffusion parameters (effective factor and effective activation energy).

If the interaction energies of atoms in the alloy are known, the theory can be used in some cases to predict the influence of a given impurity on the diffusion coefficient.

Chapter V

Motion of Microparticles in the Crystal Lattice Field of Ordered Alloys

27. Investigation of the Motion of a Nearly-Free Microparticle in a Completely Ordered Crystal

Many properties of alloys are determined not only by the arrangement and motion of atoms in the crystal lattice, but also by the state of the conduction electrons. For example, in the study of electrical conductivity, galvanomagnetic, magnetic, optical and other properties of alloys it is necessary to know the effect of the energy spectrum and wave function of electrons on ordering and change in composition. Strictly speaking, this type of problem should be solved from the standpoint of the many-electron theory, but this theory has not yet been sufficiently developed to be applied for solution of this problem. In this chapter the calculations will therefore be made by the single-electron approximation, and the results for alloys will be only qualitative ones. If it is found that the treatment of a system of interacting electrons in the crystal lattice of an alloy may be reduced to the problem of the motion of noninteracting quasi-particles in a lattice field, the results obtained may also become applicable to the present case (without the use of the poorly founded single-electron approximation). As mentioned in Chap. I, along with ordered alloys possessing metallic properties, there also exist ordered nonmetallic solid solutions (for example, ferrites), and also disordered nonmetallic crystals (either entirely disordered or only at sites of a definite type). The motion of conduction electrons in such solids may be treated by the one-electron approximation. The method presented in this chapter may be applied to this case as well.

In this section we shall consider the motion of a microparticle (called an electron in the following) in the field of the crystal lattice atoms of a completely ordered multicomponent crystal of the stoichiometric composition of arbitrary structure. Thus, we shall assume the potential energy of an electron to be small and treat it as a small perturbation (nearly free electron approximation). The Brillouin zones for a number of pure metals and several completely ordered compounds and alloys have been examined in [260–262]. A similar problem for the special case of an ordered two-dimensional square lattice has been investigated in [263], which also gives some qualitative considerations for metals with a face-centered cubic lattice. The zone structure of a binary alloy with simple and face-centered cubic lattices has been investigated in [264]. An investigation of the change in the energy spectrum and wave function of an electron during the transition from a completely ordered alloy to a pure metal with the same arrangement of atoms has been presented for the general case of a multicomponent alloy of arbitrary structure in [265].

Let us consider an alloy (or nonmetallic crystal of a specified type) by regarding its crystal lattice as geometrically perfect, i.e., neglecting various types of distortions of the cell shape, and the thermal vibrations of the atoms.

The unit cell of the crystal will be characterized by the vectors $\boldsymbol{a}_1, \boldsymbol{a}_2, \boldsymbol{a}_3$. In each cell we shall choose one site of a definite type as the origin, and the position of atoms forming the cell will be determined by the vectors $\boldsymbol{h}_{\varkappa_\alpha}$, drawn from the site taken as $\boldsymbol{\varkappa}_\alpha$ the origin of a given cell to all other sites with a cell number, occupied by atoms of species α. The common origin will be chosen from any of such origin sites. Then the position of any atom is determined by the vector $\boldsymbol{R}_s + \boldsymbol{h}_{\varkappa_\alpha}$, where $\boldsymbol{R}_s = s_1\boldsymbol{a}_1 + s_2\boldsymbol{a}_2 + s_3\boldsymbol{a}_3$ (s_1, s_2, s_3 are integers and s is the cell number). The potential energy of the electron in such a crystal may be written as

$$V(\boldsymbol{r}) = \sum_{s=1}^{N^0} \sum_{\alpha=1}^{\zeta} \sum_{\varkappa_\alpha=1}^{\lambda_\alpha} V_\alpha\left(\boldsymbol{r} - \boldsymbol{R}_s - \boldsymbol{h}_{\varkappa_\alpha}\right), \tag{27.1}$$

where N^0 is the number of unit cells in the crystal; ζ is the number of components; λ_α is the number of atoms of species α in the cell; and $V_\alpha(\boldsymbol{r} - \boldsymbol{R}_s - \boldsymbol{h}_{\varkappa_\alpha})$ is the potential energy of an electron in the field of an atom of species α located at the point $\boldsymbol{R}_s + \boldsymbol{h}_{\varkappa_\alpha}$. We shall expand the potential energy $V(\boldsymbol{r})$, which is a periodic function of the coordinates, in a Fourier series

$$V(\boldsymbol{r}) = \sum_{g} v_g e^{2\pi i (\boldsymbol{g}\boldsymbol{r})}. \tag{27.2}$$

Here $\boldsymbol{g}$ is a reciprocal lattice vector having the form

$$\boldsymbol{g} = g_1\boldsymbol{d}_1 + g_2\boldsymbol{d}_2 + g_3\boldsymbol{d}_3, \tag{27.3}$$

where g_1, g_2, g_3 are integers and $\boldsymbol{d}_1, \boldsymbol{d}_2, \boldsymbol{d}_3$ are the basis vectors of the reciprocal lattice related to $\boldsymbol{a}_1, \boldsymbol{a}_2, \boldsymbol{a}_3$ by

$$(\boldsymbol{a}_i \boldsymbol{d}_j) = \delta_{ij} \qquad (i, j = 1, 2, 3).$$

The expansion coefficients v_g, as is well known, may be written as

$$v_g = \frac{1}{\Delta_0} \int\limits_{\Delta_0} V(\boldsymbol{r})\, e^{-2\pi i (\boldsymbol{g}\boldsymbol{r})}\, d\tau, \tag{27.4}$$

where the integration extends over the volume Δ_0 of the unit cell and $v_{-g} = v_g^*$. Substituting (27.1) into (27.4), changing the variable $\boldsymbol{r} \to \boldsymbol{r}' = \boldsymbol{r} - \boldsymbol{R}_S - \boldsymbol{h}_{\varkappa_\alpha}$ and noting that $e^{-2\pi i (\boldsymbol{g}\boldsymbol{R}_s)} = 1$, we find

$$v_g = \sum_{\alpha=1}^{\zeta} v_{\alpha g} \sum_{\varkappa_\alpha = 1}^{\lambda_\alpha} e^{-2\pi i (\boldsymbol{g}\boldsymbol{h}_{\varkappa_\alpha})}, \tag{27.5}$$

where

$$v_{\alpha g} = \frac{1}{\Delta_0} \int V_\alpha(\boldsymbol{r})\, e^{-2\pi i (\boldsymbol{g}\boldsymbol{r})}\, d\tau \tag{27.6}$$

The integral in (27.6) is taken over the unit cell of the crystal (it is practically equal to the integral over infinite limits) and the prime in the new variable is discarded.

It is known [261, 262] that the equations of the planes bounding the Brillouin zone are

$$(\boldsymbol{k}\boldsymbol{g}) + \pi g^2 = 0 \tag{27.7}$$

(where $\boldsymbol{k}$ is the wave vector of the electron). These equations should be written only for those vectors $\boldsymbol{g}$ for which $v_g \neq 0$. They will be called the reflection planes. The electron energy E changes discontinuously across these planes,

$$\Delta E = 2\,|v_g|. \tag{27.8}$$

The reflection planes may be divided into principal and auxiliary planes. The principal reflection planes are those for which ΔE does not become zero during a transition to a case of a single pure component with the same arrangement of atoms (as in an alloy or compound). The auxiliary planes are those for which

ΔE (i.e., v_g) does become zero. The auxiliary planes occur only in an ordered crystal. In the case of a crystal consisting of one pure component α, Eq. (27.5) takes the form

$$v_g^{(\text{pure metal})} = v_{\alpha g} \sum_{\varkappa=1}^{\mu} e^{-2\pi i\,(g h_\varkappa)} \tag{27.9}$$

where μ is the number of sites forming the basis, and $h_\varkappa$ is a vector determining the position of the site with number $\varkappa$ in the cell. Since $v_{\alpha g} \neq 0$, Eq. (27.9) for v_g (pure metal) becomes 0 only in the case when

$$L_g = \sum_{\varkappa=1}^{\mu} e^{-2\pi i\,(g,\, h_\varkappa)} \tag{27.10}$$

is equal to zero. Thus the auxiliary planes will be obtained for those reciprocal lattice vectors g, for which

$$v_g \neq 0, \quad L_g = 0. \tag{27.11}$$

To obtain an equation for the auxiliary planes in the k-space, one should substitute all the vectors g satisfying conditions (27.11) into Eq. (27.7). The principal planes correspond to those g vectors for which

$$L_g \neq 0, \tag{27.12}$$

and their equations are obtained by substituting these values of g into (27.7).

Obviously the magnitude of $v_{\alpha g}$, which can be determined by Eq. (27.6) is a Fourier series expansion coefficient of the potential energy of an electron in the Bravais lattice (having one atom per unit cell), constructed with identical atoms of species α and having the same fundamental vectors a_1, a_2, a_3 as the complex lattice of a compound or alloy. For such a Bravais lattice, the reflection planes are obtained for all vectors g, since all $v_{\alpha g} \neq 0$.

We shall now consider the examples of completely ordered binary alloys with a β-brass type lattice and a $AuCu_3$ type face lattice. In these cases we shall divide the reflection planes, and consequently also the boundaries of the Brillouin zone formed by them, into principal and auxiliary planes.

In the case of an alloy with a β-brass type, body-centered cubic lattice we choose the vectors $\boldsymbol{a}_1, \boldsymbol{a}_2, \boldsymbol{a}_3$ along three mutually perpendicular edges of a cubic cell. Then $a_1 = a_2 = a_3 = a$, $d_1 = d_2 = d_3 = \frac{1}{a}$, $\mu = 2$, $\lambda_1 = \lambda_2 = 1$, $\boldsymbol{h}_{x_1} = \boldsymbol{h}_1 = 0$, $\boldsymbol{h}_{x_2} = \boldsymbol{h}_2 = \frac{1}{2}(\boldsymbol{a}_1 + \boldsymbol{a}_2 + \boldsymbol{a}_3)$. Consequently, according to (27.5) and (27.10)

$$v_g = v_{1g} + (-1)^{g_1+g_2+g_3} v_{2g}, \tag{27.13}$$

$$L_g = 1 + (-1)^{g_1+g_2+g_3} \tag{27.14}$$

Reflection planes occur for all $\boldsymbol{g}$ vectors of the lattice, reciprocal to the simple cubic lattice with constant a. The first four Brillouin zones formed by these planes are illustrated in Fig. 111. The principal planes of this group are obtained for those $\boldsymbol{g}$ vectors for which the sum $g_1 + g_2 + g_3$ is even, and the auxiliary planes are obtained if this sum is odd. Thus a transition from a pure metal with a body-centered cubic lattice to a completely ordered alloy with a β-brass type lattice in the $\boldsymbol{k}$-space (inside the first Brillouin zone, which is a dodecahedron (Fig. 111b) bounded by the principal reflection planes) produces an additional zone surface with a shape of a cube whose edge is $2\frac{\pi}{a}$ (Fig. 111a).

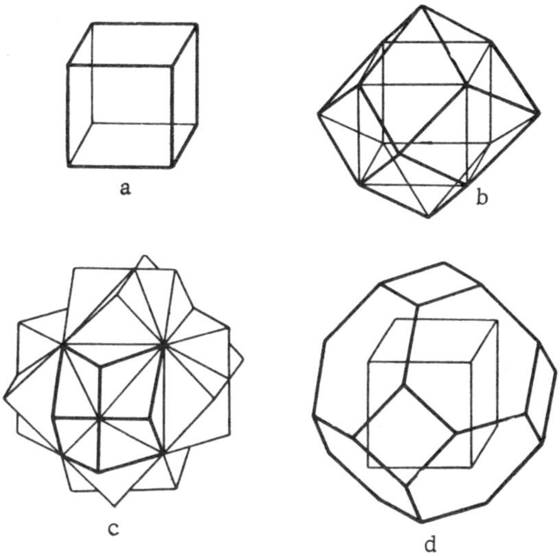

Fig. 111. Brillouin zone for a cubic lattice.

In the case of an alloy with a $AuCu_3$ face-centered lattice, the vectors $\boldsymbol{a}_1, \boldsymbol{a}_2, \boldsymbol{a}_3$ and $\boldsymbol{d}_1, \boldsymbol{d}_2, \boldsymbol{d}_3$ may be chosen just as in the previous case. Furthermore, in this case $\mu = 4$, $\lambda_1 = 1$, $\lambda_2 = 3$,

$h_1=0, h_2=\frac{1}{2}(a_1+a_2), h_3=\frac{1}{2}(a_1+a_3), h_4=\frac{1}{2}(a_2+a_3)$ $(h_{x_1}=h_1, h_{x_2}=h_2, h_3, h_4)$.

Then

$$v_g=v_{1g}+v_{2g}\left(e^{-\pi i(g_1+g_2)}+e^{-\pi i(g_1+g_3)}+e^{-\pi i(g_2+g_3)}\right), \quad (27.15)$$

$$L_g=1+e^{-\pi i(g_1+g_2)}+e^{-\pi i(g_1+g_3)}+e^{-\pi i(g_2+g_3)}. \quad (27.16)$$

In this manner reflection planes are again obtained for all g vectors of the lattice reciprocal to the simple cubic lattice and having a cubic cell edge equal to a. The principal planes ($L_g \neq 0$), according to (27.16), are obtained for those g for which g_1, g_2, g_3 have the same parity, and the auxiliary planes ($L_g=0$) are obtained when the numbers g_1, g_2, g_3 have different parity (i.e., two of them are even and one odd or vice versa). Thus for a pure metal with a face-centered cubic lattice, the first Brillouin zone, which is bounded by the principal reflection planes of type (111) and (200), is the polyhedron illustrated in Fig. 111d. This polyhedron represents the outer surface of the fourth Brillouin zone for a simple cubic lattice. When passing to a completely ordered alloy with a $AuCu_3$ type lattice inside this zone, there emerge auxiliary planes of the three inner zones, which are the first three Brillouin zones for the simple cubic lattice (Fig. 111a, b, c).

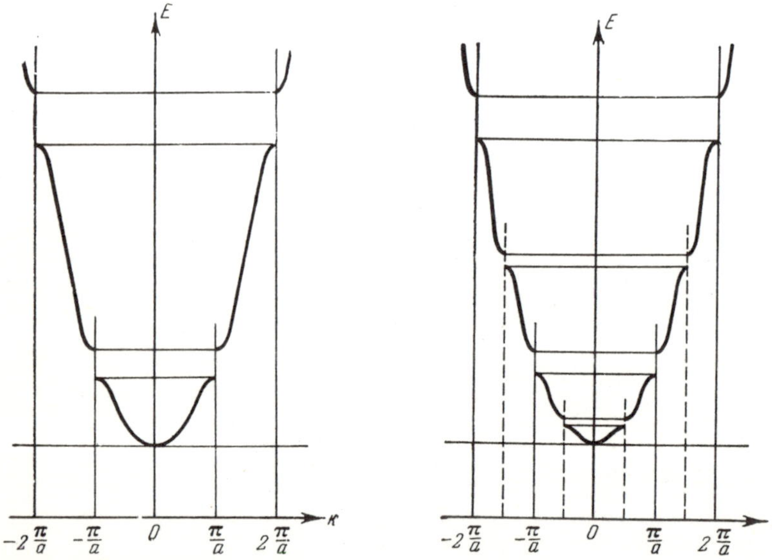

Fig. 112. Approximate form of electron energy as a function of the wave number in a one-dimensional model in the nearly-free electron approximation.
a—pure metal with a lattice constant a; b—completely ordered alloy with a lattice constant $2a$.

Thus the k-space of a completely ordered alloy is divided into a smaller number of zones than that of a pure metal with the same arrangement of atoms. The emergence of additional energy discontinuities across the zone boundaries may be illustrated by a one-dimensional problem, in which the electron potential energy is a function of only a single coordinate. The approximate form of the curve of the electron energy E as a function of the wave number k is illustrated in Fig. 112. Fig. 112a corresponds to a pure metal, in which the electron potential energy has period a, and Fig. 112b corresponds to a completely ordered alloy, in which the electron potential energy has the period $2a$.

28. Motion of a Tightly Bound Microparticle in a Completely Ordered Crystal

We now consider the problem of the motion of an electron in the crystal lattice of a completely ordered crystal, using the tightly bound electron approximation [266].

A completely ordered alloy, unlike a pure metal, contains several species of atoms with different energy levels and different wave functions of the valence electron (in an isolated atom). We shall find the wave function of the electron to the zeroth approximation and its energy to a first approximation. We shall assume that the ground states of the valence electron in different isolated atoms are close to one another, i.e., the difference between them will be regarded as small compared to the distance to other energy levels of these atoms. Each unit cell of the alloy will be regarded as a molecule, formed by the atoms of the basis. Let the unit cell of the alloy, consisting of ζ components, contain μ atoms forming the basis. Let $E_\varkappa$ and $\psi_\varkappa$ ($\varkappa = 1, \ldots, \mu$) denote the ground states and the corresponding wave functions of the valence electron in an isolated atom (number $\varkappa$) of the basis, wherein the levels are assumed to be nondegenerate and the wave functions real. If the basis contains several identical atoms, the eigenvalues $E_\varkappa$ and the forms of the functions $\psi_\varkappa$ corresponding to them will be identical.

In the treatment of the almost degenerate problem of the motion of an electron in the field of a set of atoms in the s-th unit cell, we shall write the wave function of the zero approximation of this problem in the form of a linear combination

$$\psi_s^0(\boldsymbol{r}) = \sum_{\varkappa=1}^{\mu} b_\varkappa \psi_\varkappa (\boldsymbol{r} - \boldsymbol{R}_s - \boldsymbol{h}_\varkappa) \tag{28.1}$$

with constant coefficients $b_\varkappa$. The wave function of an electron moving in the crystal field of the entire crystal in the zeroth approximation may then be written as a linear combination of functions (28.1)

$$\Psi(r) = \sum_{s=1}^{N^0} C_s \psi_s^0(r) = \sum_{s=1}^{N^0} \sum_{\varkappa=1}^{\mu} C_{s\varkappa} \psi_\varkappa(\boldsymbol{r} - \boldsymbol{R}_s - \boldsymbol{h}_\varkappa), \tag{28.2}$$

where

$$C_{s\varkappa} = C_s b_\varkappa. \tag{28.3}$$

Translational symmetry requires that the wave function $\Psi(\boldsymbol{r})$ have the form

$$\Psi(\boldsymbol{r}) = e^{i(\boldsymbol{kr})} u_{\boldsymbol{k}}(\boldsymbol{r}), \tag{28.4}$$

where $u_{\boldsymbol{k}}(\boldsymbol{r})$ is a function, having the periodicity of the crystal lattice and corresponding to a state with the electron wave vector $\boldsymbol{k}$.* From this we obtain

$$C_s = C_0 e^{i(\boldsymbol{kR}_s)}. \tag{28.5}$$

To determine the energy E of the electron in the first approximation we substitute the wave function (28.2) into the Schrödinger equation:

$$(H - E) \sum_{s=1}^{N^0} \sum_{\varkappa=1}^{\mu} C_{s\varkappa} \psi_{s\varkappa} = 0, \tag{28.6}$$

where

$$\psi_{s\varkappa} = \psi_\varkappa(\boldsymbol{r} - \boldsymbol{R}_s - \boldsymbol{h}_\varkappa), \tag{28.7}$$

$$H = -\frac{\hbar^2}{2m}\Delta + V(\boldsymbol{r}), \tag{28.8}$$

and m is the mass of an electron. Now, the potential energy of the electron $V(\boldsymbol{r})$ according to (27.1) may be rewritten as

---

*In the tightly-bound electron approximation, $\boldsymbol{k}$ denotes the reduced wave vector.

$$V(\boldsymbol{r}) = \sum_{s=1}^{N^0} \sum_{\varkappa=1}^{\mu} V_{s\varkappa}. \tag{28.9}$$

Here

$$V_{s\varkappa} = V_\varkappa (\boldsymbol{r} - \boldsymbol{R}_s - \boldsymbol{h}_\varkappa) \tag{28.10}$$

denotes the potential energy of an electron in the field of an atom at site number $\varkappa$ of the sth unit cell. We multiply (28.6) by the wave function $\psi_{s'\varkappa'}$ and integrate over the fundamental region of the crystal, taking into account the fact that the slight overlap of the wave functions of different atoms enables one to neglect the integrals of the product of such functions. Moreover, normalizing the functions $\psi_{s\varkappa}$, we obtain the following system of equations:

$$\sum_{s=1}^{N^0} \sum_{\varkappa=1}^{\mu} C_{s\varkappa} H_{s\varkappa s'\varkappa'} = E C_{s'\varkappa'}. \tag{28.11}$$

Here

$$H_{s\varkappa s'\varkappa'} = \int \psi_{s'\varkappa'} H \psi_{s\varkappa}\, d\tau = E_\varkappa \delta_{ss'} \delta_{\varkappa\varkappa'} + \int \psi_{s'\varkappa'} (V - V_{s\varkappa}) \psi_{s\varkappa}\, d\tau, \tag{28.12}$$

where $\delta_{ss'}$ and $\delta_{\varkappa\varkappa'}$ are Kronecker symbols. Substituting (28.3) and (28.5) into (28.11), we find

$$\sum_{\varkappa=1}^{\mu} \sum_{s=1}^{N^0} H_{s\varkappa s'\varkappa'} e^{i\boldsymbol{k}(\boldsymbol{R}_s - \boldsymbol{R}_{s'})} b_\varkappa = E b_{\varkappa'}. \tag{28.13}$$

Next we introduce the notation

$$M_{\varkappa\varkappa'}(\boldsymbol{k}) = \sum_{s=1}^{N^0} H_{s\varkappa s'\varkappa'} e^{i\boldsymbol{k}(\boldsymbol{R}_s - \boldsymbol{R}_{s'})}. \tag{28.14}$$

In the nearest neighbor approximation (usually employed for calculations by the method of tightly bound electrons), the only integrals in (28.12) regarded as nonvanishing are those which contain wave functions referring to the same or to neighboring atoms. Therefore, the magnitudes of $M_{\varkappa\varkappa'}(\boldsymbol{k})$ are independent of the position of the s'th cell. Locating the origin at the first site of cell number s', setting $R_{s'}=0$ in (28.14), numbering the cells by s_1, s_2, s_3 (which determine the vector $\boldsymbol{R}_s = s_1\boldsymbol{a}_1 + s_2\boldsymbol{a}_2 + s_3\boldsymbol{a}_3$), and using Eq. (28.12), we obtain

$$M_{\varkappa\varkappa'}(\boldsymbol{k}) = \sum_{s_1 s_2 s_3} \Big[E_\varkappa \delta_{s0}\delta_{\varkappa\varkappa'} + \int \psi_{000\varkappa'} (V - V_{s_1 s_2 s_3 \varkappa}) \psi_{s_1 s_2 s_3 \varkappa}\, d\tau \Big] \cdot e^{i(s_1 \boldsymbol{k}\boldsymbol{a}_1 + s_2 \boldsymbol{k}\boldsymbol{a}_2 + s_3 \boldsymbol{k}\boldsymbol{a}_3)}. \tag{28.15}$$

Substituting (28.14) into (28.13) and taking into account the fact that $M_{\varkappa\varkappa'}(\boldsymbol{k})$ is independent of s', the system of equations (28.11) reduces to μ equations

$$\sum_{\varkappa=1}^{\mu} M_{\varkappa\varkappa'}(\boldsymbol{k})\, b_\varkappa - E b_{\varkappa'} = 0 \qquad (\varkappa' = 1, 2, \ldots, \mu). \tag{28.16}$$

The condition for the existence of a nonvanishing solution of this system of μ equations

$$\begin{vmatrix} M_{11}-E & M_{12} & \ldots & M_{1\mu} \\ M_{21} & M_{22}-E & \ldots & M_{2\mu} \\ \cdot & \cdot & \cdot & \cdot \\ M_{\mu 1} & M_{\mu 2} & \ldots & M_{\mu\mu}-E \end{vmatrix} = 0 \tag{28.17}$$

gives a μth order equation for the determination of the dependence of the electron energy E on the wave vector $\boldsymbol{k}$. Eq. (28.17) has μ roots. This means that for a given $\boldsymbol{k}$, i.e. for given coefficients $M_{\varkappa\varkappa'}(\boldsymbol{k})$, μ values of energy will exist. The reduced wave vector $\boldsymbol{k}$, as is well known, may vary over the limits of the first Bravais zone for a given structure of the ordered alloy. Therefore in the first zone this calculation gives the μ branches of the energy. To remove this multiplicity one may choose a large range of variation of the wave vector, where the energy may be regarded as a single-valued function of $\boldsymbol{k}$.

We shall consider as an example a one-dimensional crystal in which the potential energies and wave function of the electrons depend on a single coordinate x. Assume that the crystal is composed of two species of regularly alternating atoms located at the same distance $\frac{a}{2}$. Then the potential energy of the electron will be a periodic function of x with a period a. The unit cell in this case contains two atoms, i.e., $\mu = 2$. Now Eq. (28.17) takes the form

$$\begin{vmatrix} M_{11}-E & M_{12} \\ M_{21} & M_{22}-E \end{vmatrix} = 0, \tag{28.18}$$

whence

$$E=\frac{M_{11}+M_{22}}{2}\pm\sqrt{\frac{(M_{11}-M_{22})^2}{4}+M_{12}M_{21}}. \qquad (28.19)$$

Determining the values of M_{11}, M_{12}, M_{21} and M_{22} from (28.15) in the nearest neighbor approximation, we obtain:

$$\left.\begin{aligned}
M_{11}&=E_1+\int\psi_{01}^2(V-V_{01})\,dx,\\
M_{22}&=E_2+\int\psi_{02}^2(V-V_{02})\,dx,\\
M_{12}&=\int\psi_{02}(V-V_{01})\psi_{01}\,dx+\int\psi_{02}(V-V_{11})\psi_{11}\,dx e^{ika},\\
M_{21}&=\int\psi_{01}(V-V_{02})\psi_{02}\,dx+\int\psi_{01}(V-V_{-12})\psi_{-12}\,dx e^{-ika}.
\end{aligned}\right\} \qquad (28.20)$$

Substituting (28.20) into (28.19), we find the following expression for the electron energy:

$$E=\frac{\mathscr{E}_{11}+\mathscr{E}_{22}}{2}\pm\sqrt{\frac{(\mathscr{E}_{11}-\mathscr{E}_{22})^2}{4}+4\mathscr{E}_{12}\mathscr{E}_{21}\cos^2\frac{ka}{2}}, \qquad (28.21)$$

where we introduce the notation

$$\left.\begin{aligned}
&\mathscr{E}_{11}=M_{11},\quad \mathscr{E}_{22}=M_{22},\\
&\mathscr{E}_{12}=\int\psi_{02}(V-V_{01})\psi_{01}\,dx=\int\psi_{02}(V-V_{11})\psi_{11}\,dx,\\
&\mathscr{E}_{21}=\int\psi_{01}(V-V_{02})\psi_{02}\,dx=\int\psi_{01}(V-V_{-12})\psi_{-12}\,dx.
\end{aligned}\right\} \qquad (28.22)$$

As is well known, the reduced wave number k varies over the range

$$-\frac{\pi}{a}<k\leqslant\frac{\pi}{a}. \qquad (28.23)$$

Each value of k in this interval corresponds according to (28.21) to two values of the electron energy $E^+(k)$ and $E^-(k)$ [taken with a plus and minus sign respectively in front of the square root sign in (28.21)]. The graph of the functions $E^+(k)$ and $E^-(k)$ for the values of the constants $\mathscr{E}_{11}=8$, $\mathscr{E}_{22}=6$, $\mathscr{E}_{12}\mathscr{E}_{21}=1$ (in arbitrary units) is illustrated in Fig. 113a. Thus in the present problem we obtain two branches of energy, similarly to the well-known problem of Born and Karman [267] for the frequency vibration spectrum of a one-dimensional chain with regularly alternating

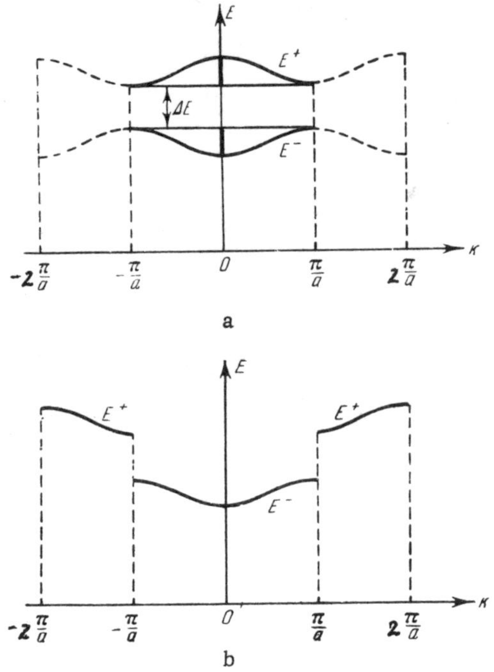

Fig. 113. The electron energy as a function of the wave number in a one-dimensional model (tight binding approximation).

atoms of different masses. If the fundamental region of the crystal contains N atoms, i.e., the number of unit cells is equal to $N^0 = \frac{N}{2}$, the reduced wave number k can take $\frac{N}{2}$ different values over the interval (28.23). Hence, the total number of energy levels, resulting from the splitting of levels E_1 and E_2 of isolated atoms of the first and second species is equal to the total number of atoms N in the fundamental region of the crystal.

As previously mentioned, the state of the electron can be characterized in several other ways. Instead of giving k over the interval (28.23) and the number of the energy branches, one may eliminate the two-valued energy and introduce a wave number k which varies over an interval twice as wide

$$-2\frac{\pi}{a} < k \leqslant 2\frac{\pi}{a} \tag{28.24}$$

Here, we consider only states corresponding to the curve E^- in its inner part where $-\frac{\pi}{a} < k \leqslant \frac{\pi}{a}$, and the curve E^+ in only the outer part. The resulting graph of the variation of E with k is given in Fig. 113b. Both methods of description are obviously entirely equivalent by virtue of the physical equivalence of points k and $k+\frac{2\pi}{a}$ on the same curve of Fig. 113a.

As follows from Eq. (28.21), there is a discontinuity in the energy spectrum at $k=\pm\frac{\pi}{a}$. The width ΔE of the spectrum is

$$\Delta E=|\mathcal{E}_{11}-\mathcal{E}_{22}|. \tag{28.25}$$

In the case of pure metal with the same arrangement of atoms as in the alloy, $\mathcal{E}_{11}=\mathcal{E}_{22}$ and ΔE becomes zero. Since $\mathcal{E}_{12}=\mathcal{E}_{21}$, it is easily shown that Eq. (28.21) in this case becomes the well-known Bloch's theoretical formula for pure metals.

Thus the qualitative difference between the energy spectrum of an electron in the one-dimensional crystal of a completely ordered alloy and the spectrum of a pure metal is (as apparent from Fig. 113a, where the bold-faced lines on the ordinate denote the regions of allowed energy values) the emergence of an additional forbidden region inside the energy band. In this sense, the energy spectrum of a tightly bound electron in the one-dimensional lattice of a completely ordered alloy is of the same type as in the nearly free electron approximation.

As a second example we shall consider the motion of an electron in the field of the body-centered cubic lattice of a completely ordered β-brass crystal. Just as in the previous case, the unit cell contains two atoms ($\mu=2$), and the equation for determination of the electron energy again has the form of (28.18), and its solution is of the form (28.19). By determining the quantities M_{11}, M_{12}, M_{21} and M_{22} from (28.15) for this structure and substituting them into (28.19), we obtain the following expression for the electron energy:

$$E=E^{\pm}=\frac{\mathcal{E}_{11}+\mathcal{E}_{22}}{2}\pm \sqrt{\frac{(\mathcal{E}_{11}-\mathcal{E}_{22})^2}{4}+64\mathcal{E}_{12}\mathcal{E}_{21}\cos^2\frac{k_xa}{2}\cos^2\frac{k_ya}{2}\cos^2\frac{k_za}{2}}, \tag{28.26}$$

where

$$\left.\begin{aligned}
\mathcal{E}_{11}&=E_1+\int(\psi_{0001})^2(V-V_{0001})\,d\tau,\\
\mathcal{E}_{22}&=E_2+\int(\psi_{0002})^2(V-V_{0002})\,d\tau,\\
\mathcal{E}_{12}&=\int\psi_{0001}(V-V_{0002})\,\psi_{0002}\,d\tau,\\
\mathcal{E}_{21}&=\int\psi_{0002}(V-V_{0001})\,\psi_{0001}\,d\tau,
\end{aligned}\right\} \tag{28.27}$$

k_x, k_y, k_z are the components of the wave vector $\boldsymbol{k}$ along the sides of the cubic cell of length equal to a. The range of variation of the wave vector will be the first Brillouin zone for an alloy with this structure, i.e., a cube whose side is $2\pi/a$. Each point of this band corresponds to two energy values: $E^+(\boldsymbol{k})$ and $E^-(\boldsymbol{k})$. As in the previous case, one may characterize the electron state by other physically equivalent methods. To accomplish this we choose a zone whose volume is as large (dodecahedron illustrated in Fig. 111b, twice bounded on the outside by the second Brillouin zone of a simple cubic lattice) and consider it as a region of variation of the wave vector. In the inner (cubic) zone the energy is assumed to be determined by the solution $E^-(\boldsymbol{k})$, and in the outer zone (the region between the surfaces of the cube and the dodecahedron) by the solution $E^+(\boldsymbol{k})$.

The electron energy changes discontinuously across the surface of the cubic zone by a magnitude which, according to (28.26), is $\Delta E = |\mathcal{E}_{11} - \mathcal{E}_{22}|$, i.e., the energy is determined by Eq. (28.25), but with the $\mathcal{E}_{11}$ and $\mathcal{E}_{22}$ that are defined by (28.27).

We note that the constant energy surface

$$E = \frac{\mathcal{E}_{11} + \mathcal{E}_{22}}{2} - \frac{|\mathcal{E}_{11} - \mathcal{E}_{22}|}{2} \tag{28.28}$$

coincides with the inner surface of this cube, whereas its outer surface is another constant energy surface

$$E = \frac{\mathcal{E}_{11} + \mathcal{E}_{22}}{2} + \frac{|\mathcal{E}_{11} - \mathcal{E}_{22}|}{2} \tag{28.29}$$

Thus the energy spectrum of an electron moving in the lattice under consideration is a system of two allowed nonoverlapping bands separated by a forbidden region of width ΔE.

In case of a pure metal (for example, of the first component) with the same arrangement of atoms as in the alloy, $\mathcal{E}_{22} = \mathcal{E}_{11}$, $\mathcal{E}_{21} = \mathcal{E}_{12}$. Therefore the energy discontinuity ΔE at the boundary of the cubic zone is equal to zero, and the expression for the electron energy (28.26) becomes the well-known Bloch formula [261] for a pure metal with a body-centered cubic lattice:

$$E = E_1 + C_1 + 8A_1 \cos\frac{k_x a}{2} \cos\frac{k_y a}{2} \cos\frac{k_z a}{2} \tag{28.30}$$

where

$$C_1 = \int (\psi_{0001})^2 (V - V_{0001})\, d\tau \tag{28.31}$$

and the exchange integral is

$$A_1 = \int \psi_{0001} (V - V_{0002}) \psi_{0002}\, d\tau. \tag{28.32}$$

We shall now find an expression for the effective mass of the electron, which can be derived in the case of small $k_x a$, $k_y a$ and $k_z a$, i.e., near the lower edge of the lower energy band. In this case Eq. (28.26) for E^- may be expanded in a power series of $k_x a$, $k_y a$ and $k_z a$, up to quadratic terms:

$$E = E^- = \frac{\mathscr{E}_{11} + \mathscr{E}_{22}}{2} - \sqrt{\frac{(\mathscr{E}_{11} - \mathscr{E}_{22})^2}{4} + 64\mathscr{E}_{12}\mathscr{E}_{21}} + \frac{8\mathscr{E}_{12}\mathscr{E}_{21}a^2}{\sqrt{\frac{(\mathscr{E}_{11} - \mathscr{E}_{22})^2}{4} + 64\mathscr{E}_{12}\mathscr{E}_{21}}} (k_x^2 + k_y^2 + k_z^2). \tag{28.33}$$

Comparing (28.33) with the formula

$$E = E_0 + \frac{\hbar^2}{2m_*} (k_x^2 + k_y^2 + k_z^2), \tag{28.24}$$

which determines the effective mass m_*, we obtain the following expression for m_*:

$$m_* = \frac{\hbar^2 \sqrt{\frac{(\mathscr{E}_{11} - \mathscr{E}_{22})^2}{4} + 64\mathscr{E}_{12}\mathscr{E}_{21}}}{16\mathscr{E}_{12}\mathscr{E}_{21}a^2} \tag{28.35}$$

Near the upper edge of the upper energy band, where, let us say, $k_x a$ is close to $\pm 2\pi$, but $k_y a$ and $k_z a$ are close to zero, the electron energy E^+, determined from (28.26) can be approximately represented as

$$E = E^+ = \frac{\mathscr{E}_{11} + \mathscr{E}_{22}}{2} + \sqrt{\frac{(\mathscr{E}_{11} - \mathscr{E}_{22})^2}{4} + 64\mathscr{E}_{12}\mathscr{E}_{21}} - \frac{8\mathscr{E}_{12}\mathscr{E}_{21}a^2}{\sqrt{\frac{(\mathscr{E}_{11} - \mathscr{E}_{22})^2}{4} + 64\mathscr{E}_{12}\mathscr{E}_{21}}} (k_x^{+2} + k_y^2 + k_z^2), \tag{28.36}$$

where

$$k_x^+ = k_x - \left(\pm \frac{2\pi}{a}\right)$$

It is apparent from (28.36) that the effective mass differs in this case only in the sign from (28.35).

It also follows from (28.26) that the electron momentum in this approximation

$$p = \frac{m}{\hbar}\,\mathrm{grad}_k E \tag{28.37}$$

becomes zero on the surface of the cubic zone.

29. Case of a Partially Ordered Crystal

In the previous sections of this chapter we have considered the motion of an electron in the crystal field of a completely ordered compound or alloy containing a perfectly periodic potential. In the case of a partially ordered crystal the problem is more complicated because the electron potential energy is no longer a periodic function of the coordinates. If, however, the influence of disorder in the alternation of atoms is small and may be considered as a small perturbation of the electron motion in a certain periodic field, the problem of determination of the energy and wave functions of an electron in a partially ordered crystal may be solved by means of the perturbation theory. The solution of this type of the perturbation theory problem for the case of disordered alloys was given in [268]. The treatment for the case of partially ordered alloys with simple cubic and lattices in the nearly free electron approximation was presented in [264]. The general case treated below for partially ordered alloys of any structure has been investigated in the nearly free electron approximation by [265], and in the tight-binding approximation by [266].

The potential energy $V(r)$ of an electron in the crystal lattice of a partially ordered alloy may be written as

$$V(r) = \sum_{s=1}^{N^0} \sum_{L=1}^{Q} \sum_{\varkappa_L=1}^{\lambda_L} V_\alpha (r - R_s - h_{\varkappa_L}), \tag{29.1}$$

where L is the type number of the site ($L = 1, \ldots, Q$), $\varkappa_L$ is the number of site type L in the unit cell ($\varkappa_L = 1, \ldots, \lambda_L$) and $h_{\varkappa_L}$ is a vector drawn from the original site of the sth cell in its site L of number $\varkappa_L$. Instead of $V_\alpha(r - R_s - h_{\varkappa_L})$ in (29.1), the potential energy of an electron in the field of an atom of the species α that occupies the given site should be substituted. The potential energy $V(r)$ may be represented in the form of a sum of two terms

$$V=\overline{V}+V'. \tag{29.2}$$

Here $\overline{V}$ is the mean potential energy

$$\overline{V}=\sum_{s=1}^{N}\sum_{L=1}^{Q}\sum_{\varkappa_L=1}^{\lambda_L}\overline{V}^{(L)}(\boldsymbol{r}-\boldsymbol{R}_s-\boldsymbol{h}_{\varkappa_L}), \tag{29.3}$$

where

$$\overline{V}^{(L)}=\sum_{\alpha=1}^{\zeta}p_\alpha^{(L)}V_\alpha \qquad (L=1,\ \ldots,\ Q) \tag{29.4}$$

is the mean potential energy of an electron in the field of type L sites. The second term in (29.2) is the randomly fluctuating part of the potential energy.

Subsequent calculations are based on the assumption that an electron in the field of atoms of a different type has potential energies V_α that are almost equal to one another, so that V' may be treated as a perturbation. Then, the exact Schrödinger equation for an electron in the lattice of a partially ordered alloy

$$\Delta\Psi+\frac{2m}{\hbar^2}(E-\overline{V}-V')\Psi=0 \tag{29.5}$$

may be solved approximately in the following manner.

In the zeroth approximation, setting $V'=0$, we obtain the equation

$$\Delta\Psi_0+\frac{2m}{\hbar^2}(E-\overline{V})\Psi_0=0. \tag{29.6}$$

The potential energy $\overline{V}$, occurring in (29.8), is a periodic function of coordinates and can be interpreted as the potential energy of an electron moving in a completely ordered crystal constituted of the approximate atoms with potential energies (29.4). Thus, the zeroth approximation problem refers to the case already examined in the previous sections of this chapter, namely to the motion of an electron in a crystal constituted of regularly alternating atoms. Thus, Ψ_0 has the form of a Bloch function $\Psi_0=e^{i\boldsymbol{kr}}u_k(\boldsymbol{r})$. Using the formulas derived in previous sections, one may find a expression for Ψ_0 and E_0 in the approximations of nearly free and tightly bound elections.

To a first approximation $V'\neq 0$, and to determine the energy $E=E_0+E'$ one must compute the correction

$$E' = \int \Psi_0^* V' \Psi_0 \, d\tau. \tag{29.7}$$

This integral equals zero for any form of the periodic function $u_k(r)$ occurring as a factor in Ψ_0. The electron energy E to a first approximation (taking into account first order quantities in the differences $V_\alpha - V_{\alpha'}$) therefore equals the zeroth approximation energy E_0.

Thus, in the treatment of the problem of the motion of an electron in the lattice of a partially ordered crystal, whose atoms establish nearly equal potential energies V_α for an electron, the zeroth approximation wave function Ψ_0 (neglecting the first order quantities in the differences $V_\alpha - V_{\alpha'}$) and the energy E in the first approximation (taking into account first order quantities in these differences) may be computed from the formulas of the theory of a completely ordered crystal constituted of the appropriate atoms with mean potential energies for the electron (29.4).

Let us consider next the binary ordered alloy with two types of sites ($\alpha = A, B, L = 1,2$). Substituting (1.5) for the probability $p_\alpha^{(L)}$ of substitution by atoms of the site in (29.4), we obtain:

$$\left.\begin{aligned} \overline{V}^{(1)} &= V^0 + \varphi^{(1)}(c_A, \eta)(V_A - V_B), \\ \overline{V}^{(2)} &= V^0 + \varphi^{(2)}(c_A, \eta)(V_A - V_B), \end{aligned}\right\} \tag{29.8}$$

where

$$V^0 = \nu V_A + (1 - \nu) V_B, \tag{29.9}$$

$$\left.\begin{aligned} \varphi^{(1)}(c_A, \eta) &= c_A - \nu + (1 - \nu)\eta, \\ \varphi^{(2)}(c_A, \eta) &= c_A - \nu - \nu\eta. \end{aligned}\right\} \tag{29.10}$$

We shall first treat the motion of an electron in a partially ordered crystal by means of the nearly free electron approximation. To calculate the energy and wave function of the electron according to the foregoing, one must replace the partially ordered alloy by a completely ordered crystal constituted of the appropriate atoms, i.e., replace V_α by $\overline{V}^{(L)}$ in the formulas of Sec. 27 (where L denotes the type of site correct for atoms of species α) or for a binary alloy

$$V_A \to \overline{V}^{(1)}, \quad V_B \to \overline{V}^{(2)}. \tag{29.11}$$

Substitution (29.11) into (27.6), we obtain instead of $v_{\alpha g}$ the quantity $\overline{v}_g^{(L)}$, which plays the role of $v_{\alpha g}$ for an alloy, formed from the appropriate atoms:

$$\bar{v}_g^{(L)} = v_g^0 + \varphi^{(L)}(c_A, \eta)(v_{Ag} - v_{Bg}) \qquad (L = 1, 2), \qquad (29.12)$$

where

$$v_g^0 = \nu v_{Ag} + (1 - \nu) v_{Bg}. \qquad (29.13)$$

Substituting (29.13) for $v_{\alpha g}$ in (27.5) and replacing the summation over the types of atoms α by a summation over the types of sites L, we find the dependence of the Fourier coefficient $\bar{v}_g$ of the potential energy of the electrons in the field of the appropriate atoms on c_A and η:

$$\bar{v}_g = \sum_{L=1}^{2} \bar{v}_g^{(L)} \sum_{\varkappa_L=1}^{\lambda_L} e^{-2\pi i g h_{\varkappa_L}} = [v_g^0 + (c_A - \nu)(v_{Ag} - v_{Bg})] L_g + $$
$$+ (v_{Ag} - v_{Bg}) \left[(1 - \nu) \sum_{\varkappa_1=1}^{\lambda_1} e^{-2\pi i g h_{\varkappa_1}} - \nu \sum_{\varkappa_2=1}^{\lambda_2} e^{-2\pi i g h_{\varkappa_2}} \right] \eta. \qquad (29.14)$$

(This $\bar{v}_g$ now plays the role of v_g.) The quantity L_g is determined by Eq. (27.10).

In the special case of auxiliary reflection planes $(L_g = 0)$ $\bar{v}_g$ takes the form

$$\bar{v}_g = (v_{Ag} - v_{Bg}) \eta \left[(1 - \nu) \sum_{\varkappa_1=1}^{\lambda_1} e^{-2\pi i g h_{\varkappa_1}} - \nu \sum_{\varkappa_2=1}^{\lambda_2} e^{-2\pi i g h_{\varkappa_2}} \right]. \qquad (29.15)$$

Consequently, $\bar{v}_g$ for the auxiliary reflection planes is proportional to the degree of long-range order η. Therefore, for disordered crystals, where $\eta = 0$ (and also for pure metals), $\bar{v}_g$ is equal to zero.

Let us obtain an expression for $\bar{v}_g$ for specific examples of crystal structures.

In the case of a crystal with a β-brass type body-centered cubic lattice for the principal reflection planes, where the sum of $g_1 + g_2 + g_3$ is even, from (29.14) and (27.14) we obtain:

$$\bar{v}_g = 2 [c_A v_{Ag} + (1 - c_A) v_{Bg}]. \qquad (29.16)$$

This quantity is independent of η and is a linear function of c_A. For the auxiliary reflection planes, where the sum $g_1 + g_2 + g_3$ is odd, from (29.15) we have

$$\bar{v}_g = \eta (v_{Ag} - v_{Bg}). \qquad (29.17)$$

In the case of the $AuCu_3$ face-centered cubic lattice for the principal reflection planes (the numbers g_1, g_2, g_3 have the same parity) according to (29.14) and (27.16) we have

$$\overline{v}_g = 4\,[c_A v_{Ag} + (1 - c_A)\,v_{Bg}], \tag{29.18}$$

i.e. we obtain an expression similar to (29.16). For the auxiliary reflection planes (the numbers g_1, g_2, g_3 have different parity), from (29.15) we again find

$$\overline{v}_g = \eta\,(v_{Ag} - v_{Bg}).$$

We note that the equation of a reflection plane (27.7) is the Wulff-Bragg condition for reflection of an electron wave of wavelength $\lambda = \frac{2\pi}{k}$ from the crystal lattice plane of a crystal with the Miller indices g_1, g_2, g_3. From this point of view one may draw an analogy between the principal reflection planes and the fundamental lines on the x-ray patterns on the one hand, and between the auxiliary reflection planes and the superlattice lines on the other hand.

Since many physical quantities in the theory of the motion of nearly free electrons in a completely ordered crystal are expressed in terms of the Fourier coefficients v_g, it becomes possible after replacing v_g by $\overline{v}_g$ and by using the formulas obtained here, to find the dependence of these quantities on the composition and the degree of long-range order in a partially ordered alloy. For example, the energy discontinuity ΔE upon crossing the reflection plane for a partially ordered alloy is, according to (27.8),

$$\Delta E = 2\,|\,v_g\,| \tag{29.19}$$

and for the auxiliary reflection planes, on the basis of (29.15), is proportional to η. For disordered crystals (and pure metals) the energy discontinuity ΔE becomes zero.

We shall now treat the problem of the motion of an electron in a partially ordered crystal with the aid of the tight binding approximation. In this approximation the problem is complicated by the fact that the energy and wave functions of an electron moving in a completely ordered crystal are not explicitly expressed in terms of the potential energies V_A and V_B. Hence we cannot immediately succeed in obtaining expressions for the energy and wave functions in partially ordered crystals by direct substitution of (29.11) into the corresponding expressions of the theory of completely ordered crystals, as may be done in the nearly free electron approximation.

By applying the method developed at the beginning of this section, obtained in the tight binding approximation, one must substitute [along with substitution (29.11) in the expression for the potential energy] the wave functions ψ_A, ψ_B and the energies E_A and E_B of an electron in isolated atoms of species A and B * by the wave functions $\tilde{\psi}_1, \tilde{\psi}_2$ and the energies $\tilde{E}_1, \tilde{E}_2$ for isolated effective atoms, in which the electron has the mean potential energies $\overline{V}^{(1)}$ and $\overline{V}^{(2)}$. Although $\tilde{\psi}_1, \tilde{\psi}_2, \tilde{E}_1$ and $\tilde{E}_2$ are not expressed explicitly in terms of $\overline{V}^{(1)}$ and $\overline{V}^{(2)}$, in the following it will be quite sufficient to know only the form of the dependence of $\tilde{\psi}_1, \tilde{\psi}_2, \tilde{E}_1$ and $\tilde{E}_2$ on the composition of the alloy and on the degree of long-range order. This dependence will be used in that approximation in which the theory of partially ordered crystals is generally constructed, i.e. taking into account only the first-order quantities in $V_A - V_B$. We shall use the perturbation theory to solve approximately the Schrödinger equation which determines the wave functions $\tilde{\psi}_1$ and $\tilde{\psi}_2$ (ground states). The potential energies $\overline{V}^{(1)}$ and $\overline{V}^{(2)}$ according to (29.8) are divided into two parts, whereby the second part $\varphi^{(1)}$ $(V_A - V_B)$ and $\varphi^{(2)}$ $(V_A - V_B)$, when V_A and V_B are assumed almost equal, may be considered as small perturbing corrections to the first term V^0. Thus, in first-order perturbation theory

$$\left.\begin{aligned} \tilde{\psi}_1 &= \tilde{\psi} + \varphi^{(1)}(c_A, \eta)\,\omega, \quad & \tilde{E}_1 &= \tilde{E} + \varphi^{(1)}(c_A, \eta)\,\chi, \\ \tilde{\psi}_2 &= \tilde{\psi} + \varphi^{(2)}(c_A, \eta)\,\omega, \quad & \tilde{E}_2 &= \tilde{E} + \varphi^{(2)}(c_A, \eta)\,\chi, \end{aligned}\right\} \tag{29.20}$$

where $\tilde{\psi}$ and $\tilde{E}$ are the wave function and energy (the same for both species of effective atoms) of the zeroth approximation, which evidently are independent of c_A and η. The quantities ω and χ may be obtained from well-known perturbation theory formulas. Thus, ω and χ are independent of c_A and η and are first-order quantities in $V_A - V_B$. Thus, for example,

$$\chi = \int \tilde{\psi}^2 (V_A - V_B)\, d\tau. \tag{29.21}$$

In (29.20) the only quantities depending on c_A and η are $\varphi^{(1)}$ and $\varphi^{(2)}$, this dependence being given by formulas (29.10). With the

---

*The formulas of Sec. 28 contain the wave functions $\psi_{s\varkappa}$ and energy $E_\varkappa$ of an electron in an atom located at the site of cell $\varkappa$. Since sites number $\varkappa$ of all the cells are occupied by atoms of a definite species, $\psi_{s\varkappa}$ and $E_\varkappa$ in these formulas are in essence replaced by the wave functions ψ_A or ψ_B and the energies E_A or E_B, depending on how these sites are substituted for the atoms.

aid of (29.20) the results obtained above for a completely ordered alloy may, in a number of cases, be used with partially ordered alloys. For this purpose it is sufficient to make the following substitutions in the corresponding formulas for completely ordered alloys:

$$\left.\begin{array}{lll}\psi_A \to \tilde{\psi}_1, & E_A \to \tilde{E}_1, & V_A \to \overline{V}^{(1)},\\ \psi_B \to \tilde{\psi}_2, & E_B \to \tilde{E}_2, & V_B \to \overline{V}^{(2)}.\end{array}\right\} \qquad (29.22)$$

As an example we shall consider a partially ordered alloy with a β-brass type body-centered cubic lattice $\left(\nu = \frac{1}{2}\right)$. Let us denote by $\tilde{\mathscr{E}}_{11}$, $\tilde{\mathscr{E}}_{22}$, $\tilde{\mathscr{E}}_{12}$ and $\tilde{\mathscr{E}}_{21}$ the quantities (28.27), written for a completely ordered alloy, i.e., the results of substitution of (29.22) into (28.27). Let us find the dependence of these quantities on c_A and η. We shall write expression (29.3) for the mean potential energy $\overline{V}$ of an electron in a partially ordered alloy of the given structure,

$$\overline{V} = \sum_{s=1}^{N^0} \sum_{L=1}^{2} \overline{V}_s^{(L)}(\varkappa) = \sum_{s=1}^{N^0} [\overline{V}_s^{(1)}(1) + \overline{V}_s^{(2)}(2)], \qquad (29.23)$$

where $\overline{V}_s^{(L)}(\varkappa)$ denote the mean potential energies $\overline{V}^{(L)}(\boldsymbol{r} - \boldsymbol{R}_s - \boldsymbol{h}_{\varkappa_L})$, occurring in (29.3), and where subscript s indicates the number of cell, the superscript L the type of site and the number in brackets—the number of the site in the cell which determines (together with s) the argument of the function $\overline{V}_s^{(L)}(\varkappa)$. Substituting (29.8) into (29.23), we obtain

$$\overline{V} = \Gamma_0 + \varphi_1\Gamma_1 + \varphi_2\Gamma_2, \qquad (29.24)$$

where

$$\left.\begin{array}{l}\Gamma_0 = \sum_{s=1}^{N^0} [V_s^0(1) + V_s^0(2)],\\ \Gamma_1 = \sum_{s=1}^{N^0} [V_{As}(1) - V_{Bs}(1)],\\ \Gamma_2 = \sum_{s=1}^{N^0} [V_{As}(2) - V_{Bs}(2)].\end{array}\right\} \qquad (29.25)$$

Taking into account the fact that in (28.7) the sites with number $\varkappa = 1$ correspond to ψ_A, E_A and V_A, and the sites with numbers $\varkappa = 2$ correspond to ψ_B, E_B and V_B and substituting (29.22) into (28.27), we find the following expression* for

$$\left.\begin{aligned}
\tilde{\mathscr{E}}_{11} &= \tilde{E}_1 + \int [\tilde{\psi}_{0001}(1)]^2 [\overline{V} - \overline{V}^{(1)}_{000}(1)]\, d\tau,\\
\tilde{\mathscr{E}}_{22} &= \tilde{E}_2 + \int [\tilde{\psi}_{0002}(2)]^2 [\overline{V} - \overline{V}^{(2)}_{000}(2)]\, d\tau,\\
\tilde{\mathscr{E}}_{12} &= \int \tilde{\psi}_{0001}(1) [\overline{V} - \overline{V}^{(2)}_{000}(2)]\, \tilde{\psi}_{0002}(2)\, d\tau,\\
\tilde{\mathscr{E}}_{21} &= \int \tilde{\psi}_{0002}(2) [\overline{V} - \overline{V}^{(1)}_{000}(1)]\, \tilde{\psi}_{0001}(1)\, d\tau.
\end{aligned}\right\} \tag{29.26}$$

Substituting (29.8), (29.20) and (29.23) into (29.26) and taking into account (29.24) and (29.25), we obtain expressions for $\tilde{\mathscr{E}}_{ij}$, containing a different form of the integral. The magnitude of $\tilde{\mathscr{E}}_{ij}$ is determined by the following three characteristics: (1) the existence of a factor of the order of the small difference $V_A - V_B$, (2) the existence of a product of weakly overlapping wave functions for different sites, (3) the existence of the square of the wave functions and any combination of potential energies weakly overlapping it. We shall regard quantities that do not possess one of these characteristics as zero-order quantities, those quantities possessing one of them as first order, and those having a combination of these criteria as higher order quantities which will be neglected. Then, the formulas for $\tilde{\mathscr{E}}_{ij}$ will contain one quantity of the zero-order in $\tilde{E}$ and three quantities of the first order

$$\left.\begin{aligned}
\mathscr{E}_{*} &= \int [\tilde{\psi}_{000}(1)]^2 [\Gamma_0 - V^0_{000}(1)]\, d\tau,\\
\chi &= \int [\tilde{\psi}_{000}(1)]^2 [V_{A000}(1) - V_{B000}(1)]\, d\tau,\\
\mathscr{E}_0 &= \int \tilde{\psi}_{000}(1)\, \tilde{\psi}_{000}(2) [\Gamma_0 - V^0_{000}(2)]\, d\tau.
\end{aligned}\right\} \tag{29.27}$$

All other integrals occurring in the expression for $\tilde{\mathscr{E}}_{ij}$ are quantities of higher order. Taking into account (29.10), we obtain the following formulas for $\tilde{\mathscr{E}}_{ij}$, which in this approximation give their explicit dependence on c_A and η.

---

*The numbers in brackets in the wave functions $\tilde{\psi}$ in (29.6), just as in $\overline{V}^{(L)}_{s}$ (which indicates the number of the site in the cell), determine the argument of these functions over the range of one (the zeroth) cell, where $s_1 = s_2 = s_3 = 0$. The subscripts 1 or 2 (characterizing the type of site) in these functions determine their form, which is different for different types of sites.

$$\left.\begin{aligned}\tilde{\mathscr{E}}_{11} &= \tilde{E} + \mathscr{E}_{*} + \left(c_A - \frac{1}{2}\right)\chi + \frac{\eta}{2}\chi,\\ \tilde{\mathscr{E}}_{22} &= \tilde{E} + \mathscr{E}_{*} + \left(c_A - \frac{1}{2}\right)\chi - \frac{\eta}{2}\chi,\\ \tilde{\mathscr{E}}_{12} &= \tilde{\mathscr{E}}_{21} = \mathscr{E}_0.\end{aligned}\right\} \qquad (29.28)$$

Substituting the obtained expressions (29.28) for $\tilde{\mathscr{E}}_{ij}$ into (28.26), we find the following expression for the energy of an electron moving in a partially ordered crystal with a β-brass type body-centered cubic lattice:

$$E = E^{\pm} = \tilde{E} + \mathscr{E}_{*} + \left(c_A - \frac{1}{2}\right)\chi \pm \pm \sqrt{\frac{\eta^2}{4}\chi^2 + 64\mathscr{E}_0^2 \cos^2\frac{k_x a}{2}\cos^2\frac{k_y a}{2}\cos^2\frac{k_z a}{2}}. \qquad (29.29)$$

Eq. (29.29) contains as parameters c_A, the concentration of atoms A in the alloys, and the degree of long-range order η and permits an investigation of the variation of the dependence of the electron energy E on wave vector $\boldsymbol{k}$ when these parameters are altered.

The energy discontinuity ΔE at the boundary of the cubic zone according to (28.25) and (29.28) in a partially ordered alloy of the given structure is equal to*

$$\Delta E = |\tilde{\mathscr{E}}_{11} - \tilde{\mathscr{E}}_{22}| = \eta|\chi|. \qquad (29.30)$$

Just as in the approximation of weakly bound electrons, the magnitude of discontinuity $\Delta E \sim \eta$ also disappears for disordered alloys.**

The general formulas obtained in this chapter enable us to draw qualitative conclusions concerning the effect of composition and long-range order on various physical phenomena.

It should be borne in mind that in derivation of the expression for the energy of an electron in a crystal lattice of

---

*The emergence of the energy discontinuity ΔE in a transition from a disordered to an ordered state of the alloy of the given structure under specific conditions could lead to existence of a change of its properties. For example, in the tight binding approximation a loss of metallic properties of the alloy during ordering is possible [266]. Such an inference is not possible, however, in the approximation of weakly bound electrons.

**The spectrum of the frequencies of natural oscillations of the crystal lattice of the alloy as in the energy spectrum of electron conduction may exhibit discontinuities during ordering [269, 270]. This subject will be considered in Sec. 40.

partially ordered alloys, the second order correction with respect to the small quantities $|V_\alpha - V_{\alpha'}|$ was neglected. This correction in the second approximation, however, may be comparable to the change of energy in the zeroth approximation caused by the variation of η (although cases are possible in which the second order correction is either small or nonexistent). Hence, conclusions which can be drawn concerning the variation of the energy of an electron with the degree of long-range order in partially ordered alloys should be considered as crude and qualitative. These conclusions enable us to judge only the possibility of the appearance of various effects associated with the influence of order on the properties of alloys.*

The change in the energy spectrum of the conduction electrons during ordering of the alloys should lead to a change of structure of the x-ray absorption and emission spectra. The shift of the K-edge of nickel and manganese at the short wave length side during ordering of the Ni_3Mn alloy was detected by Karal'nik [271].

*Formulas for the energy of an electron in completely ordered alloys have, of course, not been touched upon in the above remarks.

Chapter VI

Scattering of Different Types of Waves by the Crystal Lattice of An Ordered Alloy

30. Formulation of the Problem of Scattering of a Wave by a Crystal Lattice of an Alloy

The most reliable methods of investigation of long- and short-range order in the arrangement of atoms of an alloy are based on scattering of different types of waves by the crystal lattice of the alloy. In this chapter the theory of x-ray and thermal neutron scattering by ordered alloys will be treated. Since scattering of thermal neutrons (and electrons) by the crystal is a manifestation of the wave properties of these particles the results obtained are expected to be similar to those for x-ray scattering. The theory of scattering enables us to determine the distribution of intensity of the scattered radiation on either the x-ray pattern or the neutron diffraction pattern, when a plane monochromatic wave is incident on the crystal. The cases of both a single crystal and a polycrystal (Debye pattern) will be examined.

The scattering of electrons will be discussed in Chapter VII in connection with the residual electrical resistance of alloys.

In the case of scattering of waves by stationary identical atoms located at the sites of a perfect crystal lattice, sharp lines appear on the x-ray pattern. These lines correspond to normal reflection from some system of atomic planes, while the diffuse background appearing on the pattern and corresponding to Compton scattering is accompanied by a change of wavelength. If the perfect periodicity of the crystal is disrupted, for example, by vibration of atoms caused by either inhomogeneous geometric lattice distortions, or by the presence of atoms of a different type in the incompletely ordered crystal, an additional diffuse scattering of waves will occur. In the following discussion we shall consider scattering by vibrations of atoms and Compton scattering. It will be assumed that all atoms of the alloy are located exactly at the sites of a geometrically perfect crystal lattice. This disregard of the geometric distortion of the lattice can be justified, if the dimensions of the different atoms in the alloy are similar and there are no distortions caused by plastic deformation of the crystal. We shall take into account only those inhomogeneities which are due to an incompletely ordered distribution of atoms on the sites of a geometrically perfect crystal lattice. In the case of neutrons, we will also consider other types of scattering: diffuse scattering caused by the presence of isotopes in the alloy, and scattering dependent on the spins of the nuclei and the neutrons.

The calculations will be carried out within the framework of the kinematic theory of scattering. We will neglect effects due to the attenuation of the incident beam as it penetrates the crystal, and also effects due to multiple scattering. Scattering probabilities for neutrons will be determined from the perturbation theory.

Two approaches are possible in calculating the intensity of the scattered radiation. One may determine the diffraction pattern produced by the secondary waves (scattered by different atoms of an alloy with given composition), the long-range and correlation parameters [20, 22, 28, 272-278] (Sec. 31-33). Here, the intensity of the diffuse scattering is a function of concentration of the components, long-range order and correlation parameters. In another approach, the diffuse scattering is related to the fluctuations in composition and in the degree of long-range order in the alloy [170, 171, 279-281] (Sec. 34). Thus, the intensity of diffuse scattering is directly related (in the microscopic theory) to the interaction energies of the alloy atoms. Application of the different methods may be convenient in various specific cases.

31. General Case of Scattering of X-Rays By Ordered Crystals

In this section we shall treat the scattering of x-rays by a multicomponent ordered alloy with a crystal lattice in which each site is a center of symmetry of the crystal.* The problem of scattering of x-rays by the crystal lattice is solved here by the first method specified at the end of Sec. 30. The intensity

*The presentation of the material in this chapter follows the treatment of [278].

of x-ray scattering I is expressed in electron units. Thus, I denotes the ratio of intensity of x-rays scattered by the crystal (at a distance much greater than the dimensions of the crystal) to the intensity of x-rays scattered under the same conditions by a classical electron. The scattering intensity expressed in electron units is determined by the well-known formula

$$I=\left|\int \rho_э(\boldsymbol{r})\, e^{i\boldsymbol{q}\boldsymbol{r}}\, d\tau\right|^2. \tag{31.1}$$

Here $\rho_э(\boldsymbol{r})$ is the probability density of electrons in the crystal, $\boldsymbol{q}=\boldsymbol{k}'-\boldsymbol{k}$, where $\boldsymbol{k}$ and $\boldsymbol{k}'$ are the wave vectors of the incident and scattered waves $\left(|\boldsymbol{k}|=\frac{2\pi}{\lambda}\right)$; and the integration is carried out over the entire volume of the crystal.

We shall subdivide the electron density $\rho_э(\boldsymbol{r})$ into terms $\rho_{s\varkappa}(\boldsymbol{r}-\boldsymbol{R}_{s\varkappa})$, each of which corresponds to an atom at a site number $\varkappa$ of the sth unit cell ($\boldsymbol{R}_{s\varkappa}=\boldsymbol{R}_s+\boldsymbol{h}_\varkappa$ is a vector drawn from the first site of the first cell to site number $\varkappa$ of the s-th cell):

$$\rho_э(\boldsymbol{r})=\sum_{s\varkappa}\rho_{s\varkappa}(\boldsymbol{r}-\boldsymbol{R}_{s\varkappa}). \tag{31.2}$$

Substituting (31.2) into (31.1) and changing the variables $\boldsymbol{r}'=\boldsymbol{r}-\boldsymbol{R}_{s\varkappa}$ in each term, we obtain

$$I=\left|\sum_{s=1}^{N^0}\sum_{\varkappa=1}^{\mu} f_{s\varkappa}e^{i\boldsymbol{q}\boldsymbol{R}_{s\varkappa}}\right|^2=\sum_{s,\,s'=1}^{N^0}\sum_{\varkappa,\,\varkappa'=1}^{\mu} f_{s\varkappa}f^*_{s'\varkappa'}e^{i\boldsymbol{q}(\boldsymbol{R}_{s\varkappa}-\boldsymbol{R}_{s'\varkappa'})}, \tag{31.3}$$

where

$$f_{s\varkappa}=\int \rho_{s\varkappa}(\boldsymbol{r}')\, e^{i\boldsymbol{q}\boldsymbol{r}'}\, d\tau' \tag{31.4}$$

and the integration is extended over the entire volume of the crystal. The quantity $f_{s\varkappa}$ is called the atomic scattering factor of an atom at site number $\varkappa$ of the s-th cell, and it is equal to the ratio of the amplitude of the wave (scattered by a given atom) to the amplitude of the wave scattered under the same conditions by a classical electron. The atomic scattering factor of a given atom in the crystal depends, generally, on the composition of the alloy, the long- and short-range order parameters and the species of the surrounding atoms. In the following, however, we shall assume that this dependence may be neglected and that $f_{s\varkappa}$ depends only on the type of atom located at site number

$\varkappa$ of the s-th cell. This assumption is sufficiently accurate for almost all atoms except for the lightest ones. In fact, the relative role of the outer electrons for heavy atoms is not very important, and the inner electron shells are only little distorted by the neighboring atoms.

We shall present an expression for the intensity of the scattered radiation in the form of a sum of terms that correspond to sharp lines and to a diffuse background. For this purpose we shall denote the quantity $f_{s\varkappa}$, which changes during a transition from one site to another, in the form of a sum of the term $\bar{f}_\varkappa$ averaged over the site of a given sublattice (L-th type)

$$\bar{f}_\varkappa = \bar{f}_\varkappa^{(L)} = \sum_{\alpha=1}^{\zeta} p_\alpha^{(L)} f_\alpha \tag{31.5}$$

(f_α is the atomic scattering factor of a type α atom) and the deviation from the mean value $f_{s\varkappa} - \bar{f}_\varkappa$

$$f_{s\varkappa} = \bar{f}_\varkappa + (f_{s\varkappa} - \bar{f}_\varkappa). \tag{31.6}$$

Substituting (31.6) into (31.3) and noting that $\boldsymbol{R}_{s\varkappa} = \boldsymbol{R}_s + \boldsymbol{h}_\varkappa$, we obtain

$$I = I_1 + I_2 + I_3, \tag{31.7}$$

where

$$I_1 = \sum_{\varkappa, \varkappa'=1}^{\mu} \bar{f}_\varkappa \bar{f}_{\varkappa'}^* e^{i\boldsymbol{q}(\boldsymbol{h}_\varkappa - \boldsymbol{h}_{\varkappa'})} \sum_{s, s'=1}^{N^0} e^{i\boldsymbol{q}(\boldsymbol{R}_s - \boldsymbol{R}_{s'})} = \left| \sum_{\varkappa=1}^{\mu} \bar{f}_\varkappa e^{i\boldsymbol{q}\boldsymbol{h}_\varkappa} \right|^2 \left| \sum_{s=1}^{N^0} e^{i\boldsymbol{q}\boldsymbol{R}_s} \right|^2, \tag{31.8}$$

$$I_2 = \sum_{s, s'=1}^{N^0} \sum_{\varkappa, \varkappa'=1}^{\mu} (f_{s\varkappa} - \bar{f}_\varkappa)(f_{s'\varkappa'}^* - \bar{f}_{\varkappa'}^*) e^{i\boldsymbol{q}(\boldsymbol{R}_{s\varkappa} - \boldsymbol{R}_{s'\varkappa'})}, \tag{31.9}$$

$$I_3 = \sum_{s, s'=1}^{N^0} \sum_{\varkappa, \varkappa'=1}^{\mu} \bar{f}_\varkappa (f_{s'\varkappa'}^* - \bar{f}_{\varkappa'}^*) e^{i\boldsymbol{q}(\boldsymbol{R}_{s\varkappa} - \boldsymbol{R}_{s'\varkappa'})} + \text{compl. conj.} \tag{31.10}$$

The last term in (31.7), i.e., I_3, is equal to zero. Indeed, for each group of terms in the sum (31.10) with a constant difference $\boldsymbol{R}_{s\varkappa} - \boldsymbol{R}_{s'\varkappa'}$ and a given $\varkappa$, the factors $\bar{f}_\varkappa e^{i\boldsymbol{q}(\boldsymbol{R}_{s\varkappa} - \boldsymbol{R}_{s'\varkappa'})}$ are identical, i.e., the expression for this group becomes

$$\bar{f}_\varkappa e^{i\boldsymbol{q}(\boldsymbol{R}_{s\varkappa} - \boldsymbol{R}_{s'\varkappa'})} \sum{}' (f_{s'\varkappa'}^* - \bar{f}_{\varkappa'}^*), \tag{31.11}$$

where $\sum'$ denotes a summation of the terms occurring in the specific group. Replacing this sum by the averaged value of the summed expression multiplied by the number of terms (which is of the order of the number atoms of the crystal) we find that (31.11), and hence I_3 are equal to zero.

We shall demonstrate that the first term I_1 in (31.7) determines the intensity of the fundamental and superlattice maxima on the x-ray pattern corresponding to the different arrangements of the atomic planes. Representing R_s by $R_s = s_1 a_1 + s_2 a_2 + s_3 a_3$ and introducing the notation

$$\alpha_j = q a_j, \tag{31.12}$$

we find the sum of the unit cells occurring in (31.8)

$$\sum_{s=1}^{N^0} e^{iqR_s} = \prod_{j=1}^{3} \sum_{s_j=1}^{N_j^0} e^{i\alpha_j s_j} = \prod_{j=1}^{3} \frac{e^{i\alpha_j}\left(1 - e^{i\alpha_j N_j^0}\right)}{1 - e^{i\alpha_j}}, \tag{31.13}$$

where N_j^0 is the number of cells in the side of crystal number j ($j = 1, 2, 3$). The squared modulus of Eq. (31.13) is

$$\left|\sum_{j=1}^{N^0} e^{iqR_s}\right|^2 = \prod_{j=1}^{3} \frac{1 - \cos \alpha_j N_j^0}{1 - \cos \alpha_j} = \prod_{j=1}^{3} \frac{\sin^2 \frac{\alpha_j N_j^0}{2}}{\sin^2 \frac{\alpha_j}{2}}. \tag{31.14}$$

This function is periodic, as it remains constant with a substitution of α_j for $\alpha_j - 2\pi g_j$, where g_j are integers. In the limiting case of an infinite crystal the ratio

$$Y(\alpha_j) = \frac{\sin^2 \frac{\alpha_j N_j^0}{2}}{\sin^2 \frac{\alpha_j}{2}} \tag{31.15}$$

in the range $-\pi < \alpha_j \ll \pi$ is equal to the δ-function of α_j, multiplied by $2\pi N_j^0$. In fact, as $\alpha_j \to 0$ $Y(\alpha_j)$ tends to $(N_j^0)^2$, i.e., in the limit of an infinite crystal, it tends to infinity. If $|\alpha_j| \gg \frac{1}{N_j^0}$, $Y(\alpha_j)$ is much smaller than $(N_j^0)^2$. In the calculation of the integral of $Y(\alpha_j)$ over the interval $\alpha_{j_1} \ll \alpha_j \ll \alpha_{j_2}$, which includes the point $\alpha_j = 0$ (but does not include the points at which α_j are a multiple of 2π), the most important region is that of small α_j. Consequently, in the calculation of such an integral in (31.15) one may replace $\sin\frac{\alpha_j}{2}$ by $\frac{\alpha_j}{2}$ and then extend the limits of integration to $-\infty$ and ∞. Then

$$\int_{\alpha_{j_1}}^{\alpha_{j_2}} Y(\alpha_j)\,d\alpha_j \approx 4\int_{-\infty}^{\infty} \frac{\sin^2 \frac{\alpha_j N_j^0}{2}}{\alpha_j^2}\,d\alpha_j = 2N_j^0 \int_{-\infty}^{\infty} \frac{\sin^2 x}{x^2}\,dx = 2\pi N_j^0. \quad (31.16)$$

In this manner, in the limiting case of an infinite crystal, the function $Y(\alpha_j)$ over the interval $-\pi < \alpha_j \leqslant \pi$ changes into the function $2\pi N_j^0 \delta(\alpha_j)$. Over the entire range of variation of the argument $-\infty < \alpha_j < \infty$, by taking into account the periodicity of the function $Y(\alpha_j)$, it can be represented in the form

$$Y(\alpha_j) = 2\pi N_j^0 \sum_{g_j=-\infty}^{\infty} \delta(\alpha_j - 2\pi g_j). \quad (31.17)$$

From Eqs. (31.8), (31.14), (31.15) and (31.17), using the fact that $N_1^0 N_2^0 N_3^0 = N^0$, we find that the quantity I_1 is

$$I_1 = 8\pi^3 N^0 \left| \sum_{\varkappa=1}^{\mu} \overline{f_\varkappa} e^{iqh_\varkappa} \right|^2 \prod_{j=1}^{3} \sum_{g_j=-\infty}^{\infty} \delta(\alpha_j - 2\pi g_j). \quad (31.18)$$

Expression (31.18) differs from zero only for those values of α_j which are multiples of 2π. According to (31.12) such values of α_j will be obtained only for certain orientations of the single crystal (given by the vectors $\boldsymbol{a}_j$) and for certain scattering angles (given for a fixed direction of the incident beam vector $\boldsymbol{q}$). Thus the quantity I_1 actually determines the intensity of the fundamental and superlattice maxima on the x-ray pattern. The question whether a given maximum is a fundamental or superstructure maximum is resolved on the basis of the structure factor (equal to the square of the modulus of the structure amplitude F)

$$|F|^2 = \left| \sum_{\varkappa=1}^{\mu} \overline{f_\varkappa} e^{iqh_\varkappa} \right|^2. \quad (31.19)$$

The mean values of the atomic scattering factor $\overline{f_\varkappa}$, averaged over sites of definite (L-th) type, according to (31.5) are expressed in terms of the probabilities of occupation of sites of this type by different atoms and hence depend on the long-range order parameters in the alloy. If for a given q corresponding to any maximum on the x-ray pattern, the structure factor above the critical temperature, i.e., in the disordered alloy, becomes zero, then the maximum is a superlattice line. If, however, in the disordered alloy the structure factor differs from zero, the maximum is a fundamental line.

We shall now consider the term I_2 in expression (31.7). We separate the sum (31.9) into two parts I_2' and I_2'', including the respective diagonal terms in which $\varkappa = \varkappa'$ and $s = s'$, and the remaining terms, in which this condition is not fulfilled

$$I_2 = I_2' + I_2'', \tag{31.20}$$

$$I_2' = \sum_{s=1}^{N^0} \sum_{\varkappa=1}^{\mu} |f_{s\varkappa} - \bar{f}_\varkappa|^2, \tag{31.21}$$

$$Y(\alpha_j) = 2\pi N_j^0 \sum_{g_j=-\infty}^{\infty} \delta(\alpha_j - 2\pi g_j). \tag{31.22}$$

Here the primes at the summation signs denote that the summation of sites characterized by the numbers $s\varkappa$ does not coincide with site $s'\varkappa'$.

Let us first transform the expression for I_2'. Summing up formula (31.21) over s for a given $\varkappa$ and replacing this sum by the product of the mean value of the summed expression over sites of a definite number (this means also of a definite type) and the number of terms, we obtain:

$$I_2' = N^0 \sum_{\varkappa=1}^{\mu} \overline{|f_{s\varkappa} - \bar{f}_\varkappa|^2} = N^0 \sum_{L=1}^{Q} \lambda_L (\overline{|f_{s\varkappa_L}|^2} - |\bar{f}_{\varkappa_L}|^2). \tag{31.23}$$

Taking into account (31.5), noting that

$$\overline{|f_{s\varkappa_L}|^2} = \sum_{\alpha=1}^{\zeta} p_\alpha^{(L)} |f_\alpha|^2, \tag{31.24}$$

and multiplying (31.24) by

$$\sum_{\alpha'=1}^{\zeta} p_{\alpha'}^{(L)} = 1, \tag{31.25}$$

we obtain

$$I'_2 = N^0 \sum_{L=1}^{Q} \lambda_L \sum_{\alpha,\alpha'=1}^{\zeta} p_\alpha^{(L)} p_{\alpha'}^{(L)} (|f_\alpha|^2 - f_\alpha f_{\alpha'}^*). \qquad (31.26)$$

Separating the sum over α, α' into two sums with $\alpha > \alpha'$ and $\alpha < \alpha'$ (terms where $\alpha = \alpha'$ are equal to zero) and substituting $\alpha \rightleftarrows \alpha'$ in the second sum, we find

$$I'_2 = N^0 \sum_{L=1}^{Q} \lambda_L \sum_{\substack{\alpha,\alpha'=1 \\ (\alpha<\alpha')}}^{\zeta} p_\alpha^{(L)} p_{\alpha'}^{(L)} |f_\alpha - f_{\alpha'}|^2. \qquad (31.27)$$

We shall now consider the term I''_2 in (31.20). We replace the summation over s' and $\varkappa'$ for a given s and $\varkappa$ by a summation over vectors ρ drawn from the site $s\varkappa$ to different sites $s'\varkappa'$. Carrying out first the summation over s for a given $\varkappa$ and ρ and noting that for all terms of this sum the factor $e^{iq(R_{s\varkappa} - R_{s'\varkappa'})} = e^{-iq\rho}$ is identical, we obtain

$$I''_2 = N^0 \sum_{\varkappa=1}^{\mu} \sum_{\rho} e^{-iq\rho} \overline{(f_{s\varkappa} - \bar{f}_\varkappa)(f_{s'\varkappa'}^* - \bar{f}_{\varkappa'}^*)}. \qquad (31.28)$$

Here the line denotes averaging over all pairs of sites $s\varkappa$ and $s'\varkappa'$, which are respectively the origin and the end of the vector ρ drawn from sites number $\varkappa$ (thus, the type of initial site L and the type of final site L' remain unchanged during the averaging).

In terms of the probabilities $p_{\alpha\alpha'}^{LL'}(\rho)$ that atoms α and α' respectively will be found* on sites L and L', the mean value occurring in formula (31.28) may be represented in the following form:

$$\overline{(f_{s\varkappa} - \bar{f}_\varkappa)(f_{s'\varkappa'}^* - \bar{f}_{\varkappa'}^*)} = \overline{f_{s\varkappa} f_{s'\varkappa'}^*} - \bar{f}_\varkappa \bar{f}_{\varkappa'}^* =$$

$$= \sum_{\alpha,\alpha'=1}^{\zeta} [p_{\alpha\alpha'}^{LL'}(\rho) - p_\alpha^{(L)} p_{\alpha'}^{(L')}] f_\alpha f_{\alpha'}^*. \qquad (31.29)$$

The quantity in brackets is the correlation parameter

$$\varepsilon_{\alpha\alpha'}^{LL'}(\rho) = p_{\alpha\alpha'}^{LL'}(\rho) - p_\alpha^{(L)} p_{\alpha'}^{(L')}. \qquad (31.30)$$

---

*The quantity is $p_{\alpha\alpha'}^{LL'}(\rho)$ corresponding to the same but different directions of the vector ρ may in general be different, if the directions joining a given pair of sites ρ are not crystallographically equivalent. Henceforth, this effect will be neglected, i.e., we set $p_{\alpha\alpha'}^{LL'}(\rho) = p_{\alpha\alpha'}^{LL'}(\rho)$.

We shall replace the summation over $\varkappa$ in (31.28) by a summation over $\varkappa_L$ and L. Since in this case each site is a center of symmetry of the crystal, after replacement of $e^{-iq\rho}$ by $\cos q\rho - i \sin q\rho$ in the sum over ρ, one may discard the term containing $\sin q\rho$. Then, from (31.28), (31.29) and (31.30) we get

$$I_2'' = N^0 \sum_{L=1}^{Q} \sum_{\varkappa_L=1}^{\lambda_L} \sum_{\rho} \cos q\rho \sum_{\alpha, \alpha'=1}^{\zeta} \varepsilon_{\alpha\alpha'}^{LL'}(\rho) f_\alpha f_{\alpha'}^*. \qquad (31.31)$$

Using the fact that I_2'' is real, one may write this quantity as

$$I_2'' = \frac{1}{2} N^0 \sum_{L=1}^{Q} \sum_{\varkappa_L=1}^{\lambda_L} \sum_{\rho} \cos q\rho \sum_{\alpha, \alpha'=1}^{\zeta} \varepsilon_{\alpha\alpha'}^{LL'}(\rho)(f_\alpha f_{\alpha'}^* + f_\alpha^* f_{\alpha'}). \qquad (31.32)$$

From the relation between the correlation parameters (17.54) it follows that the correlation parameters for $\alpha = \alpha'$ may be written as

$$\varepsilon_{\alpha\alpha}^{LL'}(\rho) = -\frac{1}{2} \sum_{\substack{\alpha'=1 \\ (\alpha' \neq \alpha)}}^{\zeta} [\varepsilon_{\alpha\alpha'}^{LL'}(\rho) + \varepsilon_{\alpha'\alpha}^{LL'}(\rho)]. \qquad (31.33)$$

Substituting (31.33) into (31.32), separating the sum over α and α' into a sum for $\alpha < \alpha'$ and a sum for $\alpha > \alpha'$, and substituting $\alpha \rightleftarrows \alpha'$ in the latter sum, we obtain

$$I_2'' = -\frac{N^0}{2} \sum_{L=1}^{Q} \sum_{\varkappa_L=1}^{\lambda_L} \sum_{\rho} \cos q\rho \sum_{\substack{\alpha, \alpha'=1 \\ (\alpha < \alpha')}}^{\zeta} [\varepsilon_{\alpha\alpha'}^{LL'}(\rho) + \varepsilon_{\alpha'\alpha}^{LL'}(\rho)] \, |f_\alpha - f_{\alpha'}|^2. \qquad (31.34)$$

We shall also give another equivalent expression for this formula, by grouping the terms in (31.34) over the coordination spheres, where the ends of the vectors ρ are located. We shall replace the sum over ρ by a sum over sites of a definite type

$$I_2'' = -\frac{1}{2} N^0 \sum_{\substack{\alpha, \alpha'=1 \\ (\alpha < \alpha')}}^{\zeta} |f_\alpha - f_{\alpha'}|^2 \sum_{L=1}^{Q} \sum_{\varkappa_L=1}^{\lambda_L} \sum_{l=1}^{\infty} \sum_{L'=1}^{Q} \times$$
$$\times [\varepsilon_{\alpha\alpha'}^{LL'}(\rho_l) + \varepsilon_{\alpha'\alpha}^{LL'}(\rho_l)] \sum_{m_{lL'}=1}^{z_{lL'}} \cos q\rho_{m_{lL'}}. \qquad (31.35)$$

Here ρ_l is the radius of the lth coordination sphere, $\boldsymbol{\rho}_{m_{lL'}}$ is a vector drawn from the central site (type L, number $\varkappa_L$) to site number $m_{lL'}$ (type L' in the lth coordination sphere), and $z_{lL'}$, is the number of sites L' in the lth coordination sphere.

It is apparent from (31.27) and (31.34) that the expressions for I_2' and I_2'' do not contain δ-functions, i.e., the term I_2 in Eq. (31.7) for the intensity of the scattered radiation corresponds to diffuse scattering and gives the background intensity on the x-ray pattern. If correlation in the crystal is insignificant and the correlation parameters may be set equal to zero, $I_2''=0$ and the intensity of diffuse scattering I_2 reduces to I_2'. Since the atomic scattering factors are monotonic functions of q, then according to (31.27) the background intensity varies monotonically with q. If however correlation in the crystal is significant, the term I_2'' should also be taken into account. The factors $\cos \mathbf{q}\boldsymbol{\rho}_{m_{lL'}}$ in I_2'' show that the diffuse background of intensity I_2' must be augmented by a part which is oscillating with variation of q.

The correlation parameters $\varepsilon_{\alpha\alpha}^{LL'}(\rho_l)$ can be found approximately as functions of composition and temperature by means of some variant of the ordering theory. These parameters may also be considered as empirical constants characterizing the state of the crystal, which can be determined from a comparison of the experimentally observed distribution of background intensity on the x-ray pattern with the theoretical formula (see below).

According to (31.7), (31.8), (31.27) and (31.35), the general formula for the intensity of the monochromatic x-ray radiation scattered by a single crystal is

$$I=8\pi^3N^0\left|\sum_{\varkappa=1}^{\mu}\bar{f}_\varkappa e^{i\mathbf{q}\mathbf{h}_\varkappa}\right|^2\prod_{j=1}^{3}\sum_{g_j=-\infty}^{\infty}\delta(\alpha_j-2\pi g_j)+$$
$$+N^0\sum_{\substack{\alpha,\,\alpha'=1\\(\alpha<\alpha')}}^{\zeta}|f_\alpha-f_{\alpha'}|^2\sum_{L=1}^{Q}\Big\{\lambda_L p_\alpha^{(L)}p_{\alpha'}^{(L)}- \tag{31.36}$$
$$-\frac{1}{2}\sum_{\varkappa_L=1}^{\lambda_L}\sum_{l=1}^{\infty}\sum_{L'=1}^{Q}[\varepsilon_{\alpha\alpha'}^{LL'}(\rho_l)+\varepsilon_{\alpha'\alpha}^{LL'}(\rho_l)]\sum_{m_{lL'}=1}^{z_{lL'}}\cos\mathbf{q}\boldsymbol{\rho}_{m_{lL'}}$$

Equation (31.36) shows that in the case of a completely ordered stoichiometric alloy, where at a given L all values of $p_\alpha^{(L)}$, except one, are equal to zero and all $\varepsilon_{\alpha\alpha}^{LL'}=0$, the intensity of this type of diffuse scattering is also equal to zero.

We shall present now an expression for the background intensity on a Debye diagram. To accomplish this, the expression for I_2 must be averaged over different orientations of the crystal. The expression I_2' does not contain the vectors $\rho_{m_{lL'}}$. Consequently, during such averaging I_2' does not change. The averaging of I_2'' involves the mean values of the quantities $\cos q\rho_{m_{lL'}}$. These mean values are

$$\overline{\cos q\rho_{m_{lL'}}} = \frac{1}{4\pi}\int_0^{2\pi}\int_0^{\pi} \cos(q\rho_{m_{lL'}} \cos\theta) \sin\theta \, d\theta \, d\varphi = \frac{\sin q\rho_{m_{lL'}}}{q\rho_{m_{lL'}}} \qquad (31.37)$$

(θ is the angle between the vectors q and $\rho_{m_{lL'}}$). Therefore, carrying out the averaging, we obtain the background intensity on a Debye diagram

$$I_{2D} = N^0 \sum_{\substack{\alpha, \alpha'=1 \\ (\alpha<\alpha')}}^{\zeta} |f_\alpha - f_{\alpha'}|^2 \sum_{L=1}^{Q} \Bigg\{ \lambda_L p_\alpha^{(L)} p_{\alpha'}^{(L)} - \frac{1}{2} \sum_{\varkappa_L=1}^{\lambda_L} \sum_{l=1}^{\infty} \sum_{L'=1}^{Q} z_{lL'} [\varepsilon_{\alpha\alpha'}^{LL'}(\rho_l) + \varepsilon_{\alpha'\alpha}^{LL'}(\rho_l)] \frac{\sin q\rho_{m_{lL'}}}{q\rho_{m_{lL'}}} \qquad (31.38)$$

where the modulus of the vector q is

$$q = \frac{4\pi}{\lambda} \sin\vartheta \qquad (31.39)$$

(2ϑ is the scattering angle, λ is the wavelength).

The formulas (31.36) and (31.38) enable us to express the intensity of the x-radiation scattered both by a single crystal and by a polycrystal in terms of the composition of the alloy, the long-range order parameter (a priori probabilities $p_\alpha^{(L)}$) and the correlation parameter.

32. Investigation of Special Cases of Scattering of X-Rays

First, we shall write the formula for the diffuse scattering of x-rays by a single crystal in the case of a binary alloy $A-B$. Since in this case $\varepsilon_{AB}^{LL'}(\rho_l) = \varepsilon_{BA}^{LL'}(\rho_l)$ (17.55), it follows from (31.36) that

$$I_2 = N^0 |f_A - f_B|^2 \sum_{L=1}^{Q} \left\{ \lambda_L p_A^{(L)} p_B^{(L)} - \right.$$

$$\left. - \sum_{\varkappa_L=1}^{\lambda_L} \sum_{l=1}^{\infty} \sum_{L'=1}^{Q} \varepsilon_{AB}^{LL'}(\rho_l) \sum_{m_{lL'}=1}^{z_{lL'}} \cos \boldsymbol{q}\boldsymbol{\rho}_{m_{lL'}} \right\} \quad (32.1)$$

For binary alloys with two types of sites ($Q = 2$) when correlation is insignificant and the term with $\varepsilon_{AB}^{LL'}(\rho_l)$ may be neglected (at temperatures substantially above the critical temperature T_0, or at temperatures below T_0, for sufficiently large degrees of long-range order) we obtain from (32.1) and (1.5) the following formula for the background intensity:

$$I_2 = N |f_A - f_B|^2 [c_A c_B - \nu(1-\nu)\eta^2]. \quad (32.2)$$

At high temperatures, for which $\eta = 0$, this formula gives a quadratic dependence of I_2 on concentrations c_A of the type $I_2 \sim c_A(1 - c_A)$, i.e., I_2 has a maximum value for $c_A = \frac{1}{2}$. In the second case when T is much less than T_0, for alloys of a different composition but annealed at the same temperature, the maximum value of the degree of long-range order η (Sec. 12) occurs with $\nu = \frac{1}{2}$ for stoichiometric alloys, $c_A = \nu = \frac{1}{2}$. For other values of ν, close to the corresponding stoichiometric compositions, $c_A = \nu$. The value of I_2, according to (32.2) and (12.11), exhibits a maximum for stoichiometric alloy $c_A = \nu$.

The formula for I_2 is greatly simplified in the case of a disordered binary alloy (when only one type of site exists). Here, $Q = 1$, $\lambda_L = \mu$, $p_A^{(L)} = c_A$, $p_B^{(L)} = c_B$, the summation over $\varkappa_L$ reduces to a multiplication by the number of terms μ, and the summation over $m_{lL'} \equiv m_l$ is carried out over all z_l sites of the l-th coordination sphere. Consequently,

$$I_2 = N |f_A - f_B|^2 \left[c_A c_B - \sum_{l=1}^{\infty} \varepsilon_{AB}(\rho_l) \sum_{m_l=1}^{z_l} \cos \boldsymbol{q}\boldsymbol{\rho}_{m_l} \right], \quad (32.3)$$

where $N = \mu N^0$.

Formulas (32.1) and (32.3) and harmonic analysis of the experimentally observed distribution of background intensity on the x-ray pattern, make possible a determination of certain combinations of correlation parameters in a binary (disordered) alloy. In some cases (ordered alloys) one may determine the correlation parameters independently [22, 278]. We now expand $\boldsymbol{q}$ in vectors of the Bravais reciprocal lattice of the disordered alloy $\boldsymbol{d}_1, \boldsymbol{d}_2, \boldsymbol{d}_3$:

$$\boldsymbol{q} = 2\pi (x\boldsymbol{d}_1 + y\boldsymbol{d}_2 + z\boldsymbol{d}_3), \tag{32.4}$$

where x, y and z are certain continuous variables, and $\rho_{m_{lL'}}$ is expanded in the vectors $\boldsymbol{a}_1, \boldsymbol{a}_2, \boldsymbol{a}_3$ of the Bravais lattice of the disordered alloy:

$$\rho_{m_{lL'}} = \nu_{1m_{lL'}}\boldsymbol{a}_1 + \nu_{2m_{lL'}}\boldsymbol{a}_2 + \nu_{3m_{lL'}}\boldsymbol{a}_3 \tag{32.5}$$

($\nu_{1m_{lL'}}, \nu_{2m_{lL'}}$ and $\nu_{3m_{lL'}}$ are integers).

Substituting (32.4) and (32.5) into (32.1) and using the relation $a_i d_j = \delta_{ij}$, we obtain

$$\begin{aligned} \frac{I_2(x, y, z)}{N^0 |f_A - f_B|^2} &= \sum_{L=1}^{Q} \lambda_L p_A^{(L)} p_B^{(L)} - \\ &- \sum_{L=1}^{Q} \sum_{\varkappa_L=1}^{\lambda_L} \sum_{l=1}^{\infty} \sum_{L'=1}^{Q} \varepsilon_{AB}^{LL'}(\rho_l) \sum_{m_{lL'}=1}^{z_{lL'}} \exp[2\pi i (x\nu_{1m_{lL'}} + y\nu_{2m_{lL'}} + z\nu_{3m_{lL'}})]. \end{aligned} \tag{32.6}$$

Formula (32.6) represents the Fourier series expansion of the function $\frac{I_2(x, y, z)}{N^0 |f_A - f_B|^2}$. Carrying out a Fourier transformation, we obtain an expression for the sum of the correlation parameters

$$\begin{aligned} \sum_{L=1}^{Q} \sum_{\varkappa_L=1}^{\lambda_L} \varepsilon_{AB}^{LL'}(\rho_l) &= \\ = -\frac{1}{N^0} \int_0^1 \int_0^1 \int_0^1 \frac{I_2(x, y, z)}{|f_A - f_B|^2} &\exp[-2\pi i (x\nu_{1m_{lL'}} + y\nu_{2m_{lL'}} + z\nu_{3m_{lL'}})] dx\, dy\, dz. \end{aligned} \tag{32.7}$$

Here the superscript L' in $\varepsilon_{AB}^{LL'}$ is determined by the type of site at which the selected vector

$$\nu_{1m_{lL'}}\boldsymbol{a}_1 + \nu_{2m_{lL'}}\boldsymbol{a}_2 + \nu_{3m_{lL'}}\boldsymbol{a}_3$$

drawn from site type L number $\varkappa_L$ will terminate.

From Eq. (32.6) we also obtain the relation

$$\sum_{L=1}^{Q} \lambda_L p_A^{(L)} p_B^{(L)} = \frac{1}{N^0} \int_0^1 \int_0^1 \int_0^1 \frac{I_2(x, y, z)}{|f_A - f_B|^2}\, dx\, dy\, dz, \qquad (32.8)$$

which enables us to choose the normalization factor on the experimental curve I_2. For a disordered alloy, Eq. (32.7) becomes

$$\varepsilon_{AB}(\rho_l) = -\frac{1}{N} \int_0^1 \int_0^1 \int_0^1 \frac{I_2(x, y, z)}{|f_A - f_B|^2} \exp\left[-2\pi i\left(x\nu_{1m_l} + y\nu_{2m_l} + z\nu_{3m_l}\right)\right] dx\, dy\, dz. \qquad (32.9)$$

The dependence of the atomic scattering factors f_A and f_B for components of the alloys on q, i.e., on x, y, z, may be regarded as known. With the aid of (32.7) or (32.9) one may therefore find certain combinations of the correlation parameters, if the experimental distribution of the intensity of I_2 of the scattered diffuse radiation is known for different values of x, y, z, i.e., for different orientations of the crystal and different scattering angles that determine q.* Therefore, in choosing different pairs of site types L and L' occurring at distances ρ_l (for which the correlation parameters are defined), it is necessary to find by means of (32.5) the numbers $\nu_{1m_lL'}$, $\nu_{2m_lL'}$ and $\nu_{3m_lL'}$ that determine the vector $\rho_{m_lL'}$ joining the sites under consideration. It is then required to evaluate numerically the integral (32.7) or (32.9). The parameters $\varepsilon_{AB}(\rho_l)$ determined in this manner enable us to evaluate the degree of concentration inhomogeneity associated with short-range order in the alloy.

Equation (32.6) shows that the ratio $\frac{I_2(x, y, z)}{|f_A - f_B|^2}$ does not change with the addition of any integers to x, y and z. Therefore in (32.7) and (32.9) the integration cannot be carried out from zero to unity, but rather over the intervals between any neighboring integers. This enables us to use the values of $I_2(x, y, z)$ corresponding to large values of q.

Let us now apply the formula to several specific structures of binary alloys. First, we consider alloys with a β-brass

---

*The experimental technique for the determination of I_2 for different x, y and z is described in [22].

type body-centered cubic lattice. In this case, according to (31.5) and (1.5), the quantities $\bar{f}_\varkappa$ are

$$\begin{aligned} \bar{f}_1 = \bar{f}^{(1)} = p_A^{(1)} f_A + p_B^{(1)} f_B = c_A f_A + c_B f_B + \frac{\eta}{2}(f_A - f_B), \\ \bar{f}_2 = \bar{f}^{(2)} = p_A^{(2)} f_A + p_B^{(2)} f_B = c_A f_A + c_B f_B - \frac{\eta}{2}(f_A - f_B). \end{aligned} \tag{32.10}$$

For this structure, and also for other cubic structures which will be considered next, the fundamental lattice vectors $\boldsymbol{a}_1$, $\boldsymbol{a}_2$, $\boldsymbol{a}_3$ lie along the edges of the cubic cells, $\boldsymbol{h}_1 = 0$, $\boldsymbol{h}_2 = \frac{1}{2}(\boldsymbol{a}_1 + \boldsymbol{a}_2 + \boldsymbol{a}_3)$. Expressing $\boldsymbol{q}$, in accordance with (31.12) and (32.4), in the form

$$\boldsymbol{q} = \alpha_1 \boldsymbol{d}_1 + \alpha_2 \boldsymbol{d}_2 + \alpha_3 \boldsymbol{d}_3 = 2\pi (x\boldsymbol{d}_1 + y\boldsymbol{d}_2 + z\boldsymbol{d}_3), \tag{32.11}$$

we find the structure factor

$$\begin{aligned} F = \sum_{\varkappa=1}^{2} \bar{f}_\varkappa e^{i\boldsymbol{q}\boldsymbol{h}_\varkappa} = (c_A f_A + c_B f_B)\,[1 + e^{\pi i\,(x+y+z)}] + \\ + \frac{\eta}{2}(f_A - f_B)\,[1 - e^{\pi i\,(x+y+z)}]. \end{aligned} \tag{32.12}$$

According to (31.18) a Bragg reflection is possible only when all numbers α_1, α_2 and α_3 are multiples of 2π, i.e., when the numbers x, y, z are integers equal to g_1, g_2, g_3 respectively. Thus, as is apparent from (32.12), if the sum $x + y + z = g_1 + g_2 + g_3$ is even, the structure amplitude is

$$F = 2\,(c_A f_A + c_B f_B). \tag{32.13}$$

The intensity of the corresponding reflection on the x-ray pattern, determined by (31.18) and proportional to $|F|^2$, is independent of the degree of long-range order (i.e., the corresponding reflection also occurs in the disordered state of the alloy). Such reflections are called fundamental. If however the sum $x + y + z = g_1 + g_2 + g_3$ is odd, then

$$F = \eta\,(f_A - f_B) \tag{32.14}$$

and the intensity of the reflection is proportional to the square of the degree of long-range order (and also to the square of the modulus of the difference between the atomic factors of atoms A and B). Such reflections disappear when the alloy passes into the disordered state, and are called superlattice reflections.

The intensity of the diffuse scattering of x-rays by alloys with a β-brass type crystal lattice is, according to (32.1),

$$I_2 = N|f_A - f_B|^2 \left\{ \left(c_A c_B - \frac{\eta^2}{4}\right) - \sum_{l'=1}^{\infty} \varepsilon_{AB}^{12}(\rho_{2l'+1}) \sum_{m_{2l'+1}=1}^{z_{2l'+1}} \cos q\rho_{m_{2l'+1}} - \right.$$

$$\left. - \frac{1}{2} \sum_{l'=1}^{\infty} \left[\varepsilon_{AB}^{11}(\rho_{2l''}) + \varepsilon_{AB}^{22}(\rho_{2l''})\right] \sum_{m_{2l''}=1}^{z_{2l''}} \cos q\rho_{m_{2l''}} \right\} \quad (32.15)$$

Here, $N = 2N^0$, the summation over l' extends over the first, third and other odd coordination spheres, while the summation over l'' is over even coordination spheres. To investigate the dependence of the background intensity on temperature and composition of the alloy by means of (32.15) we shall limit ourselves to the case where the correlation in the alloy is not large. Here we need consider correlation only in the first coordination sphere and set the correlation parameters for more remote coordination spheres equal to zero. Using Eq. (17.56) for the correlation parameters ε_{AB}^{12} associated with the first coordination sphere, (obtained with the aid of the Kirkwood method and taking into account terms of the order $\left(\frac{w}{kT}\right)^2$ in the free energy), we find that in this appoximation

$$I_2 = N|f_A - f_B|^2 \left(c_A c_B - c_A^2 c_B^2 \frac{w}{kT} \sum_{m-1}^{z} \cos q\rho_m\right), \quad (32.16)$$

for a disordered alloy. Here m enumerates sites of the first coordination sphere and z is the coordination number for this sphere ($z = 8$).*

At sufficiently high temperatures, when $\frac{w}{kT} \to 0$ the second term in (32.16), related to short-range order, becomes very small, and the expression for I_2 becomes

$$I_2 = N|f_A - f_B|^2 c_A c_B. \quad (32.17)$$

The diffuse scattering is a monotonic function of the scattering

*Equation (32.16) becomes invalid near the critical point, because one must take into account correlation beyond the first coordination sphere in this region (Sec. 34).

angle only, owing to the dependence of the atomic factors f_A and f_B on q. The dependence of I_2 on the composition of the alloy is determined by the factor $c_A c_B = c_A(1-c_A)$.

The second term in (32.16), which contains the factor $\sum_{m=1}^{z} \cos q\rho_m$, leads to the appearance of an oscillatory part of the background related to correlation in the alloy. The vectors ρ_m drawn from the central site to the site of the first coordination sphere may be written in the form

$$\rho_m = \frac{1}{2}(\pm a_1 \pm a_2 \pm a_3). \tag{32.18}$$

Using (32.11), we find that the product $q\rho_m$ becomes

$$q\rho_m = \pi(\pm x \pm y \pm z). \tag{32.19}$$

Equation (32.16) for I_2 may therefore be written as

$$I_2 = N|f_A - f_B|^2 \left\{ c_A c_B - 2c_A^2 c_B^2 \frac{w}{kT} [\cos\pi(x+y+z) + \right. \tag{32.20}$$

$$\left. + \cos\pi(x+y-z) + \cos\pi(x-y+z) + \cos\pi(-x+y+z)] \right\} =$$

$$= N|f_A - f_B|^2 \left(c_A c_B - 8c_A^2 c_B^2 \frac{w}{kT} \cos\pi x \cos\pi y \cos\pi z \right).$$

Equation (32.19) shows that for values of q corresponding to structure reflections, when the sum $x+y+z$ is even (and hence the sum $\pm x \pm y \pm z$ is also even), $\cos q\rho_m = +1$ for all values of m. For values of q corresponding to superlattice reflections, when the sum $x+y+z$ is odd, $\cos q\rho_m = -1$ for all values of m. As follows from (32.16), this means that for ordered alloys ($w > 0$) in the case of scattering of monochromatic radiation from a single crystal, the background of the x-ray pattern due to an incompletely ordered arrangement of atoms is concentrated near the superlattice reflection and weakens near the fundamental reflections. On the other hand, for disordered single-phase alloys ($w < 0$), the background intensity increases near the fundamental reflections and decreases near the positions that would correspond to superlattice reflections.*

---

*For diffuse background on a Debye diagram, the specified correspondence between the positions of maximum of the background intensity and the positions does not usually hold.

The magnitude of the oscillatory part of the background at high temperatures [when Eq. (32.16) is applicable] is inversely proportional to the absolute temperature of annealing. The concentration dependence of this part of the background is determined by the factor $c_A^2 c_B^2 = c_A^2 (1 - c_A)^2$, which also shows a maximum at $c_A = \frac{1}{2}$.

We now consider the case of scattering of x-rays by alloys with a $AuCu_3$ type face-centered cubic lattice. In this case, $\mu = 4$, $\boldsymbol{h}_1 = 0$, $\boldsymbol{h}_2 = \frac{1}{2}(\boldsymbol{a}_1 + \boldsymbol{a}_2)$, $\boldsymbol{h}_3 = \frac{1}{2}(\boldsymbol{a}_1 + \boldsymbol{a}_3)$, $\boldsymbol{h}_4 = \frac{1}{2}(\boldsymbol{a}_2 + \boldsymbol{a}_3)$, and the magnitudes $\bar{f}_\varkappa$ become

$$\left.\begin{aligned} \bar{f}_1 = \overline{f^{(1)}} = p_A^{(1)} f_A + p_B^{(1)} f_B = \\ = c_A f_A + c_B f_B + \frac{3}{4}\eta (f_A - f_B), \\ \bar{f}_2 = \bar{f}_3 = \bar{f}_4 = \bar{f}^{(2)} = p_A^{(2)} f_A + p_B^{(2)} f_B = \\ = c_A f_A + c_B f_B - \frac{1}{4}\eta (f_A - f_B). \end{aligned}\right\} \tag{32.21}$$

The structure factor according to (32.11) and (32.21) is

$$F = \sum_{\varkappa=1}^{4} \bar{f}_\varkappa e^{iq h_\varkappa} = (c_A f_A + c_B f_B)\,[1 + e^{\pi i (x+y)} + e^{\pi i (x+z)} + e^{\pi i (y+z)}] + \\ + \frac{1}{4}\eta (f_A - f_B)\,[3 - e^{\pi i (x+y)} - e^{\pi i (x+z)} - e^{\pi i (y+z)}]. \tag{32.22}$$

For those reflections in which all numbers $x = g_1$, $y = g_2$, $z = g_3$ are either all even or all odd, the structure factor is

$$F = 4(c_A f_A + c_B f_B). \tag{32.23}$$

If however two numbers of the set of numbers $x = g_1$, $y = g_2$, $z = g_3$ are even, but one is odd, or alternatively two numbers are odd and one is even, then

$$F = \eta (f_A - f_B). \tag{32.24}$$

In the first case the intensity of the regular reflection is independent of η and the reflection is a fundamental reflection. In the second case the intensity is proportional to η^2 and the reflection is a superlattice reflection.

Considering correlation in the first coordination sphere only and determining the correlation parameter for this sphere by means of the Kirkwood approximation (allowing for terms of the order $\left(\frac{w}{kT}\right)^2$ in the expansion of the free energy) and using (32.3) and (17.65), we find that the intensity of the diffuse scattering by a disordered alloy with a face-centered cubic lattice, just as for alloys with a body-centered cubic lattice, is determined by Eq. (32.16). Thus, the vectors ρ_m correspond to a face-centered cubic lattice and $z=12$. The background intensity in the case of alloys with a face-centered cubic lattice may be represented in the form

$$I_2=N|f_A-f_B|^2\Big\{c_Ac_B-2c_A^2c_B\frac{w}{kT}[\cos\pi(x+y)+\cos\pi(x+z)+ \\ +\cos\pi(y+z)+\cos\pi(x-y)+\cos\pi(x-z)+\cos\pi(y-z)]\Big\}= \\ =N|f_A-f_B|^2\Big[c_Ac_B-4c_A^2c_B^2\frac{w}{kT}(\cos\pi x\cos\pi y+ \\ +\cos\pi x\cos\pi z+\cos\pi y\cos\pi z)\Big]. \quad (32.25)$$

The foregoing discussion for alloys with a body-centered cubic lattice also applies here. (We note that for alloys with a face-centered cubic lattice $\sum_{m=1}\cos q\rho_m$ for the fundamental reflection is equal to 12, and for a superlattice reflection is equal to -4). For alloys with a AuCu type face-centered tetragonal lattice

$$\left.\begin{aligned}\bar{f}_1=\bar{f}_2=\bar{f}^{(1)}=c_Af_A+c_Bf_B+\frac{1}{2}\eta(f_A-f_B),\\ \bar{f}_3=\bar{f}_4=\bar{f}^{(2)}=c_Af_A+c_Bf_B-\frac{1}{2}\eta(f_A-f_B).\end{aligned}\right\} \quad (32.26)$$

Using $\boldsymbol{h}_1=0$, $\boldsymbol{h}_2=\frac{1}{2}(\boldsymbol{a}_1+\boldsymbol{a}_2)$, $\boldsymbol{h}_3=\frac{1}{2}(\boldsymbol{a}_1+\boldsymbol{a}_3)$, $\boldsymbol{h}_4=\frac{1}{2}(\boldsymbol{a}_2+\boldsymbol{a}_3)$, we find the following expression for the structure factor

$$F=(c_Af_A+c_Bf_B)[1+e^{\pi i(x+y)}+e^{\pi i(x+z)}+e^{\pi i(y+z)}]+ \\ +\frac{1}{2}\eta(f_A-f_B)[1+e^{\pi i(x+y)}-e^{\pi i(x+z)}-e^{\pi i(y+z)}]. \quad (32.27)$$

If all the integers $x=g_1$, $y=g_2$, $z=g_3$ are either even or odd, then

$$F=4(c_Af_A+c_Bf_B) \quad (32.28)$$

and the corresponding reflection is a fundamental one. If the numbers x and y are even but z is odd, or x and y are odd but z is even,

$$F=2\eta(f_A-f_B). \quad (32.29)$$

In this case we have a superlattice reflection. If, however, x and z or y and z are of the same sign, and y or x, respectively have the opposite sign, then $F=0$ and the corresponding regular reflection does not exist. In the disordered state, AuCu type alloys, just as $AuCu_3$ type alloys, have a face-centered cubic lattice. Diffuse scattering for this case has already been considered above.

As a last example we shall consider an alloy with a Fe_3Al type crystal lattice. For this structure $\mu=16$,

$$\boldsymbol{h}_1=0,\quad \boldsymbol{h}_2=\frac{1}{2}(\boldsymbol{a}_1+\boldsymbol{a}_2),\quad \boldsymbol{h}_3=\frac{1}{2}(\boldsymbol{a}_1+\boldsymbol{a}_3),\quad \boldsymbol{h}_4=\frac{1}{2}(\boldsymbol{a}_2+\boldsymbol{a}_3),$$

$$\boldsymbol{h}_5=\frac{1}{2}(\boldsymbol{a}_1+\boldsymbol{a}_2+\boldsymbol{a}_3),\qquad \boldsymbol{h}_6=\frac{1}{2}\boldsymbol{a}_1,\quad \boldsymbol{h}_7=\frac{1}{2}\boldsymbol{a}_2,\quad \boldsymbol{h}_8=\frac{1}{2}\boldsymbol{a}_3,$$

$$\boldsymbol{h}_9=\frac{1}{4}(\boldsymbol{a}_1+\boldsymbol{a}_2+\boldsymbol{a}_3),\qquad \boldsymbol{h}_{10}=\frac{1}{4}(\boldsymbol{a}_1+3\boldsymbol{a}_2+3\boldsymbol{a}_3),$$

$$\boldsymbol{h}_{11}=\frac{1}{4}(3\boldsymbol{a}_1+\boldsymbol{a}_2+3\boldsymbol{a}_3),\qquad \boldsymbol{h}_{12}=\frac{1}{4}(3\boldsymbol{a}_1+3\boldsymbol{a}_2+\boldsymbol{a}_3),$$

$$\boldsymbol{h}_{13}=\frac{1}{4}(3\boldsymbol{a}_1+\boldsymbol{a}_2+\boldsymbol{a}_3),\qquad \boldsymbol{h}_{14}=\frac{1}{4}(\boldsymbol{a}_1+3\boldsymbol{a}_2+\boldsymbol{a}_3),$$

$$\boldsymbol{h}_{15}=\frac{1}{4}(\boldsymbol{a}_1+\boldsymbol{a}_2+3\boldsymbol{a}_3),\qquad \boldsymbol{h}_{16}=\frac{3}{4}(\boldsymbol{a}_1+\boldsymbol{a}_2+\boldsymbol{a}_3).$$

Here, the sites corresponding to the vectors $\boldsymbol{h}_1, \ldots, \boldsymbol{h}_8$ are always occupied by atoms A. Atoms A and B are on the remaining sites, where in an ordered state the sites with vectors $\boldsymbol{h}_9, \ldots, \boldsymbol{h}_{12}$ are correct for atoms A, and sites with vectors $\boldsymbol{h}_{13}, \ldots, \boldsymbol{h}_{16}$ for atoms B. Therefore the quantities $\bar{f}_\varkappa$ are

$$\left.\begin{aligned}
&\bar{f}_1=\bar{f}_2=\ldots=\bar{f}_8=f_A,\\
&\bar{f}_9=\bar{f}_{10}=\bar{f}_{11}=\bar{f}_{12}=\bar{f}^{(1)}=c_Af_A+c_Bf_B+\frac{1}{2}\eta\,(f_A-f_B),\\
&\bar{f}_{13}=\bar{f}_{14}=\bar{f}_{15}=\bar{f}_{16}=\bar{f}^{(2)}=c_Af_A+c_Bf_B-\frac{1}{2}\eta\,(f_A-f_B),
\end{aligned}\right\}\quad (32.30)$$

where c_A, c_B are the concentrations of atoms A and B; η is the degree of long-range order for one-half of the sites, which may be occupied by either atoms A or B.

The structure factor in the case under consideration is

$$\begin{aligned}
F=\Big[f_A+(c_Af_A+c_Bf_B)\,e^{\frac{\pi i}{2}(x+y+z)}\Big]&[1+e^{\pi ix}+e^{\pi iy}+e^{\pi iz}+\\
&+e^{\pi i(x+y)}+e^{\pi i(x+z)}+e^{\pi i(y+z)}+e^{\pi i(x+y+z)}]+\qquad (32.31)\\
&+\frac{1}{2}\eta\,(f_A-f_B)\,e^{\frac{\pi i}{2}(x+y+z)}[1+e^{\pi i(x+y)}+e^{\pi i(x+z)}+e^{\pi i(y+z)}-\\
&-e^{\pi ix}-e^{\pi iy}-e^{\pi iz}-e^{\pi i(x+y+z)}].
\end{aligned}$$

If all the integers $x=g_1$, $y=g_2$, $z=g_3$ are even and their sum $x+y+z$ is a multiple of four, then

$$F=8\,(f_A+c_Af_A+c_Bf\ ) \tag{32.32}$$

and the corresponding reflection is a fundamental one. If however x, y and z are even and their sum is not a multiple of four (but, of course, it is a multiple of two), then

$$F=8\,(f_A-c_Af_A-c_Bf_B)=8c_B\,(f_A-f_B). \tag{32.33}$$

The corresponding structure factor is a smaller quantity. As in the previous case, the intensity of the reflection is independent of the degree of long-range order, but the structure factor is proportional to $|f_A-f_B|^2$, and the intensity of such a type of reflection vanishes when $f_A=f_B$. Furthermore when x, y and z are the odd integers g_1, g_2 and g_3, then

$$F=4e^{\frac{\pi i}{2}(g_1+g_2+g_3)}\,(f_A-f_B)\,\eta. \tag{32.34}$$

The structure factor in this case is proportional to η^2 and the reflection of the superlattice type. Lastly, if the numbers x, y and z have different signs, the corresponding reflections do not exist.

In the determination of the intensity of diffuse scattering by alloys with a Fe_3Al type crystal lattice, one should take into account the fact that sites at the corners of the cells are permanently occupied by atoms of one type. These sites form an ideal periodic sublattice and do not participate in the formation of the diffuse background. One may therefore consider only sites at the centers of the cells, which have an NaCl type lattice in the ordered state, and a simple cubic lattice in the disordered state. Consequently, for a disordered alloy at temperatures appreciably greater than the critical temperature, formula (32.16) is valid as for the previously considered structures. In the latter, however, the vectors ρ_m corresponded to a simple cubic lattice, and $z=6$. Therefore, the expression for I_2, in this case becomes

$$I_2=N\,|f_A-f_B|^2\left[c_Ac_B-2c_A^2c_B^2\frac{w}{kT}(\cos\pi x+\cos\pi y+\cos\pi z)\right]. \tag{32.35}$$

In the same manner, one may treat the scattering of x-rays by other structures with the aid of the general formulas. The substantial redistribution of diffuse scattering intensity in reciprocal lattice space is related to the establishment of correlation in the alloy. The redistribution, which can be determined from the formulas given in this section, has been recently experimentally investigated for several alloys (Sec. 3). Thus, for instance, in the disordered $AuCu_3$ alloy, as shown by Fig. 17 [22], a sharp increase in the intensity of diffuse scattering has been detected close to those points in the reciprocal lattice space that correspond to the positions of the superlattice reflections. The latter would appear during a transition of the alloy into an ordered state. Such an enhancement of intensity of diffuse scattering for ordered alloys with a $AuCu_3$ type face-centered cubic lattice follows qualitatively from (32.25), where correlation only in the first coordination sphere is taken into account. The analysis of the experimentally determined intensity distribution of diffuse scattering by means of a Fourier transform according to (32.9), enabled Cowley [22] to determine the correlation parameters for the first ten coordination spheres. These correlation parameters are given in Table 2.

33. Scattering of Slow Neutrons By Ordered Alloys

We shall consider the problem of scattering of slow neutrons by the crystal lattice of an ordered multi-component alloy of any composition, taking into account correlation in all the coordination spheres [278]. A similar problem, but neglecting correlation, has been considered by [282], and stoichiometric alloys were discussed in [283].

Let us consider elastic scattering of slow neutrons caused by inhomogeneities that are in turn associated with an irregular arrangement of different type atoms and isotopes on the sites of a geometrically perfect lattice. Unlike the case of scattering of x-rays, neutrons possessing a magnetic moment may experience magnetic scattering, if uncompensated magnetic moments of the electron shells exist in the atoms. This effect may be important for alloys of transition metals. In the following we will examine alloys of metals in which the magnetic scattering of neutrons by the electron shells is insignificant and will be neglected. We shall restrict ourselves to the case where both the capture of neutrons by nuclei and the scattering by thermal vibrations of the lattice will be neglected.

Neutron scattering differs from x-ray scattering by the fact that the neutrons are scattered by atomic nuclei, and not by the electron shells (in the absence of magnetic scattering). Noting the small radius of action of the nuclear forces, it may be assumed that the potential energy $V(r)$ of the neutron in the crystal lattice differs from zero only at points occupied by nuclei and may be expressed by means of δ-functions [284, 285]

$$V(r)=\sum_{s=1}^{N^0}\sum_{\varkappa=1}^{\mu}[A_{s\varkappa}+B_{s\varkappa}(K_{s\varkappa x}S_x+K_{s\varkappa y}S_y+K_{s\varkappa z}S_z)]\,\delta(r-R_{s\varkappa}). \quad (33.1)$$

Here $A_{s\varkappa}$ and $B_{s\varkappa}$ are constants describing the interaction between a neutron and a nucleus at site number $\varkappa$ of the sth unit cell. $A_{s\varkappa}$ is determined by that part of the interaction energy which is independent of the spin directions of the neutrons and of the nucleus, and $B_{s\varkappa}$ is that part of the interaction energy which is dependent on the orientation of these spins. $K_{s\varkappa x}$, $K_{s\varkappa y}$, $K_{s\varkappa z}$ represent the momentum matrices of the nucluei, S_x, S_y, S_z are the momentum matrices of the neutron. In the case of an alloy, the constants A and B are different for atoms of different elements and for different isotopes. These constants can be determined experimentally. As shown in [284], which treats the case of a pure metal, the probability of elastic scattering of slow monochromatic neutrons by the crystal lattice of a single crystal into an element of solid angle $d\Omega$ per unit time is

$$dW=\frac{m_n k\,d\Omega}{4\pi^2\hbar^3\tau}\left[\left|\sum_{s=1}^{N^0}\sum_{\varkappa=1}^{\mu}A_{s\varkappa}e^{iqR_{s\varkappa}}\right|^2+\frac{1}{4}\sum_{s=1}^{N^0}\sum_{\varkappa=1}^{\mu}B_{s\varkappa}^2 j_{s\varkappa}(j_{s\varkappa}+1)\right]. \quad (33.2)$$

Here m_{n} is the neutron mass, $q=\frac{p'-p}{\hbar}=k'-k$, p' and p are the momenta of the scattered and primary neutrons, τ is the crystal volume, $j_{s\varkappa}$ is the quantum number of the momentum of the nucleus located at site $s\varkappa$. It is obvious that (33.2) may be used with alloys as well as with pure metals.

Let us transform the first sum in the square brackets of formula (33.2). For this we write the quantity $A_{s\varkappa}$ as

$$A_{s\varkappa}\equiv\widetilde{A}_{s\varkappa}+A_{s\varkappa}-\widetilde{A}_{s\varkappa}, \quad (33.3)$$

where $\widetilde{A}_{s\varkappa}$ is the mean value of $A_{s\varkappa}$ for isotopes of the element α, whose atom occupies site $s\varkappa$:

$$\widetilde{A}_{s\varkappa}=\widetilde{A}_{\alpha}=\sum_{\beta}c_{\alpha\beta}A_{\alpha\beta}. \quad (33.4)$$

Here $A_{\alpha\beta}$ is the value of the constant under consideration for isotope of species β and element α, and $c_{\alpha\beta}$ is the relative atomic concentration of isotope β of element $\alpha\left(\sum_{\beta} c_{\alpha\beta}=1\right)$. Substituting (33.3) into (33.2), one may write the first term in the square brackets in (33.2) as

$$\mathfrak{A}=\left|\sum_{s=1}^{N^0}\sum_{\varkappa=1}^{\mu} A_{s\varkappa}e^{iqR_{s\varkappa}}\right|^2=\mathfrak{A}_1+\mathfrak{A}_2+\mathfrak{A}_3, \tag{33.5}$$

where

$$\mathfrak{A}_1=\sum_{s,\,s'=1}^{N^0}\sum_{\varkappa,\,\varkappa'=1}^{\mu}\widetilde{A}_{s\varkappa}\widetilde{A}_{s'\varkappa'}e^{iq(R_{s\varkappa}-R_{s'\varkappa'})}, \tag{33.6}$$

$$\mathfrak{A}_2=\sum_{s,\,s'=1}^{N^0}\sum_{\varkappa,\,\varkappa'=1}^{\mu}\left(A_{s\varkappa}-\widetilde{A}_{s\varkappa}\right)\left(A_{s'\varkappa'}-\widetilde{A}_{s'\varkappa'}\right)e^{iq(R_{s\varkappa}-R_{s'\varkappa'})}, \tag{33.7}$$

$$\mathfrak{A}_3=\sum_{s,\,s'=1}^{N^0}\sum_{\varkappa,\,\varkappa'=1}^{\mu}\widetilde{A}_{s\varkappa}\left(A_{s'\varkappa'}-\widetilde{A}_{s'\varkappa'}\right)e^{iq(R_{s\varkappa}-R_{s'\varkappa'})}+\text{complex conjugate} \tag{33.8}$$

Equation (33.6) for $\mathfrak{A}_1$ has the same form as Eq. (33.3) for I, if in this formula the values of $f_{s\varkappa}$ and $f^*_{s'\varkappa'}$ are replaced by the quantities averaged over the isotopes $\widetilde{A}_{s\varkappa}$ and $\widetilde{A}_{s'\varkappa'}$, respectively. Therefore, transforming the expression for $\mathfrak{A}_1$ just as was done in the derivation of (31.36), we find

$$\begin{aligned}\mathfrak{A}_1=8\pi^3N^0\left|\sum_{\varkappa=1}^{\mu}\bar{\widetilde{A}}_{s\varkappa}e^{iqh_{\varkappa}}\right|^2\prod_{j=1}^{3}\sum_{g_j=-\infty}^{\infty}\delta(\alpha_j-2\pi g_j)+\\+N^0\sum_{\substack{\alpha,\,\alpha'=1\\(\alpha<\alpha')}}^{\zeta}(\widetilde{A}_{\alpha}-\widetilde{A}_{\alpha'})^2\sum_{L=1}^{Q}\left\{\lambda_L p_{\alpha}^{(L)}p_{\alpha'}^{(L)}-\right.\\\left.-\frac{1}{2}\sum_{\varkappa_L=1}^{\lambda_L}\sum_{l=1}^{\infty}\sum_{L'=1}^{Q}\left[\varepsilon_{\alpha\alpha'}^{LL'}(\rho_l)+\varepsilon_{\alpha'\alpha}^{LL'}(\rho_l)\right]\sum_{m_{lL'}=1}^{z_{lL'}}\cos q\rho_{m_{lL'}}\right\},\end{aligned} \tag{33.9}$$

where $\bar{\widetilde{A}}_{s\varkappa}$ is the mean value of the quantity $\widetilde{A}_{s\varkappa}$ for all sites number $\varkappa$ (of the specific type L):

$$\bar{\widetilde{A}}_{s\varkappa}=\sum_{\alpha=1}^{\zeta}p_{\alpha}^{(L)}\widetilde{A}_{\alpha}=\sum_{\alpha=1}^{\zeta}\sum_{\beta}p_{\alpha}^{(L)}c_{\alpha\beta}A_{\alpha\beta}. \tag{33.10}$$

Let us transform the quantity $\mathfrak{A}_2$. The summation in (33.7) extends over sites $s\varkappa$ (the diagonal part of the sum) and over pairs of sites $s\varkappa$ and $s'\varkappa'$ (off-diagonal part). We shall group the terms of the off-diagonal part of the sum, so that in each group the vector $\boldsymbol{R}_{s\varkappa}-\boldsymbol{R}_{s'\varkappa'}$ joining these sites will be fixed; the atoms at the origins of these vectors will be atoms of the same element α; and the atoms at the ends of these vectors will be atoms of the same element α'. For each such group the constant factor

$$e^{iq(\boldsymbol{R}_{j\varkappa}-\boldsymbol{R}_{s'\varkappa'})}$$

can be put in front of the summation, i.e., the group of terms under consideration takes the form

$$e^{iq(\boldsymbol{R}_{s\varkappa}-\boldsymbol{R}_{s'\varkappa'})}\sum_{ss'\varkappa\varkappa'}{}'\left(A_{s\varkappa}-\widetilde{A}_{s\varkappa}\right)\left(A_{s'\varkappa'}-\widetilde{A}_{s'\varkappa'}\right), \tag{33.11}$$

where the prime in the summation sign denotes that the summation extends only over terms occurring in this group. We replace the summation in (33.11) by the mean value of the summed expression which was multiplied by the number of terms. Noting the randomness of the distribution of isotopes over sites occupied by atoms of the given element, the averaging of each factor in the sum (33.11) over the types of isotopes may be carried out independently. The mean value of the summed expression will thus become zero. In the expression for $\mathfrak{A}_2$ only terms with $s=s'$ and $\varkappa=\varkappa'$ will remain

$$\mathfrak{A}_2=\sum_{s=1}^{N^0}\sum_{\varkappa=1}^{\mu}\left(A_{s\varkappa}-\widetilde{A}_{s\varkappa}\right)^2. \tag{33.12}$$

Replacing the summation over $\varkappa$ by a summation over sites of the given type L and over the types of sites, changing the order of summation and replacing the sum over s by the product of the mean value of summed expression over sites of the given type and the number of cells N^0, we obtain

$$\mathfrak{A}_2=N^0\sum_{L=1}^{Q}\sum_{\varkappa_L=1}^{\lambda_L}\overline{\overline{\left(A_{s\varkappa}-\widetilde{A}_{s\varkappa}\right)^2}}=N^0\sum_{L=1}^{Q}\sum_{\varkappa_L=1}^{\lambda_L}\left[\overline{\overline{(A_{s\varkappa}^2)}}-\overline{(\widetilde{A}_{s\varkappa})^2}\right]. \tag{33.13}$$

Substituting A from (33.4) and also a similar expression for $\overline{(A_{s\varkappa})^2}$ into (33.13), averaging over the types of atoms α, that replace sites of type L [in analogy to the procedure in (33.10)], and multiplying the first term in (33.13) by $\sum_{\beta'} c_{\alpha\beta'} = 1$, we obtain

$$\mathfrak{A}_2 = N^0 \sum_{L=1}^{Q} \lambda_L \sum_{\alpha=1}^{\zeta} p_\alpha^{(L)} \sum_{\beta, \beta'} c_{\alpha\beta} c_{\alpha\beta'} \left(A_{\alpha\beta}^2 - A_{\alpha\beta} A_{\alpha\beta'}\right). \qquad (33.14)$$

Let us separate the sum over β and β' into two sums. In one of the sums $\beta < \beta'$, while in the other $\beta > \beta'$. In the second sum we shall interchange the summation indices $\beta \rightleftarrows \beta'$, noting that $N^0\mu = N$ and

$$\sum_{L=1}^{Q} \lambda_L p_\alpha^{(L)} = \mu c_\alpha, \qquad (33.15)$$

where c_α is the relative atomic concentration of element α. Then

$$\mathfrak{A}_2 = N \sum_{\alpha=1}^{\zeta} c_\alpha \sum_{\substack{\beta, \beta' \\ (\beta < \beta')}} c_{\alpha\beta} c_{\alpha\beta'} (A_{\alpha\beta} - A_{\alpha\beta'})^2. \qquad (33.16)$$

We shall now consider the third term $\mathfrak{A}_3$ in (33.5). Separating the sum (33.8) into a group of terms in the same manner as was done in the transformation of $\mathfrak{A}_2$, and replacing the sum of each group of terms by the mean value of the summed expression over the types of isotopes, multiplied by the number of terms in the group, we show that $\mathfrak{A}_3 = 0$.

In the last term of (33.2) containing the constant $B_{s\varkappa}$, the sum over all sites of the lattice $s, \varkappa$ may be replaced by the mean value of the summed expression over the types of isotopes and over the types of atoms multiplied by the number of sites N:

$$\frac{1}{4} \sum_{s=1}^{N^0} \sum_{\varkappa=1}^{\mu} B_{s\varkappa}^2 j_{s\varkappa} (j_{s\varkappa} + 1) = \frac{1}{4} N \sum_{\alpha=1}^{\zeta} c_\alpha \sum_{\beta} c_{\alpha\beta} B_{\alpha\beta}^2 j_{\alpha\beta} (j_{\alpha\beta} + 1). \qquad (33.17)$$

Using Eqs. (33.5), (33.9), (33.16) and (33.17), we obtain, in place of (33.2), the following expression for the probability of slow neutron scattering:

$$dW = \frac{m_\eta k N^0 \, d\Omega}{4\pi^2 \hbar^3 \tau} \left\{ 8\pi^3 \left| \sum_{\varkappa=1}^{\mu} \bar{\bar{A}}_{s\varkappa} e^{iqh_\varkappa} \right|^2 \prod_{j=1}^{3} \sum_{g_j=-\infty}^{\infty} \delta(\alpha_j - 2\pi g_j) + \right.$$

$$+ \sum_{\substack{\alpha, \alpha' = 1 \\ (\alpha < \alpha')}}^{\zeta} (\tilde{A}_\alpha - \tilde{A}_{\alpha'})^2 \sum_{L=1}^{Q} \left[\lambda_L p_\alpha^{(L)} p_{\alpha'}^{(L)} - \frac{1}{2} \sum_{\varkappa_L=1}^{\lambda_L} \sum_{l=1}^{\infty} \sum_{L'=1}^{Q} (\varepsilon_{\alpha\alpha'}^{LL'}(\rho_l) + \right.$$

$$+\varepsilon_{\alpha\alpha}^{LL'}(\rho_l)\Big)\sum_{m_{lL'}=1}^{z_{lL'}}\cos q\rho_{m_{lL'}}\Bigg]+\mu\sum_{\alpha=1}^{\zeta}c_\alpha\sum_{\substack{\beta,\beta\\(\beta<\beta')}}c_{\alpha\beta}c_{\alpha\beta'}(A_{\alpha\beta}-A_{\alpha\beta'})^2+$$

$$+\frac{1}{4}\mu\sum_{\alpha=1}^{\zeta}c_\alpha\sum_{\beta}c_{\alpha\beta}B_{\alpha\beta}^2 j_{\alpha\beta}(j_{\alpha\beta}+1)\Bigg\}. \quad (33.18)$$

In formula (33.18) the first term determines the intensity of the regular reflection, and the remaining terms the intensity of the diffuse background. Thus, the first terms in (33.18) occurring in $\mathfrak{A}_1$ coincide, except for the constant factor, with expression (31.36) for the intensity of scattered x-rays, if the quantities f_α in the latter are replaced by $\widetilde{A}_{s\varkappa}$ (and hence, f_α by $\widetilde{A}_\alpha$). Thus, the specified terms give that part of the scattering of neutrons which is similar to scattering of x-rays and the role of the atomic factor of atom type α is played by the quantities $\widetilde{A}_\alpha$. The latter are proportional to the amplitudes of the coherent scattering of neutrons $f_\alpha^{(\eta)}$ introduced in Sec. 4. The last two terms in (33.18) are not analogous to the scattering of x-rays and give the additional part of the diffuse background specific for the case of scattering of neutrons. This background is caused by the random distribution of isotopes and by the existence of interaction energy depending on the direction of the spins of both the neutrons and the nuclei. This part of the background is a linear function of the concentration of components of the alloy c_α. Inasmuch as the quantities $A_{\alpha\beta}$ (and hence, $\widetilde{A}_\alpha$), as well as $B_{\alpha\beta}$ are constants (i.e., unlike the atomic factors of the scattering of x-rays they are independent of the scattering angle), the part of the background that is not related to correlation in the alloy will not depend on the scattering angle. Owing to the existence of terms containing $\cos q\rho_{m_{lL'}}$, the dependence on this angle reveals only a part of the background related to the correlation in the alloy.

In the case of disordered alloys, the scattering probabilities dW_Φ corresponding to the background on the neutron diffraction pattern take, according to Eq. (33.18), the following form

$$dW_\Phi=\frac{m_\eta kN\,d\Omega}{4\pi^2\hbar^3\tau}\Bigg\{\sum_{\substack{\alpha,\alpha'=1\\(\alpha<\alpha')}}^{\zeta}(\widetilde{A}_\alpha-\widetilde{A}_{\alpha'})^2\Bigg[c_\alpha c_{\alpha'}-$$

$$-\sum_{l=1}^{\infty}\varepsilon_{\alpha\alpha'}(\rho_l)\sum_{m_l=1}^{z_l}\cos q\rho_{m_l}\Bigg]+\sum_{\alpha=1}^{\zeta}c_\alpha\sum_{\substack{\beta,\beta'\\(\beta<\beta')}}c_{\alpha\beta}c_{\alpha\beta'}(A_{\alpha\beta}-A_{\alpha\beta'})^2+$$

$$+\frac{1}{4}\sum_{\alpha=1}^{\zeta} c_\alpha \sum_{\beta} c_{\alpha\beta} B_{\alpha\beta}^2 j_{\alpha\beta}(j_{\alpha\beta}+1)\Bigg\}. \tag{33.19}$$

For binary disordered alloys, when correlation is not important and the terms containing $\varepsilon_{\alpha\alpha'}(\rho_l)=\varepsilon_{AB}(\rho_l)$ may be neglected, the background intensity on the neutron diffraction pattern is represented by the sum of a term proportional to $c_A(1-c_A)$, i.e., having a maximum for $c_A=\frac{1}{2}$, and a term which is a linear function of c_A. For binary ordered alloys with two types of sites with a rather large degree of order, and when the correlation is insignificant, the background intensity will contain a term having a minimum value at the stoichiometric composition $c_A=\nu$ in analogy with the case of scattering of x-rays. This term will appear in addition to the term linear in c_A.

Similarly to the case of the scattering of x-rays, harmonic analysis can be used to determine several combinations of the correlation parameters, or, in some cases, the individual correlation parameters for binary alloys. This can be accomplished when the distribution of background intensity on a neutron diffraction pattern and the values of $\tilde{A}_\alpha$ are known.

To determine the intensity of diffuse scattering of neutrons by polycrystals one should average the probability of diffuse scattering over the different orientations of the crystal. In this manner one obtains a formula analogous to (31.38) but with the addition of the last two terms in Eq. (33.18). The latter are independent of the orientation of the crystal relative to the directions of the primary and scattered beams of neutrons, and will not change during the averaging step.

Using the general formula (33.18) for the probability of scattering of slow monochromatic neutrons by a single crystal of the alloy, one may investigate the scattering of neutrons by alloys with different specific structures. Equation (33.18), for the probability of scattering of neutrons, differs from (31.36), for the intensity of the scattering of x-rays, by a constant factor, by existence of the two additional terms and by the presence of $\tilde{A}_\alpha$ instead of f_α. The formulas obtained in Sec. 32 for the different special cases of scattering of x-rays, can be used to obtain corresponding formulas for the probability of scattering of slow neutrons. For this purpose, the formulas for the structure factor F should be changed by substituting $f_\alpha \to \tilde{A}_\alpha$, while the formulas for the background intensity should also contain the last terms occurring in the braces of (33.18) multiplied by N^0. The obtained expression should then be multiplied by $\frac{m_n k d\Omega}{4\pi^2 \hbar^3 \tau}$.

The theoretical formulas for the intensity of diffuse scattering show that the background intensity of the neutron diffraction pattern should increase during a transition from an almost completely ordered into a disordered state. This result is confirmed by the experimental data on neutron scattering by the alloys FeCo (Fig. 26) and Ni_3Mn (Fig. 28). If no correlation exists in the alloy, the intensity of diffuse scattering of thermal neutrons, according to the above theory, should not depend on the angle of scattering. The dependence on the angle of scattering appears when correlation in the alloy is taken into account. Therefore, the appearance of a diffuse maximum in the distribution of the background intensity on the lower curve of Fig. 28 (corresponding to scattering by a disordered alloy of Ni_3Mn) is the evidence of the existence of short-range order in this alloy.

The methods of investigation of scattering expounded above may also be applied to the case of scattering of some other waves, for example, electron waves, by the crystal lattice of an alloy. The application of such a method to determine the residual resistivity of alloys caused by the scattering of electrons by inhomogeneities in the crystal lattice will be examined in Chapter VII.

34. Application of the Fluctuation Method to Problems of Scattering

As already mentioned in Sec. 30, two approaches are possible in the treatment of scattering of waves by alloys. The first method of treatment has been presented in Secs. 31-33. In the formulas obtained, the intensity of the scattered radiation is expressed in terms of the a priori probabilities of substitution of lattice sites by atoms of a different type and in terms of the correlation parameter. These formulas are therefore suitable for determination of the intensity of scattering, if the composition of the alloy, the long-range order and the correlation parameters are known. These formulas can also be used for solving the reverse problem—the determination of the correlation parameters (or their combinations) from the experimentally known distribution of diffuse scattering intensity, and for the determination of long-range parameters from the intensity of the superlattice lines. However, a theoretical calculation of the correlation parameters for the various coordination spheres as a function of temperature, composition the alloy and constants characterizing the interaction energy of atoms in the crystals is usually difficult. Consequently, this method of treatment is less suitable for determination of the intensity of radiation scattering as a function of the specified quantities. Relations of this type have been determined in the nearest neighbor approximation only for high temperatures, at which the correlation in the solution is not large (Sec. 32). The very interesting case where the temperature of the solution is of the same order as the critical temperature in the ordered state has not been considered.

For an investigation of the dependence of the intensity of scattering on temperature, composition and interatomic interaction constant, another method of treatment, originally introduced in the problem of scattering of light by Einstein [286], is much more convenient. In the case of scattering of x-rays or slow neutrons by an alloy, the crystal is considered as a periodic state constituted of the effective atoms, upon which the fluctuations of composition and of the degree of long-range order are imposed. The periodic structure accounts for the emergence of regular reflections and the fluctuations cause the diffuse scattering of waves. The probability of fluctuations depends substantially on temperature, composition and energy states, which enables us to relate in a simple manner these quantities to the intensity of the diffuse scattering.

The probabilities of fluctuations and the associated intensities of diffuse scattering of waves by alloys may be determined either by means of phenomenological treatment, or by means of a treatment based on the specific atomic model of the alloy.

1. Phenomenological theory

The treatment of scattering of x-rays and thermal neutrons is a more complicated problem than the scattering of light, because the wavelength of the x-radiation and the de Broglie wavelengths for thermal neutrons are of the same order of magnitude as the lattice constant in the crystal. The situation is greatly simplified if the direction of the scattered wave corresponding to diffuse scattering is close either to the direction of the wave corresponding to a regular reflection or to the direction of the incident beam. In these cases the phase differences between waves scattered by different atoms begin to differ noticeably from a number which is a multiple of π only for those interatomic distances that are substantially greater than the lattice constant. The fluctuations of composition and of order which are related to the diffuse scattering at those angles encompass therefore a great number of lattice constants. This makes it possible to carry out a phenomenological treatment not based on any specific atomic model of the crystal and using the thermodynamic theory of ordering.

The phenomenological treatment of x-ray scattering by crystals was first carried out by Landau [170], who considered scattering by fluctuations in the degree of long-order of single-component crystals of the molecular type, at a temperature close to the second-order phase transition temperature. Furthermore, by means of the phenomenological theories [171] we shall consider scattering of waves by solid solutions, where fluctuations in the degree of long-range order are accompanied by fluctuations in composition, and where the correlation between these fluctuations is significant (Sec. 13). This type of calculation can be carried out for those cases in which the thermodynamic potential of the solid solution is known, i.e., for temperatures near the critical temperature or the critical temperature of disorder, and for ideal, dilute and almost completely ordered solid solutions. The calculations will be carried out for the case of x-ray scattering. However, as was shown in Sec. 33, the results obtained are also applicable to the case of scattering of slow neutrons whose probabilities of scattering are determined by means of the rule given at the end of Sec. 33.

We shall consider alloys whose crystal lattice in the ordered state is subdivided into sites of only two types, and where the number of sites of the first and second types are the same. In spite of this limitation, apparently, all ordered solutions known at the present time corresponding to the stoichiometric composition *AB* are embraced. The calculation will be applicable to disordered alloys of any structure.

The amplitude of the monochromatic radiation (in electron units) scattered by a single crystal according to the kinematic theory of scattering and in accordance with (31.1) is

$$A = \int \rho_э(r) e^{iqr} d\tau. \tag{34.1}$$

The calculation of (34.1) is made much easier if we use the assumptions of Secs. 30 and 31 (that all atoms are located precisely at the sites of a perfect periodic lattice and that the distribution of the electron density of the given atom is independent on the type of atoms surrounding it). The electron density $\rho_э(r)$ in a solid solution is a function which changes in an extremely complicated manner in space. However, since the exponential factor in the integral formula (34.1) remains constant along a plane perpendicular to the vector q, during the integration one may replace the function $\rho_э(r)$ by a function averaged over the type of atoms along the specified plane. Such an averaged function $\bar{\rho}_э(r)$, according to (31.2), may be written as

$$\begin{aligned}\bar{\rho}_э(r) &= \sum_{s,\,\varkappa_1} \bar{\rho}_{s1}(r - R_{s\varkappa_1}) + \sum_{s,\,\varkappa_2} \rho_{s2}(r - R_{s\varkappa_2}) = \\ &= p_A^{(1)\prime} \sum_{s,\,\varkappa_1} \rho_A(r - R_{s\varkappa_1}) + p_B^{(1)\prime} \sum_{s,\,\varkappa_1} \rho_B(r - R_{s\varkappa_1}) + \\ &+ p_A^{(2)\prime} \sum_{s,\,\varkappa_2} \rho_A(r - R_{s\varkappa_2}) + p_B^{(2)\prime} \sum_{s,\,\varkappa_2} \rho_B(r - R_{s\varkappa_2}),\end{aligned} \tag{34.2}$$

where $p_A^{(1)\prime}$, $p_A^{(2)\prime}$, $p_B^{(1)\prime}$ and $p_B^{(2)\prime}$ denote the probability of substitution of sites of the first and second types in the specified plane by atoms A and B; due to the fluctuations these probabilities differ from the probabilities averaged over the crystal: $p_A^{(1)}$, $p_A^{(2)}$, $p_B^{(1)}$ and $p_B^{(2)}$. Let us introduce the periodic functions

$$\left.\begin{aligned}\rho_A(r) &= \sum_{s\varkappa} \rho_A(r - R_{s\varkappa}), \\ \rho_B(r) &= \sum_{s\varkappa} \rho_B(r - R_{s\varkappa}),\end{aligned}\right\} \tag{34.3}$$

which represent the electron density of the crystal when all sites of the lattice of the given alloy are filled with atoms A or B, respectively. Let us also introduce the function

$$\begin{aligned}\rho'(r) = \frac{1}{2} \sum_{s,\,\varkappa_1} [\rho_A(r - R_{s\varkappa_1}) - \rho_B(r - R_{s\varkappa_1})] - \\ - \frac{1}{2} \sum_{s,\,\varkappa_2} [\rho_A(r - R_{s\varkappa_2}) - \rho_B(r - R_{s\varkappa_2})],\end{aligned} \tag{34.4}$$

possessing the symmetry of the ordered crystal, which is lower than the symmetry of the functions $\rho_A(r)$ and $\rho_B(r)$. Expressing the probabilities $p_A^{(1)\prime}$, $p_A^{(2)\prime}$, $p_B^{(1)\prime}$ and $p_B^{(2)\prime}$ in terms of the atomic concentrations c'_A and c'_B in the plane under consideration and the degree of long-range order η' in this plane, then from formulas similar to (1.5) (for $\nu=\frac{1}{2}$) and from (34.2), (34.3) and (34.4) we obtain

$$\bar{\rho}_3(\boldsymbol{r})=c'_A\rho_A(\boldsymbol{r})+c'_B\rho_B(\boldsymbol{r})+\eta'\rho'(\boldsymbol{r}). \tag{34.5}$$

The functions $\rho_A(\boldsymbol{r}), \rho_B(\boldsymbol{r})$ and $\rho'(\boldsymbol{r})$ are independent of c'_A and η'. As a consequence of the fluctuations in composition and in the degree of long-range order in the crystal, the values of c_A, c_B and η, will differ from the mean values of c'_A, c'_B and η' averaged over the entire crystal. The former are functions of coordinates of the plane.

The periodic functions $\rho_A(\boldsymbol{r})$, $\rho_B(\boldsymbol{r})$ and $\rho'(\boldsymbol{r})$ can be expanded in a Fourier series

$$\left.\begin{aligned}\rho_A(\boldsymbol{r})=\sum_i\lambda_{Ai}e^{2\pi i\boldsymbol{g}_i\boldsymbol{r}},\quad \rho_B(\boldsymbol{r})=\sum_i\lambda_{Bi}e^{2\pi i\boldsymbol{g}_i\boldsymbol{r}},\\ \rho'(\boldsymbol{r})=\sum_l\lambda'_l e^{2\pi i\boldsymbol{g}'_l\boldsymbol{r}}.\end{aligned}\right\} \tag{34.6}$$

Here, $\boldsymbol{g}_i$ and $\boldsymbol{g}'_l$ are lattice vectors reciprocal to the lattices of the disordered [having the same symmetry as the function $\rho_A(\boldsymbol{r})$ and $\rho_B(\boldsymbol{r})$] and of the ordered crystals, respectively. According to (34.4) the integral

$$\int\rho'(\boldsymbol{r})\,e^{-2\pi i\boldsymbol{g}_i\boldsymbol{r}}\,d\tau \tag{34.7}$$

taken over the entire volume V of the ordered alloy is equal to zero. This means that in the expansion of $\rho'(\boldsymbol{r})$ there exist no terms in which $\boldsymbol{g}'_l$ coincides with any $\boldsymbol{g}_i$.

Substituting (34.5) and (34.6) into (34.1) and separating the quantities c'_A, c'_B and η' into their mean values and a fluctuating part: $c'_A=c_A+\Delta c_A$, $c'_B=c_B+\Delta c_B$, $\eta'=\eta+\Delta\eta$, we obtain

$$\begin{aligned}A=8\pi^3\sum_i(c_A\lambda_{Ai}+c_B\lambda_{Bi})\,\delta_V(\boldsymbol{q}_i)+8\pi^3\eta\sum_l\lambda'_l\,\delta_V(\boldsymbol{q}'_l)+\\+\sum_i(\lambda_{Ai}-\lambda_{Bi})\int\Delta c_A e^{i\boldsymbol{q}_i\boldsymbol{r}}\,d\tau+\sum_l\lambda'_l\int\Delta\eta e^{i\boldsymbol{q}'_l\boldsymbol{r}}\,d\tau.\end{aligned} \tag{34.8}$$

Here we have noted that $\Delta c_B=-\Delta c_A$, and introduced the following notation:

$$\begin{aligned}\boldsymbol{q}_i&=2\pi\boldsymbol{g}_i+\boldsymbol{q}=2\pi\boldsymbol{g}_i+\boldsymbol{k}'-\boldsymbol{k},\\ \boldsymbol{q}'_l&=2\pi\boldsymbol{g}_l+\boldsymbol{q}=2\pi\boldsymbol{g}'_l+\boldsymbol{k}'-\boldsymbol{k};\end{aligned} \tag{34.9}$$

$\delta_V(q)$ in the limiting case of an infinite crystal represents the δ-function in the space of three measurements, i.e., the product of three δ-functions for the separate components of the vector q *.

The intensity of the scattered radiation is proportional to the square of the modulus of the amplitude A averaged over the different configuration of the system. In the calculation of the mean value $|A|^2$ one should note that all terms, except the squares of the moduli of the individual terms, are equal to zero. In fact, the product of the δ-functions of different arguments is equal to zero, the different Fourier components of the fluctuation in composition and degree of long-range order (as shown in Sec. 13) are statistically independent, and consequently their products vanish after statistical averaging. The product of the δ-function and these Fourier components (containing the Fourier components to the first power) obviously also disappears as the result of the statistical averaging. The square of the "δ-function" defined for a finite volume of the crystal V (as shown in [170]) is equal to this "δ-function" multiplied by $\frac{V}{8\pi^3}$ **.

As a result we obtain the following expression for the scattered radiation intensity expressed in electron units

$$I = 8\pi^3 V \sum_i |c_A \lambda_{Ai} + c_B \lambda_{Bi}|^2 \delta_V(q_i) +$$

$$+ 8\pi^3 V \eta^2 \sum_l |\lambda_l'|^2 \delta_V(q_l) + \sum_i |\lambda_{Ai} - \lambda_{Bi}|^2 \overline{\left| \int \Delta c_A e^{iq_i r} d\tau \right|^2} +$$

---

*Thus, δ-functions in (34.8) are obtained in the limit of an infinite crystal. For a finite crystal of volume V there obviously should occur the following functions with sharp maxima:

$$\delta_V(q) = \frac{1}{8\pi^3} \int_V e^{iqr} d\tau.$$

**Representing the square modulus of the "δ-function" in the form of a double integral

$$|\delta_V(q)|^2 = \frac{1}{64\pi^6} \int \int e^{iq(r-r')} d\tau \, d\tau'$$

and making the change of variables

$$r_1 = \frac{r + r'}{2}, \quad r_2 = r - r_1',$$

we obtain

$$|\delta_V(q)|^2 = \frac{1}{64\pi^6} \int \int e^{iqr_2} d\tau_1 \, d\tau_2 = \frac{V}{8\pi^3} \frac{1}{8\pi^3} \int e^{iqr_2} d\tau_2 = \frac{V}{8\pi^3} \delta_V(q).$$

$$+\sum |\lambda_l'|^2 \overline{\left|\int \Delta\eta e^{iq_l' r}\, d\tau\right|^2}, \qquad (34.10)$$

where the horizontal bar over the last terms denote the statistical average over all configurations (averaged over the statistical ensemble).

Using the phenomenological theory we shall calculate the intensity of the scattered radiation near the fundamental or superlattice reflection. Thus, one of the quantities q_i or q_l is small, and in the expression for I one may retain the squared modulus only of that Fourier component of composition, or of the degree of long-range order, for which the corresponding q_i or q_l' is small. The squares of these remaining Fourier components, as may be shown, are in order of magnitude $q_i^2 a^2$ or $q_l'^2 a^2$ times smaller respectively (a is the lattice constant) and will be henceforth neglected. Noting next that according to (34.6), (34.3), (34.4) and (31.4)

$$\lambda_{Ai} = \frac{N}{V} f_A, \quad \lambda_{Bi} = \frac{N}{V} f_B, \quad \lambda_l' = \frac{N}{2V}(f_A - f_B)$$

[the atomic factors f_A and f_B, according to (31.4), correspond to the vector $q = -2\pi g_i$ or $q = -2\pi g_l'$], we find that the intensity of the scattered radiation near the fundamental corresponding to the vector g_i of the reciprocal lattice is

$$I = 8\pi^3 \frac{N^2}{V}\left[|c_A f_A + c_B f_B|^2 \delta_V(q_i) + \frac{|f_A - f_B|^2}{8\pi^3 V} \overline{\left|\int \Delta c_A e^{iq_i r}\, d\tau\right|^2}\right]. \qquad (34.11)$$

Similarly, the intensity of radiation near the superlattice reflection corresponding to the vector g_l' of the reciprocal lattice is defined by the formula

$$I = 2\pi^3 \frac{N^2}{V} |f_A - f_B|^2 \left[\eta^2 \delta_V(q_l') + \frac{1}{8\pi^3 V} \overline{\left|\int \Delta\eta e^{iq_l' r}\, d\tau\right|^2}\right]. \qquad (34.12)$$

In formulas (34.11) and (34.12) the first terms containing the δ-function determine the intensity of the regular reflection, and the second term the intensity of the diffuse scattering, i.e., the background on the x-ray pattern and the neutron diffraction pattern. It is not difficult to show that the expression for the intensity of the regular reflection agrees with (31.18) for I_1, applicable to binary alloys with the same number of sites of the first and second type.

The intensity of the diffuse scattering is determined by the distribution of fluctuations in composition and in the degree of long-range order within the crystal. The integrals occurring in (34.11) and (34.12) are, according to (13.15), proportional to the corresponding Fourier components of composition and degree of long-range order

$$\int \Delta c_A e^{iq_i r}\, d\tau = \begin{cases} V c^*_{Aq_i} & \text{for } q_{iz} > 0, \\ V c_{A-q_i} & \text{for } q_{iz} < 0; \end{cases} \quad (34.13)$$

$$\int \Delta \eta e^{iq'_l r}\, d\tau = \begin{cases} V \eta^*_{q'_l} & \text{for } q_{lz} > 0, \\ V \eta_{-q'_l} & \text{for } q_{lz} < 0. \end{cases} \quad (34.14)$$

Since

$$\overline{|c_{Aq_i}|^2} = \overline{|c_{A-q_i}|^2}, \quad \overline{|\eta_{q'_l}|^2} = \overline{|\eta_{-q'_l}|^2}, \quad (34.15)$$

expressions (34.11) and (34.12), taking into account (34.13) and (34.14), may be written in the following form near the fundamental reflection:

$$I = 8\pi^3 \frac{N^2}{V}\left[|c_A f_A + c_B f_B|^2 \delta_V(q_i) + \frac{|f_A - f_B|^2}{8\pi^3} V \overline{|c_{Aq_i}|^2}\right], \quad (34.16)$$

and near the superlattice reflection as

$$I = 2\pi^3 \frac{N^2}{V} |f_A - f_B|^2 \left[\eta^2 \delta_V(q'_l) + \frac{V}{8\pi^3} \overline{|\eta_{q'_l}|^2}\right]. \quad (34.17)$$

We now apply the formulas obtained to the special cases where the expressions for the thermodynamic potential, obtained from the phenomenological theory, are known.

We shall first consider the case of temperatures close to the second-order phase transition temperature. The thermodynamic expressions for η and $\overline{|\eta_\varkappa|^2}$, valid in this temperature range, have been given in Chap. II. Substituting (11.9) and (13.22) (for $\varkappa = q'_l$) into (34.17) and using (11.12), we find that the intensity of the x-ray radiation scattered by cubic crystals below the critical temperature is

$$I = 2\pi^3 \frac{N^2}{V} |f_A - f_B|^2 \frac{a'' T_0}{A_4^{0''}} \left\{ \frac{T_0 - T}{T_0} \delta_V(q'_l) + \right.$$

$$\left. + \frac{k}{16\pi^3 \Delta C''_p} \frac{1}{2\frac{T_0 - T}{T_0}\left[1 - \frac{\Delta C''_p}{T_0 \frac{\partial^2 \varphi_0}{\partial c_A^2}}\left(\frac{dT_0}{dc_A}\right)^2\right] + \frac{a}{a'' T_0} q_l'^2} \right\} (T \ll T_0), \quad (34.18)$$

where the quantities a'', $A_4^{0''}$, φ_0 and $\Delta C_p''$ refer to a unit volume of the crystal. Above the critical temperature there remains only the diffuse part of the scattered radiation. As follows from (34.17), (13.21) and (11.12), in this case

$$I=\frac{1}{8}\frac{N^2}{V}|f_A-f_B|^2\frac{a''T_0}{A_4^{0''}}\frac{k}{\Delta C_p''}\frac{1}{\frac{T-T_0}{T_0}+\frac{a}{a''T_0}q_l'^2}\ (T>T_0). \quad (34.19)$$

Thus, in the case of scattering from an ordered crystal sharp superlattice reflections are formed, whose shape is described by δ-functions, and the intensity is proportional to T_0-T. These reflections vanish at the critical temperature. In addition, near the temperature T_0 (both below and above it) for small values of q_l', i.e., near the superlattice line, one should observe an exceedingly intense background on the x-ray pattern caused by the sharp increase in magnitude in fluctuations of the degree of long-range order. The maximum of the background intensity according to (34.18) and (34.19) occurs at $q_l'=0$ and is inversely proportional to the different $|T-T_0|$. For given nonzero q_l', the background intensity is greatest at $T=T_0$ and its decrease upon departure from the temperature T_0 is proportional to $|T-T_0|$, i.e., one should observe a break on the curve of the dependence of I on T (for constant q_l') at the maximum point for $T=T_0$. For alloys whose composition corresponds to the maximum critical temperature, $\frac{dT_0}{dc_A}=0$, the background intensity will decrease upon departure from T_0 towards $T<T_0$ twice as fast as towards $T>T_0$. In other cases the ratio of the rate of decrease of the background intensity is one half as great.

All parameters occurring in (34.18) and (34.19), apart from the quantity $\frac{a}{a''T_0}$, may be determined if we know the concentration dependence of the critical temperature, the increase in the specific heat at $T=T_0$, the concentration dependence of the chemical potentials of the components for $T\approx T_0$ $(T>T_0)$ and the temperature dependence of the degree of long-range order near the transition point. The single unknown parameter $\frac{a}{a''T_0}$ is determined from a comparison of the calculated background intensity with experimental data for definite values of T and q_l', after which one may calculate the value of the intensity of the scattered radiation for different scattering angles (near the superlattice reflection) and temperatures (near the critical temperature) for all superlattice reflections. The parameters $\frac{\alpha}{a''T_0}$ can also be computed by means of the statistical theory of ordering, as will be shown below.

Special features of x-ray scattering should occur near the critical point, where the curve of the second-order phase transition points changes into a "decay" curve. At this point, according to (11.30), the expression in the square brackets in the denominator of (34.18) becomes zero. Near the critical point this expression may be represented in the form of an expansion (13.24). Then the formula for the intensity of the scattered radiation near the superlattice reflection for $T < T_0$ takes the form

$$I = 2\pi^3 \frac{N^2}{V} |f_A - f_B|^2 \frac{a''T_0}{A_4^{0''}} \left\{ \frac{T_0 - T}{T_0} \delta_V(q_l') + \right.$$

$$\left. + \frac{k}{16\pi^3 \Delta C_p''} \frac{1}{2\frac{T_0 - T}{T_0}\left[g_1(T - T_k) + g_2(T_0 - T_k) + \frac{\alpha}{a''T_0} q_l'^2\right.} \right\} \quad (34.20)$$

$$\text{for} \quad (T \ll T_0),$$

where g_1 and g_2 are constants of Eq. (13.24).

Thus, near the critical point as the temperature is lowered from T_0, the intensity of the diffuse scattering decreases much more slowly than it does far from the critical point. In the case of disordered alloys, for $T > T_0$, the intensity of the diffuse scattering, as the transition temperature is approached, increases according to (34.19).

For second-order phase transitions, the fluctuations in composition of the alloy, and consequently also the background intensity near the fundamental reflection do not become anomalously large. However, near the critical point of the transition of the ordering curve to a "decay" curve the magnitude of the fluctuation in composition in the ordered alloy also increases (according to (13.26)) greatly. Equations (34.16) and (13.26) show that the intensity of the scattered radiation near the fundamental reflections is

$$I = 8\pi^3 \frac{N^2}{V} \left[|c_A f_A + c_B f_B|^2 \delta_V(q_i) + \right.$$

$$\left. + \frac{|f_A - f_B|^2}{8\pi^3} \frac{kT}{\frac{\partial^2 \varphi_0}{\partial c_A^2}} \frac{2\frac{T_0 - T}{T_0} + \frac{\alpha}{a''T_0} q_i^2}{2\frac{T_0 - T}{T_0}[g_1(T - T_k) + g_2(T_0 - T_k)] + \frac{\alpha}{a''T_0} q_i^2} \right]. \quad (34.21)$$

Thus, for very small q_i at $T \ll T_0$, when the following condition is fulfilled

$$\frac{\alpha}{a'' T_0} q_i^2 \ll \frac{T_0 - T}{T_0},$$

one should observe an intense diffuse scattering of x-rays by fluctuations also near the fundamental reflection. Above the transition temperature, and also at $T = T_0$ and somewhat below the transition temperature when

$$\frac{\alpha}{a'' T_0} q_i^2 \gg \frac{T_0 - T}{T}$$

the intensity of diffuse scattering close to the fundamental reflections is small.

We shall now consider almost completely ordered, dilute and ideal solid solutions.

For almost completely ordered solutions of the structure considered here (with the same number of sites of the first and second type), it follows from (13.27) and (13.28) $\left(\text{for } \nu = \frac{1}{2}\right)$, that the mean square moduli of the Fourier components of fluctuation in the degree of long-range order and composition are determined by

$$\frac{1}{4} \overline{|\eta_\varkappa|^2} = \overline{|c_{A\varkappa}|^2} = \frac{\Omega_0}{V} \left(c_A c_B - \frac{1}{4} \eta^2\right). \tag{34.22}$$

Substituting these values for $\overline{|\eta_\varkappa|^2}$ and $\overline{|c_{A\varkappa}|^2}$ into (34.16) and (34.17) and using $\frac{\Omega_0}{V} = \frac{1}{N}$, we find that the intensity of the scattered radiation is

$$I = 8\pi^3 \frac{N^2}{V} \left[\sum_i |c_A f_A + c_B f_B|^2 \delta_V(q_i) + \frac{\eta^2}{4} \sum_l |f_A - f_B|^2 \delta_V(q'_l) + \right.$$
$$\left. + N |f_A - f_B|^2 \left(c_A c_B - \frac{1}{4} \eta^2\right) \right]. \tag{34.23}$$

The expression for the intensity of the diffuse scattering, determined by the last term in (34.25), agrees with (32.2), which was obtained by assumming that no correlation exists in the alloy. This corresponds to the fact that correlation is of little significance in almost completely ordered alloys.

In the case of dilute solutions, according to (34.16) and (13.30), the intensity of the scattered radiation is

$$I = 8\pi^3 \frac{N^2}{V} \sum_i |c_A f_A + c_B f_B|^2 \delta_V(q_i) + N|f_A - f_B|^2 c_A. \quad (34.24)$$

Lastly, from (34.16) and (13.34) we obtain the intensity of diffuse scattering for ideal solutions:

$$I = 8\pi^3 \frac{N^2}{V} \sum_i |c_A f_A + c_B f_B|^2 \delta_V(q_i) + N|f_A - f_B|^2 c_A(1 - c_A). \quad (34.25)$$

The expression for the intensity of diffuse scattering, determined by the last term of Eq. (34.25), agrees with (32.3), if the correlation parameters in the latter $\varepsilon_{AB}(\rho_l)$ are set equal to zero. This corresponds to the absence of correlations in ideal solutions. Correlation also disappears for $c_A \to 0$, in accordance with which the last term of (34.24) agrees with Eq. (32.3) for small values of c_A and $\varepsilon_{AB}(\rho_l) = 0$.

2. Microscopic theory

In the first part of the present section we have considered the scattering of x-rays by fluctuations extending over a large number of lattice constants. This enabled us to give a phenomenological treatment without the use of any specific atomic model of the alloy. However, in view of the fact that the wavelength of x-rays is of the same order of magnitude as the lattice constant, these fluctuations are due to diffuse scattering at angles similar to the direction of the regular reflection or to the incident beam. In order to find the intensity of scattering at arbitrary angles it is necessary to consider fluctuations whose dimensions are of the same magnitude as the lattice constant. Calculation of the probabilities of such fluctuations cannot be performed on the basis of thermodynamic considerations, and one should adopt a specific statistical model of a solution which, of course, should contain a number of simplifying assumptions. The use of such a model also enables us to interrelate the intensity of the diffuse scattering with the interatomic interaction constants. The calculation will be made [279] using the ordinary statistical model of an alloy described in Chap. III.

The crystal lattice of an alloy whose set of sites forms a Bravais lattice may be partitioned into n geometrically identical sublattices which are equally translated with respect to one another (containing sites of a given type) so that each atom interacts only with the neighboring atoms of the other sublattices and does not interact with atoms located at the sites of its own sublattice. Thus, by choosing a sufficiently large n one may take into account the interaction with atoms lying in a remote coordination sphere. Some of the different sublattices may in general contain sites of the same type.

The arrangement of atoms on the lattice sites of the binary alloy $A-B$ may be characterized by a set of quantities p_{Ait} equal to unity or zero, depending on whether atom A or B is located on site number t of the i-th sublattice. To determine the intensity of the diffuse scattering we shall start with Eq. (31.9), replacing μ by n, $\varkappa$ by i, s by t and N^0 by the total number of sites in the sublattice N_0. The atomic scattering factor f_{ti} of an atom at site number t of the i-th sublattice, may be written in terms of the quantity

$$f_{ti} = p_{Ait} f_A + (1 - p_{Ait}) f_B. \tag{34.26}$$

The mean value of the atomic scattering factor $\overline{f_i}$ over the sites of the given sublattice for a binary alloy, according to (31.5), is

$$\overline{f_i} = p_{Ai} f_A + (1 - p_{Ai}) f_B, \tag{34.27}$$

where p_{Ai} denotes the probability of substitution of a site of th ith sublattice by atom A.

From Eqs. (31.9), (34.36) and (34.27) it follows that the intensity of scattering of the diffuse radiation is

$$I_2 = |f_A - f_B|^2 \left| \sum_{i=1}^{n} \sum_{t=1}^{N_0} (p_{Ait} - p_{Ai}) e^{iq\mathbf{R}_{ti}} \right|^2. \tag{34.28}$$

Introducing the quantities

$$p_{iq} = \frac{1}{N_0} \sum_{t=1}^{N_0} (p_{Ait} - p_{Ai}) e^{iq\mathbf{R}_{ti}}, \tag{34.29}$$

we represent the expression for the intensity of the diffuse scattering in the form

$$I_2 = N_0^2 |f_A - f_B|^2 \left| \sum_{i=0}^{n} p_{iq} \right|^2. \tag{34.30}$$

As was shown in Sec. 32, $\frac{I_2}{|f_A - f_B|^2}$ is a periodic function of q. In the following, we may therefore restrict ourselves to a consideration of q over a range corresponding to the smallest values of q.

Let us denote the atomic planes perpendicular to vector q by the index λ. If q is not perpendicular to any system of atomic planes, the crystal may be divided into "infinitely thin" layers, perpendicular to q, whose number will also be denoted by the index λ. The numbers of sites belonging to one of the sublattices are the same for different parallel planes containing sites of this sublattice. We shall denote this number by ν'. In the summation over sites lying in one plane (layer), the factor $e^{iqR_{ti}}$ in (34.29) remains constant. Therefore, in the sum (34.29) the magnitude p_{Ait} of the mean concentration p^{λ}_{Ai} of atoms A on the sites of the i-th sublattice in planes λ can be replaced by the appropriate value, 1 or 0. If the Oz axis is parallel to the vector q, Eq. (34.29) takes the following form:

$$p_{iq} = \frac{\nu'}{N_0} \sum_{\lambda_i=1}^{N_0/\nu'} (p^{\lambda}_{Ai} - p_{Ai})\, e^{iqz_{i\lambda}}, \tag{34.31}$$

where $z_{i\lambda}$ is the section intersected by the plane λ_i and taken from the origin of the oz axis.

Thus, according to (34.30) and (34.31) the intensity of the diffuse scattering is determined by the deviation of concentration p^{λ}_{Ai} in the different planes from the mean concentration p_{Ai}, i.e., by the distribution of fluctuations in concentration of atoms A on sites of the different sublattices. The probability of some fluctuations, according to (13.1), is proportional to $e^{-\frac{R}{kT}}$. The minimum work R necessary to produce such fluctuation according to Eq. (13.2), applicable to the system under consideration, may be represented in the form of an expansion of the thermodynamic potential Φ in powers of the deviation of the concentration at sites of the different sublattices in different atomic planes from the equilibrium concentrations

$$R = \frac{1}{2} \sum_{i} \sum_{\lambda} \frac{\partial^2 \Phi}{\partial p^{\lambda^2}_{Ai}} (p^{\lambda}_{Ai} - p_{Ai})^2 + \\ + \sum_{\substack{i,\, j \\ (i<j)}} \sum_{\lambda_1 \lambda_2} \frac{\partial^2 \Phi}{\partial p^{\lambda_1}_{Ai}\, \partial p^{\lambda_2}_{Aj}} (p^{\lambda_1}_{Ai} - p_{Ai})(p^{\lambda_2}_{Aj} - p_{Aj}) +$$

$$+\sum_{i}\sum_{\substack{\lambda_1,\lambda_2\\(\lambda_1<\lambda_2)}}\frac{\partial^n\Phi}{\partial p_{Ai}^{\lambda_1}\,\partial p_{Ai}^{\lambda_2}}\left(p_{Ai}^{\lambda_1}-p_{Ai}\right)\left(p_{Ai}^{\lambda_2}-p_{Ai}\right). \qquad (34.32)$$

It may be shown that the second derivative of the thermodynamic potential with respect to the concentration p_{Ai}^{λ} of atoms A in planes which do not contain sites between which the interactions are taken into account, is equal to zero. Consequently, in the previously adopted division of the crystal into sublattices, the last term in (34.32) vanishes, but the second term contains only quantities corresponding to those planes whose atoms interact with one another. Furthermore, we note that the derivatives with respect to concentration (for the atomic planes), occurring in (34.32), are related to the derivatives with respect to the usual concentration (for the entire crystal) by

$$\frac{\partial^2\Phi}{\partial p_{Ai}^{\lambda 2}}=\frac{\nu'}{N_0}\frac{\partial^2\Phi}{\partial p_{Ai}^{2}},\qquad \frac{\partial^2\Phi}{\partial p_{Ai}^{\lambda_1}\,\partial p_{Aj}^{\lambda_2}}=\frac{\nu'}{N_0}\frac{z''}{z'}\frac{\partial^2\Phi}{\partial p_{Ai}\,\partial p_{Aj}}, \qquad (34.33)$$

where z' is the total number of sites of the j-th sublattice adjoining a given site of the i-th sublattice, and z'' is the number of sites of the j-th sublattice lying in plane λ_2 neighboring the site of the i-th sublattice located in plane λ_1.

To determine the intensity of diffuse scattering according to (34.30) and (34.31) one must know the Fourier components of the quantities p_{Ai}^{λ}. Therefore, it is more convenient to calculate directly the mean values of the squares and the products of the Fourier components of the quantities $p_{Ai}^{\lambda_1}$, $p_{Aj}^{\lambda_2}$ (but not these quantities themselves). The Fourier series expansion of the difference $p_{Ai}^{\lambda}-p_{Ai}$ is

$$p_{Ai}^{\lambda}-p_{Ai}=\sum_{K}\left(r_{iK}e^{-iKz_{i\lambda}}+r_{iK}^{*}e^{iKz_{i\lambda}}\right), \qquad (34.34)$$

where K takes values satisfying the conditions of periodicity, which lie over the range $0<K<\frac{2\pi}{\Delta z}$, where Δz is the distance between the neighboring atomic planes (layers).

Substituting (34.33) and (34.34) into (34.32) and noting that

$$K\left(z_{j\lambda_2}-z_{i\lambda_1}\right)=K\rho_{\xi ij},$$

where the vector K is directed along the Oz axis, and $\rho_{\xi ij}$ is a vector drawn from the site of the i-th sublattice (located in plane λ_1) to one of the sites (number ξ) of the j-th sublattice in plane λ_2, we find:

$$R=\sum_i \frac{\partial^2\Phi}{\partial p_{Ai}^2}\sum_K r_{iK}r_{iK}^* +$$
$$+\frac{1}{z}\sum_{\substack{i,j\\(i<j)}}\frac{\partial^2\Phi}{\partial p_{Ai}\partial p_{Aj}}\sum_{\xi=1}^{z}\sum_K \cos \rho_{\xi ij}K\left(r_{iK}r_{jK}^*+r_{iK}^*r_{jK}\right). \quad (34.35)$$

The expression for minimum work Eq. (34.35) separates into terms R_K corresponding to different K. This means that the fluctuations of different Fourier components of the quantity $p_{Ai}^{\lambda}-p_{Ai}$ are statistically independent. The expression for the probability distribution of fluctuations of the Fourier component in this case separates into a product of factors corresponding to the individual components. The probability of fluctuations of the q-th Fourier components, which according to (34.30), (34.31) determine the intensity of the diffuse scattering, is

$$w\sim\exp\left(-\frac{R_q}{kT}\right)=$$
$$=\exp\left[-\sum_{i=1}^{n}a_{ii}\left(r_{iq}'^2+r_{iq}''^2\right)-\sum_{\substack{i,j=1\\(i\neq j)}}^{n}a_{ij}\left(r_{iq}'r_{jq}'+r_{iq}''r_{jq}''\right)\right]. \quad (34.36)$$

Here

$$a_{ii}=\frac{1}{kT}\frac{\partial^2\Phi}{\partial p_{Ai}^2},\qquad a_{ij}=\frac{1}{z'kT}\frac{\partial^2\Phi}{\partial p_{Ai}\partial p_{Aj}}\sum_{\xi=1}^{z'}\cos\rho_{\xi ij}q\quad(i\neq j), \quad (34.37)$$

and r_{iq}' and r_{iq}'' denote the real and imaginary parts of the complex quantity r_{iq}. With the aid of the expression obtained for the probability (34.36) one may determine in the usual manner ([107] Sec. 109) the mean values of the squares and the product of the Fourier components, referred to the different sublattices

$$\frac{1}{2}\overline{\left(r_{iq}r_{jq}^*+r_{iq}^*r_{jq}\right)}=b_{ij}. \quad (24.38)$$

Here the set of quantities b_{ij} forms a matrix reciprocal to $\|a_{ij}\|$ which is defined by formula (34.37). Equation (34.38) is applicable both for $i=j$ and for $i\neq j$.

As follows from (34.31) and (34.34), $p_{iq}=r_{iq}$. Therefore, the intensity of the diffuse scattering can be expressed, from (34.30) and (34.38), as follows

$$I_2=N_0^2|f_A-f_B|^2\sum_{i,j=1}^{n}b_{ij}. \quad (34.39)$$

Thus, to determine the intensity of the diffuse scattering it is necessary to know the thermodynamic potential of an alloy as a function of the variables p_{Ai}, calculate from (34.37) the matrix $\|a_{ij}\|$, and then determine the sum of the matrix elements of the reciprocal matrix $\|b_{ij}\|$.

If n is large, the inversion of the matrix $\|a_{ij}\|$ is, in general, a difficult problem. In a number of cases, however, we may directly compute the sum of all matrix elements of the reciprocal matrix $\sum_{i,j=1}^{n} b_{ij}$, without making an inversion of the matrix $\|a_{ij}\|$. This is found to be possible when the sum of the matrix elements of each row (or column) of the matrix $\|a_{ij}\|$ is independent of the number of the row (column) and is equal to A

$$\sum_{i=1}^{n} a_{si} = A. \tag{34.40}$$

We shall demonstrate that in this case the sum of all matrix elements of the reciprocal matrix $\|b_{ij}\|$ is equal to

$$\sum_{i,j=1}^{n} b_{ij} = \frac{n}{A}. \tag{34.41}$$

In fact, summing the equations

$$\sum_{s=1}^{n} b_{js} a_{si} = \delta_{ij} \tag{34.42}$$

which interrelate the matrix elements of the direct and reciprocal matrices over all i and j, we obtain

$$\sum_{i,j,s=1}^{n} b_{js} a_{si} = n. \tag{34.43}$$

Summing (34.43) over i and noting (34.40), we obtain (34.41).

We shall now apply these general formulas to some special cases.

We first consider diffuse scattering of x-rays by disordered alloys. In order to take into account the interactions of pairs of atoms at great distances from one another, we choose, as was done in Sec. 19, a polyhedron in the crystal which contains a sufficiently large number of unit cells. In this manner we can neglect the interaction between an atom at the center of the polyhedron and outside atoms. Then, if each site of the n sites of the polyhedron is regarded as belonging to its own sublattice, each atom will not interact with atoms of its own sublattice but will interact more strongly with one atom of each of the other sublattices. When calculating the intensity of diffuse scattering one may, therefore, apply the foregoing method. Since the number of sites z' of any sublattice neighboring with one of the sites of another sublattice equals unity, only a single term for a_{ij} is retained in Eq. (34.37). In a disordered solid solution the derivatives $\frac{\partial^2\Phi}{\partial p_{Ai}\,\partial p_{Aj}}$, computed for equilibrium values of the concentrations p_{Ai} and p_{Aj} and given T and P, depend only on the distance between the nearest neighbor sites of the ith and jth sublattices (and also on the direction of the vector connecting them) and do not depend on the numbers of these sublattices. Furthermore, in a Bravais type crystal lattice, all sites are equivalent and the systems of vectors drawn from a given site to all other sites are equivalent for all the original sites. Thus, all diagonal matrix elements a_{ii}, defined by (34.37), are identical. The same off-diagonal matrix elements occur in different rows of the matrix a_{ij}, although they are arranged in a different order. Hence, the sum of the matrix elements of each row is independent of the number of the row, and in order to calculate the sum of the matrix elements in the reciprocal matrix we may use (34.41), substituting in place of A the sum

$$A = \sum_{j=1}^{n} a_{ij} = \frac{1}{kT}\frac{\partial^2\Phi}{\partial p_{Ai}^2} + \frac{1}{kT}\sum_{\substack{j=1\\(j\neq i)}}^{n}\frac{\partial^2\Phi}{\partial p_{Ai}\,\partial p_{Aj}}\cos \boldsymbol{\rho}_{ij}\mathbf{q}. \tag{34.44}$$

The summation of j in (34.44) extends over all sites interacting with a given site of the ith sublattice. This sum may be separated into a sum over the coordination spheres surrounding the central sites and a sum inside each sphere. Denoting the number of the coordination sphere by the subscript l, the number of the site in the lth sphere by the subscript m, the vector drawn from the central site in the mth cell of the lth coordination sphere by $\boldsymbol{\rho}_{m_l}$, and introducing the notation

$$x_0 = \frac{1}{N_0 kT}\frac{\partial^2\Phi}{\partial p_{A1}^2}, \quad x_l = \frac{1}{N_0 kT}\frac{\partial^2\Phi}{\partial p_{A1}\,\partial p_{Am_l}}, \tag{34.45}$$

we write the expression for A in the following form

$$A = N_0 x_0 + N_0 \sum_l x_l \sum_{m_l=1}^{z_l} \cos \boldsymbol{\rho}_{m_l}\mathbf{q}, \tag{34.46}$$

where z_l is the coordination number of the lth coordination sphere.* From Eqs. (34.39), (34.41) and (34.46) we obtain the following expression for the intensity of diffuse scattering by a disordered alloy

$$I_2 = N|f_A - f_B|^2 \frac{1}{x_0 + \sum_{l=1}^{\infty} x_l \sum_{m_l=1}^{z_l} \cos \rho_{m_l} q}. \tag{34.47}$$

In order to apply formula (34.47), it is necessary to know the thermodynamic potential of the alloy. At sufficiently high temperatures the correlation in the alloy can be neglected. The expression for the thermodynamic potential (taking into account interaction of atoms at any distance) may be obtained by the same method used to find the free energy in Sec. 16, where only the interaction with atoms of the first coordination sphere was considered. As a result we obtain the thermodynamic potential as a function of the variables p_{Ai}

$$\begin{aligned} \Phi = -\frac{N_0}{2} \sum_{j=1}^{n} \sum_{\substack{m_l=1 \\ (m_l \neq j)}}^{n} \{ & p_{Aj} p_{Am_l} v_{AA}(\rho_l) + [p_{Aj}(1 - p_{Am_l}) + \\ & + (1 - p_{Aj}) p_{Am_l}] v_{AB}(\rho_l) + (1 - p_{Aj})(1 - p_{Am_l}) v_{BB}(\rho_l)\} \\ & + N_0 kT \sum_{m_l=1}^{n} [p_{Am_l} \ln p_{Am_l} + (1 - p_{Am_l}) \ln (1 - p_{Am_l})]. \end{aligned} \tag{34.48}$$

Here, in the first sum the summation over m is extended over all $n-1$ sites around the central site belonging to the jth sublattice, and the summation over j covers all sublattices. Substituting the expression for the thermodynamic potential (34.48) into (34.45), differentiating and replacing p_{Ai} and c_A we find that at high temperatures

$$x_0 = \frac{1}{c_A(1 - c_A)}, \quad x_l = \frac{w_l}{kT}, \tag{34.49}$$

*With identical distances between the central and m_lth sites, the quantities may be different at certain orientations of their vectors. In such a case we will assume that the coordination sphere is subdivided into several coordination spheres with same radius, and that the value of x_l is identical in each of them. Here, z_l will denote the coordination numbers of such spheres.

where $w_l = 2v_{AB}(\rho_l) - v_{AA}(\rho_l) - v_{BB}(\rho_l)$ is the ordering energy for the lth coordination sphere. It follows from (34.47) and (34.49) that the intensity of the diffuse scattering in an alloy far above the critical temperature is

$$I_2 = N\,|f_A - f_B|^2 \frac{c_A(1-c_A)}{1 + c_A(1-c_A)\sum_{l=1}^{\infty}\frac{w_l}{kT}\sum_{m_l=1}^{z_l}\cos\rho_{m_l}q}. \tag{34.50}$$

In the case when the only important interaction is that with atoms of the first coordination sphere, the expansion of (34.50), up to terms of the first order in $\frac{w}{kT}$, coincides with (32.16).

Equation (34.50) is valid only at high temperatures, when the conditions $\frac{|w_l|}{kT} \ll 1$ are fulfilled for all coordination spheres. An expression for the intensity of diffuse scattering, valid at any temperature, can be obtained, if the concentration of one of the alloy components is sufficiently small. The thermodynamic potential of the alloy for such a case has been determined in [287] and [169]. According to [169], at $c_A \ll 1$ Φ takes the form

$$\Phi = -\frac{1}{2}N\sum_{m_l=2}^{n} v_{BB}(\rho_l) - N_0\sum_{j=1}^{n} p_{Aj}\sum_{\substack{m_l=1\\(m_l\neq j)}}^{n}[v_{AB}(\rho_l) - v_{BB}(\rho_l)] +$$
$$+ kTN_0\sum_{m_l=1}^{n}\left[p_{Am_l}\ln p_{Am_l} + (1-p_{Am_l})\ln(1-p_{Am_l})\right] \tag{34.51}$$
$$- kTN_0\sum_{\substack{j,\,m_l=1\\(j<m_l)}}^{n} p_{Aj}p_{Am_l}\left(e^{-\frac{w_l}{kT}} - 1\right)$$

Equations (34.45) and (34.51) show that

$$x_0 = \frac{1}{c_A(1-c_A)}, \quad x_l = 1 - e^{-\frac{w_l}{kT}}. \tag{34.52}$$

Substituting (34.52) into (34.47), we find that up to and including terms of the second order in c_A the intensity of diffuse scattering of x-rays by weak solutions is

$$I_2 = N\,|f_A - f_B|^2 \frac{c_A(1-c_A)}{1 + c_A \sum_{l=1}^{\infty}\left(1 - e^{-\frac{w_l}{kT}}\right) \sum_{m_l=1}^{z_l} \cos \rho_{m_l} q}. \qquad (34.53)$$

At high temperatures (34.53) agrees with (34.50), written for small values of c_A.

It is evident from (34.50) and (34.53) that both with an increase of temperature and a decrease of concentration of one of the alloy components, when the solution becomes ideal or dilute (correlation becomes insignificant), the expression for the background intensity becomes $I_2 = N\,|f_A - f_B|^2 c_A(1-c_A)$ and agrees with (34.25).

If the only significant interaction is that with atoms of the first coordination sphere ($w_1 = w$, $w_l = 0$ at $l \neq 1$), the dependence of I_2 on scattering angles, orientations of the single crystal and wavelength is determined by the sum $\sum_{m=1}^{z} \cos \rho_m q$. The expressions for this sum in the case of various specific structures have been considered in Sec. 32. For weak solutions, according to (34.53), nonuniformity of the background is more pronounced for disordered alloys ($w < 0$) than for ordered alloys (with the same $\frac{|w|}{kT}$).

The case when correlation in the disordered alloy is significant can be treated by means of the nearest neighbor approximation. Correlation in this case is taken into account not only in filling neighboring sites, but also in the more remote coordination spheres. We first consider alloys whose crystal lattices may be divided into two sublattices with the same number of sites, and where each site is surrounded only by sites of another sublattice. Such a type of structure occurs, for example, in disordered alloys with a body-centered cubic lattice or a simple cubic lattice. We introduce the notation

$$a_1 = \frac{2}{NkT} \frac{\partial^2 \Phi}{\partial p_{A1}^2}, \qquad a_2 = \frac{2}{zNkT} \frac{\partial^2 \Phi}{\partial p_{A1} \partial p_{A2}}, \qquad (34.54)$$

where p_{A1} and p_{A2} are the concentrations of atoms A in the specified sublattices. According to (34.39), (34.41), (34.40) and (34.37), the intensity of diffuse scattering in this case is

$$I_2 = N\,|f_A - f_B|^2 \frac{1}{a_1 + a_2 \sum_{m=1}^{z} \cos \rho_m q}. \qquad (34.55)$$

The background intensity may be determined from (34.55) and (34.54), if the thermodynamic potential of the alloy is known as a function of the variables p_{A1}, p_{A2}. We shall use Eq. (17.5) for Φ* in the form of an expansion in powers of $\frac{w}{kT}$ (replacing $p_A^{(1)}$, $p_A^{(2)}$ by p_{A1}, p_{A2} in the formulas of Sec. 17). Retaining in the expansion those terms that are proportional to $\left(\frac{w}{kT}\right)^4$, and replacing (after differentiation) p_{A1} and p_{A2} by c_A, from (34.54), (17.5), (17.23-17.25) we obtain

$$a_1 = \frac{1}{F} + zF\left(\frac{w}{kT}\right)^2 - zF(1-4F)\left(\frac{w}{kT}\right)^3 - F\Big[-\frac{z}{12}(7-72F+ \\ +180F^2) + \frac{y}{2}F(1-6F)\Big]\left(\frac{w}{kT}\right)^4 + \ldots, \qquad (34.56)$$

$$a_2 = \frac{w}{kT} - \frac{1}{2}(1-4F)\left(\frac{w}{kT}\right)^2 + \frac{1}{6}(1-6F)^2\left(\frac{w}{kT}\right)^3 - \\ -(1-4F)\left[\frac{1}{24}(1-12F)^2 + \left(\frac{y}{z} - 1\right)F^2\right]\left(\frac{w}{kT}\right)^4 + \ldots, \qquad (34.57)$$

where $F = c_A(1-c_A)$. The terms proportional to $\left(\frac{w}{kT}\right)^5$ and $\left(\frac{w}{kT}\right)^6$ should give, as may be shown, small corrections to expansions (34.56) and (34.57). At high temperatures $\left(\frac{w}{kT} \to 0\right)$ the expression for the scattering intensity determined by (34.55), (34.56) and (34.57) taking into account terms proportional to $\frac{w}{kT}$ agrees with (34.50), the latter being applicable to the case where the interaction is considered only in the first coordination sphere. At a low concentration of one of the components of the alloy, the expression obtained taking into account terms proportional to c_A^2 agrees with (34.53). Here, we set $w_1 = w$, $w_l = 0$ for $l \neq 1$ and expand $1 - e^{-\frac{w}{kT}}$ in powers of $\frac{w}{kT}$, taking into account terms proportional to $\left(\frac{w}{kT}\right)^4$.

In the structures under consideration the superlattice reflections are realized for values of $\mathbf{q}$ for which the conditions $\mathbf{q}\boldsymbol{\rho}_m = \pi(2n_m+1)$ are fulfilled (here, n_m are integers). Near these values of the vector $\mathbf{q}$ one may expand $\cos \boldsymbol{\rho}_m\mathbf{q}$ in powers of

*In the problems considered in this chapter, the difference between the thermodynamic potential and the free energy may be neglected.

the components of the vectors q'_l (where q'_l represents the addition to q until we reach a value at which a given (lth) superlattice reflection is attained). Limiting ourselves to quadratic terms in this expansion, from (34.55) we obtain

$$I_2 = N\,|f_A - f_B|^2 \frac{1}{a_1 - za_2 + \frac{a_2}{2}\sum_{m=1}^{z}(\rho_m q'_l)^2}. \tag{34.58}$$

It has been mentioned, in the first article of this section, that the intensity of diffuse scattering near the critical temperature increases sharply close to the superlattice lines. Such a result may be obtained from the statistical theory based on (34.58). Let us consider, for example, an alloy of stoichiometric composition $\left(c_A = \frac{1}{2},\ F = \frac{1}{4}\right)$. The ratio $\frac{w}{kT_0}$, where T_0 is the critical temperature, for such an alloy is determined by (17.39). Substituting this expression into (34.56) and (34.57) for a_1 and a_2, carrying out an expansion in powers of $\frac{1}{z}$ and, as usual, omitting terms containing $\frac{1}{z^4}$ and higher powers of $\frac{1}{z}$, we find that at $T = T_0$, $a_1 = za_2$. Near the critical temperature the difference $a_1 - za_2$ is proportional to $T - T_0$. The expression for the intensity of diffuse scattering by an alloy of stoichiometric composition slightly above the critical temperature may therefore be written in the form

$$I_2 = N\,|f_A - f_B|^2\,\beta \frac{1}{\frac{T - T_0}{T_0} + \gamma q'_l}. \tag{34.59}$$

Here, the expansions of the quantities β and γ in powers of $\frac{1}{z}$ for alloys with a β-brass type crystal lattice and with a simple cubic lattice have the form

$$\left.\begin{aligned} \beta &= \frac{1}{4}\left[1 + \frac{1}{z} + \frac{5}{3}\frac{1}{z^2} + 3\left(\frac{y}{z} + 1\right)\frac{1}{z^3} + \dots\right], \\ \gamma &= \frac{1}{6}\left[1 + \frac{2}{z} + \frac{14}{3}\frac{1}{z^2} + \left(4\,\frac{y}{z} + \frac{32}{3}\right)\frac{1}{z^3} + \dots\right]\rho^2, \end{aligned}\right\} \tag{34.60}$$

where $\rho = |\rho_m|$ is the distance between nearest neighbor atoms. It may seem from Eqs. (34.19), (34.59) and (34.60) that the quantity $\frac{\alpha}{a''T_0}$, which appeared as a parameter in the thermodynamic theory, is equal in order of magnitude, according to the statistical theory, to $\frac{1}{4}\rho^2$. An estimate of the parameter β yields $\beta \approx \frac{1}{3}$.

In the case of disordered alloys with a face-centered cubic lattice, in calculating the scattering intensity using the nearest neighbor approximation, the crystal may be divided into four simple cubic sublattices, each of which contains one of the sites of the cubic cell of the original lattice. Thus, each site has as neighbors only the sites of the other sublattices, and the formulas introduced above can be applied. The intensity of the diffuse scattering is again determined by Eq. (34.55), but the dependence of a_1 and a_2 on the composition of the alloy has a somewhat different form. Using Eqs. (17.5), (17.60)-(17.64) which determine the thermodynamic potential of an alloy with a face-centered cubic lattice, we find from (34.54), that

$$a_1 = \frac{1}{F} + 12F\left(\frac{w}{kT}\right)^2 - 12F\left(\frac{w}{kT}\right)^3 + \qquad (34.61)$$

$$a_2 = \frac{w}{kT} - \frac{1}{2}(1-4F)\left(\frac{w}{kT}\right)^2 + \frac{1}{6}(1+12F-60F^2)\left(\frac{w}{kT}\right)^3 + \qquad (34.62)$$

These expansions may be used at temperatures somewhat higher than the critical temperature, because for this structure the value of $\frac{w}{kT_0}$ is higher than unity. A calculation of the intensity of the diffuse scattering by ternary disordered alloys has been made in [280].

We shall now consider the case of ordered alloys. Here the sum of the matrix elements belonging to different rows of the matrix (34.37) are not generally the same, so that the sum of the matrix elements of the inverse matrix $\sum_{i,j-1}^{n} b_{ij}$ cannot be evaluated by (34.41), but an inversion of matrix a_{ij} must be made. However, the previous method of calculation is possible in the case of solutions with a stoichiometric composition AB whose sites of the first type are surrounded by sites of the second and first types (in the same manner as the second type sites are surrounded by sites of the first and second types). Such structures occur in the following types of crystals: NaCl (or Fe_3Al), β-brass, AuCu, etc. Crystal lattices of the structures under consideration may be divided into n sublattices, as was done in the previous section for disordered alloys. Thus, $\frac{n}{2}$ sublattices consist of sites of the first type and just as many sublattices of the sites of the second type. In the case of alloys with stoichiometric composition AB, the derivatives $\frac{\partial^2\Phi}{\partial p_{Ai}^2}$ for different i (corresponding to both sites of the first type and sites of the second type) in the statistical model adapted for the alloy are identical while for the structures under consideration the sum of the matrix elements belonging to any row of matrix (34.37) is independent of the number of the row. Therefore, $\sum_{i,j=1}^{n} b_{ij}$ may be calculated from formula (34.41). Thus, in the case under consideration, as in the case of disordered alloys, the intensity of diffuse scattering is determined by (34.47). In this formula, however, sites of different types, even those located at equal distances, must now be considered as belonging to different coordination spheres, so that z_l is equal to the number of sites of a given type located at a specific distance from the central site (cf. footnote to Eq. 34.46)

The quantities x_0 and x_l for different coordination spheres occurring in formula (34.47) for I_2 may be calculated from formulas (34.45) in the case of almost completely ordered alloys, when the expression for the thermodynamic potential of the alloy is determined from (19.5). Differentiating, replacing the symbol for the ordering energy w^{ij} for sites located at a distance of the radius of the lth coordination sphere ρ_l by w_l and assuming that the first sublattice consists of sites of the first type, we obtain

$$x_0 = \frac{1}{p_{A1}(1 - p_{A1})} = \frac{4}{1 - \eta^2}, \tag{34.63}$$

$$x_l = 1 - e^{-\frac{w_l}{kT}}, \tag{34.64}$$

if the l-th coordination sphere around a site of the first sublattice consists of sites of second type, and

$$x_l = e^{\frac{w_l}{kT}} - 1, \tag{34.65}$$

if this coordination sphere consists of sites of the second type.

In the special case of almost completely ordered solutions AB, where each site is surrounded only by sites of the other type (NaCl, β-brass type crystals), treated in the nearest neighbor approximation, it follows from (34.47), (34.63) and (34.65) that the intensity of the diffuse scattering is

$$I_2 = \frac{1}{4} N |f_A - f_B|^2 \frac{1-\eta^2}{1+\frac{1-\eta^2}{4}\left(e^{\frac{w}{kT}}-1\right)\sum\limits_{m=1}^{z} \cos \rho q}. \tag{34.66}$$

As shown in Sec. 19, when $T \to 0$, the degree of long-range order exponentially tends to unity as described by $1-\eta \approx 2e^{-\frac{z}{2}\frac{w}{kT}}$. Therefore, the second term in the denominator of (34.66), leading to a concentration of the background near the superlattice lines, decreases exponentially for $T \to 0$ as $e^{-\frac{z-2}{2}\frac{w}{kT}}$. Thus, the nonuniformity of the background disappears with a decrease in temperature in the case of almost completely ordered alloys. Here, (34.66) agrees with the last term of (34.23) (in which we must set $c_A = \frac{1}{2}$), obtained from thermodynamic considerations. As the temperature is lowered, there is a decrease not only in the nonuniformity of the distribution of diffuse scattering, but also in the absolute value of the scattering intensity, because $I_2 \sim 1-\eta \sim e^{-\frac{z}{2}\frac{w}{kT}}$. In a completely ordered crystal the intensity of the diffuse scattering of the type under consideration is obviously equal to zero.

For almost completely ordered alloys with a AuCu type crystal lattice and stoichiometric composition, Eqs. (34.47), (34.63), (34.64) and (34.65) show that (in the nearest neighbor approximation) the intensity of the diffuse scattering is

$$I_2 = \frac{1}{4} N |f_A - f_B|^2 \times$$
$$\times \frac{1-\eta^2}{1+\frac{1-\eta^2}{4}\left[\left(1-e^{-\frac{w}{kT}}\right)\sum\limits_{m_1=1}^{4} \cos \rho_{m_1} q + \left(e^{\frac{w}{kT}}-1\right)\sum\limits_{m_2=1}^{8} \cos \rho_{m_2} q\right]}. \tag{34.67}$$

Here the subscript m_1 enumerates the vectors ρ_{m_1} drawn from the central site to neighboring sites of the same type, and the subscript m_2 enumerates the vector ρ_{m_2} drawn to neighboring sites of the other type. It is seen from (34.67) that, even if we neglect the usual small tetragonal character of the AuCu crystals (with different lengths of the vectors ρ_{m_1} and ρ_{m_2}), the [001] direction perpendicular to planes containing one type of site, and the [010] and [100] directions lying in these planes will not be equivalent with respect to scattering of x-rays.

If the composition of the alloy differs from the stoichiometric composition AB, the sum of the matrix elements of any row of the matrix a_{ij} depends on the number of the row, and formula (34.41) cannot be applied. In the case when the ordered solution contains an equal number of sites of the first and second types, each site being surrounded only by sites of the other type, then in order to apply the general formula (34.49), one should partition the crystal into sublattices of the sites of the same type. Matrix a_{ij} in this case is a matrix of rank two. For almost completely ordered solid solutions (whose composition is close to stoichiometric), its matrix elements are determined by means of Eqs. (34.37) and (19.15). By determining the inverse matrix a_{ij} and then the intensity of diffuse scattering from (34.39), we approximately obtain (when the degree of long-range order is close to its maximum value)

$$I_2 = N|f_A - f_B|^2 \Big\{ c_A(1-c_A) - \frac{\eta^2}{4} - \left(c_A^2 - \frac{\eta^2}{4}\right)\left[(1-c_A)^2 - \frac{\eta^2}{4}\right]\left(e^{\frac{w}{kT}} - 1\right)\sum_{m=1}^{z} \cos \rho_m q \Big\}. \tag{34.68}$$

The general case when the degree of long-range order is not close to unity has been treated in the nearest neighbor approximation in [279].

For alloys with a face-centered cubic lattice in the nearest neighbor approximation the lattice should be divided into four sublattices. It is not difficult to invert the resulting matrix of the fourth order $\|a_{ij}\|$ in the case of almost completely ordered alloys, since the off-diagonal matrix elements in this case are much smaller than the diagonal matrix element. As a result, for $AuCu_3$ type almost completely ordered alloys, Eq. (19.17) leads to the following expression for thermodynamic potential

$$I_2 = N|f_A - f_B|^2 \Big\{ c_A(1-c_A) - \frac{3}{16}\eta^2 - \frac{1}{2}\left(c_A - \frac{\eta}{4}\right)\left(1 - c_A + \frac{\eta}{4}\right)\Big[\left(c_A + \frac{3}{4}\eta\right)\left(1 - c_A - \frac{3}{4}\eta\right)\left(e^{\frac{w}{kT}} - 1\right) + \left(c_A - \frac{\eta}{4}\right)\left(1 - c_A + \frac{\eta}{4}\right)\left(1 - e^{-\frac{w}{kT}}\right)\Big]\sum_{m=1}^{12} \cos \rho_m q \Big\}. \tag{34.69}$$

For AuCu type almost ordered alloys, using Eq. (19.19) we obtain the thermodynamic potential from

$$I_2 = N|f_A - f_B|^2 \left\{ c_A(1-c_A) - \frac{\eta^2}{4} - \left(1 - e^{-\frac{w}{kT}}\right) \times \right.$$
$$\times \left[\left(c_A(1-c_A) - \frac{\eta^2}{4}\right)^2 + (1-4c_A(1-c_A))\frac{\eta^2}{4}\right] \sum_{m_1=1}^{4} \cos \rho_{m_1} q -$$
$$\left. - \left(e^{\frac{w}{kT}} - 1\right)\left(c_A^2 - \frac{\eta^2}{4}\right)\left[(1-c_A)^2 - \frac{\eta^2}{4}\right] \sum_{m_2-1}^{8} \cos \rho_{m_2} q \right\}. \quad (34.70)$$

The formulas obtained for I_2 enabled us to determine the ordering (or disordering) energy w_l from the experimentally found distribution of background intensity for the various coordination spheres. Harmonic analysis may be applied for this purpose. We shall start with Eq. (34.47) for the intensity of diffuse scattering, which is applicable for disordered alloys and ordered alloys of the stoichiometric composition AB. Expanding the vector q in terms of the reciprocal lattice vector according to (32.4) and the vector ρ_{m_l} in terms of vectors of the Bravais lattice of the disordered alloy according to (32.5) (for brevity the index L is omitted), we obtain

$$\frac{N|f_A - f_B|^2}{I_2(x, y, z)} =$$
$$= x_0 + \sum_{\nu_{1m_l}=-\infty}^{\infty} \sum_{\nu_{2m_l}=-\infty}^{\infty} \sum_{\nu_{3m_l}=-\infty}^{\infty} x_l \exp\left[2\pi i\left(x\nu_{1m_l} + y\nu_{2m_l} + z\nu_{3m_l}\right)\right],$$
$$(34.71)$$

where the integers $\nu_{\alpha m_l}$ for each α ($\alpha = 1, 2, 3$) run through all integral values from $-\infty$ to ∞. A Fourier transformation yields

$$x_0 = N \int_0^1 \int_0^1 \int_0^1 \frac{|f_A - f_B|^2}{I_2(x, y, z)} \, dx\, dy\, dz, \quad (34.72)$$

$$x_l = N \int_0^1 \int_0^1 \int_0^1 \frac{|f_A - f_B|^2}{I_2(x, y, z)} \cos 2\pi\left(x\nu_{1m_l} + y\nu_{2m_l} + z\nu_{3m_l}\right) dx\, dy\, dz. \quad (34.73)$$

Here the numbers ν_{1m_l}, ν_{2m_l} and ν_{3m_l} correspond to any site in the lth coordination sphere.

The quantities x_l with the aid of Eqs. (34.49), (34.52), (34.64), (34.65), (34.56), (34.57), (34.61), (34.62), applicable to the different special cases, are expressed in terms of the ordering energy. Thus the ordering energy for different coordination spheres can be determined from the experimental distribution of background intensity for different orientations of the crystal and different scattering angles (x, y, z). In contrast to the method presented in Sec. 32, we take here the Fourier transform of the ratio $\frac{|f_A - f_B|^2}{I_2}$, and not of $\frac{I_2}{|f_A - f_B|^2}$, and determine directly the ordering energies w_l, rather than the correlation parameters whose relation to w_l is established only for the simplest special cases by the statistical theory.

The results given above pertain to the case of scattering by alloys, in which there is no geometrical distortion of the crystal lattice. Such distortions, due to the different dimensions of atoms in the substitutional alloys, defects arising from plastic deformation, and also such defects as vacant lattice sites and interstitial atoms also lead to diffuse scattering and may in a number of cases (for example, when atomic radii differ appreciably) greatly alter the distribution pattern of the background intensity. Therefore, the problem of further development of the theory entails noting the geometric distortions of the crystal lattice in ordered alloys, and also the construction of a theory of diffuse scattering by the thermal vibrations of atoms. To solve the last problem it is necessary to first consider the very complicated subject of thermal vibrations in a partially ordered (not ideally periodic) crystal (for example, [270], reviewed in Sec. 40) and apply the results to scattering by thermal vibrations. Since the theory of the electrical resistivity of alloys, which will be considered in the next chapter, is based on the results of scattering of waves in a crystal lattice, the direction of development of the theory of scattering indicated above also makes it possible to take into account the influence of geometric distortions and thermal vibrations of atoms on the electrical resistivity of alloys.

Chapter VII

Theory of Residual Electrical Resistivity of Alloys

35. Determination of the Scattering Probability of Electrons by the Crystal Lattice of an Alloy

It has been shown in Sec. 6 that order in the arrangement of atoms of different species on the lattice sites strongly influences electrical resistivity of alloys. Therefore, an investigation of the electrical resistivity, and especially of the residual electrical resistivity, coupled with a study of x-ray and neutron scattering is a sufficiently sensitive method for analyzing the ordering of atoms in alloys.

As is well known, electrical resistivity is produced in alloys by the scattering of electrons by imperfections in the crystal lattice. We shall consider only the electrical resistivity due to disorder in the alternation of atoms; geometric distortion of the crystal lattice and thermal motion will be neglected. We shall present a theory of residual electrical resistivity of alloys with an undistorted crystal lattice.

The first papers on the theory of residual electrical resistivity of alloys employed a one-electron model of a metal. The electrical resistivity of disordered alloys neglecting correlation has been considered in [268]. The residual electrical resistivity of binary ordered alloys of any composition (having an arbitrary Bravais lattice in the disordered state) has been considered in [288]. In [289], correlation was taken into account for binary and ordered stoichiometric alloys with a simple cubic lattice. These authors made assumptions that are usually employed within the framework of one-electron approximation. These assumptions concern the independence of the electron energy on the direction of the wave vector, and the possibility of using a mean free path which is independent of direction, etc. As shown in [290], these assumptions are not necessary for obtaining the dependence of residual electrical resistivity on composition, degree of long-range order and correlation parameter. Therefore, in constructing a theory that does not have as its goal the determination of the numerical value of electrical resistivity, such assumptions are not required. Moreover, not even the one-electron approximation is essential to obtain these dependences, and one may construct a many-electron theory of residual electrical resistivity in alloys. From the standpoint of the many-electron theory of metals a calculation of the residual electrical resistivity has been made in [291] for binary ordered alloys, taking into account correlation in the first coordination sphere. Reference [292] dealt with ternary disordered alloys taking into account correlation in all coordination spheres, while [293] considered the general case of multicomponent ordered alloys of the substitutional type, and [294] described interstitial ordered alloys.

In this chapter we shall present the many-electron theory of residual electrical resistivity for ordered substitutional alloys. In such a theory one does not endeavor to determine the numerical value of the resistivity, but attempts to find its dependence on composition and on parameters describing the long-range order and correlation in the alloy. It is therefore possible to perform the calculations using the smallest possible number of models. In addition to accounting for the translational symmetry properties of the crystal, it is also assumed that the potential energies of the electrons in the field of atoms of one type are very similar, that the electric field intensity in the metal is small, that Ohm's law applies and that the number of conduction electrons is independent of composition and of the order parameters.

We shall determine the probability (per unit time) of the transition of a system of electrons from one state to another caused by a pertubation produced by an imperfectly ordered arrangement of atoms on the sites of the crystal lattice of the alloy. We shall calculate the square of the modulus of the matrix element of the perturbation energy which is proportional to this probability.

Let us consider the system of N_e conduction electrons in the crystal lattice of multicomponent ordered alloy. The Hamiltonian of such a system is

$$H = -\frac{\hbar^2}{2m}\sum_{i=1}^{N_e}\Delta_i + V(r_1, r_2, \ldots, r_{N_e}) + \sum_{\substack{i, j=1 \\ (i<j)}}^{N_e}\frac{e^2}{|r_i - r_j|}, \quad (35.1)$$

where $-\frac{\hbar^2}{2m}\Delta_i$ is the kinetic energy operator of the ith electron, the last term represents the energy of interaction between the conduction electrons, and the potential energy $V=V(r_1,\ldots,r_{N_e})$ of these electrons in the field of the atoms forming the crystal lattice has the form

$$V=\sum_{i=1}^{N_e}\sum_{s=1}^{N^0}\sum_{\varkappa=1}^{\mu}V_{s\varkappa}(r_i-R_{s\varkappa}). \qquad (35.2)$$

Here $V_{s\varkappa}(r_i-R_{s\varkappa})$ is the potential energy of the ith electron in the field of the atom located at site number $\varkappa$ of the sth unit cell.

As the zeroth approximation we take the state of the system of electrons located in a field of a completely ordered crystal lattice consisting of the effective atoms. Thus the potential energy of the electrons in the field of each effective atom occupying a given type of site is equal to the mean potential energy of an electron in the field of atoms located at the sites of this type. The Hamiltonian of the zeroth approximation may be written as

$$H_0=-\frac{\hbar^2}{2m}\sum_{i=1}^{N_e}\Delta_i+\overline{V}(r_1, r_2, \ldots, r_{N_e})+\sum_{\substack{i, j=1\\(i<j)}}^{N_e}\frac{e^2}{|r_i-r_j|}, \qquad (35.3)$$

where

$$\overline{V}=\overline{V}(r_1, \ldots, r_{N_e})=\sum_{i=1}^{N_e}\sum_{s=1}^{N'}\sum_{\varkappa=1}^{\mu}\overline{V}_{\varkappa}(r_i-R_{s\varkappa}), \qquad (35.4)$$

and

$$\overline{V}_{\varkappa}(r_i-R_{s\varkappa})=\sum_{\alpha=1}^{\zeta}p_{\alpha}^{(L)}V_{\alpha}(r_i-R_{s\varkappa}). \qquad (35.5)$$

Here $p_{\alpha}^{(L)}$ is the probability that the type L sites are occupied by a type α atom, and $V_{\alpha}(r_i-R_{s\varkappa})$ is the potential energy of the ith electron in the field of a type α atom located at site number $\varkappa$ of the sth unit cell. The Hamiltonian H_0 defined by (35.3) is the Hamiltonian of the system of electrons in the periodic field having translational symmetry. Therefore, the eigenfunctions $\psi_n^0(r_1, \ldots, r_{N_e})$ of this Hamiltonian having in the zero approximation the wave function of the nth stationary phase, may be represented as*

*See, for example, Ref. [295]. The functions ψ_n^0 are antisymmetric. The spin part of the function is regarded as included in u_n and is unimportant for the following considerations.

$$\psi_n^0(\boldsymbol{r}_1, \ldots, \boldsymbol{r}_{N_e}) = \sum_P (-1)^P P \exp\left(i \sum_{i=1}^{N_e} \boldsymbol{k}_i \boldsymbol{r}_i\right) u_n(\boldsymbol{r}_1, \ldots, \boldsymbol{r}_{N_e}). \quad (35.6)$$

Here n is the set of quantum numbers of this state, P denotes the permutation operator of the electrons, $(-1)^P = +1$ for an even number of translations corresponding to a given permutation and equals -1, if the number of translations is odd; $\boldsymbol{k}_i$ are vectors characterizing the nth state of the electron system. The function $u_n(\boldsymbol{r}_1, \ldots, \boldsymbol{r}_{N_e})$ remains constant when the coordinates of all the electrons are shifted by the lattice vector $\boldsymbol{R}_s = s_1\boldsymbol{a}_1 + s_2\boldsymbol{a}_2 + s_3\boldsymbol{a}_3$ (s_1, s_2, s_3 are integers and $\boldsymbol{a}_1$, $\boldsymbol{a}_2$, $\boldsymbol{a}_3$ are basis vectors of the lattice)

$$u_n(\boldsymbol{r}_1 + \boldsymbol{R}_s, \ldots, \boldsymbol{r}_{N_e} + \boldsymbol{R}_s) = u_n(\boldsymbol{r}_1, \ldots, \boldsymbol{r}_{N_e}). \quad (35.7)$$

The perturbing energy is the difference of the potential energies

$$V' = V - \overline{V} = \sum_{i=1}^{N_e} \Delta V(\boldsymbol{r}_i), \quad (35.8)$$

where

$$\Delta V(\boldsymbol{r}_i) = \sum_{s=1}^{N'} \sum_{\varkappa=1}^{\mu} [V_{s\varkappa}(\boldsymbol{r}_i - \boldsymbol{R}_{s\varkappa}) - V_\varkappa(\boldsymbol{r}_i - \boldsymbol{R}_{s\varkappa})]. \quad (35.9)$$

Under the influence of perturbation, the system of electrons makes a transition from state n to another state m with the wave function

$$\psi_m^0(\boldsymbol{r}_1, \ldots, \boldsymbol{r}_{N_e}) = \sum_P (-1)^P P \exp\left(i \sum_{i=1}^{N_e} \boldsymbol{k}_i' \boldsymbol{r}_i\right) u_m(\boldsymbol{r}_1, \ldots, \boldsymbol{r}_{N_e}). \quad (35.10)$$

The matrix element of the perturbing energy corresponding to such a transition is

$$V'_{nm} = \int \psi_n^{0*} V' \psi_m^0 \, d\tau_1 \ldots d\tau_{N_e} = \sum_{i=1}^{N_e} \int \Phi_{nm}(\boldsymbol{r}_i) \Delta V(\boldsymbol{r}_i) \, d\tau_i, \quad (35.11)$$

where

$$\Phi_{nm}(\boldsymbol{r}_i) = \int \psi_n^{0*}(\boldsymbol{r}_1, \ldots, \boldsymbol{r}_{N_e}) \psi_m^0(\boldsymbol{r}_1, \ldots, \boldsymbol{r}_{N_e}) d\tau_1 \ldots d\tau_{i-1} d\tau_{i+1} \ldots d\tau_{N_e} \quad (35.12)$$

and the integration is performed over the unit cell of the crystal. We shall demonstrate that the function $\Phi_{nm}(r_i)$ is multiplied by a constant factor $e^{iq'a_j}$ when r_i is replaced by $r_i + a_j$ $(j=1, 2, 3)$

$$\Phi_{nm}(r_i + a_j) = e^{iq'a_j}\Phi_{nm}(r_i), \tag{35.13}$$

where

$$q' = \sum_{i=1}^{N_e}(k'_i - k_i). \tag{35.14}$$

In fact, the function

$$\begin{aligned}\Phi_{nm}(r_i + a_j) = \int \psi_n^{0*}(r_1, \ldots, r_i + a_j, \ldots, r_{N_e}) \times \\ \times \psi_m^0(r_1, \ldots, r_i + a_j, \ldots, r_{N_e})\, d\tau_1 \ldots d\tau_{i-1}\, d\tau_{i+1} \ldots d\tau_{N_e}\end{aligned} \tag{35.15}$$

because periodicity does not change during a change of the variables of integration $r_{i'} \to r_{i'} + a_j$(where $i' \neq i$), i.e.,

$$\begin{aligned}\Phi_{nm}(r_i + a_j) = \int \psi_n^{0*}(r_1 + a_j, \ldots, r_i + a_j, \ldots, r_{N_e} + a_j) \times \\ \times \psi_m^0(r_1 + a_j, \ldots, r_i + a_j, \ldots, r_{N_e} + a_j)\, d\tau_1 \ldots d\tau_{i-1}\, d\tau_{i+1} \ldots d\tau_{N_e}.\end{aligned} \tag{35.16}$$

Using (35.6) and (35.10)

$$\left.\begin{aligned}\psi_n^{0*}(r_1 + a_j, \ldots, r_{N_e} + a_j) = e^{-i\sum\limits_{i=1}^{N_e} k_i a_j}\psi_n^{0*}(r_1, \ldots, r_{N_e}), \\ \psi_m^0(r_1 + a_j, \ldots, r_{N_e} + a_j) = e^{i\sum\limits_{i=1}^{N_e} k'_i a_j}\psi_m^0(r_1, \ldots, r_{N_e}),\end{aligned}\right\} \tag{35.17}$$

and (35.16), we obtain (35.13). As is well known, the function having the property (35.13) may be written as

$$\Phi_{nm}(r_i) = e^{iq'r_i}U_{nm}(r_i), \tag{35.18}$$

where $U_{nm}(r_i)$ is a periodic function with a period equal to the lattice vector a_j.

Noting the symmetry of the wave function ψ_n^0 and ψ_m^0 with respect to permutation coordinates of individual electrons, it is readily shown that the form of the function $\Phi_{nm}(\boldsymbol{r}_i)$ is independent of the number of electrons i. Since the form of the function $\Delta V(\boldsymbol{r}_i)$ is also independent of i, the summation of (35.11) may be replaced by a multiplication by N_e, and the subscript i in $\boldsymbol{r}_i$ and $d\tau_i$ may be discarded

$$V'_{nm} = N_e \int \Phi_{nm}(\boldsymbol{r}) \Delta V(\boldsymbol{r})\, d\tau. \tag{35.19}$$

Substituting (35.9) and (35.18) into (35.19) we obtain:

$$V'_{nm} = N_e \sum_{s=1}^{N} \sum_{\varkappa=1}^{\mu} \int e^{i\boldsymbol{q}'\boldsymbol{r}} U_{nm}(\boldsymbol{r}) [V_{s\varkappa}(\boldsymbol{r} - \boldsymbol{R}_{s\varkappa}) - \overline{V}_{\varkappa}(\boldsymbol{r} - \boldsymbol{R}_{s\varkappa})]\, d\tau. \tag{35.20}$$

In each term of the sum in (35.20) we introduce the new variable

$$\boldsymbol{r}' = \boldsymbol{r} - \boldsymbol{R}_{s\varkappa}, \tag{35.21}$$

where $\boldsymbol{R}_{s\varkappa} = \boldsymbol{R}_s + \boldsymbol{h}_\varkappa$ ($\boldsymbol{h}_\varkappa$ is a vector drawn from the first site of the sth cell to its site number $\varkappa$). Denoting next the variable of integration $\boldsymbol{r}'$ by $\boldsymbol{r}$, we obtain the matrix element of the perturbing energy

$$V'_{nm} = N_e \sum_{s=1}^{N'} \sum_{\varkappa=1}^{\mu} e^{i\boldsymbol{q}'\boldsymbol{R}_{s\varkappa}} (V_{s\varkappa}^{nm} - \overline{V}_{\varkappa}^{nm}), \tag{35.22}$$

where

$$\left.\begin{aligned} V_{s\varkappa}^{nm} &= \int e^{i\boldsymbol{q}'\boldsymbol{r}} U_{nm}(\boldsymbol{r} + \boldsymbol{h}_\varkappa) V_{s\varkappa}(\boldsymbol{r})\, d\tau, \\ \overline{V}_{\varkappa}^{nm} &= \int e^{i\boldsymbol{q}'\boldsymbol{r}} U_{nm}(\boldsymbol{r} + \boldsymbol{h}_\varkappa) \overline{V}_{\varkappa}(\boldsymbol{r})\, d\tau. \end{aligned}\right\} \tag{35.23}$$

Let us substitute (35.5) into (35.23) for $\overline{V}_{\varkappa}^{nm}$

$$\overline{V}_{\alpha}^{nm} = \sum_{\alpha=1}^{\zeta} p_{\alpha}^{(L)} V_{\alpha}^{nm}, \tag{35.24}$$

where

$$V_{\alpha}^{nm} = \int e^{i\boldsymbol{q}'\boldsymbol{r}} U_{nm}(\boldsymbol{r} + \boldsymbol{h}_\varkappa) V_{\alpha}(\boldsymbol{r})\, d\tau. \tag{35.25}$$

We shall find the square of the modulus of the matrix element V'_{nm}. From (35.22) we obtain

$$\frac{1}{N_e^2} |V'_{nm}|^2 = \sum_{s, s'=1}^{N^0} \sum_{\varkappa, \varkappa'=1}^{\mu} e^{i\boldsymbol{q}'(\boldsymbol{R}_{s\varkappa} - \boldsymbol{R}_{s'\varkappa'})} (V_{s\varkappa}^{nm} - \overline{V}_{\varkappa}^{nm}) (V_{s'\varkappa'}^{nm*} - \overline{V}_{\varkappa'}^{nm*}). \tag{35.26}$$

Comparing the expression (35.26) with (31.9) for the intensity I_2 of diffuse scattering of x-rays, we demonstrate that the right-hand side of (35.26) goes over into the right-hand side of (31.9) upon the substitution of

$$V^{nm}_{s\varkappa} \to f_{s\varkappa}, \quad \bar{V}^{nm}_{\varkappa} \to \bar{f}_{\varkappa}, \quad q' \to q. \tag{35.27}$$

Such a substitution is obviously equivalent to

$$V^{nm}_{\alpha} \to f_{\alpha}, \quad q' \to q. \tag{35.28}$$

One may, therefore, without carrying out a new calculation, immediately write the final expression for $\frac{1}{N_e^2}|V'_{nm}|^2$, having substituted for the atomic scattering factors f_α the quantity V^{nm}_α, and replacing q by q' in those terms of Eq. (31.36) that do not contain δ functions, and which determine the intensity I_2 of diffuse scattering of x-rays. This yields

$$\frac{1}{N_e^2}|V'_{nm}|^2 = N^0 \sum_{\substack{\alpha,\alpha'=1\\(\alpha<\alpha')}}^{\zeta} |V^{nm}_{\alpha} - V^{nm}_{\alpha'}|^2 \sum_{L=1}^{Q} \Big\{\lambda_L p^{(L)}_{\alpha} p^{(L)}_{\alpha'} - \\ -\frac{1}{2}\sum_{\varkappa_L=1}^{\lambda_L}\sum_{l=1}^{\infty}\sum_{L'=1}^{Q}\left[\varepsilon^{LL'}_{\alpha\alpha'}(\rho_l)+\varepsilon^{LL'}_{\alpha'\alpha}(\rho_l)\right]\sum_{m_{lL'}=1}^{z_{lL'}}\cos q'\rho_{m_{lL'}}\Big\}. \tag{35.29}$$

In the derivation of (35.29), V_α and $V_{\alpha'}$ are assumed to be nearly equal. If all V_α for different values of α should be identical (pure metal), the function U_{nm} occurring in (35.25) would have a periodicity corresponding to the lattice of a pure metal, i.e., would have the periods of the vector $\boldsymbol{h}_\varkappa$, and the value $U_{nm}(\boldsymbol{r}+\boldsymbol{h}_\varkappa)$ for different $\boldsymbol{h}_\varkappa$ would appear to be identical. Therefore, in an ordered alloy with the same arrangement of atoms, formed of atoms with nearly equal V_α, these values differ by a magnitude of the order $|V_\alpha - V_{\alpha'}|$. From (35.29), discarding terms of the third and higher orders, we may assume that V^{nm}_α and $V^{nm}_{\alpha'}$ are independent of $\boldsymbol{h}_\varkappa$, i.e., assume these quantities equal to their values at $\boldsymbol{h}_\varkappa = \boldsymbol{h}_1 = 0$:

$$V^{nm}_{\alpha} = \int e^{iq'r} U_{nm}(\boldsymbol{r}) V_\alpha(\boldsymbol{r})\, d\tau. \tag{35.30}$$

We can therefore neglect the dependence of the wave functions of the zero approximation on both the composition and the order parameters.

Formula (35.29) gives an expression for the magnitude $\frac{1}{N_e^2}|V'_{nm}|^2$, which relates the composition of the multicomponent alloy, long-range order parameters $(p_\alpha^{(L)})$ and correlation parameters $\varepsilon_{\alpha\alpha'}^{LL'}(\rho_l)$ for all coordination spheres. In the case when correlation in the alloy is not significant, terms containing the correlation parameter in (35.29) may be discarded and the dependence of $\frac{1}{N_e^2}|V'_{nm}|^2$ on composition and long-range parameters (neglecting correlation) can be found.

Let us rewrite formula (35.29) as

$$\frac{1}{N_e^2}|V'_{nm}|^2 = N^0 \sum_{\substack{\alpha,\alpha'=1\\(\alpha<\alpha')}}^{\zeta} |V_\alpha^{nm} - V_{\alpha'}^{nm}|^2 f_{\alpha\alpha'}^{nm}, \tag{35.31}$$

where

$$f_{\alpha\alpha'}^{nm} = \sum_{L=1}^{Q} \lambda_L p_\alpha^{(L)} p_{\alpha'}^{(L)} - \frac{1}{2}\sum_{L=1}^{Q} \sum_{\varkappa_L=1}^{\lambda_L} \sum_{l=1}^{\infty} \sum_{L'=1}^{Q} [\varepsilon_{\alpha\alpha'}^{LL'}(\rho_l) + \varepsilon_{\alpha'\alpha}^{LL'}(\rho_l)] \sum_{m_{lL'}=1}^{z_{lL'}} \cos q'\rho_{m_{lL'}}. \tag{35.32}$$

The quantities $f_{\alpha\alpha'}^{nm}$ are, in general, dependent on the quantum numbers of states n and m, because they contain the vector q'. We shall give the expression for $f_{\alpha\alpha'}^{nm}$ as a function of composition, long-range parameters and correlation parameters for the case of binary alloys $A-B$. Thus, in the sum over α and α' there remains only one term, and since in this case $\varepsilon_{AB}^{LL'}(\rho_l) = \varepsilon_{BA}^{LL'}(\rho_l)$, $f_{\alpha\alpha'}^{nm} = f_{AB}^{nm}$ becomes

$$f_{AB}^{nm} = \frac{|V'_{nm}|^2}{N_e^2 N^0 |V_A^{nm} - V_B^{nm}|^2} = \sum_{L=1}^{Q} \lambda_L p_A^{(L)} p_B^{(L)} - \sum_{L=1}^{Q} \sum_{\varkappa_L=1}^{\lambda_L} \sum_{l=1}^{\infty} \sum_{L'=1}^{Q} \varepsilon_{AB}^{LL'}(\rho_l) \sum_{m_{lL'}=1}^{z_{lL'}} \cos q'\rho_{m_{lL'}}. \tag{35.33}$$

In particular, if correlation is insignificant, then

$$f_{AB}^{nm} = f_{AB} = \sum_{L=1}^{Q} \lambda_L p_A^{(L)} p_B^{(L)} \tag{35.34}$$

and is independent of the quantum numbers of the initial and final states n and m.

The obtained expression (35.31) for the quantity $\frac{1}{N_e^2}|V'_{nm}|^2$ makes possible a determination of the probability w_{nm} of a transition of the system of electrons (per unit time) from state n to state m as a function of composition of the multicomponent alloy, the long-range order parameters and the correlation parameters up to quantities of the third order $|V_\alpha - V_{\alpha'}|$. In fact, the proportionality factor K_{nm} in the formula

$$w_{nm} = K_{nm} \frac{1}{N_e^2} |V'_{nm}|^2 \qquad (35.35)$$

does not vanish in the case where all V_α coincide, when it is obviously independent of c_α, $p_\alpha^{(L)}$ and $\varepsilon_{\alpha\alpha'}^{LL'}$. Consequently, its expansion in powers of $|V_\alpha - V_{\alpha'}|$ begins with a term that is independent of these quantities.

36. Derivation of Equations for the Residual Electrical Resistivity

To determine the mean value of the electric current density we shall use the matrix density method. Let us construct a density matrix from the many-electron wave functions (35.6) of the zeroth approximation problem. W_{nm} denotes the matrix elements of density matrix in the presense of an electric field, and W_{nm}^0 their values in the absence of an electric field. As is well known, the mean value of the component of the electric current density in the direction of the field may be written as

$$j = \sum_{nm} W_{nm} j_{nm}, \qquad (36.1)$$

where j_{nm} are the matrix elements of the current density. In the absence of an electric field, the mean electric current density is equal to zero

$$\sum_{nm} W_{nm}^0 j_{nm} = 0. \qquad (36.2)$$

Hence

$$j = \sum_{nm} \Delta W_{nm} j_{nm}, \qquad (36.3)$$

where

$$\Delta W_{nm} = W_{nm} - W_{nm}^0. \qquad (36.4)$$

In order to determine the residual electrical resistivity ρ_0, it is necessary to find the mean current density j from formula (36.3), for which the quantities ΔW_{nm} must be known. In the calculation of ΔW_{nm} we restrict ourselves merely to the determination of their dependence on the composition and order parameters, without seeking to determine the numerical value of ρ_0. While performing the calculation, we shall allow for the fact that $|V_\alpha - V_{\alpha'}|$ is small, and discard terms of third and higher orders in this difference. As is customary, restricting ourselves to the range of applicability of Ohm's law, we retain only linear terms in the electric field intensity F.

The quantities W_{nm} may be determined from the steady-state condition

$$\frac{dW_{nm}}{dt} = 0. \tag{36.5}$$

We shall consider the fact that the system of electrons is in a state of dynamic equilibrium and that the quantities W_{nm} change both as a result of the action of the electric field and scattering of electrons by imperfections. The change of W_{nm} per unit time due to the action of the electric field is denoted by $\left(\frac{\partial W_{nm}}{\partial t}\right)_{\text{field}}$, and that caused by scattering by $\left(\frac{\partial W_{nm}}{\partial t}\right)_{\text{alloy}}$. Then the steady state condition (36.5) may be rewritten as

$$\left(\frac{\partial W_{nm}}{\partial t}\right)_{\text{field}} + \left(\frac{\partial W_{nm}}{\partial t}\right)_{\text{alloy}} = 0. \tag{36.6}$$

The quantity $\left(\frac{\partial W_{nm}}{\partial t}\right)_{\text{field}}$ is a function of the field intensity F. By expanding this function as a power series in F and noting that the expansion must not contain any term that is independent of the field intensity, we obtain:

$$\left(\frac{\partial W_{nm}}{\partial t}\right)_{\text{field}} = \beta_{nm} F. \tag{36.7}$$

The change of W_{nm} per unit time, caused by scattering of electrons, depends on the transition probability (35.35) w_{nm}, and may be represented as

$$\left(\frac{\partial W_{nm}}{\partial t}\right)_{\text{alloy}} = \sum_{n'm'} \gamma^{nm}_{n'm'} w_{n'm'}. \tag{36.8}$$

Substituting (35.31) and (35.35) into (36.8), we find

$$\left(\frac{\partial W_{nm}}{\partial t}\right)_{\text{alloy}} = \sum_{\substack{\alpha, \alpha'=1 \\ (\alpha < \alpha')}}^{\zeta} \sum_{n'm'} \theta_{\alpha\alpha'}^{nmn'm'} f_{\alpha\alpha'}^{n'm'}, \tag{36.9}$$

where

$$\theta_{\alpha\alpha'}^{nmn'm'} = N^0 \gamma_{n'm'}^{nm} K_{n'm'} \left| V_{\alpha}^{n'm'} - V_{\alpha'}^{n'm'} \right|^2. \tag{36.10}$$

Since the quantities $\left(\frac{\partial W_{nm}}{\partial t}\right)_{\text{alloy}}$ depend not only on the transition probability, but also on the matrix element $W_{n''m''} = W_{n''m''}^0 + \Delta W_{n''m''}$, then $\gamma_{n'm'}^{nm}$ and, consequently, also $\theta_{\alpha\alpha'}^{nmn'm'}$ are functions of $\Delta W_{n''m''}$. For $\Delta W_{n''m''} = 0$, i.e., in the absence of a field, the quantity $\left(\frac{\partial W_{nm}}{\partial t}\right)_{\text{alloy}}$, according to the steady state condition, must become zero. Hence, also $\theta_{\alpha\alpha'}^{nmn'm'}$ for $\Delta W_{n''m''} = 0$ must become zero, i.e., in the expansion of $\theta_{\alpha\alpha'}^{nmn'm'}$ in powers of $\Delta W_{n''m''}$, the free term is absent. Carrying out this expansion and restricting ourselves to linear terms in $\Delta W_{n''m''}$, we obtain

$$\theta_{\alpha\alpha'}^{nmn'm'} = \sum_{n''m''} M_{\alpha\alpha'}^{nmn'm'n''m''} \Delta W_{n''m''}. \tag{36.11}$$

Substituting (36.7), (36.9) and (36.11) into the equilibrium condition (36.6), we obtain a system of equations for the determination of $\Delta W_{n''m''}$

$$\sum_{\substack{\alpha, \alpha'=1 \\ (\alpha < \alpha')}}^{\zeta} \sum_{n'm'} f_{\alpha\alpha'}^{n'm'} \sum_{n''m''} M_{\alpha\alpha'}^{nmn'm'n''m''} \Delta W_{n''m''} = -\beta_{nm} F. \tag{36.12}$$

The quantities β_{nm} occurring in Eq. (36.7), and also the quantities $\gamma_{n'm'}^{nm}$ and $K_{n'm'}$ occurring in Eq. (36.10) do not become zero, as all the differences $|V_\alpha - V_{\alpha'}|$ tend to zero; they become independent of composition, long-range parameters and correlation parameters. Therefore in the approximation being considered in Eqs. (36.12) one may regard β_{nm} and $M_{\alpha\alpha'}^{nmn'm'n''m''}$ as independent of composition and of these parameters. The quantities depending on composition, long-range order parameters and correlation parameters are functions of $f_{\alpha\alpha'}^{n'm'}$ having the form (35.32).

From Eqs. (36.12) one may obtain the dependence of ΔW_{nm} on composition, long-range order parameters and correlation parameters only in several special cases, when the functions $f_{\alpha\alpha'}^{nm}$ are independent of the quantum numbers of the states n and m. In these cases, according to (36.3), the mean current density and the residual electrical resistivity may be determined. Several cases of this type will be considered in the following two sections.

37. Residual Electrical Resistivity of Binary Alloys

We shall consider the special case of a binary ordered alloy AB. In this case the functions $f_{\alpha\alpha'}^{nm}$ have the form (35.33), and in the absence of correlation the form (35.34). As already mentioned, in alloys where correlation is insignificant, the functions $f_{\alpha\alpha'}^{nm}$ according to (35.34) are independent of n and m. In alloys in which correlation is important, the functions f_{AB}^{nm} may be regarded as independent of n and m, if the following condition is fulfilled:

$$q'\rho_{m_{lL'}} \ll 1, \tag{37.1}$$

i.e., for coordination spheres which are not too remote and for sufficiently small values of q'. In the one-electron approximation, a condition similar to the "smallness" of q' is fulfilled for all quantum conditions in the case of "bad" metals, i.e., metals with a small number of conduction electrons N_e. When (37.1) is fulfilled, all the cosines in (35.33) may be replaced by unity. Then, for f_{AB}^{nm} we obtain

$$f_{AB}^{nm} = f_{AB} = \sum_{L=1}^{Q} \lambda_L p_A^{(L)} p_B^{(L)} - \sum_{L=1}^{Q} \sum_{\varkappa_L=1}^{\lambda_L} \sum_{l} \sum_{L'=1}^{Q} \varepsilon_{AB}^{LL'}(\rho_l)\, z_{lL'}. \tag{37.2}$$

Using expression (37.2) for f_{AB}, we shall reevaluate the role of correlation in alloy, which in real metals (where condition (37.1) is not always fulfilled) may give a substantially smaller effect.

In the case of binary alloys, when the function $f_{AB}^{nm} = f_{AB}$ is independent of n and m, the system of equations (36.12) becomes

$$\sum_{n''m''} M_{AB}^{nmn''m''} \Delta W_{n''m''} = -\beta_{nm} \frac{F}{f_{AB}}, \tag{37.3}$$

where

$$M_{AB}^{nmn''m''} = \sum_{n'm'} M_{AB}^{nmn'm'n''m''}. \tag{37.4}$$

The solution of the system of linear inhomogeneous equations (37.3) is

$$\Delta W_{n''m''} = b_{n''m''} \frac{F}{f_{AB}}. \tag{37.5}$$

Replacing n'', m'' by n, m and substituting (37.5) into (36.3), we obtain

$$j = \frac{F}{f_{AB}} \sum_{nm} b_{nm} j_{nm}. \tag{37.6}$$

Hence, the residual electrical resistivity of the alloy ρ_0 is

$$\rho_0 = A f_{AB}, \tag{37.7}$$

where the constant $A = \frac{1}{\sum_{nm} b_{nm} j_{nm}}$ is independent of composition, long-range order parameters and correlation parameters.

We shall first consider the dependence of the residual electrical resistivity ρ_0 of the alloy on composition and the degree of long-range order neglecting correlation. In this case f_{AB} is determined from (35.34) and the derivation of (37.7) for ρ_0 is not based on the inequality (37.1). According to (37.7) ρ_0 is given by

$$\rho_0 = A \sum_{L=1}^{Q} \lambda_L p_A^{(L)} p_B^{(L)}. \tag{37.8}$$

For an alloy with two types of sites ($Q = 2$), taking into account (1.5), we obtain for ρ_0 a dependence on c_A and η previously found in [288] using the one-electron approximation

$$\rho_0 = A' [c_A (1 - c_A) - \nu (1 - \nu) \eta^2], \tag{37.9}$$

where $A' = A\mu$. In the case of disordered alloys ($\eta = 0$), from (37.9) we obtain a parabolic dependence of ρ_0 on c_A

$$\rho_0 = A' c_A (1 - c_A) \tag{37.10}$$

with a maximum at $c_A = \frac{1}{2}$, which has been found in [268] using the one-electron approximation. Such a parabolic concentration dependence of the residual electrical resistivity is in good agreement with the experimental curve for the alloy Ag-Au, illustrated in Fig. 34.

For ordered alloys with an equal number of sites of the first and second types $\left(\nu=\frac{1}{2}\right)$, Eq. (37.9) gives

$$\rho_0=A'\left[c_A(1-c_A)-\frac{1}{4}\eta^2\right], \tag{37.11}$$

and for $AuCu_3$ type alloys $\left(\nu=\frac{1}{4}\right)$ we have

$$\rho_0=A'\left[c_A(1-c_A)-\frac{3}{16}\eta^2\right]. \tag{37.12}$$

It is apparent from Eqs. (37.11) and (37.12) that in conformity with the experimental results expounded in Sec. 6, the residual electrical resistivity of an alloy of given composition decreases with an increase of the degree of long-range order η. In particular, for a completely ordered alloy ($c_A=\nu$, $\eta=1$) the residual electrical resistivity ρ_0 is equal to zero.

The formulas obtained also explain the experimentally discovered minima in the concentration curve of electrical resistivity. In fact, in a series of alloys of different compositions, annealed until equilibrium is established at the same temperature, a different degree of long-range order was found, increasing as the composition of the alloy approached the stoichiometric composition. Therefore, if in Eqs. (37.9) or (37.11) and (37.12) we note the indicated relation $\eta(c_A)$, the residual electrical resistivity ρ_0 of the specified series of quenched alloys (assuming that η at $c_A=\nu$ is not small) must be a minimum in the region of stoichiometric composition. Such types of minima of the residual electrical resistivity of alloys of the system Au–Cu are clearly marked in Fig. 35. In the concentration and temperature curves of ρ_0 for the alloys Au–Cu (see Fig. 35 and 40), in which the transition to the ordered state is a first-order phase transition, discontinuities of the variation of ρ_0 due to discontinuous changes of η are marked at a temperature and compositions corresponding to this transition [61]. It is also apparent from (37.11) that in β-brass type alloys, where the order-disorder transition is a second-order phase transition, the electrical resistivity at the transition point should change continuously. Since in the ordered state near the critical temperature T_0, the quantity η^2 is according to (11.9) proportional to T_0-T, the graph of $\rho_0(T)$ should contain a break at $T=T_0$. Such a type of break is noticeable on the curve of the temperature dependence of the electrical resistivity of β-brass illustrated in Fig. 39.

For stoichiometric alloys ($c_A = \nu$), we obtain from (37.9) the dependence of ρ_0 on η

$$\rho_0 = A'\nu(1-\nu)(1-\eta^2). \tag{37.13}$$

The proportionality of the residual electrical resistivity of such alloys to $1-\eta^2$ has been corroborated experimentally [296].

We shall now consider alloys in which correlation is important. Eqs. (37.2) and (37.7) [when condition (37.1) is fulfilled] give

$$\rho_0 = A \sum_{L=1}^{Q} \left[\lambda_L p_A^{(L)} p_B^{(L)} - \sum_{\varkappa_L=1}^{\lambda_L} \sum_{l} \sum_{L'=1}^{Q} z_{lL'} \varepsilon_{AB}^{LL'}(\rho_l)\right]. \tag{37.14}$$

The correlation for different coordination spheres $\varepsilon_{AB}^{LL'}(\rho_l)$ or some of their combinations may be determined from experimental data (from the distribution of intensity of diffuse scattering of x-rays or neutrons, using the method presented in Chapter VI). The long-range order parameters may also be determined from x-ray or neutron diffraction data on the intensity of the superlattice reflections. For a determination of long-range order and the correlation parameters one may use the formulas of the statistical theory of ordering. These will be given below for some alloys.

In the case of disordered alloys, taking into account correlation in the first coordination sphere only, Eq. (37.14) yields

$$\rho_0 = A'[c_A(1-c_A) - z\varepsilon_{AB}], \tag{37.15}$$

where z is the coordination number of the first coordination sphere and $\varepsilon_{AB} = \varepsilon_{AB}(\rho_1)$. Substituting for the correlation parameter ε_{AB} its expression in (17.56) (at $\eta = 0$) or (17.65) found by the Kirkwood method, we rewrite formula (37.15) as

$$\rho_0 = A'\left[c_A(1-c_A) - zc_A^2(1-c_A)^2 \frac{w}{kT}\right], \tag{37.16}$$

where w is the ordering energy. This formula is applicable at temperatures much greater than the critical temperature, because at lower temperatures it is impossible to consider only linear terms in $\frac{w}{kT}$ in the expression for ε_{AB}, and the correlation in the filling should be noted not only for neighboring sites.

Eq. (37.16) shows that the establishment of correlation in the first coordination sphere for alloys with a small number of conduction electrons reduces the residual electrical resistivity of ordered alloys ($w > 0$) and enhances the residual electrical resistivity of the disordered alloys ($w < 0$). The second term in (37.16) due to short-range order (correlation) increases in absolute value with a decrease of annealing temperatures T of the alloy. Taking correlation at this temperature into account, gives a larger correction to the electrical resistivity the closer the composition to the composition $c_A = \frac{1}{2}$. We note that formula (37.16) contains a rather large correlation correction to the residual electrical resistivity. Thus, for example, at $c_A = \frac{1}{2}$ and $\frac{w}{kT} = \frac{1}{z}$ (the temperature T is several times greater than the critical temperature) the establishment of correlation reduces the residual resistance of the alloy by twenty-five percent.

It should be emphasized that annealing leading to the establishment of short-range order should not lead to so great a change in the residual electrical resistivity. In fact, annealing of the alloy at a given temperature leads only to a change of the short-range order which corresponds to another temperature (for example, the melting point), and does not produce a complete short-range order. From this point of view one may expect a greater effect on the residual electrical resistivity to be caused not by annealing, but by plastic deformation, because the latter may greatly disrupt the short-range order of the alloy. Moreover, in real alloys the role of correlation may be appreciably smaller because of violation of conditions (37.1). In alloys with a sufficiently large number of conduction electrons it is no longer possible to replace the cosine by unity in Eq. (35.33) and the effect of correlation may be more complicated: the coefficient at $\varepsilon_{AB}^{LL'}(\rho_l)$ may differ from $z_{lL'}$ and even become negative. Therefore, the residual electrical resistivity in such alloys may not change in the same direction (during the establishment of correlation) as it should change in the case considered above for "bad" metals, when correlation is taken into account only in the first coordination sphere. Moreover, since a change of annealing temperature changes the relative role of correlation in the different coordination spheres (which may cause either an increase or a decrease in the electrical resistivity) it is possible that the effect of correlation on ρ_0 may be different in various temperature ranges. This possibly explains the experimental data given in Sec. 6 [74] concerning the dependence of ρ_0 on the annealing temperature of the alloy $AuCu_3$ (see Fig. 42), where an increase of the degree of short-range order above 485° C is accompanied by a decrease in ρ_0, while above 485° C it is accompanied by an increase.

For alloys in which condition (37.1) is not fulfilled, we were unable to determine the dependence of ρ_0 on the correlation parameters within the framework of the many-electron theory developed above. This can be done by the one-electron theory. This calculation has been performed for alloys with a simple cubic lattice in [297] and for alloys with a face-centered cubic lattice in [298]. In particular, for disordered alloys $AuCu_3$ the residual electrical resistivity has been determined by noting correlation in the first ten coordination spheres at three different temperatures (405, 460, 550° C). Thus, the experimental values of the correlation parameters cited in Sec. 3 (Table 2) were analyzed by Cowling. The calculation led to a monotonic decrease of ρ_0 with an increase of annealing temperatures, which is in qualitative agreement with the experimental results of [74] over the temperature range 393 to 485° C, but does not correspond to the behavior of the temperature curve above 485° C (see Fig. 42).

In the case of ordered alloys with an equal number of sites of the first and second types, where sites of one type are surrounded only by sites of the other type, Eq. (37.14) leads to the following expression for ρ_0 taking into account the correlation in the first coordination sphere:

$$\rho_0 = A'\left[c_A(1-c_A) - \frac{1}{4}\eta^2 - z\varepsilon_{AB}^{12}\right]. \tag{37.17}$$

Substituting for ε_{AB}^{12} the expression (19.14) for $\varepsilon_{AB}^{12}(\rho_1)$, obtained for an almost completely ordered alloy, we find

$$\rho_0 = A'\left[c_A(1-c_A) - \frac{1}{4}\eta^2 - z\left(c_A - \frac{\eta}{2}\right)\times\right.$$
$$\left.\times\left(1 - c_A - \frac{\eta}{2}\right)\left(e^{\frac{w}{kT}} - 1\right)\right]. \tag{37.18}$$

For alloys with stoichiometric composition $\left(c_A = \frac{1}{2}\right)$ we find

$$\rho_0 = \frac{A'}{4}\left[1 - \eta^2 - z(1-\eta)^2\left(e^{\frac{w}{kT}} - 1\right)\right]. \tag{37.19}$$

In almost completely ordered stoichiometric alloys of the structure under consideration, according to (19.16), as the temperature decreases the quantity $1-\eta$ tends exponentially to zero

$$1-\eta \approx 2\exp\left(-\frac{z}{2}\frac{w}{kT}\right).$$

Consequently, the correlation correction in (37.19) tends exponentially to zero at $T\to 0$. Thus, the residual electrical resistivity ρ_0 tends exponentially to zero $\left(\text{as } e^{-\frac{z}{2}\frac{w}{kT}}\right)$.

For binary $AuCu_3$ ordered alloys with a face-centered cubic lattice, taking into account (1.5), one may obtain from (37.14) an expression for the residual electrical resistivity, allowing for correlation in the first coordination sphere

$$\rho_0 = A'\left[c_A(1-c_A) - \frac{3}{16}\eta^2 - 6\left(\varepsilon_{AB}^{12} + \varepsilon_{AB}^{22}\right)\right], \qquad (37.20)$$

where

$$\varepsilon_{AB}^{12} = \varepsilon_{AB}^{12}(\rho_1), \quad \varepsilon_{AB}^{22} = \varepsilon_{AB}^{22}(\rho_1).$$

In the case of almost completely ordered alloys of this structure with a stoichiometric composition $\left(c_A = \frac{1}{4}\right)$, substituting the expression (19.14) into (37.20) for ε_{AB}^{12} and ε_{AB}^{22} and allowing for (1.5), we find for the first coordination sphere

$$\rho_0 = \frac{3}{16}A'\left[1-\eta^2 - 2\left(3e^{\frac{w}{kT}} - 2 - e^{-\frac{w}{kT}}\right)(1-\eta)^2\right]. \qquad (37.21)$$

In almost completely ordered $AuCu_3$ type alloys with a face-centered lattice of the stoichiometric composition (when $\frac{w}{kT} \gg 1$), Eq. (19.18) yields

$$1-\eta \approx \frac{4}{\sqrt{3}}e^{-2\frac{w}{kT}}.$$

Therefore, as the annealing temperature is lowered (in the case of ordering of the alloy) the residual electrical resistivity tends exponentially to zero, and the correlation correction plays an ever decreasing role.

We note again that the formulas obtained for the residual electrical resistivity give in a number of cases the same type of dependence on composition, long-range order parameters and correlation parameters as those given in Chapter VI for diffuse scattering of x-rays and neutrons. This is explained by the fact that the electrical resistivity is caused by the scattering of electron waves by inhomogeneities of the crystal lattice, caused by an incompletely ordered arrangement of atoms.

38. Residual Electrical Resistivity of Ternary Alloys

The system of equations (36.12) may be solved with respect to $\Delta W_{n''m''}$, in a more general case than in Sec. 37, for multicomponent ordered alloys in which the concentrations of all components—except two—are small. In particular, this may be done for ternary alloys with a low concentration of one of the components.

For ternary alloys $A-B-C$, when condition (37.1) is fulfilled, Eq. (36.12) becomes

$$f_{AB}\sum_{n''m''} M_{AB}^{nmn''m''}\,\Delta W_{n''m''}+f_{AC}\sum_{n''m''} M_{AC}^{nmn''m''}\,\Delta W_{n''m''}+ \\ +f_{BC}\sum_{n''m''} M_{BC}^{nmn''m''}\,\Delta W_{n''m''}=-\beta_{nm}F, \tag{38.1}$$

where, according to (35.32),

$$f_{AB}=\sum_{L=1}^{Q}\lambda_L p_A^{(L)}p_B^{(L)}-\frac{1}{2}\sum_{L=1}^{Q}\sum_{\varkappa_L=1}^{\lambda_L}\sum_{l}\sum_{L'=1}^{Q}\left[\varepsilon_{AB}^{LL'}(\rho_l)+\varepsilon_{BA}^{LL'}(\rho_l)\right]z_{lL'}, \tag{38.2}$$

$$f_{AC}=\sum_{L=1}^{Q}\lambda_L p_A^{(L)}p_C^{(L)}-\frac{1}{2}\sum_{L=1}^{Q}\sum_{\varkappa_L=1}^{\lambda_L}\sum_{l}\sum_{L'=1}^{Q}\left[\varepsilon_{AC}^{LL'}(\rho_l)+\varepsilon_{CA}^{LL'}(\rho_l)\right]z_{lL'}, \tag{38.3}$$

$$f_{BC}=\sum_{L=1}^{Q}\lambda_L p_B^{(L)}p_C^{(L)}-\frac{1}{2}\sum_{L=1}^{Q}\sum_{\varkappa_L=1}^{\lambda_L}\sum_{l}\sum_{L'=1}^{Q}\left[\varepsilon_{BC}^{LL'}(\rho_l)+\varepsilon_{CB}^{LL'}(\rho_l)\right]z_{lL'} \tag{38.4}$$

and

$$M_{\alpha\alpha'}^{nmn''m''}=\sum_{n'm'} M_{\alpha\alpha'}^{nmn'm'n''m''} \qquad (\alpha,\ \alpha'=A,\ B,\ C). \tag{38.5}$$

For alloys where the concentration c_C is appreciably lower than c_A and c_B, the functions f_{AC} and f_{BC} are small compared with

f_{AB}, and Eqs. (38.1) may be solved by the method of successive approximations. In the zeroth approximation, setting f_{AC} and f_{BC} equal to zero, we obtain, similarly to the case of binary alloys,

$$\Delta W^0_{n''m''} = b_{n''m''} \frac{F}{f_{AB}}. \tag{38.6}$$

As a first approximation we set

$$\Delta W_{n''m''} = \Delta W^0_{n''m''} + \delta_{n''m''}, \tag{38.7}$$

where $\delta_{n''m''}$ is a small correction. Substituting (38.7) and (38.6) into (38.1) and discarding second and higher order quantities, we obtain the following system of equations for $\delta_{n''m''}$:

$$\sum_{n''m''} M_{AB}^{nmn''m''} \delta_{n''m''} = -\frac{F}{f_{AB}^2}\Big[f_{AC} \sum_{n''m''} M_{AC}^{nmn''m''} b_{n''m''} + f_{BC} \sum_{n''m''} M_{BC}^{nmn''m''} b_{n''m''}\Big]. \tag{38.8}$$

Solving this system of equations, we find

$$\delta_{n''m''} = \frac{F}{f_{AB}^2}\left(h_{AC}^{n''m''} f_{AC} + h_{BC}^{n''m''} f_{BC}\right), \tag{38.9}$$

where $h_{AC}^{n''m''}$ and $h_{BC}^{n''m''}$ are coefficients which in this approximation are independent of composition, long-range order parameters and correlation parameters. Substituting (38.6) and (38.9) into (38.7) and replacing n'', m'' by n and m, we find the first approximation of ΔW_{nm}

$$\Delta W_{nm} = \left(b_{nm}\frac{1}{f_{AB}} + h_{AC}^{nm}\frac{f_{AC}}{f_{AB}^2} + h_{BC}^{nm}\frac{f_{BC}}{f_{AB}^2}\right) F. \tag{38.10}$$

Equations (38.10) and (36.3) show that the conductivity of the alloy σ takes the form

$$\sigma = \frac{1}{f_{AB}} \sum_{nm} b_{nm} j_{nm} + \frac{f_{AC}}{f_{AB}^2} \sum_{nm} h_{AC}^{nm} j_{nm} + \frac{f_{BC}}{f_{AB}^2} \sum_{nm} h_{BC}^{nm} j_{nm}. \tag{38.11}$$

Discarding small quadratic terms in c_C, the following expression for the residual electrical resistivity $\rho_0 = \frac{1}{\sigma}$ is obtained:

$$\rho_0 = A_1 f_{AB} + A_2 f_{AC} + A_3 f_{BC}, \tag{38.12}$$

where A_1, A_2, A_3 are constants. It is obvious from physical considerations that these coefficients must be positive. In the case where $c_C=0$, $f_{AC}=f_{BC}=0$, these formulas become identical with (37.7) for binary alloys.

Let us consider the special case of disordered alloys. The functions f_{AB}, f_{AC} and f_{BC}, according to (38.2)-(38.4), are

$$\left.\begin{aligned} f_{AB} &= \mu[c_A c_B - \sum_l z_l \varepsilon_{AB}(\rho_l)], \\ f_{AC} &= \mu\,[c_A c_C - \sum_l z_l \varepsilon_{AC}(\rho_l)], \\ f_{BC} &= \mu\,[c_B c_C - \sum_l z_l \varepsilon_{BC}(\rho_l)]. \end{aligned}\right\} \quad (38.13)$$

Substituting (38.13) into (38.12), we find that the residual electrical resistivity of a disordered alloy with a low concentration of C atoms is

$$\begin{aligned} \rho_0 = A_1'\,[c_A c_B - \sum_l z_l \varepsilon_{AB}(\rho_l)] + A_2'\,[c_A c_C - \sum_l z_l \varepsilon_{AC}(\rho_l)] + \\ + A_3'\,[c_B c_C - \sum_l z_l \varepsilon_{BC}(\rho_l)], \end{aligned} \quad (38.14)$$

where $A_1'=\mu A_1$, $A_2'=\mu A_2$ and $A_3'=\mu A_3$. If correlation in a disordered alloy is insignificant and all $\varepsilon_{AB}(\rho_l)$, $\varepsilon_{AC}(\rho_l)$ and $\varepsilon_{BC}(\rho_l)$ are equal to zero, Eq. (38.14) gives

$$\rho_0 = A_1' c_A c_B + A_2' c_A c_C + A_3' c_B c_C, \quad (38.15)$$

which agrees with the expression for electrical resistance obtained in the single-electron approximation [268] for alloys of arbitrary composition. Nordheim [268] has used the additional, generally unsupported, assumptions mentioned in Sec. 35 regarding the possibility of introducing a mean free path that is independent of direction, etc. Within the framework of these assumptions for "bad" metals (with a small number of current carriers), Eq. (38.14) can be obtained in the one-electron approximation for any composition of the alloy. If, however, we reject these assumptions, this formula may also be obtained in the single-electron approximation (without introducing the concept of mean free path) only in the case of small concentrations of C atoms [290].

Neglecting correlation, we can calculate the correction $\Delta\rho_0$ to the residual electrical resistivity of the binary alloy

$A-B$ from (33.15). This correction accounts for the addition of the small amount of impurity of the C component, is proportional to c_C and depends linearly on the composition of the binary alloy $A-B$. For instance, if the concentration c_A remains constant when a different quantity of C atoms is added, then

$$\Delta\rho_0 = [A_3' + (A_2' - A_1' - A_3')\,c_A]\,c_C. \tag{38.16}$$

Equation (38.16) shows that if $c_A \ll 1$ [in this case, $A_1' c_A c_B$ is much greater than the other two terms in (38.15)], then $\Delta\rho_0$ is always positive. This result should be expected, since the replacement of a certain number of atoms B of the almost pure metal B by atoms C (the number of A atoms remaining constant), will obviously increase the number of scattering centers that will scatter electrons independently of one another. On the other hand, if $c_B \ll 1$ (and $c_A =$ const), i.e., if the atoms C replace atoms D with an increase of c_C in an almost pure metal A, then $\Delta\rho_0$ may be either positive or negative.

In a series of alloys with constant concentration of alloying atoms C ($c_C =$ const), the dependence of residual resistivity of the alloys on c_A (or c_B) can be represented according to (38.15) not by a parabolic relation of the type $c_A(1-c_A)$ as in the case of a binary alloy, but by the sum of this quadratic function and the linear function (38.16) of c_A. Thus, the curve of ρ_0 as a function of c_A becomes asymmetric and the displacement of the maximum is proportional to c_C.

If correlation in the disordered alloy cannot be neglected, the residual electrical resistivity must be determined from (38.14). We shall consider the case of rather high temperatures, when the following conditions hold

$$kT \gg |w_{AB}|, \quad kT \gg |w_{AC}|, \quad kT \gg |w_{BC}|, \tag{38.17}$$

and where w_{AB}, w_{AC} and w_{BC} are the ordering energies of the corresponding binary alloys with the same geometric arrangement of atoms as in a ternary alloy. Then the correlation parameters for the first coordination sphere are determined by (20.39) and by formulas obtained from the latter by cyclic permutation of the subscripts A, B and C. The correlation parameters for the remaining coordination spheres in the nearest neighbor approximation are proportional to $\frac{w_{\alpha\alpha'}}{kT}$ and, on the basis of (38.17), higher power terms may be discarded. Substituting the expression obtained in this manner for

$\varepsilon_{AB}(\rho_1)=\varepsilon_{AB}$, $\varepsilon_{AC}(\rho_1)=\varepsilon_{AC}$ and $\varepsilon_{BC}(\rho_1)=\varepsilon_{BC}$ into (38.14), we find that for small c_C ($c_C \ll c_A$, $c_C \ll c_B$) the residual electrical resistivity is

$$\begin{aligned}\rho_0 = A_1'\left[c_A(1-c_A) - zc_A^2(1-c_A)^2\frac{w_{AB}}{kT}\right] + \\ + \left[A_3' + (A_2' - A_1' - A_3')c_A\right]c_C + \\ + zc_A(1-c_A)\left\{A_1'\left[(4c_A-1)\frac{w_{AB}}{2kT} + (1-2c_A)\frac{w_{AC}-w_{BC}}{2kT}\right] + \right. \\ \left. + (A_2' - A_3')\left[(1-2c_A)\frac{w_{AB}}{2kT} - \frac{w_{AC}-w_{BC}}{2kT}\right]\right\}c_C.\end{aligned} \tag{38.18}$$

Equation (38.18) shows that the correlation correction in the expression for ρ_0 becomes negligibly small when the concentration of atoms A or B is sufficiently small; however, it may be appreciable when c_A and c_B are of the same order of magnitude. As may be seen from this formula, the correlation correction at sufficiently high temperatures is inversely proportional to the annealing temperature T of the alloy. Taking correlation into account, the variation $\Delta\rho_0$ of the residual resistivity when alloying atoms C are added to the alloy $A-B$ (with $c_A=$ const) will no longer be a linear function of c_A, determined by (38.16). We must add to this linear function, in accordance with (38.18), the product of a different linear function and the factor $c_A(1-c_A)$.

We shall now consider the case of ternary ordered alloys $A-B-C$ with two types of sites ($Q=2$) and a small concentration of C atoms. For alloys with a body-centered cubic β-brass type lattice, taking into account correlation in the first coordination sphere, Eq. (38.12) and the expressions for f_{AB}, f_{AC} and f_{BC} (38.2)-(38.4) lead to the following equation for ρ_0:

$$\begin{aligned}\rho_0 = A_1\left[p_A^{(1)}p_B^{(1)} + p_A^{(2)}p_B^{(2)} - 8(\varepsilon_{AB}^{12} + \varepsilon_{BA}^{12})\right] + \\ + A_2\left[p_A^{(1)}p_C^{(1)} + p_A^{(2)}p_C^{(2)} - 8(\varepsilon_{AC}^{12} + \varepsilon_{CA}^{12})\right] + \\ + A_3\left[p_B^{(1)}p_C^{(1)} + p_B^{(2)}p_C^{(2)} - 8(\varepsilon_{BC}^{12} + \varepsilon_{CB}^{12})\right],\end{aligned} \tag{38.19}$$

where $\varepsilon_{\alpha\alpha'}^{LL'} = \varepsilon_{\alpha\alpha'}^{LL'}(\rho_1)$, noting that $\varepsilon_{\alpha\alpha'}^{LL'} = \varepsilon_{\alpha'\alpha}^{L'L}$. If correlation in the alloy is unimportant, we may set $\varepsilon_{\alpha\alpha'}^{LL'}=0$, and Eq. (38.19) will yield

$$\begin{aligned}\rho_0 = A_1(p_A^{(1)}p_B^{(1)} + p_A^{(2)}p_B^{(2)}) + A_2(p_A^{(1)}p_C^{(1)} + p_A^{(2)}p_C^{(2)}) + \\ + A_3(p_B^{(1)}p_C^{(1)} + p_B^{(2)}p_C^{(2)}).\end{aligned} \tag{38.20}$$

In a similar manner, for alloys with a face-centered cubic $AuCu_3$ type lattice and taking into account correlation in the first coordination sphere, we obtain from (38.12)

$$\begin{aligned}\rho_0 = A_1 [p_A^{(1)} p_B^{(1)} + 3p_A^{(2)} p_B^{(2)} - 12 (\varepsilon_{AB}^{12} + \varepsilon_{BA}^{12} + 2\varepsilon_{AB}^{22})] + \\ + A_2 [p_A^{(1)} p_C^{(1)} + 3p_A^{(2)} p_C^{(2)} - 12 (\varepsilon_{AC}^{12} + \varepsilon_{CA}^{12} + 2\varepsilon_{AC}^{22})] + \\ + A_3 [p_B^{(1)} p_C^{(1)} + 3p_B^{(2)} p_C^{(2)} - 12 (\varepsilon_{BC}^{12} + \varepsilon_{CB}^{12} + 2\varepsilon_{BC}^{22})].\end{aligned} \tag{38.21}$$

Neglecting correlation, the expression for residual electrical resistivity of these alloys becomes

$$\begin{aligned}\rho_0 = A_1 (p_A^{(1)} p_B^{(1)} + 3p_A^{(2)} p_B^{(2)}) + A_2 (p_A^{(1)} p_C^{(1)} + 3p_A^{(2)} p_C^{(2)}) + \\ + A_3 (p_B^{(1)} p_C^{(1)} + 3p_B^{(2)} p_C^{(2)}).\end{aligned} \tag{38.22}$$

For ternary alloys just as for binary alloys, the role of correlation in the formulas given above is somewhat overestimated.

In conclusion we note that the many-electron theory leads to the same dependence of the residual electrical resistivity on composition and order parameters as the single-electron theory.

Chapter VIII

Magnetic, Galvanomagnetic, Optical and Mechanical Properties of Alloys

39. Magnetic Galvanomagnetic and Optical Properties of Ordered Alloys

As mentioned in the first chapter, changes in composition and order in the configuation of atoms on the crystal lattice sites greatly influence the magnetic and galvanomagnetic properties of alloys. Let us consider first several theories of ferromagnetic alloys. The most thoroughly developed subjects in this theory are the dependence of Curie temperature and saturation magnetization (and consequently also the mean atomic magnetic moments) on composition and order parameters of the alloy. A qualitative explanation of the fractional nature of the mean atomic magnetic moments in alloys is given by Dorfman [299]. In the series of papers by Stoner, Slater, Pauling and others [300-304] ferromagnetic alloys have been examined on the basis of the one-electron model of ferromagnetism. A more systematic theory of ferromagnetism of alloys has been constructed on the basis of the many-electron model of a crystal. Rudnitskiy [305] has pointed out the possibility of extending the theory of ferromagnetism of pure metals to alloys by replacing the exchange integral by its mean value. Vonsovskiy [306, 140] has determined the dependence of Curie temperature θ on the composition of the alloy and the degree of order, and has investigated the temperature dependence of the magnetization of the alloy. A simple derivation of the concentration dependence of θ in disordered and ordered alloys has been given by Komar [123], who has also made a detailed comparison of theory with experimental data. On the basis of the assumption concerning the additive effect of neighboring atoms on a given atom, Akulov [307] has considered the dependence of the saturation magnetization and Curie temperature on both composition and order. A quantum mechanical explanation of the fractional nature of the mean atomic magnetic moments, both for pure ferromagnetic metals and alloys, has been given by Vonsovskiy and Vlasov [308].

We shall first consider the dependence of Curie temperature of a ferromagnetic alloy on both composition and degree of order and also the dependence of the magnetization of the alloy on temperature, using the work of Vonsovskiy [306] as a basis. In this paper, the calculations were made using the exchange model of a ferromagnetic substance [309, 310]. In the calculation of the partition function near the Curie temperature θ, the energy spectrum of the system for a given magnetization can be approximated by a discrete energy level—by a "center of gravity" of energy. In this approximation, as is well known [140], the temperature dependence of the spontaneous magnetization M of a pure ferromagnetic metal (within one domain) is determined by the equation

$$M = M_0 \tanh\left(\frac{M}{M_0}\frac{Az}{2kT}\right), \tag{39.1}$$

where M_0 is the spontaneous saturation magnetization (at the absolute zero of temperature), A is the exchange integral for a pair of nearest neighbor atoms and z is the coordination number. As shown by Vonsovskiy, Eq. (39.1) remains valid also for ferromagnetic alloys, if the exchange integral A is replaced by its mean value $\bar{A}$ over all pairs of neighboring atoms of the alloy

$$M = M_0 \tanh\left(\frac{M}{M_0}\frac{\bar{A}z}{2kT}\right), \tag{39.2}$$

where

$$\bar{A} = \frac{2}{zN}(N_{AA}A_{AA} + N_{BB}A_{BB} + N_{AB}A_{AB}), \tag{39.3}$$

A_{AA}, A_{BB} and A_{AB} are exchange integrals for pairs of neighboring atoms AA, BB and AB in the alloy $A-B$.

Near the Curie point (as $T \to \theta - 0$), when $\frac{M}{M_0} \ll 1$, in the expansion of $\tanh\left(\frac{M}{M_0}\frac{\bar{A}z}{2kT}\right)$ one may retain only the first terms. Retaining only the first term of the expansion, we find an expression for the Curie temperature

$$\theta = \frac{\bar{A}z}{2k} = \frac{1}{kN}(N_{AA}A_{AA} + N_{BB}A_{BB} + N_{AB}A_{AB}). \tag{39.4}$$

Substituting (14.7) and (14.8) for N_{AA} and N_{BB}, we obtain

$$\theta = \frac{1}{2kN}\,[z\,(N_A A_{AA} + N_B A_{BB}) + N_{AB}\,(2A_{AB} - A_{AA} - A_{BB})]. \quad (39.5)$$

Expressing N_A, N_B, N_{AB} in terms of the atomic concentration c_A, c_B, the degree of long-range order η and the correlation parameters ε_{AB}^{11}, ε_{AB}^{12}, ε_{AB}^{22} and noting the relation between the number of pairs of atoms of a definite type in the probabilities of the appearance of these pairs, and also taking into account Eqs. (1.11) and (1.5), after simple transformations, we obtain

$$\theta = \frac{1}{2k}\{z\,(c_A A_{AA} + c_B A_{BB}) + [zc_A c_B + \nu(1-\nu)(z - z_{11} - z_{22})\,\eta^2 + $$
$$+ 2\nu z_{12}\varepsilon_{AB}^{12} + \nu z_{11}\varepsilon_{AB}^{11} + (1-\nu)\,z_{22}\,\varepsilon_{AB}^{22}]\,(2A_{AB} - A_{AA} - A_{BB})\}, \quad (39.6)$$

where z_{11}, z_{12} are the respective numbers of sites of the first and second types neighboring on a given site of the first type, z_{22} is the number of sites of second type neighboring on a site of the second type.

In the particular case of disordered alloys with a random distribution of atoms on the lattice sites (when the degree of long-range order and all the correlation parameters are equal to zero), as follows from (39.6), the Curie temperature can be determined by

$$\theta = \frac{z}{2k}(c_A^2 A_{AA} + c_B^2 A_{BB} + 2c_A c_B A_{AB}) =$$
$$= \frac{z}{2k}[A_{BB} + 2\,(A_{AB} - A_{BB})\,c_A - (2A_{AB} - A_{AA} - A_{BB})\,c_A^2]. \quad (39.7)$$

Thus, in this case θ is a function of the square of the concentration c_A. Such dependence of θ on c_A (from which we obtain a linear relation when $2A_{AB} - A_{AA} - A_{BB} \approx 0$) is observed experimentally for many alloys (see Sec. 9, where the solid line in Figs. 70-75 illustrates the theoretical curve for disordered alloys).

If short-range order is significant in disordered alloys (where only one type of site occurs), the formula for θ becomes

$$\theta = \frac{z}{2k}[c_A^2 A_{AA} + c_B^2 A_{BB} + 2c_A c_B A_{AB} + \varepsilon_{AB}\,(2A_{AB} - A_{AA} - A_{BB})]. \quad (39.8)$$

It follows from (39.8) and (39.7) that in ordered alloys (where $\varepsilon_{AB} > 0$) the establishment of short-range order (for example, as a result of high-temperature annealing) should lead to an increase of the Curie temperature, if the concentration curve $\theta(c_A)$ is convex away from the abscissa. In disordered alloys (where $\varepsilon_{AB} < 0$) an inverse type of relation should exist between the influence of short-range order on θ and the direction of convexity of the curve.

For ordered alloys with a β-brass type crystal lattice, $\nu = \frac{1}{2}$, $z_{11} = z_{22} = 0$, $z_{12} = z$ and the expression for the Curie temperature becomes

$$\theta = \frac{z}{2k}\Big[c_A^2 A_{AA} + c_B^2 A_{BB} + 2c_A c_B A_{AB} + \left(\frac{1}{4}\eta^2 + \varepsilon_{AB}\right)(2A_{AB} - A_{AA} - A_{BB})\Big]. \quad (39.9)$$

For alloys with a $AuCu_3$ type crystal lattice, in which $\nu = \frac{1}{4}$, $z_{11} = 0$, $z_{12} = z = 12$, $z_{22} = 8$, from Eq. (39.6) we obtain

$$\theta = \frac{z}{2k}\Big[c_A^2 A_{AA} + c_B^2 A_{BB} + 2c_A c_B A_{AB} + \left(\frac{1}{16}\eta^2 + \frac{1}{2}\varepsilon_{AB}^{12} + \frac{1}{2}\varepsilon_{AB}^{22}\right)(2A_{AB} - A_{AA} - A_{BB})\Big]. \quad (39.10)$$

For alloys with a AuCu type crystal lattice, where $\nu = \frac{1}{2}$, $z_{11} = z_{22} = 4 = \frac{z}{3}$, $z_{12} = 8 = \frac{2z}{3}$,

$$\theta = \frac{z}{2k}\Big[c_A^2 A_{AA} + c_B^2 A_{BB} + 2c_A c_B A_{AB} + \left(\frac{1}{12}\eta^2 + \frac{2}{3}\varepsilon_{AB}^{12} + \frac{1}{6}\varepsilon_{AB}^{11} + \frac{1}{6}\varepsilon_{AB}^{22}\right)(2A_{AB} - A_{AA} - A_{BB})\Big]. \quad (39.11)$$

It is apparent from (39.9), (39.10) and (39.11) that the establishment of long-range order, just as in the establishment of short-range order in a disordered alloy, should lead to an increase of the Curie temperature, if the curve of the dependence of θ on c_A in the disordered alloy becomes convex away from the abscissa; and to a decrease of Curie temperature, if this curve becomes convex toward the abscissa. As is apparent from Figs. 70, 74, 75, this conclusion is supported by experimental data on the dependence of θ on c_A and η in the alloys Ni-Mn, Fe-Pt and Fe-Pd. In the alloy Ni-T (see Fig. 73) the increase of θ during ordering is comparatively small. This occurs in accordance with the fact that the coefficient of c_A^2 in the dependence of the Curie temperature of the disordered alloy on c_A, as is apparent from Fig. 73 (θ is almost a linear function of c_A) is small.

It follows from these formulas for θ, that for definite values of the exchange integrals the ordering may lead to the emergence of ferromagnetism in an alloy which was not ferromagnetic in the disordered state. An example of such an alloy is Ni_3Mn. The vanishing of ferromagnetism during ordering is also possible.

The dependence of the rate $\frac{M}{M_0}$ on temperature, composition of the alloy and order parameters may be determined from Eq. (39.2), if we note that $\bar{A}=\frac{2k\theta}{z}$ and use the formulas given above for θ. In particular, for disordered alloys where short-range order is also absent, it follows from (39.2) and (39.7) that $\frac{M}{M_0}$ changes with temperature according to the same law as in the case of pure metals [see (39.1)]. In the more general case, owing to the fact that the degree of long-range order and the correlation parameters depend on temperature, the temperature dependence of the equilibrium values of $\frac{M}{M_0}$ becomes more complex.

At low temperatures, where $T \ll \theta$, Eq. (39.1) even in the case of pure metals leads to a qualitatively incorrect dependence of $\frac{M}{M_0}$ on T. In this temperature region the correct temperature dependence of magnetization of pure metals is obtained by the Bloch theory [311]. This theory was extended to the case of alloys by Vonsovskiy [306]. Just as in the case of pure metals, the deviation from saturation magnetization at sufficiently low temperatures was also found to be proportional to $T^{3/2}$ (if it is assumed that the order parameters in this temperature range are independent of temperature).

We shall now consider the dependence of the mean atomic magnetic moment of ferromagnetic alloys on composition and degree of long-range order, following the work of Vonsovskiy and Vlasov [308]. As is well known, the ratio of the saturation magnetization M_0 to the total number of atoms per unit volume is called the mean atomic magnetic moment μ. In [308] the mean atomic magnetic moments have been determined with the help of the $s-d$ exchange model of a ferromagnetic substance, proposed by Vonsovskiy [312]. In this model, owing to the exchange interaction of the inner d electrons and the outer electrons, the latter are magnetized and contribute to the magnetization of the crystal. Consequently, even for pure metals the mean atomic magnetic moments must be nonintegral. The free energy of a ferromagnetic substance has been computed by assuming that the interaction between the outer and inner electrons is small. Then, from the condition of the minimum free energy with respect to variation of magnetization of the inner and outer electrons, it was possible to find the equilibrium value of magnetization and, consequently, the mean atomic magnetic moment.

To determine the exchange energy in the interaction between the outer electron and the inner electrons the method presented in Chapter V was employed [266]. According to this method, in calculating the wave functions and electron energies, the disordered alloy is replaced by a pure metal, consisting of effective atoms with a mean potential energy of the electron, and the partially ordered alloy is replaced by a completely ordered alloy consisting of appropriately selected effective atoms. The calculation of the mean atomic magnetic moment in the case of disordered alloys yields the following formula:

$$\mu = \overline{\mu}_d + \varkappa \overline{\mu}_s, \tag{39.12}$$

where $\overline{\mu}_d$ and $\overline{\mu}_s$ are the mean atomic moments of the inner and outer electrons, respectively, calculated from the displacement rule, and $\varkappa$ is the ratio of two quadratic functions of c_A:

$$\varkappa = \frac{\alpha + \beta c_A + \gamma c_A^2}{\alpha' + \beta' c_A + \gamma' c_A^2}. \tag{39.13}$$

The quantities $\alpha, \beta, \gamma, \alpha', \beta', \gamma'$ in this approximation are independent of composition. Thus, the mean atomic magnetic moment of the alloy cannot be computed from a simple displacement rule, which is in agreement with experimental findings [140].

For ordered alloys with stoichiometric composition and a β-brass type body-centered cubic lattice, the following formula has been obtained for the parameter $\varkappa$, which, according to (39.12), determines the mean atomic magnetic moment

$$\varkappa = A + B\eta^2. \tag{39.14}$$

It should be mentioned that the fact that the mean atomic magnetic moments for pure metals are nonintegral and the deviation of their concentration dependence from a straight line (corresponding to the displacement rule) may be due not only to $s-d$ exchange interaction, but also to other causes.

In [308] the energy spectrum of the d electron in an ordered ferromagnetic alloy with a β-brass type crystal lattice has also been examined. Just as for nonferromagnetic alloys considered in Chapter V, a discontinuity appears here in the energy spectrum during ordering. However, for ferromagnetic alloys the magnitude of the discontinuity is different for electrons with "right" and "left" spins and depends linearly on the magnetization of the s electrons.

We shall now consider nonferromagnetic alloys.

The subject of the effect of ordering of alloys on the Hall constant has been qualitatively examined in [264, 263] by means of the nearly-free electron approximation. In the tight binding approximation, which enables us to obtain a stronger effect of ordering on the Hall effect, the variation of the Hall constant during ordering of the alloy has been considered in [313]. Thus, the original formula showed the general definition of the Hall constant [262]

$$R=-\frac{4\pi^3}{ec'}\frac{I}{I_x I_y}, \tag{39.15}$$

where e is the absolute value of the electron charge ($e>0$), c' is the velocity of light and

$$I=\int\limits_S \frac{\partial E}{\partial k_y}\left(\frac{\partial E}{\partial k_y}\frac{\partial^2 E}{\partial k_x^2}-\frac{\partial E}{\partial k_x}\frac{\partial^2 E}{\partial k_x \partial k_y}\right)\frac{dS}{|\operatorname{grad}_k E|}, \tag{39.16}$$

$$\left.\begin{aligned} I_x &= \int\limits_S \left(\frac{\partial E}{\partial k_x}\right)^2 \frac{dS}{|\operatorname{grad}_k E|}, \\ I_y &= \int\limits_S \left(\frac{\partial E}{\partial k_y}\right)^2 \frac{dS}{|\operatorname{grad}_k E|}. \end{aligned}\right\} \tag{39.17}$$

Here E is the energy of a conduction electron, which is a function of the components of the wave vector k_x, k_y, k_z, and $\operatorname{grad}_k E$ is taken over the wave vector space and the integration is extended over the surface S, filled by the electron region in the k-space (over the Fermi surface). Thus, the magnetic moment is parallel to the z axis, the electric field lies in the xy plane and current flows along the x axis.

For alloys with a β-brass type body-centered lattice, in which the number of conduction electrons is independent of composition, calculations show that the Hall constant should not depend on either composition or the degree of long-range order, i.e., it should have the same value in both the disordered and completely ordered states. For several alloy structures, the Hall constant is strongly dependent on both c_A and η. For example, in alloys with $AuCu_3$ type face-centered cubic lattice, the ordering may lead to a change of sign of R. In the disordered state $R < 0$, i.e., the Hall effect is normal, but in a completely ordered state $R > 0$, i.e., ordering leads to the appearance of an anomalous Hall effect. These results are in agreement with the experimental data of Frank, presented in Sec. 9, for β-brass, which has a comparatively slight change of R during ordering (see Fig. 80, where one must eliminate the temperature dependence of R which is not related to ordering), and also the data of Komar and Sidorov, which revealed a change in the sign of R during ordering of a $AuCu_3$ alloy (Fig. 79).

In a similar manner one may find an explanation [314, 315] of the experimental results presented in Sec. 9 pertinent to the relative change of the electrical resistivity $\frac{\Delta\rho}{\rho^0}$ in a weak transverse magnetic field as a function of c_A and η. It is well known that in pure metals $\frac{\Delta\rho}{\rho^0}$ must be proportional to the square of the ratio $\frac{H}{\rho^0}$.

$$\frac{\Delta\rho}{\rho^0} \sim \left(\frac{H}{\rho^0}\right)^2, \tag{39.18}$$

where H is the magnetic field intensity. In ordered alloys, with a change in the degree of long-range order there is the possibility of a departure from relation (39.18), due to the dependence of the electron energy spectrum on c_A and η. This departure has been detected in the experiments of Komar, who has shown that in $AuCu_3$ alloys the variation $\frac{\Delta\rho}{\rho^0}$ during ordering is stronger than predicted by formula (39.18), if we note the dependence of the electrical resistivity in the absence of a magnetic field on η. The observed maximum on the curve of $\frac{\Delta\rho}{\rho^0}$ versus c_A for ordered Au-Cu alloys close to composition corresponding to $AuCu_3$ (see Fig. 81), is associated to an appreciable extent with the existence of a minimum ρ_0 in this composition.

We shall now consider the optical properties of alloys [360]. To investigate the reflecting and absorbing ability of alloys in the infrared spectral region, the well-known theory of Drude-Zener [295] can be extended to the case of alloys. In this theory we neglect the optical quantum transitions of electrons and take into account only the accelerating action of the slowly varying field of a light wave, as a result of which we obtain the following formulas for the specific electrical resistivity σ and the dielectric constants ε:

$$\sigma = \frac{N_e e^2 \Gamma}{m_* (\omega^2 + \Gamma^2)}, \tag{39.19}$$

$$\varepsilon = 1 - \frac{4\pi\sigma}{\Gamma}. \tag{39.20}$$

Here N_e is the number of conduction electrons, $\Gamma = \frac{1}{\tau}$, where τ is the relaxation time for the conduction electrons, m_* is their effective mass and ω is the frequency of the light waves. In the case of binary disordered alloys, in which the electrical resistivity (for constant current) obeys the Matthiessen rule and correlation is unimportant, we may assume that

$$\Gamma = c_A \Gamma_A + c_B \Gamma_B + c_A c_B \Gamma', \tag{39.21}$$

where Γ_A and Γ_B are the reciprocal relaxation times in pure metals A and B, and Γ' is a coefficient that is independent of composition. The first two terms in (39.21) are due to scattering by lattice vibrations, and the last term corresponds to the residual electrical resistance.

Substituting (39.21) into (39.19) and (39.20), we find for disordered alloys

$$\sigma = \frac{N_e e^2 (c_A \Gamma_A + c_B \Gamma_B + c_A c_B \Gamma')}{m_* [\omega^2 + (c_A \Gamma_A + c_B \Gamma_B + c_A c_B \Gamma')^2]}, \tag{39.22}$$

$$\varepsilon = 1 - \frac{4\pi\sigma}{c_A \Gamma_A + c_B \Gamma_B + c_A c_B \Gamma'}. \tag{39.23}$$

The effective mass m_* in (39.22) may be assumed as independent of composition.

Using familiar relations of the optics of metals (see, for example, [317]), with the help of (39.22) and (39.23) we can determine the dependence of the reflection coefficient and the absorption coefficient on the concentration c_A. Thus, for example, the calculation carried out in [316] for the reflection coefficient of the alloy Ag–Au in the infrared region, has shown that a minimum should occur near the point $c_A = \frac{1}{2}$ on the curve of the concentration dependence of this coefficient.

For binary ordered alloys investigated at sufficiently low temperatures, when the scattering by lattice vibrations is negligible and correlation in the alloy is insignificant, the relaxation time Γ^{-1} is determined by the formula [compare (37.9)]

$$\Gamma = [c_A c_B - \nu(1-\nu)\eta^2]\Gamma'. \tag{39.24}$$

In this case the formulas for σ and ε take the form

$$\sigma = \frac{N_e e^2 [c_A c_B - \nu(1-\nu)\eta^2]\Gamma'}{m^* \{\omega^2 + [c_A c_B - \nu(1-\nu)\eta^2]^2 \Gamma'^2\}}, \tag{39.25}$$

$$\varepsilon = 1 - \frac{4\pi\sigma}{[c_A c_B - \nu(1-\nu)\eta^2]\Gamma'}, \tag{39.26}$$

where the effective mass m^* may be assumed to be independent of the concentration of the components and of the degree of order.

From expressions (39.25) and (39.26), giving the dependence of σ and ε on c_A and η, it is apparent that changes of slope should be observed at the critical temperature on the curves of the optical constants versus the annealing temperature of the alloy at second-order phase transition points, and step-changes should be observed at first-order phase transition points.

The effect of ordering on the photoeffect in alloys with a β-brass type body-centered cubic lattice has been theoretically investigated by Sokolov [318]. It was discovered that ordering should lead to a displacement of the red boundary of the photoeffect. A break should be observed in the curve of the temperature dependence of the photoelectric current in alloys with this structure.

40. Mechanical Properties of Ordered Alloys

From the experimental data given in Sec. 9, it is apparent that ordering exerts a strong influence on the elastic properties of alloys.

The theory of the elastic properties of β-brass type ordered alloys has been developed by Samoylovich [319] for the case of stoichiometric alloys. Orlov made calculations for disordered alloys [320] and for the general case of ordered alloys of any composition [321], using a more accurate expression for the energy of a deformed crystal.

In these papers the determinations of the moduli of elasticity $c_{11}-c_{12}$ and c_{44} associated with the deformations that are not accompanied by volume changes (longitudinal elongation accompanied by a transverse compression, and shear). As is well known, the moduli c_{ij} are coefficients which determine the dependence of the free energy of a deformed crystal on the components e_{mn} of the deformation tensor. For a cubic crystal this dependence has the form

$$F' = F^0 + \frac{1}{2} c_{11}\left(e_{xx}^2 + e_{yy}^2 + e_{zz}^2\right) + c_{12}\left(e_{xx}e_{yy} + e_{xx}e_{zz} + e_{yy}e_{zz}\right) + \\ + \frac{1}{2} c_{44}\left(e_{xy}^2 + e_{xz}^2 + e_{yz}^2\right), \qquad (40.1)$$

where F' is the free energy per unit volume and F^0 is independent of e_{mn}. From (40.1) it follows that the magnitude c_{44} is expressed in terms of the second derivative of the free energy with respect to e_{xy} in the following manner:

$$c_{44} = \frac{\partial^2 F'}{\partial e_{xy}^2}. \qquad (40.2)$$

To find the difference $c_{11}-c_{12}$, it is necessary to consider the deformation, during which the crystal is elongated along one of the cubic axes (ox) and is compressed along the two other axes (oy and oz) in such a manner that its volume remains unchanged. Thus, obviously,

$$e_{yy} = e_{zz} = -\frac{1}{2} e_{xx}.$$

Substituting $-\frac{1}{2} e_{xx}$ for e_{yy} and e_{zz} in (40.1), and differentiating with respect to e_{xx}, we find

$$c_{11} - c_{12} = \frac{2}{3}\left(\frac{\partial^2 F'}{\partial e_{xx}^2}\right)_{e_{yy}=e_{zz}=-\frac{1}{2}e_{xx}} \qquad (40.3)$$

In the papers cited, the dependence of entropy of the components of the elastic deformation tensor was disregarded and the free energy in formulas (40.2) and (40.3) was replaced by the energy of the elastically deformed crystal E.

From formulas (40.2 and (40.3) it is apparent that a calculation of the quantity c_{44} and $c_{11}-c_{12}$ requires a determination of the energy of a deformed crystal during both a shear deformation and a longitudinal elongation accompanied by a transverse compression (with constant volume). Then, introducing the notation φ for the shear deformation and ε for a relative elongation in the second deformation, the quantities $c_{11}-c_{12}$ and c_{44} may be calculated from

$$c_{11}-c_{12}=\frac{2}{3}\frac{d^2E}{d\varepsilon^2}, \qquad c_{44}=\frac{d^2E}{d\varphi^2}. \tag{40.4}$$

In [319-321] a calculation of the energy of a deformed crystal was made on the basis of the model of an alloy employed by Mott [322]. In this model (in contrast with the model adopted in Chapter III) the energy of the crystal does not reduce to a sum of the interaction energies of pairs of atoms, and explicitly takes into account the different forms of interaction energy in the system comprised of ions and conduction electrons.

The conduction electrons are considered as a gas of Fermi particles filling the crystal lattice.

In β-brass the sites of the crystal lattice are occupied by Cu^+ and Zn^{++} ions of different valence. If the conduction electrons are uniformly distributed over the β-brass stoichiometric crystal (assuming that the copper atom gives one electron to the conduction band, and the zinc atom—two), in the Wigner-Seitz polyhedra (cells) surrounding the Zn^{++} and Cu^+ ions there should be found charges equal to $\frac{1}{2}e$ and $\frac{1}{2}e$ respectively (e is the absolute value of the electron charge). As a consequence there arises in the crystal an ionic bond leading to an electrostatic interaction energy. The electrostatic interaction energy of point ions arranged with complete order in the sites of the undeformed β-brass type body-centered crystal lattice, is defined by

$$E_e=-\frac{\alpha Q^2}{2a}, \tag{40.5}$$

where Q is the ionic charge, a is the lattice constant and α is the Madelung constant, which is equal to 2.035 for the structure under consideration. For $Q|=\frac{1}{2}e$, formula (40.5) leads to a very large electrostatic interaction energy, as a consequence of which the calculated ordering energy is found to substantially exceed the experimentally measured ordering energy of β-brass. This energy, however, is appreciably reduced, if we take into account the fact that the mean density of electric charge is not distributed uniformly in the crystal, but reaches a large value near the ions, while near the Cu^+ ions these values are correspondingly smaller than near the Zn^{++} ions. This type of screening of the ionic charges leads to an appreciable lowering of the electrostatic energy. In [322, 319, 321] this screening was taken into account by means of the Thomas-Fermi method. The Wigner-Seitz polyhedra are approximately replaced by spheres, on the surface of which the potential is equal to zero. As a result it was found that the energy, as before, is determined by formula (40.5). However, Q should not be replaced in this formula by $\pm\frac{1}{2}e$, but by a finite value of charge equal in the case of a completely ordered stoichiometric alloy to

$$Q' = \pm\frac{e}{2}\frac{q_э r_0}{\sinh q_э r_0}, \tag{40.6}$$

where r_0 is the radius of the selected sphere and $q_э$ is a screening constant

$$q_э^2 = \frac{4m^* e^2}{h^2}\left(\frac{3n_0}{\pi}\right)^{1/3}, \tag{40.7}$$

m^* is the effective mass of the conduction electron and n_0 is the number of conduction electrons per unit volume.

Using the value $\frac{1}{q_э} =$ (0.37) A calculated by Mott, for completely ordered β-brass, we find

$$Q' = \pm 0.075\ e. \tag{40.8}$$

In the case of a partially ordered stoichiometric alloy, Q was replaced by $\bar{Q}$, equal to the mean screening charge for sites of each type

$$\bar{Q} = \eta Q'. \tag{40.9}$$

For nonstoichiometric alloys the number of conduction electrons and, consequently, $q_э$ and Q' depend on the concentration of the component of the alloy.

Besides the electrostatic energy, the interaction of the electronic shells of ions plays an essential role in an undeformed crystal. The energy of this interaction in the nearest neighbor approximation may be determined with the help of a formula, similar to (14.6), for the energy in the statistical model of an alloy

$$E_i = -N_{AA}\chi_{AA} - N_{BB}\chi_{BB} - N_{AB}\chi_{AB}, \tag{40.10}$$

where χ_{AA}, χ_{BB} and χ_{AB} are the interaction energies of the electronic shells of pairs of neighboring ions AA, BB and AB, which in this approximation are independent of composition and order, while the number of pairs N_{AA}, N_{BB} and N_{AB} do depend on c_A, η and the correlation parameters. The ordering energies of β-brass calculated with the help of formulas (40.5), (40.8)-(4.10) turned out to be of the same order of magnitude but somewhat larger than the ordering energy found experimentally.

To determine the quantity $c_{11} - c_{12}$ and c_{44} with the help of the same model of an alloy, the energy of the deformed crystal was determined for the deformation indicated above. As a consequence of the fact that such deformations are accompanied by a displacement of the lattice sites, there is a change in the previously examined electrostatic energy and energy of overlap of the ionic shells. Deformation of the crystal leads to the fact that the spheres which have replaced the Wigner-Seitz polyhedra are transformed into ellipsoids. Thus, opposite charges arise in the surface layers of the ellipsoids, causing the ellipsoids to acquire quadrupole moments. Moreover, the total charges of the ellipsoids change during deformation. Consequently, the energy of a unit volume of the deformed crystal may be represented in the form

$$E = E_e + E_i + E_Q + E_{eQ} + E_s, \tag{40.11}$$

where E_e is the interaction energy of charges of the atomic spheres, E_i is the interaction energy of the inner electron shells, E_Q is the mutual interaction energy of the quadrupole moments, E_{eQ} is their interaction energy with charges and E_s is the energy of the ellipsoids. These energies have been calculated for a deformed crystal, and their dependence on composition and degree of long-range order has been found.

In the calculation of the second derivatives of the energy of the deformed crystal with respect to deformation [319, 321] the dependence of the degree long-range order on deformation was neglected. The expressions obtained by this method for the moduli c_{ij} are, in general, applicable to the case where these quantities are measured by means of oscillation at high frequencies. Here, the period of oscillation is much less than the relaxation time of ordering, and the degree of long-range order cannot change during the deformation. These results are also approximately valid if the degree of long-range order is a weak function of the deformation. Taking into account this dependence should lead to a discontinuous change of certain elasticity constants at the temperature T_0 of the second-order phase transition.

As a result of calculations [321], the following expressions have been obtained for the elasticity moduli $c_{11}-c_{12}$ and c_{44}:

$$\left.\begin{aligned} c_{11}-c_{12} &= A+A'\eta^2, \\ c_{44} &= B+B'\eta^2. \end{aligned}\right\} \tag{40.12}$$

Here A, A', B and B' are coefficients independent of the degree of long-range order η, but dependent on temperature, composition of the alloy, and the correlation parameters.

Thus, the elastic constants are quadratic functions of the degree of long-range order. It should be noted that the results obtained are of a qualitative character, because in the calculation of the energy of the deformed crystal we have neglected the change of the energy of the conduction electrons caused by the change in their energy spectrum during deformation. As has already been mentioned in Chapter II, the magnitude of η^2 at small η is proportional to the difference T_0-T. Therefore, at the transition point from the disordered state (where $\eta\equiv 0$) into the ordered state, the curves of the temperature dependence of c_{11}, c_{12} and c_{44} should show a break (with the previously specified method of measurement of the moduli by means of high-frequency vibration). This break is due to the appearance of additional terms proportional to T_0-T at $T<T_0$. Along with the moduli c_{ij} one often uses the elastic coefficients s_{ij}, which are the expansion coefficients of the elastic part of the free energy with respect to the stress tensor components. For cubic crystals the quantities $s_{11}-s_{12}$ and s_{44} are related to $c_{11}-c_{12}$ and c_{44} by the simple relations

$$s_{11}-s_{12}=\frac{1}{c_{11}-c_{12}}, \quad s_{44}=\frac{1}{c_{44}}. \tag{40.13}$$

Since at small values of η the expressions for $s_{11}-s_{12}$ and s_{44}, obtained from formulas (40.12) and (40.13), can be expanded in a power series in η and can be restricted to quadratic terms, these quantities, and also the Young's moduli, shear moduli, etc., that are related to c_{ij} or s_{ij} contain quadratic terms in η

near the transition temperature (but not linear terms for β-brass type alloys), i.e., linear terms in $T_0 - T$. Therefore, breaks should be observed at the critical point on the temperature dependence curves of these quantities. These breaks are very apparent on the experimental curves of the temperature dependence of s_{11}, s_{12}, s_{44}, and of the reciprocal values of Young's modulus (see Figs. 61 and 62). Thus, the temperature dependence in the ordered phase, in accordance with the theoretical conclusion, is well approximated by a linear function of $T_0 - T$. An estimate of the order of magnitude of the changes of $c_{11} - c_{12}$ and c_{44} during ordering, carried out in [319], is also in satisfactory agreement with the experimental data.

Kholodenko [323] has investigated the change of the elastic moduli of alloys during a passage through the second-order phase transition point in terms of the thermodynamic theory of ordering. It has been shown that at the second-order phase transition point some elastic constants experience discontinuous changes. Expressions were found which relate the changes in the elasticity constants to the changes of several other thermodynamic quantities.

Knowing the dependence of the free energy of the crystal on the interatomic distances, one may determine not only the elastic constants, but also the lattice constant. Using the elastic sphere model, Pines [324] has investigated the concentration dependence of the lattice constants of solid solutions. In this model, distortions of the crystal lattice around an impurity atom are defined as distortions caused by an elastic sphere placed in a spherical cavity of a somewhat different radius, cutout from the elastic body. In [324] this method was used to investigate the departures from Vegard's Law, i.e., the linear variation of the lattice parameter with the concentration.

The dependence of the lattice parameter of a binary substitutional alloy with two types of sites in the ordered state on its composition and order parameters has been determined by Aptekar' and Finkel'shtein [325]. These authors used the statistical model of a solid solution in the nearest neighbor approximation. The configurational part of the energy of the alloy (14.6), when subjected to the transformations used in the derivation of formula (39.6) may be given by

$$E_i = -\frac{Nz}{2}\left[c_A v_{AA} + c_B v_{BB} + (c_A c_B + \Lambda) w\right], \qquad (40.14)$$

where the ordering energy w is defined by formula (14.10),

$$\Lambda = \nu(1-\nu)\left(1-\frac{z_{11}}{z}-\frac{z_{22}}{z}\right)\eta^2 + 2\nu\frac{z_{12}}{z}\varepsilon_{AB}^{12} + \nu\frac{z_{11}}{z}\varepsilon_{AB}^{11} + (1-\nu)\frac{z_{22}}{z}\varepsilon_{AB}^{22} \tag{40.15}$$

and the notation corresponds to that adopted in (39.6). Thus, in contrast with the assumptions made in the previous sections, the energies v_{AA}, v_{BB} and v_{AB} are regarded not as constants, but as functions of the interatomic separation.

In addition to the energy of pairwise interaction described by formula (40.14) in [325] the other parts of the energy (electron, thermal etc.) whose dependence on composition is assumed to be linear, are also taken into account:

$$E' = c_A E'_A + c_B E'_B, \tag{40.16}$$

where the quantities E'_A, E'_B are the corresponding parts of the energy of pure metals A and B, the number of atoms is equal to the total number of atoms in the alloy. In a similar manner in the expression for the entropy of the alloy, in addition to the configurational term $k \ln W$ (where W is the number of physically distinguishable interchanges of atoms A and B on sites of the first and second type) other parts of the entropy, whose dependence on composition is also assumed to be linear, have been taken into account (for example, that related to the thermal vibrations of the crystal lattice):

$$S = k \ln W + c_A S_A + c_B S_B, \tag{40.17}$$

where S_A and S_B are the entropies of pure metals A and B (with the same number of atoms N as the alloy).

By means of Eqs. (40.14), (40.16) and (40.17) the expression for the free energy of the alloy may be represented as

$$F = (c_A^2 - \Lambda) F_A + (c_B^2 - \Lambda) F_B + 2(c_A c_B + \Lambda) F_{AB} - kT \ln W, \tag{40.18}$$

where

$$\left.\begin{aligned} F_A &= -\frac{Nz}{2} v_{AA} + E'_A - TS_A, \\ F_B &= -\frac{Nz}{2} v_{BB} + E'_B - TS_B, \\ F_{AB} &= -\frac{Nz}{2} v_{AB} + \frac{E'_A + E'_B}{2} - T\frac{S_A + S_B}{2} = \frac{F_A + F_B}{2} - \frac{Nz}{4} w. \end{aligned}\right\} \tag{40.19}$$

The quantities F_A, F_B and F_{AB} are functions of the lattice constant a'. These functions take their minimum values at $a' = a_A$,

$a' = a_B$ and $a' = a_{AB}$, respectively. Expansions of F_A, F_B and F_{AB} in powers of $a' - a_A$, $a' - a_B$ and $a' - a_{AB}$ near their minimum values have the form

$$\left.\begin{aligned} F_A(a') &= F_A(a_A) + \frac{9}{2} N\Omega_A^0 K_A \left(\frac{a' - a_A}{a_A}\right)^2, \\ F_B(a') &= F_B(a_B) + \frac{9}{2} N\Omega_B^0 K_B \left(\frac{a' - a_B}{a_B}\right)^2, \\ F_{AB}(a') &= F_{AB}(a_{AB}) + \frac{9}{2} N\Omega_{AB}^0 K_{AB} \left(\frac{a' - a_{AB}}{a_{AB}}\right)^2, \end{aligned}\right\} \tag{40.20}$$

where Ω_A^0, Ω_B^0 and Ω_{AB}^0 are the atomic volumes, which determine the values of the lattice constance a_A, a_B and a_{AB} while K_A, K_B and K_{AB} are the corresponding bulk moduli:

$$K_A = V\left(\frac{\partial^2 F_A}{\partial V^2}\right)_{a'=a_A}, \quad K_B = V\left(\frac{\partial^2 F_B}{\partial V^2}\right)_{a'=a_B}, \quad K_{AB} = \left(\frac{\partial^2 F_{AB}}{\partial V^2}\right)_{a'=a_{AB}} \tag{40.21}$$

(V is the volume of the crystal). If we restrict ourselves to a consideration of solutions with almost almost equal atomic volume Ω_A^0, Ω_B^0 and Ω_{AB}^0, the ratios $\frac{\Omega_A^0}{a_A^2}$, $\frac{\Omega_B^0}{a_B^2}$ and $\frac{\Omega_{AB}^0}{a_{AB}^2}$ in formulas (40.20) may be replaced by a mean ratio $\frac{\Omega^0}{a_0^2}$. Then, substituting expressions (40.20) into (40.19), we obtain a formula determining the dependence of the free energy of a solid solution on the lattice constant:

$$\begin{aligned} F = F_0 + \frac{9}{2} N \frac{\Omega^0}{a_0^2} \{ & c_A^2 K_A (a' - a_A)^2 + c_B^2 K_B (a' - a_B)^2 + \\ & + 2c_A c_B K_{AB} (a' - a_{AB})^2 + \\ & + \Lambda [2K_{AB}(a' - a_{AB})^2 - K_A (a' - a_A)^2 - K_B (a' - a_B)^2]\}, \end{aligned} \tag{40.22}$$

where

$$\begin{aligned} F_0 = (c_A^2 - \Lambda) F_A(a_A) + (c_B^2 - \Lambda) F_B(a_B) + \\ + 2(c_A c_B + \Lambda) F_{AB}(a_{AB}) - kT \ln W. \end{aligned} \tag{40.23}$$

The lattice constant of the crystal corresponding to a given external pressure P, as is well known, is determined from the condition

$$\frac{\partial F}{\partial V} = -P. \tag{40.24}$$

To calculate the lattice constant at atmospheric pressure, which is practically the same as the lattice constant of the crystal at zero pressure, we set $P=0$ in (40.24). Then the lattice constant can be computed from

$$\frac{\partial F}{\partial a'} = 0. \tag{40.25}$$

Substituting (40.22) into (40.25) and solving the resulting equation, we find a formula determining the dependence of the lattice constant on the composition of the alloy and the order parameters

$$a' = a_A + \Delta a',$$

$$\Delta a' = \frac{c_B^2 K_B (a_B - a_A) + 2c_A c_B K_{AB}(a_{AB} - a_A) + \Lambda [2K_{AB}(a_{AB} - a_A) - K_B (a_B - a_A)]}{c_A^2 K_A + c_B^2 K_B + 2c_A c_B K_{AB} + \Lambda (2K_{AB} - K_A - K_B)}. \tag{40.26}$$

Next, taking the second derivative of (40.22) with respect to a', we may find an expression for the bulk modulus of the alloy

$$K = c_A^2 K_A + c_B^2 K_B + 2c_A c_B K_{AB} + \Lambda (2K_{AB} - K_A - K_B). \tag{40.27}$$

It must be emphasized that in the calculation of K we have neglected the dependence of the long-range parameters and correlation parameters on volume. Consequently, Eq. (40.27) is applicable only in those cases when in the measurement of K the specified parameters do not vary with volume. This remark does not apply to the formula for a', since the first derivative $\frac{\partial \eta}{\partial V}$ is multiplied by the derivative $\frac{\partial F}{\partial \eta}$, which is equal to zero at equilibrium. Eqs. (40.26) and (40.27) enable us to describe the variation of the lattice constant and the bulk modulus of a number of alloys with composition and order parameter. It is well known from experiment that in disordered alloys Au-Ag the magnitude of K is a linear function of concentration.

Eq. (40.27) leads to such a relation, if

$$K_{AB} = \frac{1}{2}(K_A + K_B).$$

Taking for K_A, K_B, a_A and a_B the values of the bulk moduli and lattice constants for the pure metals Au and Ag, setting $2a_{AB}-a_A-a_B=-0.022$ Å and restricting ourselves to the case when $\Lambda=0$, we may obtain a relation for the lattice constant of the Ag-Au alloys as a function of composition. This relation is in good agreement with experimental data (Fig. 114) [326].

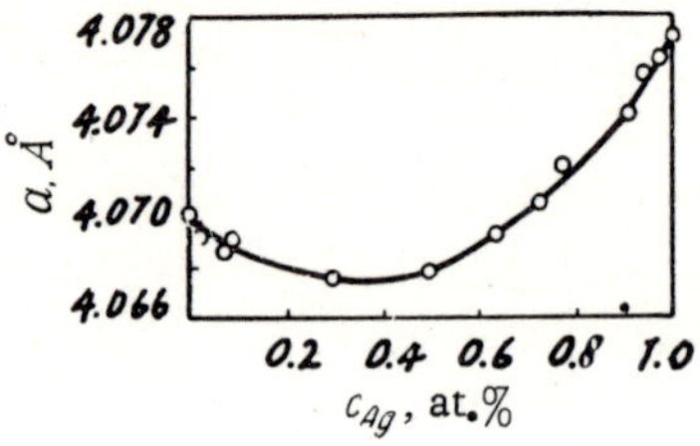

Fig. 114. Lattice constant of the alloy Ag–Au as a function of the concentration of silver. Solid curve-theoretical.

In [325] for the Cu-Au alloy it was assumed that $a_{AB}=\frac{1}{2}(a_A+a_B)$, $K_{AB}=\frac{1}{2}(K_A+K_B)$. Then formula (40.26) reduces to a simpler dependence of a' on c_A and Λ

$$\left.\begin{aligned} a' &= a_A + c_B(a_B - a_A) + A(c_Ac_B - \Lambda), \\ \text{where} \quad A &= \frac{K_B - K_A}{2(c_AK_A + c_BK_B)}(a_B - a_A). \end{aligned}\right\} \qquad (40.28)$$

In order to explain the experimentally observed decrease of the lattice constant of the Cu_3Au alloy during a transition from the disordered state to an almost completely ordered state (see Fig. 83), the value of A was taken to be $A=0.1056$ Å. Then, noting that $a_B-a_A=0.462$ Å, and applying formula (40.28) for the disordered alloy Cu_3Au, in which short-range order is also absent ($\Lambda=0$), we obtain the value $a'=3.7435$ Å which is in good agreement with the experimental value of $a'=3.7427$ Å. It must be noted, however, that the assumption of the equality $K_{AB}=\frac{1}{2}(K_A+K_B)$ does not explain the change of the bulk modulus during ordering. Taking into account the fact that Λ, according to (40.15), is related not only to the degree of long-range order, but also to the correlation parameters, one may determine from Eqs. (40.26) and (40.27) the dependence of the lattice constant and bulk modulus not only on a but on also on the correlation parameters.

Using a similar method, Aptekar' [327] has investigated the stability of CuAu and CuPt crystal lattices with respect to deformations occurring without a change in volume. The dependence of the tetragonal character of a CuAu type lattice on the degree of long-range order has been examined by Wilson [328].

The variation of the lattice constant of an alloy of nontransition metals of cubic structure with composition has also been investigated by Orlov [329], who used a different model of the alloy. In this paper he neglected the dependence of the entropy of the alloy on its volume, and the crystal lattice constant a was determined from the condition minimizing the total energy E per unit volume of the crystal, considered as a function of the lattice constant a' of the deformed crystal:

$$\frac{dE(a')}{da'} = 0. \tag{40.29}$$

The energy of the alloy as a function of its composition has been determined with the help of a simple model similar to that used by Orlov in calculating the elasticity moduli. In this case, however, the conduction electrons were treated as an electron gas uniformly distributed in the alloy. By investigating the dependence of the lattice constant on composition, we were able to show, in particular, the possibility of an increase of the lattice constant of the alloy with the addition of impurity atoms with a smaller ionic radius than the solvent ions. This conclusion agrees with experiment for a number of alloys. For disordered alloys we have explained the experimentally observable deviations from the linear dependence of the lattice constant on the composition of the alloy. It has also been shown that during ordering the lattice constant of a cubic crystal decreases. This result is in agreement with measurements of the lattice constant of the alloy AuCu (see Fig. 83). The concentration dependence of the lattice constant of alloys has also been examined in [331].

As already explained in this section, ordering of an alloy may considerably influence its elastic properties. Consequently, the velocity of propagation of elastic vibrations (sound), which, as is well known, is related to the elastic constants, should also change markedly during ordering. Moreover, ordering may exert a strong influence on the absorption of sound. In fact, the periodic changes of temperature and pressure arising when an elastic wave passes through the crystal lead to a periodic change in the degree of long-range order, which corresponds to a change in temperature and pressure if equilibrium has been established. In real crystals, because of the existence of a finite relaxation time, the deviations of η from the equilibrium value (in the absence of a soundwave) will be retarded with respect to the changes in temperature and pressure. Since a change in the degree of long-range order is accompanied by irreversible processes of approach to the equilibrium value of η (in the case of temperature and pressure changes) caused by displacement of the atoms to new equilibrium positions, energy should be dissipated, i.e., absorption of sound should occur. The dependence of the degree of long-range order on the magnitude of the elastic deformation and the diffuse reaction in ordered alloys was first considered by Gorskiy [332]. This subject has also been considered by Zener [333].

A method for calculating the absorption of the elastic vibrations, due to the relaxation of internal parameters, has been developed by Mandel'shtam and Leontovich [334]. For the case in which relaxation of only one internal parameter is significant, the following expression has been obtained for the length of the wave vector k:

$$k = \omega \sqrt{\frac{1 - i\omega\tau}{c_0^2 - c_\infty^2 i\omega\tau}}. \tag{40.30}$$

Here ω is the frequency of the vibration, τ is the relaxation time, c_0 is the velocity of propagation of the elastic wave under consideration during an "infinitely slow" process ($\omega\tau \ll 1$) when the internal parameter can take its equilibrium value during a time considerably shorter than the period of vibration, c_∞ is the velocity of the wave during an "infinitely fast" process ($\omega\tau \gg 1$) when the internal parameter in general cannot change during a time equal to the period of vibration. It is obvious that during the propagation of elastic vibrations in a single crystal the velocities c_0 and c_∞ depend on the orientation of the direction of propagation relative to the crystal axes.

It is apparent from Eq. (40.30) that for the case in which relaxation of the internal parameter ($\tau \neq 0$) occurs, k is a complex number. This means that a wave propagating through the crystal is attenuated. The absorption coefficient of sound α, determining the weakening of intensity of the elastic wave (which occurs according to the law $e^{-2\alpha x}$), equals the imaginary part of k. For the case when $c_\infty - c_0 \ll c_0$, from (40.30) we obtain the following expression for α:

$$\alpha = \frac{c_\infty - c_0}{c_\infty^2} \frac{\omega^2\tau}{1 + \omega^2\tau^2}. \tag{40.31}$$

Stepanov [335] has used formula (40.31) for an investigation of the influence of relaxation of the degree of long-range order on absorption of sound in β-brass. In particular, he emphasized the possibility of significant dependence of both the frequency (at which the logarithmic damping decrement is a maximum) and of the maximum value of the decrement on the orientation of the propagation of the wave relative to the axes of the single crystal.

Landau and Khalantnikov [336] have investigated in detail the temperature dependence of the absorption coefficient of sound (related to the relaxation mechanism under consideration) near the second-order phase transition point. To determine the temperature dependence of the relaxation time τ a kinetic equation was used

$$\frac{d\eta}{dt} = -\gamma \frac{\partial \Phi}{\partial \eta}. \tag{40.32}$$

It was assumed that near the second-order phase transition point the kinetic coefficient γ does not have a singularity, and the derivative $\frac{\partial \Phi}{\partial \eta}$ is calucated for a given value of η, which differs from the equilibrium value η_p. Expanding $\frac{\partial \Phi}{\partial \eta}$ in a power series in $\left(\frac{\partial \Phi}{\partial \eta}\right)_{\eta=\eta_p}$, using for the thermodynamic potential its expansion (11.2) in powers of η (obtained from the thermodynamic considerations of second-order phase transitions) and also formulas (11.5) and (11.8), we find

$$\frac{\partial \Phi}{\partial \eta} = \left(\frac{\partial \Phi}{\partial \eta}\right)_{\eta=\eta_p} + \left(\frac{\partial^2 \Phi}{\partial \eta^2}\right)_{\eta=\eta_p} (\eta - \eta_p) =$$
$$= (A_2 + 3A_4\eta^2)(\eta - \eta_p) = 2a(T_0 - T)(\eta - \eta_p). \tag{40.33}$$

Substituting (40.33) into (40.32) we obtain:

$$\frac{d\eta}{dt} = -2\gamma a (T_0 - T)(\eta - \eta_p). \tag{40.34}$$

Comparing this equation with the one resulting from (7.1) (where one must replace Q by η),

$$\frac{d\eta}{dt} = -\frac{1}{\tau}(\eta - \eta_p), \tag{40.35}$$

we find that the relaxation time τ is

$$\tau = \frac{1}{2\gamma a (T_0 - T)}. \qquad (40.36)$$

Thus, as we approach the critical temperature T_0 in the case of a second-order phase transition, the relaxation time of the degree of long-range order increases sharply.

In the neighborhood of the second-order phase transition temperature, which occurs in Eq. (40.31), the velocities of sound c_∞ and c_0 can be determined simply. The velocity c_∞ corresponding to a constant value of η during the passage of the wave is, obviously, equal to the velocity of sound in the disordered phase, where $\eta \equiv 0$. The velocity c_0 is equal to the velocity of the elastic wave with a small frequency ($\omega\tau \ll 1$) in the ordered phase and may be calculated from the usual formulas from the equilibrium values of the elastic constants. At sufficiently low frequencies of oscillation, when $\omega\tau \ll 1$ (at temperatures which are not too close to T_0), it becomes apparent from (40.31) and (40.36) that the absorption coefficient α increases in proportion to $\frac{1}{T_0 - T}$ as the second-order phase transition temperature is approached. In this temperature range α is proportional to the square of the vibration frequency. For a given ω, even if ω is small, at temperatures sufficiently close to T_0 the condition $\omega\tau \ll 1$ ceases to be fulfilled and the temperature dependence of α becomes more complicated. For a temperature at which $\omega\tau = 1$, Eq. (40.31) (for a given ω) takes a maximum value

$$\alpha_{\max} = \frac{c_\infty - c_0}{2c_\infty^2}\,\omega. \qquad (40.37)$$

With a further increase of temperature, when $\omega\tau \gg 1$, the absorption coefficient of sound decreases in proportion to $T_0 - T$ and is independent of ω. Lastly, above the transition temperature when $\eta \equiv 0$, the absorption of sound associated with the relaxation mechanism under consideration, in general, does not exist. In the disordered phase (just as in the ordered phase) there exist of course other relaxation mechanisms caused by thermal conductivity, diffusion associated with macrofluxes of atoms of different types arising during a change of pressure and temperature, etc., which also lead to absorption of sound.

It should be emphasized that this discussion of the temperature dependence of relaxation time τ and the absorption coefficient of sound α, during which the kinetic coefficient γ is regarded as constant, is applicable only in the neighborhood of the second-order phase transition point. In fact, in solids one should expect a strong temperature dependence of the quantity γ. Therefore, as we depart from the temperature T_0 where the relative variation $T_0 - T$ is not as great as the relative variation of γ, the temperature dependence of the relaxation time may be determined mainly by the temperature dependence of γ and will be of the exponential rather than of the hyperbolic type. Thus, the relaxation time will not decrease but will increase with a decrease in temperature (passing through a certain minimum). Such an increase of τ according to formula (40.31) for a definite range of frequencies of oscillation may lead to the appearance of a second maximum on the curve of the temperature dependence of α at lower temperatures.

Ordering influences not only the propagation of elastic vibrations (acoustic waves) but also the spectrum of the characteristic (thermal) vibrations of atoms of the crystal lattice. Lifshits and Stepanov [270] have investigated this subject by means of a special method of perturbation theory [337]. An ordered solid solution with two types of sites has been considered, the number of sites of the first type being equal to the number of sites of the second type. On sites with a certain degree of long-range order η are distributed atoms of two types with concentrations c_A and c_B, differing only in their masses, which are equal to m_A and m_B, respectively. The problem of the determination of the spectral density of the vibration frequencies has been solved by means of perturbation theory, in which the relative difference of masses of the atoms was the small parameter:

$$\varepsilon_m = \frac{m_A - m_B}{\bar{m}}, \tag{40.38}$$

where $\bar{m}$ is the mean mass determined by

$$\bar{m} = c_A m_A + c_B m_B.$$

As the zeroth approximation they chose a completely ordered alloy, formed from the effective atoms with the following masses

$$\bar{m}_1 = p_A^{(1)} m_A + p_B^{(1)} m_B = \bar{m} + \frac{1}{2}\eta (m_A - m_B) \tag{40.39}$$

on the sites of the first type and

$$\bar{m}_2 = p_A^{(2)} m_A + p_B^{(2)} m_B = \bar{m} - \frac{1}{2}\eta (m_A - m_B) \tag{40.40}$$

on sites of the second type. The calculations were carried out in the second approximation (taking into account magnitudes of the order ε_m^2) and made possible a determination of the spectral density of the frequencies $g(z)$, where z is the square of the vibration frequency

$$g(z) = g^0(z) + \frac{d}{dz}\xi^0(z). \tag{40.41}$$

Here, $g^0(z)$ is the spectral density of a completely ordered alloy, formed from the effective atoms, and

$$\xi^0(z) = \varepsilon_m^2\left(c_A c_B - \frac{\eta^2}{4}\right) z^2 g^0(z) \fint_0^{z_r^0} \frac{g^0(z')\,dz'}{z - z'}, \tag{40.42}$$

where z_r^0 is the square of the limiting frequency, and $\fint$ denotes the principal value of the integral. The second term in formula (40.41) represents a second-order correction caused by fluctuating inhomogeneities in the arrangement of atoms A and B on the sites of the crystal lattice. Eq. (40.42) enables us to calculate this correction neglecting correlation. We note that dependence of ξ^0 on concentration and degree of long-range order is the same as for the residual electrical resistivity of the alloy [see Eq. (37.11)]. In particular, for a completely ordered alloy $\left(c_A = c_B = \frac{1}{2},\ \eta = 1\right)$ this correction becomes zero and the spectral density is determined only by the first term of formula (40.41).

For partially ordered alloys we shall first consider the spectral density in the zeroth approximation $g^0(z)$, corresponding to a completely ordered alloy of the effective atoms. In this approximation for three-dimensional crystals and for sufficiently small values of ε_m, no forbidden intervals are obtained in the frequency spectrum. However, in several special cases such discontinuities may occur. This occurs, for example, in crystals with a NaCl type cubic lattice in the approximation which accounts only for the interaction between nearest-neighbor atoms. In this case the spectral density in the zeroth approximation $g^0(z)$ has a discontinuity, the width of the forbidden frequency spectrum being determined by

$$\Delta z \sim \varepsilon_m z_0 \eta, \tag{40.43}$$

where z_0 is the square of the frequencies bounding the discontinuity $(z_0 = z_0^-, z_0^+)$. Thus, the width of the discontinuity is proportional to the degree of long-range order, which is similar to the results obtained in Chapter V for the energy spectrum of the conduction electrons of an ordered alloy. The width of the discontinuity Δz is a maximum in a completely ordered alloy $\left(\eta = 1,\ c_A = \frac{1}{2}\right)$. Near values of z_0 equal to z_0^- or z_0^+ bounding the discontinuity, the spectral density $g^0(z)$ in the case under consideration has a singularity of the type

$$g^0(z) \sim \sqrt{\frac{\varepsilon_m \eta z_0}{2\,|z_0 - z|}}. \tag{40.44}$$

The second term in formula (40.41), which takes into account the fluctuation inhomogeneity, is equal to zero for a completely ordered alloy, and increases during disordering. Near the frequency bounding the discontinuity of the zeroth approximation spectrum, the fluctuation terms Δg, according to [270], take the form

$$\Delta g \sim \frac{\varepsilon_m}{\eta}\left(c_A c_B - \frac{\eta^2}{4}\right)\left(\frac{\varepsilon_m \eta z_0}{2\,|z_0 - z|}\right)^{3/2}. \tag{40.45}$$

From Eqs. (40.44) and (40.45) it follows that when

$$\frac{\varepsilon_m \eta z_0}{2\,|z_0 - z|} \sim 1$$

the magnitude of Δg is appreciably smaller than $g^0(z)$, if η is not small (i.e., in this case one may use the zeroth approximation formulas), and may be of the order $g^0(z)$ for $\eta \ll 1$.

References

1. SYKES, C. and H. WILKINSON. J. Inst. Metals, 61:223, 1937.
2. Metals Handbook. Am. Soc. Metals, 1948.
3. HIRABAYASHI. J. Phys. Soc. Japan, 6:129, 1951.
4. FOWLER, R. H. and E. A. GUGGENHEIM, Statistical Thermodynamics. Cambridge University Press, 1939.
5. GORTER, Ye. V. Usp. Fiz. Nauk, 57:279, 1955.
6. BRADLEY, A. J. and J. W. RODGERS. Proc. Roy. Soc., A144:340, 1934.
7. JONES, F. W. and C. SYKES. Proc. Roy. Soc., A161:440, 1947.
8. CHIPMAN, D. and B. E. WARREN. J. Appl. Phys., 21:415, 1941.
9. HOWARTH, F. E. Phys. Rev., 54:693, 1938.
10. LEECH, P. and C. SYKES. Phil. Mag., 27:742, 1939.
11. ELLIS, W. E. and E. S. GREINER. Trans. Am. Soc. Metals, 29:415, 1941.
12. KRIVOGLAZ, M. A. Fiz. Metal. i Metalloved., 1:393, 1955.
13. OWENS, E. A. and I. G. EDMUNDS. Proc. Phys. Soc. (London), 50:389, 1938.
14. MULDAWER, L. J. Appl. Phys., 22:663, 1951.
15. AGEYEV, N. V. and D. N. SHOYKHET. Izv. Sektora Fiz.-Khim. Analiza, 13:65, 1940.
16. AGEYEV, N. V. and D. N. SHOYKHET. Ann. Phys., 23:90, 1935.
17. AGEYEV, N. V. Chemistry of Metal Alloys. Moscow-Leningrad, Izv-vo AN SSSR, 1941.
18. SYKES, C. and F. W. JONES. Proc. Roy. Soc., A166:376, 1938.
19. SYKES, C. and H. EVANS. J. Inst. Metals, 58:255, 1936.
20. WILCHINSKY, Z. W. J. Appl. Phys., 15:806, 1944.
21. KOMAR, A. P. and N. N. BUYNOV. Zh. Eksperim. i Teor. Fiz., 17:555, 1947.
22. COWLEY, J. M. Journ. Appl. Phys., 21:24, 1950.
23. BRADLEY, A. J. and A. H. JAY. Proc. Roy. Soc., A136:210, 1932.
24. ROBERTS, B. W. and VINEYARD, G. H. Journ. Appl. Phys., 27:203, 1956.

25. GUINIER, A. J. Proc. Phys. Soc., 57:310, 1945.
26. ROBERTS, B. W., see WARREN, B. E. and B. L. AVERBACH, Modern Research Techniques in Physical Metallurgy. 1953. p. 95.
27. NORMAN, N. and B. E. WARREN. Journ. Appl. Phys., 22:483, 1951.
28. FLINN, P. A., B. L. AVERBACH and P. S. RUDMAN. Acta Cryst. 7:153, 1954.
29. FLINN, P. A., B. L. AVERBACH and M. COHEN. Acta Met., 1:664, 1953.
30. WALKER, C. B. Journ. Appl. Phys., 23:118, 1952.
31. RUDMAN, P. S., P. A. FLINN and B. L. AVERBACH. Journ. Appl. Phys., 24:365, 1953.
32. KONOBEYEVSKIY, S. T. Zh. Eksperim. i Teor. Fiz., 13:200, 1943.
33. LYUBOV, B. Ya. and B. I. MAKSIMOV. Zh. Tekhn. Fiz., 23:1202, 1953.
34. WALKER, C. B. and A. GUINER. Acta Met., 1:568, 1953.
35. WALKER, C. B., J. BLIN and A. GUINIER. Comptes Rend., 235:254, 1952.
36. BAGARYATSKIY, Yu. A. Zh. Tekhn. Fiz., 20:424, 1950.
37. BAGARYATSKIY, Yu. A. Dokl. Akad. Nauk SSSR, 77:261, 1951.
38. BAGARYATSKIY, Yu. A. Dokl. Akad. Nauk SSSR, 87:559, 1952.
39. YELISTRATOV, A. M. Izv. Akad. Nauk SSSR, Ser. Fiz., 15:60, 1951.
40. YELISTRATOV, A. M. Dokl. Akad. Nauk SSSR, 87:581, 1952.
41. YELISTRATOV, A. M., S. D. FINKEL'SHTEYN and T. Yu. GOL'SHTEYN. Dokl. Akad. Nauk SSSR, 88:669, 1953.
42. YELISTRATOV, A. M. Dokl. Akad. Nauk SSSR, 101:473, 1955.
43. YELISTRATOV, A. M. Dokl. Akad. Nauk SSSR, 101:69, 1955.
44. GERMER, L. H., F. E. HAWORTH and J. J. LANDER. Phys. Rev., 61:614, 1942.
45. HUGHES, D., Neutron Optics. Foreign Literature Press, 1955 (In Russian).
46. PALEVSKY, H. and D. J. HUGHES. Phys. Rev. 92:202, 1953.
47. NIX, F. C., H. C. BEYER and J. R. DUNNING. Phys. Rev., 58:1031, 1940.
48. NIX, F. C. and G. CLEMENT. Phys. Rev., 68:159, 1945.
49. SHULL, C. G. and S. SIEGEL. Phys. Rev., 75:1008, 1949.

50. SHULL, C. G. and M. K. WILKINSON. Phys. Rev., 97:304, 1955.
51. LYASHCHENKO, B. G., D. F. LITVIN, I. M. PUZEY and Yu. G. ABOV. Kristallografiya, 2:64, 1957.
52. SYKES, C. Proc. Roy. Soc., A148:422, 1935.
53. SYKES, C. and F. W. JONES. J. Inst. Metals, 59:257, 1936; Proc. Roy. Soc., A157:213, 1936.
54. MOSER, H. Physik. Z., 37:737, 1936.
55. KHOMYAKOV, K. G., V. A. KHOLLER and V. A. TROSHKINA. Vestn. Mosk. Univ., No. 6:43, 1950.
56. WEBELER, B., R. WEBELER and F. TRUMBORE. Acta Met., 1:374, 1953.
57. Entsiklopediya Metallofiziki (Encyclopedia of Metal Physics). ONTI, 1937. p. 302.
58. KURNAKOV, N. S., S. F. ZHEMCHUZHNYY and M. I. ZASEDATELEV. Izv. St. Peterburgskogo Politekhnicheskogo Instituta, 22:487, 1914; Zh. Eksperim. i Teor. Fiz., 47:871, 1915; J. Inst. Metals, 15:305, 1916.
59. TAMMANN, G. Z. Anorg. Allgem. Chem., 107:1, 1919.
60. JOHANSSON, C. and J. LINDE. Ann. Phys., 25:1, 1936; Ann. Phys., 28:449, 1937.
61. BUTYLENKO, A. K., V. M. DANILENKO, Yu. V. MIL'MAN, and others. Zh. Eksperim. i Teor. Fiz., 23:731, 1952.
62. POSPISIL, V. Ann. Phys., 18:497, 1933.
63. JOHANSSON, C. and J. LINDE. Ann. Phys., 82:449, 1927.
64. KURNAKOV, N. S. and V. A. NEMILOV. Z. Anorg. Allgem. Chem., 168:339, 1928.
65. SVENSSON, B. Ann. Phys., 14:699, 1932.
66. NIKS, F. and V. SHOKLI. Usp. Fiz. Nauk, 20:344, 536, 1938.
67. SMIRNOV, A. A. Zh. Eksperim. i Teor. Fiz., 4:229, 1934; Phys. Z. Sowjetunion, 5:599, 1934.
68. SMIRNOV, A. A. Dokl. Akad. Nauk SSSR, No. 3:172, 1953; No. 4:250, 1954; Izv. Kievsk. Politekhn. Inst., 15:57, 1954; D. L. Dexter. Phys. Rev., 87:786, 1952.
69. FRANK, V. Kgl. Danske Videskab. Selskab., Mat. Fys. Medd., 30:3, 1955.
70. SIEGEL, S. Phys. Rev., 57:537, 1940.
71. McGEARY, R. and S. SIEGEL. Phys. Rev., 65:347, 1944.
72. LIFSHITS, B. G., B. V. MOLOTILOV, N. N. MYULLER and N. A. SAVOST'YANOVA. Fiz. Metal. i Metalloved., 3:477, 1956.
73. ROSENBLATT, D. B., R. SMOLUCHOWSKI and G. J. DIENES. J. Appl. Phys., 26:1044, 1955.
74. DAMASK, A. C. J. Phys. Chem. Solids, 1:23, 1956.

75. SIEGEL, S. J. Chem. Phys., 8:860, 1940.
76. LORD, N. W. J. Chem. Phys., 21:692, 1953.
77. DUGDALE, R. A. Phil. Mag., 1:537, 1956.
78. DUGDALE, R. A. and A. GREEN. Phil. Mag., 45:163, 1954.
79. BURNS, F. P. and S. L. QUIMBY. Phys. Rev., 97:1567, 1955.
80. KUCZYNSKI, G., R. HOCHMAN and M. DOYAMA. J. Appl. Phys., 26:871, 1955.
81. KAMEL, R. J. Inst. Metals, 84:55, 1955.
82. KADYKOVA, G. N. and Ya. P. SELISSKIY. Fiz. Metal. i Metalloved., 3:497, 1956.
83. GUTOVSKIY, I. G. and Ya. P. SELISSKIY. Fiz. Metal. i Metalloved., 2:375, 1956.
84. BRAGG, W. L. and E. J. WILLIAMS. Proc. Roy. Soc., A145:699, 1934.
85. GORSKIY, V. S. Phys. Z. Sowjetunion, 8:443, 1935.
86. ANASTASEVICH, V. S. and Ya. I. FRENKEL'. Zh. Eksperim. i Teor. Fiz., 9:586, 1939.
87. DIENES, G. J. Acta Met., 3:549, 1955.
88. VINEYARD, G. H. Phys. Rev., 102:981, 1956.
89. SEITZ, F. Discussions Faraday Soc., 5:271, 1949; (Effect of radiations on semiconductors and insulators). Foreign Literature Press, 1954. p. 9. (In Russian).
90. SIEGEL, S. Modern Research Techniques in Physical Metallurgy (Cleveland. American Soc. of Metals). 1953.
91. BROOM, T. Advan. Phys., 3:26, 1954.
92. ZAKHAROV, A. I. Usp. Fiz. Nauk, 57:525, 1955.
93. GLEN, J. W. Advan. Phys., 4:381, 1955.
94. GLICK, H. L., F. C. BROOKS, W. F. WITZIG and W. E. JOHNSON. Phys. Rev., 87:1074, 1952.
95. GLICK, H. L. and W. F. WITZIG. Phys. Rev., 91:236, 1953.
96. BLEWITT, T. H. and R. R. COLTMAN. Phys. Rev., 85:384, 1952.
97. SIEGEL, S. Phys. Rev., 75:1823, 1949.
98. COLTMAN, R. R. and T. H. BLEWITT. Phys. Rev., 91:236, 1953.
99. BRINKMAN, J. C. E. DIXON and C. J. MEECHAN. Acta Met., 2:1, 1954.
100. DIXON, C. E., J. MEECHAN and J. A. BRINKMAN. Phil. Mag., 44:449, 1953.
101. MARTIN, A. B., S. B. AUSTERMAN, R. R. EGGLESTON, J. F. McGEE and M. TORPINIAN. Phys. Rev., 81:664, 1951.

102. MARTIN, A. B., S. B. AUSTERMAN, R. R. EGGLESTON, J. F. McGEE and M. TORPINIAN. Bull. Am. Phys. Soc., 25(6):21, 1950; DIENES, G. J. Ann. Rev. Nuclear Sci., 2:187, 1953.
103. ARONIN, L. R. J. Appl. Phys., 25:344, 1954.
104. EGGLESTON, R. and F. BOWMAN. J. Appl. Phys., 24:229, 1953.
105. BILLINGTON, D. S. and S. SIEGEL. Metal Progr., 58:847, 1950.
106. SIEGEL, S. Phys. Rev., 57:537, 1940.
107. ROHL, H. Ann. Phys., 18:155, 1933.
108. ARTMAN, R. J. Appl. Phys., 23:475, 1952. Elasticity and non-elasticity of Metal. Foreign Literature Press, 1954. p. 181. (In Russian).
109. GOOD, W. A. Phys. Rev., 60:605, 1941.
110. NOWACK, L. Z. Metallk., 22:94, 1930.
111. SACHS, G. Praktische Metallk., 3:68, 1935.
112. HARKER, D. Trans. Am. Soc. Metals, 32:210, 1944.
113. SACHS, G. and J. WEERTS. Z. Phys., 67:507, 1931.
114. KADYKOVA, G. N. and Ya. P. SELISSKIY. Fiz. Metal. i Metalloved., 3:486, 1956.
115. HERMAN, M. and N. BROWN. J. of Metals, 8, Sct. 2:604, 1956.
116. DAHL, O. Z. Metallk., 28:133, 1936.
117. CLAREBROUGH, L. M. and J. F. NICKOLAS. Australian J. Sci. Res., A3:284, 1950.
118. DEHLINGER, U. and L. GRAF. Z. Phys., 64:359, 1930.
119. LINDE, J. O. Ann. Phys., 30:151, 1937.
120. LIFSHITS, B. G. and M. P. RAVDEL'. Dokl. Akad. Nauk SSSR, 93:1033, 1953.
121. VOLKENSHTEYN, N. V. and A. P. KOMAR. Zh. Eksperim. i Teor. Fiz., 11:723, 1941.
122. KOMAR, A. P. and N. V. VOLKENSHTEYN. Izv. Sektora Fiz.-Khim. Analiza, Inst. Obshch. Neorgan. Khim., Akad. Nauk SSSR, 26:105, 1942.
123. KOMAR, A. P. Tr. Inst. Metal. Akad. Nauk SSSR, Ural'sk. Filial, 12:50, 1949; Izv. Akad. Nauk SSSR, Ser. Fiz., 11:497, 1947.
124. KAYA, S. and M. NAKAYAMA. Proc. Phys. Math. Soc. Japan, 22:126, 1940.
125. KAYA, S. and A. KUSSMANN. Z. Phys., 72:293, 1931.
126. FALLOT, M. Ann. Phys. (Paris), 6:305, 1936.
127. BITTER, R. Introduction to Ferromagnetism. London, 1937.
128. KUSSMAN, A. and NITKA. Phys. Z., 39:373, 1938.

129. BRADLEY, A. J. and J. W. RODGERS. Proc. Roy. Soc., A144:340, 1934.
130. DEKHTYAR, M. V. Fiz. Metal. i Metalloved., 3:54, 1956.
131. IVANOVSKIY, V. I. Fiz. Metal. i Metalloved., 4:70, 1957.
132. JELLINGHAUS, W. Z. Tech. Phys., 17:33, 1936.
133. HULTGREN, R. and R. JAFFEE. J. Appl. Phys., 12:501, 1941.
134. KHANSEN, M., Struktura Binarnykh Splavov (Structure of binary alloys). Moscow, Gostekhizdat, 1941.
135. MARIAN, V. Ann. Phys. (Paris), 7:459, 1937.
136. FALLOT, M. Ann. Phys. (Paris), 10:391, 1938.
137. HEUSLER, O. Ann. Physik, 19:155, 1934.
138. GRABBE, E. M. Phys. Rev., 57:728, 1940.
139. BECKER, R. and W. DORING. Ferromagnetismus. Berlin, 1939.
140. VONSOVSKIY, S. V. and Ya. S. SHUR, Ferromagnetizm (Ferromagnetism). Moscow-Leningrad, Gostekhizdat, 1948.
141. BOZORTH, R. M. Ferromagnetism. New York, 1951.
142. GLAZER, A. A. and Ya. S. SHUR. Fiz. Metal. i Metalloved., 3:568, 1956.
143. SEEMANN, H. J. and E. VOGT. Ann. Phys., 2:976, 1929.
144. SVENSSON, B. Ann. Physik, 14:699, 1932.
145. KOMAR, A. P. Doctoral dissertation. Urals Branch of the Academy of Sciences of the USSR, 1942.
146. KOMAR, A. P. and S. K. SIDOROV. Dokl. Akad. Nauk SSSR, 23:144, 1939.
147. KOMAR, A. P. and S. K. SIDOROV. Zh. Tekhn. Fiz., 11:711, 1947.
148. KOMAR, A. P. Dokl. Akad. Nauk SSSR, 27:554, 1940.
149. KOMAR, A. P. Zh. Eksperim. i Teor. Fiz., 11:717, 1941.
150. SIDOROV, S. K. Zh. Eksperim. i Teor. Fiz., 16:503, 1946.
151. SIDOROV, S. K. Zh. Eksperim. i Teor. Fiz., 16:629, 1946.
152. BUYNOV, N. N. Zh. Eksperim. i Teor. Fiz., 17:41, 1947.
153. BETTERIDGE, W. J. Inst. Met., 75:559, 1949.
154. JONES, D. M. and E. A. OWEN. Proc. Phys. Soc., B67:297, 1954.
155. SELISSKIY, Ya. P. Fiz. Metal. i Metalloved., 4:191, 1957.
156. GERTSRIKEN, S. D. and I. Ya. DEKHTYAR. Tr. Fiz. Fakul'teta Kievsk. Gos. Univ., 1952. p. 209.
157. KUPER, A. B., D. LAZARUS, J. R. MANNING and C. T. TOMIZUKA. Phys. Rev., 104:1936, 1956.
158. TAKAGI, Y. and T. SATO. Proc. Phys. Soc. Japan, 21:251, 1939.

159. ERENFEST, P. Proc. Kon. Akad. Amsterdam, 36:153, 1933.
160. KRIVOGLAZ, M. A. and A. A. SMIRNOV. Vopr. Fiz. Metal. i Metalloved. Akad. Nauk Ukr. SSR, No. 8:65, 1957.
161. LANDAU, L. D. Zh. Eksperim. i Teor. Fiz., 7:19, 1937; Phys. Z. Sowjetunion, 11:26, 1937.
162. LANDAU, L. D. Zh. Eksperim. i Teor. Fiz., 7:627, 1937; Phys. Z. Sowjetunion, 11:545, 1937.
163. LANDAU, L. D. Phys. Z. Sowjetunion, 8:113, 1935.
164. LIFSHITS, Ye. M. Zh. Eksperim. i Teor. Fiz., 11:255, 1941; J. Phys., 6:61, 1942.
165. LIFSHITS, Ye. M. Zh. Eksperim. i Teor. Fiz., 11:269, 1941.
166. LIFSHITS, Ye. M. Zh. Eksperim. i Teor. Fiz., 14:353, 1944.
167. LANDAU, L. and Ye. LIFSHITS, Statisticheskaya Fizika (Statistical physics). Gostekhizdat, 1951.
168. Internationale Tabellen zur Bestimmung von Kristallstrukturen. Vol. 1. 1935; International Tables for X-ray crystallography. Vol. 1. 1952.
169. KRIVOGLAZ, M. A. Dokl. Akad. Nauk SSSR, 117:213, 1957; Zh. Fiz. Khim., 31:1930, 1957.
170. LANDAU, L. D. Zh. Eksperim. i Teor. Fiz., 7:1232, 1937; Phys. Z. Sowjetunion, 12:123, 1937.
171. KRIVOGLAZ, M. A. Zh. Eksperim. i Teor. Fiz., 31:625, 1956.
172. BOROVSKIY, I. B. and K. P. GUROV. Fiz. Metal. i Metalloved., 4:187, 1957.
173. NEWELL, G. F. and E. W. MONTROLL. Rev. Mod. Phys., 25:353, 1953.
174. RUMER, Yu. B. Usp. Fiz. Nauk, 53:245, 1954.
175. ter HAAR, D. Elements of Statistical Mechanics. New York, Rinehart and Co., 1954.
176. KRAMERS, H. A. and G. H. WANNIER. Phys. Rev., 60:252, 1941.
177. MONTROLL, E. W. J. Chem. Phys., 9:706, 1941.
178. LASSETTRE, E. N. and J. P. HOWE. J. Chem. Phys., 9:747, 1941.
179. ASHKIN, J. and W. E. LAMB. Phys. Rev., 64:159, 1943.
180. ONSAGER, L. Phys. Rev., 65:117, 1944.
181. KAUFMAN, B. Phys. Rev., 76:1232, 1949.
182. KAC, M. and J. C. WARD. Phys. Rev., 88:1332, 1952.
183. YANG, C. N. Phys. Rev., 85:809, 1952.
184. ISING, E. Z. Phys., 31:253, 1925.

185. GORSKIY, V. S. Z. Phys., 50:64, 1928.
186. BRAGG, W. L. and E. J. WILLIAMS. Proc. Roy. Soc., A151:540, 1935.
187. BRAGG, W. L. and E. J. WILLIAMS. Proc. Roy. Soc., A152:231, 1935.
188. NESTERENKO, Ye. G. and A. A. SMIRNOV. Dokl. Akad. Nauk SSSR, No. 3:184, 1951; Vopr. Fiz. Metal. i Metalloved. Akad. Nauk Ukr. SSSR, No. 3:152, 1952.
189. GUGGENHEIM, E. A. Mixtures. Oxford, 1952. VII.
190. KIRKWOOD, J. G. J. Chem. Phys., 6:70, 1938.
191. BETHE, H. A. and J. G. KIRKWOOD. J. Chem. Phys., 7:578, 1939.
192. LIFSHITS, I. M. Zh. Eksperim. i Teor. Fiz., 9:481, 1939.
193. CHANG, T. S. Proc. Roy. Soc., 173:48, 1939.
194. CHANG, T. S. J. Chem. Phys., 9:169, 1941.
195. GUGGENHEIM, E. Proc. Roy. Soc., A148:304, 1935.
196. RUCHBROOKE. Proc. Roy. Soc., A166:296, 1938.
197. LI, Y. Y. Phys. Rev., 76:972, 1949.
198. YANG, C. N. J. Chem. Phys., 13:66, 1945.
199. YANG, C. N. and Y. Y. LI. Chinese J. Phys., 7:59, 1947.
200. LI, Y. Y. J. Chem. Phys., 17:447, 1949.
201. MIRTSKHULAVA, I. A. Zh. Eksperim. i Teor. Fiz., 19:407, 1949.
202. BETHE, H. A. Proc. Roy. Soc., A150:552, 1935.
203. PEIERLS, R. Proc. Roy. Soc., A154:207, 1936.
204. CHANG, T. S. Proc. Cambr. Phil. Soc., 35:265, 1939.
205. Van der WAERDEN. Z. Phys., 118:473, 1941.
206. TREFFTZ, E. Z. Phys., 127:371, 1950.
207. SMIRNOV, A. A. and I. A. STOYANOV. Fiz. Metal. i Metalloved., 2:524, 1956.
208. KRIVOGLAZ, M. A. and Z. A. MATYSINA. Vopr. Fiz. Met. i Metalloved. Akad. Nauk Ukr. SSR, No. 6:114, 1955.
209. KRAMERS, H. A. and G. H. WANNIER. Phys. Rev., 60:252, 1941.
210. Ter HAAR, D. Physica, 18:386, 1952; 19:611, 1953.
211. DEMPSEY, E. and D. ter HAAR. Physica, 20:437, 1954; 20:667, 1954; 22:1, 1956.
212. KVASNIKOV, I. A. Dokl. Akad. Nauk SSSR, 110:755, 1956; 113:544, 777, 1957.
213. ZERNIKE, F. Physica, 7:565, 1940.
214. COWLEY, J. M. Phys. Rev., 77:669, 1950.
215. KIKUCHI, R. Phys. Rev., 81:988, 1951; J. Chem. Phys., 19:1230, 1951; KURATA, M., R. KIKUCHI and T. WATARI. J. Chem. Phys., 21:434, 1953.

216. HIJMANS, J. and J. De BOER. Physica, 21:471, 485, 499, 1955; 22:408, 429, 1956.
217. CHANG, T. S. J. Chem. Phys., 9:174, 1941.
218. PINES, B. Ya. Zh. Eksperim. i Teor. Fiz., 11:725, 1941.
219. LESNIK, A. G. Vopr. Fiz. Metal. i Metalloved. Akad. Nauk Ukr. SSR, No. 5:104, 1954.
220. STEPANOV, P. Ye. Zh. Eksperim. i Teor. Fiz., 9:1352, 1939; 10:103, 1940.
221. CHANG, T. S. Proc. Cambr. Phil. Soc., 35:265, 1939
222. KHOLODENKO, L. Zh. Eksperim. i Teor. Fiz., 20:1083 1950.
223. KRIVOGLAZ, M. A. and A. A. SMIRNOV. Usp. Fiz. Nauk 55:391, 1955.
224. FRENKEL', Ya. I. Z. Phys., 35:652, 1926; Kineticheskaya Teoriya Zhidkostey (Kinetic theory of liquids). Moscow-Leningrad, Izd. Akad. Nauk SSSR, 1945.
225. SMIGELSKAS, A. D. and E. O. KIRKENDALL. Trans. Am. Inst. Min. Met. Engrs., 171:130, 1947.
226. CORREA da SILVA, L. C. and R. F. MEHL. Trans. Am. Inst. Min. Met. Engrs., 191:155, 1951; BUCKLE, H. and J. BLIN. J. Inst. Metals, 80:385, 1951-1952; BARNES, R. S. Proc. Phys. Soc., B65:512, 1952; SEITH, W. and A. KOTTMANN. Naturwiss., 39:40, 1952; Angew. Chem., 64:376, 1952.
227. PINES, B. Ya. and Ya. Ye. GEGUZIN. Zh. Tekhn. Fiz., 23:1559, 1953.
228. PINES, B. Ya. Usp. Fiz. Nauk, 52:501, 1954.
229. BARRER, R. Diffusion in Solids. Foreign Literature Press, 1948 (In Russian).
230. JOST, W. Diffusion in Solids, Liquids, Gases. New York, 1952.
231. ARKHAROV, V. I. Okisleniye Metallov (Oxidation of metals). Metallurgizdat, 1945.
232. BUGAKOV, V. Z. Diffuziya v Metallakh i Splavakh (Diffusion in metals and alloys). Moscow, Gostekhizdat, 1949.
233. Le CLAIRE, A. D. Progr. Metal Phys., 1:306, 1949; Usp. Fiz. Metal., 1:224, 1956.
234. GROH, J. and G. HEVESY. Ann. Phys., 63:85, 1920.
235. ZABRUBSKIY, A. M. Izv. Akad. Nauk SSSR, Otd. Mat. i Yestestv. Nauk, 1937. p. 903.
236. GERTSRIKEN, S. D. and I. Ya. DEKHTYAR. Vopr. Fiz. Met. i Metalloved. Akad. Nauk Ukr. SSR, No. 2:108, 1950.
237. GRUZIN, P. L., Yu. V. KORNEV and G. V. KURDYUMOV. Dokl. Akad. Nauk SSSR, 80:49, 1951.

238. FRENKEL', Ya. I. and M. I. SERGEYEV. Zh. Eksperim. i Teor. Fiz., 9:189, 1939.
239. ONSAGER, L. Phys. Rev., 37:405, 1931; 38:2265, 1931.
240. DARKEN, L. S. Trans. Am. Inst. Min. Met. Engrs., 175:148, 1948.
241. BARDEEN, J. and C. HERRING. Atom Movements. Am. Soc. for Metals, 1951.
242. SEITZ, F. Phys. Rev., 74:1513, 1948.
243. SEITZ, F. Phase Transformations in Solids. Symposium held at Cornell University 77; 1951.
244. JOHNSON, W. A. Trans. Am. Inst. Min. Met. Engrs., 147:331, 1942.
245. WAGNER, C. and F. ENGELHARDT. Z. Physik. Chem., A159:241, 1932.
246. LeCLAIRE, A. D. and R. S. BARNES. J. Metals, 3:1060, 1951.
247. KONOBEYEVSKIY, S. T. Zh. Eksperim. i Teor. Fiz., 13:200, 1943.
248. LYUBOV, B. Ya. and N. S. FASTOV. Dokl. Akad. Nauk SSSR, 84:939, 1952.
249. KRIVOGLAZ, M. A. and A. A. SMIRNOV. Zh. Eksperim. i Teor. Fiz., 24:409, 1953.
250. KRIVOGLAZ, M. A. Dokl. Akad. Nauk Ukr. SSR, No. 5:344, 1953.
251. KRIVOGLAZ, M. A. and A. A. SMIRNOV. Vopr. Fiz. Metal. i Metalloved. Akad. Nauk Ukr. SSR, No. 4:95, 1953.
252. WILLIAMSON, G. K. and R. E. SMALLMAN. Acta Cryst., 6:361, 1953.
253. SMIRNOV, A. A. Zh. Tekhn. Fiz., 24:1804, 1954.
254. KRIVOGLAZ, M. A. and A. A. SMIRNOV. Zh. Eksperim. i Teor. Fiz., 24:673, 1953.
255. JACK, K. H. Proc. Roy. Soc., A195:34, 1948; PETCH, N. J. J. Iron and Steel Inst., 145:111, 1942.
256. SMIRNOV, A. A. Dokl. Akad. Nauk Ukr. SSSR, No. 5:351, 1950; Vopr. Fiz. Metal. i Metalloved. Akad. Nauk Ukr. SSR, No. 3:143, 1952.
257. KRIVOGLAZ, M. A. Vopr. Fiz. Metal. i Metalloved. Akad. Nauk Ukr. SSR, No. 7:95, 1956.
258. SMIRNOV, A. A. Zh. Tekhn. Fiz., 23:56, 1953.
259. KRIVOGLAZ, M. A. and A. A. SMIRNOV. Dokl. Akad. Nauk SSSR, 96:495, 1954.
260. MOTT, N. and H. JONES. The Theory of the Properties of Metals and Alloys. Oxford, 1936.

261. BETE, G. and A. ZOMMERFEL'D. Elektronnaya Teoriya Metallov (Electron theory of metals). GONTI, 1938.
262. VIL'SON, A., Kvantovaya Teoriya Metallov (Quantum theory of metals). Gostekhizdat, 1941.
263. RUDNITSKIY, V. Ye. Zh. Eksperim. i Teor. Fiz., 9:1069, 1939.
264. MUTO, T. Sci. Pap. Inst. Phys. Chem. Res. (Tokyo), 34:377, 1938.
265. SMIRNOV, A. A. Dokl. Akad. Nauk Ukr. SSR, No. 1:67, 1955.
266. SMIRNOV, A. A. Zh. Eksperim. i Teor. Fiz., 17:730, 1947.
267. BORN, M. and T. KARMAN. Phys. Z., 13:397, 1912.
268. NORDHEIM, L. Ann. Phys., 9:607, 1931.
269. MEN', A. N. and A. N. ORLOV. Dokl. Akad. Nauk SSSR, 40:753, 1953.
270. LIFSHITS, I. M. and G. I. STEPANOVA. Zh. Eksperim. i Teor. Fiz., 31:156, 1956.
271. KARAL'NIK, S. M. Fiz. Metal. i Metalloved., 3:503, 1956.
272. LAUE, M. Ann. Phys., 78:167, 1925; Röntgenstrahlinterferenzen. Leipzig. 1948.
273. OBRAZTSOV, Yu. N. Zh. Eksperim. i Teor. Fiz., 8:593, 1938.
274. LIFSHITS, I. M. Zh. Eksperim. i Teor. Fiz., 8:959, 1938.
275. LIFSHITS, I. M. Zh. Eksperim. i Teor. Fiz., 9:500, 1939.
276. WARREN, B. E. B. L. AVERBACH and B. W. ROBERTS. J. Appl. Phys., 22:1493, 1951.
277. BAGARYZTSKIY, Yu. A. Dokl. Akad. Nauk SSSR, 43:35, 1953.
278. DANILENKO, V. M., M. A. KRIVOGLAZ, Z. A. MATYSINA and A. A. SMIRNOV. Fiz. Metal. i Metalloved., 4:28, 1957.
279. KRIVOGLAZ, M. A. Zh. Eksperim. i Teor. Fiz., 32:1368, 1957.
280. KRIVOGLAZ, M. A. Fiz. Metal. i Metalloved., 5:203, 1957.
281. KRIVOGLAZ, M. A. Vopr. Fiz. Metal. i Metalloved. Akad. Nauk Ukr. SSR, No. 8:199, 1957.
282. SMIRNOV, A. A. and B. V. PADUCHEV. Zh. Eksperim. i Teor. Fiz., 21:541, 1951.
283. SMIRNOV, A. S. and S. V. VONSOVSKIY. J. Phys., 5:263, 1941.
284. POMERANCHUK, I. Zh. Eksperim. i Teor. Fiz., 8:894, 1938.

285. AKHIYEZER, A. and I. POMERANCHUK. Nekotoryye Voprosy Teorii Yadra (Some problems concerning the nuclear theory). Gostekhizdat, 1948.
286. EINSTEIN, A. Ann. Phys., 33:1275, 1910.
287. ZHUKHOVITSKIY, A. A., B. N. FINKEL'SHTEYN and I. S. KULIKOV. Dokl. Akad. Nauk SSSR, 84:227, 1951.
288. SMIRNOV, A. A. Zh. Eksperim. i Teor. Fiz., 17:743, 1947.
289. RYZHANOV, S. G. Zh. Eksperim. i Teor. Fiz., 9:4, 1939.
290. KRIVOGLAZ, M. A. and A. A. SMIRNOV. Vopr. Fiz. Metal. i Metalloved. Akad. Nauk Ukr. SSR, No. 7:115, 1956.
291. KRIVOGLAZ, M. A. and Z. A. MATYSINA. Zh. Eksperim. i Teor. Fiz., 28:61, 1955.
292. KRIVOGLAZ, M. A., Z. A. MATYSINA and A. A. SMIRNOV. Fiz. Metal. i Metalloved., 1:385, 1956.
293. DYKHNE, A. M., Z. A. MATYSINA and A. A. SMIRNOV. Fiz. Metal. i Metalloved., 5:220, 1957.
294. SMIRNOV, A. A. and I. A. STOYANOV. Ukr. Fiz. Zh., 1:333, 1956.
295. SEITZ, F., Sovremennaya Teoriya Tverdogo Tela (Modern theory of solids). Gostekhizdat, 1949.
296. KOMAR, A. P. Zh. Eksperim. i Teor. Fiz., 17:753, 1947.
297. MURAKAMI, T. J. Phys. Soc. Japan, 8:458, 1953.
298. GIBSON, J. B. J. Phys. Chem. Solids, 1:27, 1956.
299. DORFMAN, Ya. G. Phys. Z. Sowjetunion, 3:399, 1933.
300. STONER, E. C. Phil. Mag., 15:1018, 1933; Phys. Soc. Rept. Progr. Phys., 11:43, 1946-1947; J. Phys. Radium, 12:372, 1951.
301. SLATER, J. C. J. Appl. Phys., 8:385, 1937.
302. PAULING, L. Phys. Rev., 54:899, 1938.
303. NIESSEN. Physica, 6:1011, 1939.
304. SMOLUCHOWSKI, R. J. Phys. Radium, 12:389, 1951.
305. RUDNITSKIY, V. Ye. Zh. Eksperim. i Teor. Fiz., 10:63, 1940.
306. VONSOVSKIY, S. V. Dokl. Akad. Nauk SSSR, 26:564, 1940; Zh. Tekhn. Fiz., 18:131, 1948.
307. AKULOV, N. S. Dokl. Akad. Nauk SSSR, 66:361, 1949.
308. VONSOVSKIY, S. V. and K. B. VLASOV. Zh. Eksperim. i Teor. Fiz., 25:327, 1953.
309. FRENKEL', Ya. I. Z. Phys., 49:31, 1928.
310. HEISENBERG, W. Z. Phys., 49:619, 1928.
311. BLOCH, F. Z. Phys., 61:206, 1930; 74:295, 1933.

312. VONSOVSKIY, S. V. Zh. Eksperim. i Teor. Fiz., 16:981, 1946; VONSOVSKIY, S. V. and Ye. A. TUROV. Zh. Eksperim. i Teor. Fiz., 24:419, 1953.
313. SMIRNOV, A. A. Zh. Tekhn. Fiz., 18:153, 1948.
314. SMIRNOV, A. A. Izv. Akad. Nauk SSSR, Ser. Fiz., 11:507, 1947.
315. SMIRNOV, A. A. Tr. Ural'sk Filiala Akad. Nauk SSSR, 12:40, 1949.
316. VONSOVSKIY, S. V., A. A. SMIRNOV and A. V= SOKOLOV. Dokl. Akad. Nauk SSSR, 80:353, 1951.
317. BORN, M. Optika (Optics). Khar'kov-Kiev, GNTIU, 1937.
318. SOKOLOV, A. V. Zh. Eksperim. i Teor. Fiz., 21:384, 1951.
319. SAMOYLOVICH, A. Zh. Eksperim. i Teor. Fiz., 14:205, 1944.
320. ORLOV, A. N. Zh. Eksperim. i Teor. Fiz., 21:1090, 1951.
321. ORLOV, A. N. Dissertation. Ural'sk. Filial Akad. Nauk SSSR, 1949; Fiz. Metal. i Metalloved., 5:212, 1957.
322. MOTT, N. F. Proc. Phys. Soc., 49:258, 1937.
323. KHOLODENKO, L. Zh. Eksperim. i Teor. Fiz., 18:812, 1948.
324. PINES, B. Ya. Zh. Eksperim. i Teor. Fiz., 11:147, 1941.
325. APTEKAR', I. L. and B. N. FINKEL'SHTEYN. Zh. Eksperim. i Teor. Fiz., 21:900, 1951.
326. SACHS, G. and J. WEERTS. Z. Phys., 60:481, 1930.
327. APTEKAR', I. L. Zh. Eksperim. i Teor. Fiz., 21:910, 1951.
328. WILSON, A. H. Proc. Cambr. Phil. Soc., 34:81, 1939.
329. ORLOV, A. N. Zh. Eksperim. i Teor. Fiz., 21:1081, 1951.
330. IZMAYLOV, S. V. and O. M. TODES. Zh. Eksperim. i Teor. Fiz., 2:254, 1932.
331. SARKISOV, E. S. Dokl. Akad. Nauk SSSR, 58:1645, 1947.
332. GORSKIY, V. S. Phys. Z. Sowjetunion, 8:457, 526, 1935.
333. ZENER, K., Elasticity and non-elasticity of metals. Foreign Literature Press, 1954 (In Russian).
334. MANDEL'SHTAM, L. I. and M. A. LEONTOVICH. Zh. Eksperim. i Teor. Fiz., 7:438, 1937; LANDAU, L. D. and Ye. M. LIFSHITS, Mekhanika Sploshnykh Sred (Mechanics of continuous media). Gostekhizdat, 1953.
335. STEPANOV, P. Ye. Dokl. Akad. Nauk SSSR, 74:217, 1950.

336. LANDAU, L. D. and I. M. KHALATNIKOV. Dokl. Akad. Nauk SSSR, 96:469, 1954.

337. LIFSHITS, I. M. Zh. Eksperim. i Teor. Fiz., 17:1017, 1947; Usp. Mat. Nauk, 7:171, 1952; Nuovo Cimento (Supplemento), 3:716, 1956; LIFSHITS, I. M. and G. I. STEPANOV. Zh. Eksperim. i Teor. Fiz., 30:938, 1956.